fourth ed.

BIOCHEMISTRY PRIMER FOR EXERCISE SCIENCE

Peter M. Tiidus, PhD
Wilfrid Laurier University
Waterloo, Ontario, Canada

A. Russell Tupling, PhD
University of Waterloo
Waterloo, Ontario, Canada

Michael E. Houston, PhD

Human Kinetics

Library of Congress Cataloging-in-Publication Data

Tiidus, Peter M., 1955-
 Biochemistry primer for exercise science / Peter M. Tiidus, A. Russell Tupling, Michael E. Houston. -- 4th ed.
    p. ; cm.
 Rev. ed. of: Biochemistry primer for exercise science / Michael E. Houston. 3rd ed. c2006.
 Includes bibliographical references and index.
 ISBN 978-0-7360-9605-8 (print) -- ISBN 0-7360-9605-1 (print)
   I. Tupling, A. Russell, 1970- II. Houston, Michael E., 1941-2008. III. Houston, Michael E., 1941-2008. Biochemistry primer for exercise science. IV. Title.
 [DNLM: 1. Exercise--physiology. 2. Metabolism--physiology. QU 120]

 612'.044--dc23
                              2012001957

ISBN: 978-0-7360-9605-8 (print)

Copyright © 2012 by Peter M. Tiidus, A. Russell Tupling, and Michael E. Houston
Copyright © 2006, 2001, 1995 by Michael E. Houston

All rights reserved. Except for use in a review, the reproduction or utilization of this work in any form or by any electronic, mechanical, or other means, now known or hereafter invented, including xerography, photocopying, and recording, and in any information storage and retrieval system, is forbidden without the written permission of the publisher.

Acquisitions Editor: Amy N. Tocco; **Developmental Editor:** Katherine Maurer; **Assistant Editors:** Steven Calderwood, Derek Campbell; **Copyeditor:** Joy Wotherspoon; **Indexer:** Michael Ferreira; **Permissions Manager:** Dalene Reeder; **Graphic Designer:** Fred Starbird; **Graphic Artist:** Denise Lowry; **Cover Designer:** Keith Blomberg; **Art Manager:** Kelly Hendren; **Associate Art Manager:** Alan L. Wilborn; **Illustrations:** © Human Kinetics; **Printer:** Total Printing Systems

Printed in the United States of America   10   9   8

The paper in this book is certified under a sustainable forestry program.

Human Kinetics
P.O. Box 5076
Champaign, IL 61825-5076
Website: www.HumanKinetics.com

In the United States, email info@hkusa.com or call 800-747-4457.
In Canada, email info@hkcanada.com.
In the United Kingdom/Europe, email hk@hkeurope.com.

For information about Human Kinetics' coverage in other areas of the world,
please visit our website: **www.HumanKinetics.com**

E5205

To my wife, Ann, my love and my best friend,
and to my sons, Erik and Tommi Tiidus,
who have grown into young men
I am justifiably proud of. —*PMT*

To my wife, Sue, and to our children,
Olivia and John. Your love and support
mean the world to me. —*ART*

Contents

Preface ix

Acknowledgments x

Tribute to Michael E. Houston xi

Part I Proteins and Enzymes: The Basis of Biochemistry — 1

Chapter 1 Amino Acids, Peptides, and Proteins — 3
Nature of Amino Acids 4
Characteristics of Peptides 10
Structure of Proteins 11
Next Stage: Proteomics—The New Systems Approach to Biomedical Health and Exercise Science 16
Summary 17
Review Questions 17

Chapter 2 Enzymes — 19
Enzymes as Catalysts 19
Rates of Enzymatic Reactions 20
Enzyme Cofactors 25
Classification of Enzymes 26
Protein Transporters 26
Oxidations and Reductions 27
Regulation of Enzyme Activity 30
Measurement of Enzyme Activity 35
Next Stage: Redox Regulation of the SERCA Pump in Vascular Physiology and Disease 35
Summary 36
Review Questions 37

Chapter 3 Gene Transcription and Protein Synthesis — 39
DNA and RNA 39
Transcription 44
Posttranscriptional Modifications of RNA 53
Translation 58
Posttranslational Processing of Polypeptides 63

Protein Degradation 64
Regulation of Gene Expression in Exercise and Training 66
Next Stage: MicroRNAs and the Adaptive Response to Exercise Training 70
Summary 71
Review Questions 72

Part II Metabolism: Regulation and Adaptation to Exercise and Training   75

Chapter 4 Energy Systems and Bioenergetics   77

Energy Requirements of Skeletal Muscle 77
Energy-Rich Phosphates 84
Energy Systems 88
Bioenergetics 99
Next Stage: Epigenetic Control of Skeletal Muscle Fiber Type and Metabolic Potential 103
Summary 104
Review Questions 105

Chapter 5 Oxidative Phosphorylation   107

Overview of Metabolism 107
Mitochondria 109
Citric Acid Cycle 114
Electron Transport Chain 120
Coupled Phosphorylation 125
Regulation of Oxidative Phosphorylation 128
Quantification of Redox Reactions 139
Oxidants and Antioxidants 142
Next Stage: Effects of Aging and Exercise Training on Muscle Mitochondria 148
Summary 150
Review Questions 151

Chapter 6 Carbohydrate and Related Metabolism   153

Carbohydrates 153
Cellular Uptake of Glucose 155
Phosphorylation of Glucose 157
Glycolysis 159
Glycogen Metabolism 165
Lactate Metabolism 176
Oxidation of Cytoplasmic NADH 184

Gluconeogenesis 186
Pentose Phosphate Pathway 196
Signaling Pathways 198
Next Stage: Carbohydrate, Exercise Performance, and Fatigue 201
Summary 202
Review Questions 203

Chapter 7 Lipid Metabolism 205

Types of Lipids 205
Fat Storage and Metabolism 209
Oxidation of Fatty Acids 218
Oxidation of Ketone Bodies 222
Synthesis of Fatty Acids 225
Fat as Fuel for Exercise 229
Metabolism During Exercise: Fat Versus Carbohydrate 231
Adipose Tissue as an Endocrine Tissue 240
Cholesterol 242
Next Stage: Exercise, 24-Hour Fat Oxidation, and Weight Loss 243
Summary 244
Review Questions 245

Chapter 8 Amino Acid and Protein Metabolism 247

Overview of Amino Acid Metabolism 247
Degradation of Amino Acids 249
Urea Cycle 252
Fate of Amino Acid Carbon Skeletons 256
Amino Acid Metabolism During Exercise 257
Additional Roles for Amino Acids 261
Next Stage: Effects of Diet and Exercise on Protein Degradation and Synthesis 262
Summary 264
Review Questions 264

Appendix: Answers to Review Questions 267
Glossary 273
References 285
Index 293
About the Authors 299

Preface

We are reminded almost constantly that a growing number of the world's population face an obesity crisis. We are told that we need to increase our physical activity while cutting back on our food intake. These bioenergetics concepts are simple, yet they are difficult for many to accomplish. Students trained in traditional exercise physiology can understand the basic energy concepts that underlie our obesity epidemic, but they may lack the molecular basics on which all human science is founded. We need educated people with not only an understanding of exercise physiology but also a thorough foundation in the thermodynamic concepts of human energy metabolism at the more detailed molecular level. This text aims to provide exercise science students with an introduction to biochemistry that will give them this greater insight on the working of metabolism and the human body's response to physical activity.

Part I of this text lays the groundwork with chapters on amino acids and proteins and then enzymes, the critical biological catalysts. It then presents the basic concepts of gene expression and protein synthesis, emphasizing how exercise and exercise training modify these processes, and how cell signaling consequent to exercise training facilitates the training response. Our understanding of these areas has expanded very rapidly during the past decade. Students of exercise biochemistry will need to familiarize themselves with new advances if they wish to more fully understand the mechanisms that regulate our responses to physical activity. Part II on metabolism, however, is still the major emphasis of this book, with all the fundamental biochemical pathways explained and then integrated with information on the exercising human. Part II also draws from the ideas discussed in part I to further develop the concepts of biochemistry as they relate to physical activity and the adaptations that occur consequent to various forms of training.

Biochemistry Primer for Exercise Science is now in its fourth edition. It has expanded over the years from a biochemistry primer for students of exercise sciences to one that integrates the basics of biochemistry into exercise physiology and expands into related and overlapping aspects of molecular biology, nutrition, and physical activity as related aspects of health and disease. The first three editions were the sole work of Dr. Michael Houston, who initially transformed his teaching notes into a textbook and then added to this groundwork with major revisions, additions, and further integration of related fields in subsequent editions. Due to Mike's untimely death in 2008, Dr. Peter Tiidus and Dr. Russell Tupling have taken over the production of a revised and updated fourth edition of this classic textbook. While it is always difficult to take over the production of a successful textbook from a talented author, both Dr. Tiidus and Dr. Tupling had the fortune of having Mike Houston as a teacher and mentor for part of their development as exercise scientists. This close relationship with Mike has helped greatly in their ability to maintain his voice as an integral component of the fourth edition of the *Biochemistry Primer*.

This edition adds and integrates new research that has appeared since the third edition into the basic understanding of exercise biochemistry. Major additions and revisions based on new research developments appear throughout this text. In particular, the explosion of understanding of the control of biochemistry and biochemical and muscular adaptations to exercise and training through signaling pathways and the regulation of gene expression are significant additions to the text. This warranted a rearrangement of some of the chapters. We have consolidated and integrated the chapters dealing with DNA, RNA, and the regulation of protein synthesis into one chapter and have moved this into the first section of the text. This move highlights the growing need for students to appreciate the basics of molecular biology in order to more fully understand metabolic and biochemical control and responses to exercise and training. Chapter 3 now further covers the basics of signaling mechanisms associated with exercise and training adaptations. These topics also appear in specific instances in later chapters to highlight our growing understanding of their role

in exercise biochemistry. Additionally, each chapter now has a Next Stage section, which highlights a new or controversial area of research related to exercise biochemistry. These sections build on and integrate material covered in each of the chapters, and they will help lead students toward newer areas of development in the broad aspects of science related to exercise biochemistry. These changes highlight the rapid research developments affecting biochemistry as it relates to exercise science and serve to update the educational needs and experiences of kinesiology and exercise science students for the 21st century.

This book provides students and professionals alike with a nonintimidating, basic understanding of the science behind biochemistry. It has been written to present essential concepts in this dynamic, complex area of scientific knowledge in an easy-to-understand manner. It will be a useful resource for students, for researchers, and for professionals who need to obtain background in an unfamiliar scientific area, or as a basic reference.

Acknowledgments

I have had the fortune of being educated by great scientists. In particular, I would like to acknowledge three of them who have been instrumental in my career development. Dr. C. David Ianuzzo first taught me exercise physiology as an undergraduate at York University and later introduced me to scientific research as my MSc supervisor. Dr. Roy J. Shephard first hired me as a very young faculty member at the University of Toronto. Most important, the late Dr. Michael E. Houston, the author of the first three editions of this text and my PhD advisor and close friend, inspired me to pursue the excellence he represented in all things. I would also like to acknowledge Cheri Houston, who first suggested that I update her late husband's outstanding textbook. Mike's extensive and comprehensive work in producing the first three editions made our revision for the fourth edition relatively easy. I cannot imagine having completed this work without Mike's manuscript to build on. Finally, I would also like to acknowledge my many colleagues and friends in the department of kinesiology and physical education at Wilfrid Laurier University who make conducting research and teaching a pleasure.

—*PMT*

I owe my development in physiology and biochemistry and as a scientist, my achievements in research and teaching, and my current position in the department of kinesiology at the University of Waterloo to the opportunities that I was given and to the mentorship that I received from Dr. Howie Green, whose passion for exercise science is unsurpassable, even today in his retirement. While training at the University of Waterloo, I was also fortunate to learn from other great teachers and scientists, including Dr. Arend Bonen, Dr. Jay Thomson, Dr. Rich Hughson, Dr. Mike Sharratt, Dr. Norm Ashton, and, of course, Dr. Mike Houston. My expertise in biochemistry and my approach to research and writing were developed further during my time as a postdoctoral fellow under Dr. David MacLennan at the University of Toronto. Today, I continue to learn and draw inspiration from my friends and colleagues at Waterloo, who make collaboration a joy. My greatest satisfaction comes from the interactions I have with students and fellows, particularly the undergraduate and graduate students and postdoctoral fellows I have supervised who challenge me to continue to move forward and strive for excellence. Finally, I would also like to acknowledge Dr. Peter Tiidus for inviting me to coauthor the fourth edition of this textbook. I am truly honored to make a contribution to Dr. Mike Houston's classic work.

—*ART*

Michael E. (Mike) Houston

February 26, 1941 – July 16, 2008

Mike Houston was the author of the first three editions of *Biochemistry for Exercise Science*. This fourth edition, which is built on his body of work, still incorporates a major portion of his third edition. Mike received his undergraduate training in biochemistry from the University of Toronto and his PhD in biochemistry from the University of Waterloo. A superb athlete and lifelong exercise fanatic, Mike was able to integrate his training in biochemistry with his love of exercise sciences and to forge a career as a teacher and scientist in the field of kinesiology. Mike was a faculty member for many years in the department of kinesiology at the University of Waterloo. He moved briefly to the University of British Columbia before finishing his career as the department chair of human nutrition, foods, and exercise at Virginia Tech in Blacksburg, Virginia.

During his career, he authored more than 100 refereed publications and taught courses on the biochemistry of exercise for almost 40 years to many undergraduate and graduate students. In 2003, Mike was presented with the Honour Award from the Canadian Society for Exercise Physiology in acknowledgment of his lifetime contribution to exercise science research and education. Mike had a warm and engaging personality and was genuinely concerned for and interested in all of his students. His many scientific contributions and personal attributes will be long remembered by his academic colleagues. He continues to live in the hearts of his many students and friends. He is deeply missed by his wife Cheri; their children Mike Junior, Beth, and Patrick; and their families.

part I

Proteins and Enzymes
The Basis of Biochemistry

Everything we do, everything we are, and everything we will become depends on the actions of thousands of different proteins. These large molecules, composed of individual units known as amino acids, regulate everything in our bodies. The arrangement of amino acids in proteins is dictated by the sequence of bases, a part of nucleotides organized into genes within our DNA. Deoxyribonucleic acid base sequence is copied during transcription as working blueprints containing RNA bases. The sequence of bases in RNA, read in groups of three during translation, dictates the sequence of amino acids in each of our thousands of different proteins. This simple yet complex story forms the basis for understanding the biochemistry of exercise science.

Part I of this book introduces the molecules and interactions that form the basis of biochemistry, including amino acids, enzymes, and the processes by which DNA is transcribed and used to make all the proteins in the body, emphasizing how exercise and exercise training modify these processes. Chapter 1 reviews acids, bases, and buffering. It also illustrates the nature of amino acids and the ways in which these are linked to make peptides and proteins. Chapter 2 affords a detailed understanding of enzymes and how these key catalytic molecules enable our complex metabolism to function. Chapter 3 reviews the essentials of gene transcription and translation, protein breakdown, and the signaling mechanisms involved in the regulation of gene expression and protein synthesis. These chapters form the basis on which subsequent chapters are built.

chapter 1

Amino Acids, Peptides, and Proteins

Proteins are the molecules responsible for what happens in cells and organisms. We can consider them the body's action molecules: They can be enzymes that catalyze all the chemical reactions in the body; receptor molecules inside cells, in membranes, or on membranes that bind specific substances; contractile proteins involved in the contraction of skeletal, smooth, and heart muscle; or transport molecules that move substances in the blood, within cells, and across membranes. Proteins are also parts of bones, ligaments, and tendons; in the form of antibodies or receptors on lymphocytes, they help protect us from disease. Peptides are like small proteins that often have specific regulatory roles. Examples of important peptides are hormones, such as insulin, and growth factors, such as insulin-like growth factors (IGFs). This chapter introduces the basic characteristics and behavior of amino acids and the peptides and proteins they join together to form. The Next Stage section highlights the study of proteomics and its relationship to exercise and training.

Twenty different amino acids are used to make proteins. The genes in our cell nuclei contain the information needed to specify which amino acids are used in making a protein and in what order. The genes in chromosomes are segments of deoxyribonucleic acid (DNA), which is a large molecule containing four different bases. The sequence of bases in a **gene** spells out the sequence of amino acids for a protein. Because of the huge size of DNA molecules, only small segments of the genes are copied at any time. The copies are in the form of new molecules known as messenger ribonucleic acid (mRNA). The copying process is known as **transcription**. The mRNA information is then used to order the binding of the 100 or more specific amino acids to make a protein. This step, known as **translation**, takes place outside of the nucleus on ribosomes.

Proteins are continually being turned over: Old protein molecules are broken down to their constituent amino acids in a process known as **degradation**, and new protein molecules are synthesized. This continual synthesis and degradation of proteins in cells is known as protein turnover (see figure 1.1). Amino acids released when a protein molecule is broken down can be reutilized. We can express the rate of turnover of an individual protein as the time taken for one-half of the protein molecules to be replaced, that is, the **half-life**. Some proteins have a short half-life, measured in minutes or hours. An example is ornithine decarboxylase (Jennissen 1995), an enzyme involved in the biosynthesis of polyamines, which has DNA stabilizing functions and is important for cell growth. Others have a half-life that may be expressed in days. For example, skeletal muscle sarco(endo)plasmic reticulum Ca^{2+}-ATPase is a Ca^{2+} transport protein that induces relaxation of muscle cells by pumping Ca^{2+} from the cytosol into the sarcoplasmic reticulum (Ferrington, Krainev, and Bigelow 1998). The total content of a particular protein in a cell can reflect the use of that protein. Consider how fast muscles atrophy when a limb is immobilized by a cast. Most of the loss of muscle can be attributed to loss of contractile proteins, due to a decrease in contractile protein synthesis that is not balanced by a decrease in contractile protein degradation. Research examining the role of specific exercise training, nutrition, and ergogenic aids in the turnover of skeletal muscle proteins is thriving, both

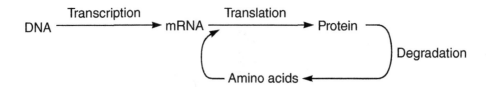

Figure 1.1 A simple outline of the lifetime of a protein, begun when its gene is transcribed, making mRNA. The mRNA information is translated into the sequence of amino acids in the protein, followed by the breakdown of the protein to its constituent amino acids.

from the perspective of basic science and in terms of the practical implications for sport and health. The regulation of protein turnover is discussed in greater detail in chapters 3 and 8.

NATURE OF AMINO ACIDS

Amino acids play key roles in the body, especially as building blocks for proteins and peptides. Of the 20 amino acids needed to make proteins and peptides, at least 9 cannot be synthesized by the body. Like vitamins and minerals, they must be obtained from the diet. As the name implies, amino acids have characteristics of simple acids, and they also contain at least one group that can act as a base.

Acids, Bases, and Buffers

Our cells operate within a narrow pH range. Our blood pH, although slightly higher than intracellular pH, is carefully regulated by the action of buffers. Deviations above and below normal pH levels lead to conditions known as **alkalosis** and **acidosis**, respectively. Severe exercise causes a temporary acidosis, first in the active muscle fibers and then in the blood. Historically, acidosis was considered to be the major cause of muscle fatigue; however, as chapter 6 discusses further, it is now understood that acidity associated with lactic acid accumulation may actually help to maintain muscle performance during intense exercise (Allen, Lamb, and Westerblad 2008). Acidosis can also be caused by starvation and diabetes, while alkalosis can occur with uncontrolled vomiting. Amino acids have groups that can behave as both acids and bases. Other important acids and bases are generated by our metabolism.

In high school, you learned about **acids** and **bases**, often by dealing with strong acids such as hydrochloric or sulfuric acids or strong bases such as sodium and potassium hydroxide. You learned that an acid is a substance that, in water, releases a proton (H^+), and that a base releases hydroxide (OH^-) ions in water. These definitions were originally advanced by Arrhenius, and they serve us well when we consider strong acids and bases. However, the range of acidity and basicity in our bodies is very narrow, hovering primarily around a pH of 7. The *pH scale* is based on the negative logarithm of the hydrogen ion concentration ($-\log [H^+]$) (where square brackets appear, read "concentration of"). A pH of 7 means that the $[H^+]$ is 10^{-7} molar (also written as 1×10^{-7} M). A pH of 7 is also neutral in that there is an equal concentration of protons and hydroxide ions. Acidity (acidosis) denotes a pH less than 7, and basicity (alkalosis) denotes a pH greater than 7. Since the pH scale is a logarithmic scale, a difference of one pH unit indicates a 10-fold difference in the concentration of protons. In biological terms, extreme acidity is found in the gastric juice of the stomach, with a pH of about 2 (i.e., gastric juice is roughly 250,000 times more acidic than the blood). In chemistry courses, you have probably used acid solutions stronger than gastric juice or bases much stronger than anything we encounter in living organisms. Therefore, biologists use a different concept for acids and bases.

The **Brønsted-Lowry** definitions for acids and bases are more effective than the complete pH scale for understanding acid and base properties of cells and tissues and fluids in our bodies. An acid is a proton donor whereas a base is a proton acceptor. If

✓ KEY POINT

A proton is simply a hydrogen atom minus its only electron. As such, it is a very tiny positively charged particle that interacts well with water molecules. You probably learned that a proton is associated with a single water molecule (hydrated) in what is described as a hydronium ion (H_3O^+). It is more likely that a proton in water interacts with four water molecules, existing as $H_9O_4^+$. For our purposes, we will just use H^+, but understand that in the human body, this naked proton is hydrated with about four water molecules.

we have pure water, with a [H⁺] of 10^{-7} M (pH = 7), adding an acid to it will result in an increase in the [H⁺] because the acid will donate more protons. If we bubble ammonia gas (NH₃) into our pure water, the [H⁺] will decrease because the ammonia will act as a base and remove some of the protons from water, generating the ammonium ion (NH_4^+); this will raise the pH.

We can define the strength of acids and bases in biological systems by using a simple system of measurement, the pK_a. We will illustrate this system by using a general acid that we designate HA. This is an acid because it can donate a proton through a process called **dissociation**. It is not a strong acid, such as HCl, and thus does not dissociate completely. Therefore, in the following equation, incomplete dissociation is shown by a double-headed arrow, indicating that an equilibrium is established in which there exist both undissociated acid (HA) and its conjugate base (A^-).

$$HA \leftrightarrow H^+ + A^-$$
(acid) (proton) (conjugate base)

Because weak acids do not dissociate completely, we can write an equilibrium expression for the reversible dissociation using square brackets to represent concentration. The **acid dissociation constant** K_a reflects the degree to which the acid is dissociated.

$$K_a = [H^+][A^-] / [HA]$$

The stronger the acid (HA), the more it is dissociated, the lower the pH, and the higher the concentration of the conjugate base (A^-). Therefore, the larger the numerical value for K_a, the stronger the acid.

Suppose we dissolve HA in a solution containing the sodium salt of A^- (i.e., NaA) such that at equilibrium, the concentration of undissociated acid, HA, is equal to the concentration of its conjugate base, A^-. Now the equilibrium equation is simplified:

$$[H^+] = K_a$$

The other terms ([HA] and [A^-]) cancel out since we are creating the circumstances in which they have the same value.

Now we will take the negative logarithm of both sides of this equation:

$$-\log [H^+] = -\log K_a$$
or
$$pH = pK_a$$

By definition, pH is the negative logarithm of [H⁺]. Thus, pK_a is the negative logarithm of K_a, the acid dissociation constant.

The pK_a for an acid is the pH of a solution in which the acid is one-half dissociated; that is, 50% of the molecules are dissociated and 50% are not dissociated. Therefore, the smaller the value of pK_a for an acid, the stronger the acid. Let's illustrate this concept by using two substances, acetic acid (CH₃COOH—a carboxylic acid) and ammonia, each dissolved in a separate beaker. We know that acetic acid will dissociate as shown in the following.

$$CH_3COOH \leftrightarrow CH_3COO^- + H^+$$
(acetic acid) (acetate)

Acetic acid is a weak acid, and we can write an equilibrium expression for it. Note that the K_a for acetic acid is 1.8×10^{-5}.

$$1.8 \times 10^{-5} = [H^+][CH_3COO^-] / [CH_3COOH]$$

Using the definition for pK_a as just shown, the pK_a for acetic acid would be $-\log 1.8 \times 10^{-5}$ or $-(\log 1.8 + \log 10^{-5})$, which turns out to be 4.74. Assume we start out with a solution of sodium acetate and acetic acid so that at equilibrium, the concentration of undissociated acetic acid equals the concentration of acetate ions. The pH where this would occur would be 4.74—that is, the pH of a solution in which the acetic acid is 50% dissociated.

When the gas ammonia is bubbled into water, some of the ammonia removes a proton from water, forming the ammonium ion (NH_4^+). Free ammonia (NH₃) would also be present in the solution. Using our description for acids as proton donors, we could say that the ammonium ion is an acid and the ammonia is its conjugate base. Therefore, we could write the dissociation of the acid NH_4^+ as follows:

$$NH_4^+ \leftrightarrow NH_3 + H^+$$

From this, we could write an equilibrium expression:

$$K_a = [NH_3][H^+] / [NH_4^+]$$

Looking up the value for the acid dissociation constant for the ammonium ion, we get the value 5.6×10^{-10}. Recall that the larger the K_a, the stronger the acid; we can see that ammonium ion is a much weaker acid than acetic acid. Of course, this means that ammonia is a much stronger base than acetate. During very severe exercise, ammonia is produced by the deamination (removal of an amino group) of adenosine monophosphate, or AMP. The ammonia can remove a proton from the cytoplasm of the exercising muscle cell to help maintain muscle cell pH, as we will see in chapter 4. Can you determine the pK_a for the ammonium ion?

Buffers are chemicals that resist changes in the pH of a solution. They play enormously important roles in our bodies given the very narrow pH range at which our metabolism operates. They do this by maintaining the [H$^+$] within a very narrow range through absorbing added protons or reducing the effects of added OH$^-$ ions. A buffer usually consists of a mixture of a weak acid and its salt, such as acetic acid and sodium acetate. Buffers also work best in the pH range just above and just below the pK$_a$ for the acid. Consider the preceding equation for the dissociation of the weak acid acetic acid. With a pK$_a$ of 4.74, acetic acid would function best as a buffer just above and below pH 4.74, where there would be a roughly equal concentration of CH$_3$COOH (undissociated acetic acid) and its conjugate base acetate (CH$_3$COO$^-$). Addition of H$^+$ ions would result in their being attached to the acetate, forming undissociated acetic acid. Addition of hydroxide ions would result in their being combined with H$^+$ ions to form water, and more acetic acid would dissociate to maintain the H$^+$ balance. Similarly, a solution with an equal concentration of NH$_3$ (ammonia) and NH$_4^+$ (ammonium ion) would buffer best at the pK$_a$ for ammonium ions (9.26).

Physiological buffers need to operate at roughly pH 7, since this is the internal pH of most cells. **Inorganic phosphate**, often denoted by the abbreviation Pi, is a mixture of dihydrogen (H$_2$PO$_4^-$) and monohydrogen (HPO$_4^{2-}$) ions. The H$_2$PO$_4^-$ is a weak acid whose dissociation is shown in the following:

$$H_2PO_4^- \leftrightarrow H^+ + HPO_4^{2-}$$

The pK$_a$ for dissociation of the dihydrogen phosphate ion is 7.2, so a mixture of the dihydrogen and monohydrogen phosphates is a good buffer in the physiological pH range. Inside cells, the free concentration of Pi (the mixture of the two phosphates) is at least 1 millimolar (1 mM or 1 mmol/L). In addition, there are other phosphate-containing compounds, such as ATP (adenosine triphosphate), whose phosphate groups can also participate in pH control of the cell. Can you predict which of the two species of Pi would dominate in a skeletal muscle cell when the pH is reduced by severe exercise? What if the pH were increased above 7 by alkalosis?

Before leaving this important topic, let's revisit the dissociation of the general weak acid, HA.

$$HA \leftrightarrow H^+ + A^-$$

From this, write the equilibrium expression:

$$K_a = [H^+][A^-] / [HA]$$

Rearrange this to produce:

$$[H^+] = K_a [HA] / [A^-]$$

Take negative logs of each side:

$$-\log[H^+] = (-\log K_a) - (\log [HA] / [A^-])$$

or

$$pH = pK_a + \log [A^-] / [HA]$$

This last equation is known as the **Henderson-Hasselbalch equation** and is a very useful way of expressing the relationship between the pH of a solution, the pK$_a$ of a weak acid, and the relative concentrations of the weak acid and its conjugate base. This equation will prove to be very useful when we talk about the acid-base characteristics of the amino acids.

Structure of Amino Acids

The genetic code provides information for 20 different amino acids that make up proteins. Figure 1.2 illustrates what is typically shown for the general structure of an amino acid. In this figure, each amino acid has a carboxyl group (a weak acid), an amino group (a weak base), a carbon atom identified as the alpha (α)-carbon (because it is adjacent to the acidic carboxyl group) (also known as carbon 2), and a variable group known as the side chain and indicated by the letter R. In fact, 19 of the 20 common amino acids have this general structure. The exception, proline, has an imino group (—HN—) instead of an amino group. What makes each amino acid unique is its side chain, or R group. Each amino acid has (a) a specific side chain (designated as an R group), (b) a name, (c) a three-letter designation, and (d) a single capital letter to represent it. Differences in the properties of the 20 common amino acids are based on their side chains.

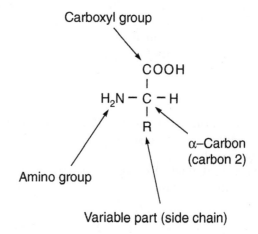

Figure 1.2 The general structure of an amino acid, showing its weak acid (carboxyl) group and weak base (amino) group. These are attached to a central or alpha (α)-carbon, along with a hydrogen atom and its variable part, given by R.

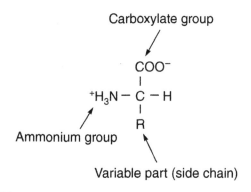

Figure 1.3 The general structure of an amino acid as it would exist at cellular pH of 7.

However, the general amino acid structure depicted in figure 1.2 never occurs exactly as shown because of the acid–base properties of the amino and carboxyl groups. Figure 1.3 shows the general structure of an amino acid as it would exist at pH 7. This is often described as the **zwitterion** form, which includes both a positively and a negatively charged group.

Figure 1.4 shows each of the 20 amino acids, including the name, three-letter short form, single-letter identification, and the overall structural formula with the specific side chain. Amino acids are typically organized on the basis of the side chain structure and how this will interact with water. Charged polar and uncharged polar side chains will interact with polar water molecules.

> **✓ KEY POINT**
>
> The 20 amino acids used to make proteins are the same ones used by all animal and plant life. Diversity in the structure of amino acids helps produce the diversity in the nature and function of proteins.

Acid–Base Properties of Amino Acids

All amino acids have at least one acid group (proton donor) and one basic group (proton acceptor). Therefore, amino acids may have both acid and base properties; that is, they are **amphoteric**. In amino acids, the major acid groups are the carboxyl (–COOH) and the ammonium (–NH$_3^+$) groups. These protonated forms can each give up a proton. The major base groups are the carboxylate (–COO$^-$) and amino (–NH$_2$) groups, which are unprotonated and can accept a proton.

Now let us look at equations describing the ionization of the protonated (acid) forms of the groups:

$$-COOH \leftrightarrow -COO^- + H^+$$

will have a K_a value and hence a pK_a.

$$-NH_3^+ \leftrightarrow -NH_2 + H^+$$

will also have a K_a and thus a pK_a.

The carboxyl group (–COOH) is a stronger acid than ammonium (–NH$_3^+$), and it will have a lower pK_a value (or its K_a value will be larger). Conversely, the amino group will be a better proton acceptor (base) than the –COO$^-$, and it will have a larger pK_a.

Most of the amino acids have two groups that can act as acids, each with a pK_a, usually identified as pK_1 (the lower value) and pK_2 (the higher value). Figure 1.5 shows the structure of the amino acid alanine when it is a zwitterion. In this form, alanine has no net charge, but it has one positive and one negative charge. The pH at which a molecule has an equal number of positive and negative charges (and therefore no net charge) is its **isoelectric point**, designated pI. For amino acids like alanine, the pI is one-half the sum of pK_1 and pK_2.

Monoamino-dicarboxylic amino acids, such as aspartic acid and glutamic acid (see figure 1.4), have two groups that can be carboxyl or carboxylate, and only one group that can be amino or ammonium. We call such groups acidic amino acids. Three pK_a values appear for these, identified from lowest to highest as pK_1, pK_2, and pK_3. As shown in figure 1.6, the isoelectric point (pI) of aspartic acid will be the pH where each molecule has one positive and one negative charge, yet no net charge. The actual value for the pI will be one-half the value of pK_1 and pK_2, that is, 2.8. Because aspartic and glutamic acids always have a negative charge at cellular pH, they are almost always described as aspartate and glutamate, respectively.

Amino acids such as lysine (see figure 1.4) have two amino/ammonium groups and only one carboxyl/carboxylate group, and are often called basic amino acids. Figure 1.7 shows the structure of lysine at its pI. Notice that lysine has no net charge, due to an equal number of positive and negative groups. The pI value for lysine will be one-half the sum of pK_2 and pK_3, that is, 9.8.

In addition to the major ionizable groups discussed so far (i.e., the amino and carboxyl groups

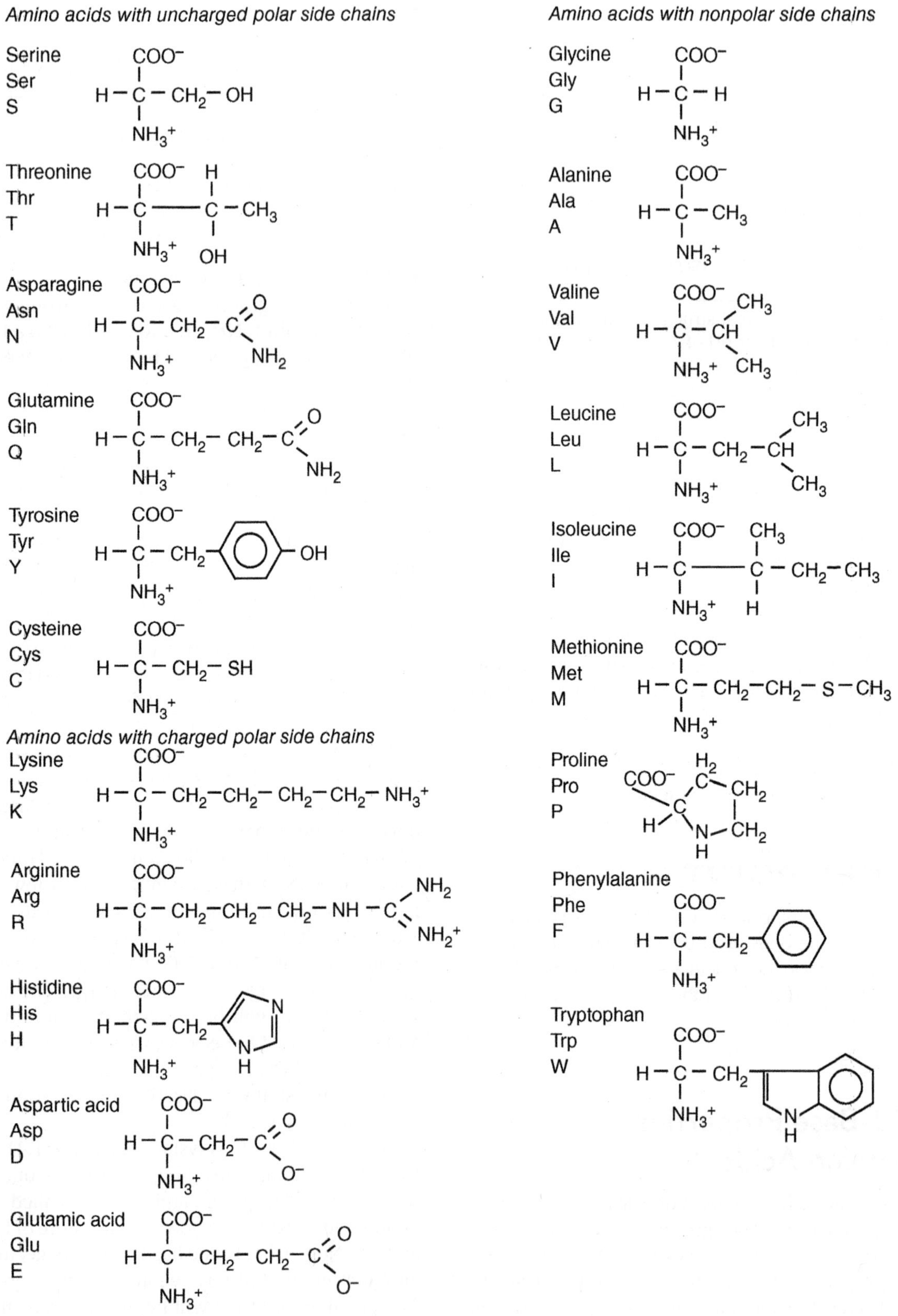

Amino acids with uncharged polar side chains

Serine
Ser
S

Threonine
Thr
T

Asparagine
Asn
N

Glutamine
Gln
Q

Tyrosine
Tyr
Y

Cysteine
Cys
C

Amino acids with charged polar side chains

Lysine
Lys
K

Arginine
Arg
R

Histidine
His
H

Aspartic acid
Asp
D

Glutamic acid
Glu
E

Amino acids with nonpolar side chains

Glycine
Gly
G

Alanine
Ala
A

Valine
Val
V

Leucine
Leu
L

Isoleucine
Ile
I

Methionine
Met
M

Proline
Pro
P

Phenylalanine
Phe
F

Tryptophan
Trp
W

Figure 1.4 Structures of the 20 amino acids used to make proteins, with their names, three-letter abbreviations, and single-letter identifiers.

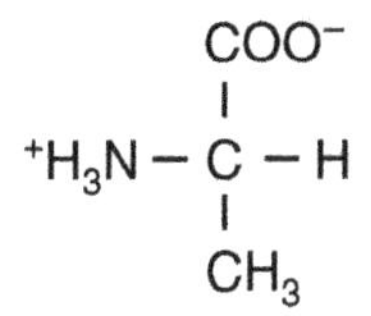

Figure 1.5 Structure of alanine at its isoelectric point.

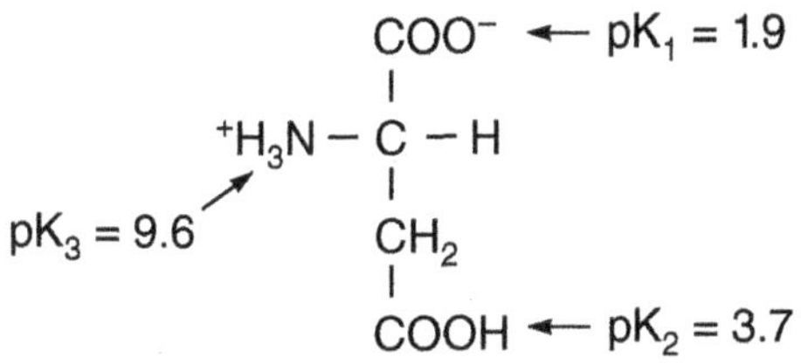

Figure 1.6 Structure of aspartate (aspartic acid) at its isoelectric point. The pK_a values for the three dissociable groups are shown in order of increasing magnitude.

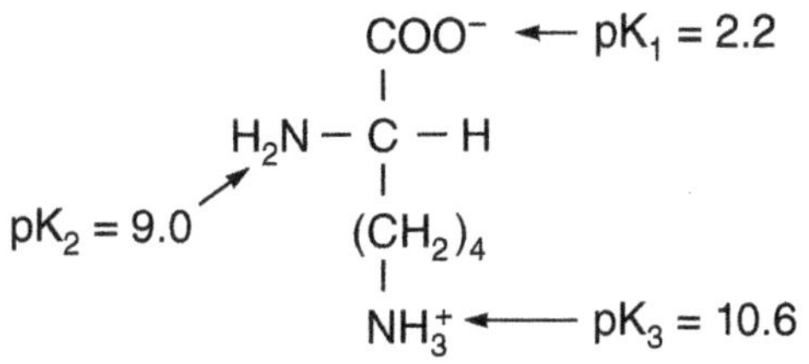

Figure 1.7 Structure of lysine at its isoelectric point. The pK_a values for the dissociable groups are shown.

attached to the α-carbon and side chain), others can be quite important (see figure 1.8). Histidine is often found at the active site of many enzymes because it influences their catalytic ability.

KEY POINT

The pH of intracellular and extracellular fluids must be carefully regulated, since even a small change in pH can add or remove a proton from an amino acid. Such a change may alter the structure and thus the function of proteins, altering our metabolism.

Stereoisomerism of Amino Acids

As shown in figure 1.4, all amino acids except glycine have four different groups attached to the α-carbon. Because of this, the α-carbon is a chiral center and the amino acid is chiral or asymmetric, with two different ways of arranging these groups; that is, the amino acid has two different configurations. Figure 1.9 shows the groups around the α-carbon to be three-dimensional; the dashed lines mean that the bonds are going into the paper, and the wedges mean that the bonds are coming out of the plane of the paper toward you. When the carboxylate group is at the top and is going into the paper, the L and D refer to the position of the ammonium group, that is, on the left side (L or *levo*) or right side (D or *dextro*). These **stereoisomers** (or space isomers) are also enantiomers—pairs of molecules that are nonsuperimposable mirror images of each other. They may be compared to a right and left hand. When you hold your left hand to a mirror, the image is that of your right hand. Similarly, holding a D-amino acid to a mirror gives an L-amino acid as the image. All naturally occurring amino acids are in the L-configuration.

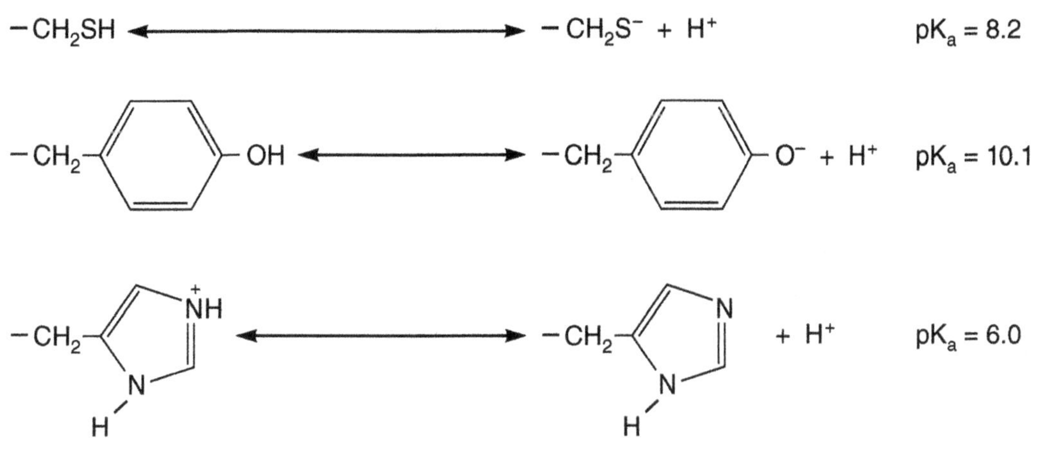

Figure 1.8 Equations to show the acid dissociation characteristics for the side chains of the amino acids cysteine (top), tyrosine (middle), and histidine (bottom). The numerical value for the pK_a for each dissociation is shown.

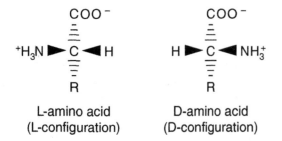

Figure 1.9 The two stereoisomeric forms for the general structure for an amino acid. The amino acids in our bodies are in the L-configuration.

CHARACTERISTICS OF PEPTIDES

Peptides are formed when amino acids join together via their ammonium and carboxylate groups in a specialized form of the amide bond known as a **peptide bond**. Because the body's amino acids are in an environment with a pH around 7, the amino group is protonated, and it is an ammonium group ($+H_3N$). In contrast, the carboxyl group is unprotonated, and it is a carboxylate group (COO^-). Figure 1.10 shows how a peptide is formed from two amino acids. Combination of the ammonium and carboxylate groups yields the peptide bond. This reaction, called condensation, eliminates a water molecule. The peptide bond is rigid and planar. Peptides are drawn by convention starting with the free ammonium group and ending with the free carboxylate group. We describe these as the **N-terminus** and the **C-terminus**, respectively.

The amino acids in a peptide are known as **amino acid residues**. The prefixes used to characterize the number of amino acid residues in a peptide are *di*—two, *tri*—three, *tetra*—four, *penta*—five, *hexa*—six, *hepta*—seven, *octa*—eight, *nona*—nine, and *deca*—ten. Thus, a nonapeptide consists of 9 amino acid residues. An *oligopeptide* is one that contains roughly 10 to 20 amino acid residues. The term **polypeptide** refers to a large peptide that contains more than 20 amino acid residues; indeed, some polypeptides have more than 4,000 amino acid residues. A protein may be a polypeptide if there is only one polypeptide in the molecule. However, many proteins contain more than one polypeptide chain. In this case, the protein is not a polypeptide, but a molecule containing two or more specific polypeptides. Since a peptide has only one N-terminus and one C-terminus, its properties are defined by the properties of the individual side chains of its constituent amino acids.

> ### ☑ KEY POINT
> Peptides are linear (as opposed to branched) molecules. This is a natural consequence of the linear sequence of bases within a gene in DNA, which is responsible for specifying the linear sequence of amino acids in the polypeptide chain. The **primary structure** of a peptide is the sequence of amino acids starting from the N-terminus.

The peptide shown in figure 1.10 is a dipeptide. We would name it glycylalanine, starting from the amino acid at the N-terminus. We could also refer to this as gly-ala, using the three-letter short form, or GA, using the single-letter amino acid designations. A very important tripeptide appears in all cells that plays a key role in maintaining our cellular oxidation–reduction (redox) status and regulates the activity of several enzymes (see chapter 2). Known as **glutathione**, it is made up of the amino acids glutamate, cysteine, and glycine. The first peptide bond between glutamate and cysteine, catalyzed by the enzyme γ-glutamylcysteine ligase, is via the side chain carboxyl of glutamate—not the usual α-carboxyl group. This is the gamma (γ)-carbon atom of glutamate. We thus designate glutathione as γ-glutamylcysteinyl-glycine (γglu-cys-gly or γECG). Unlike the majority of peptides that are synthesized on ribosomes following an RNA template, enzymatic synthesis of glutathione is independent of nucleic acid.

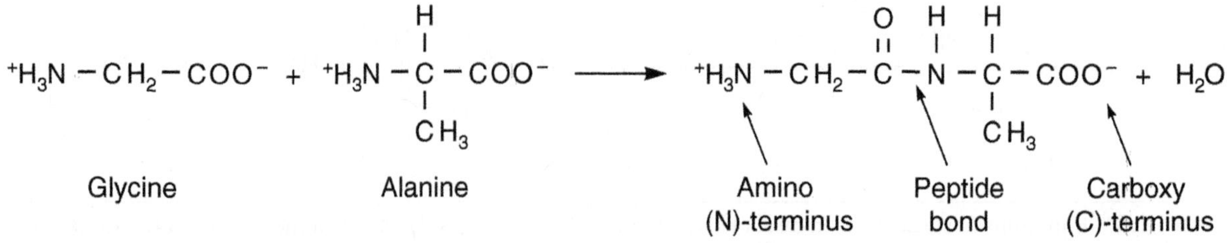

$$^+H_3N-CH_2-COO^- + {}^+H_3N-\underset{CH_3}{\underset{|}{C}}H-COO^- \longrightarrow {}^+H_3N-CH_2-\underset{}{\overset{O}{\overset{\|}{C}}}-\underset{}{\overset{H}{\overset{|}{N}}}-\underset{CH_3}{\underset{|}{C}}H-COO^- + H_2O$$

Glycine   Alanine   Amino (N)-terminus   Peptide bond   Carboxy (C)-terminus

Figure 1.10 The combination of two amino acids, glycine and alanine, to form the dipeptide glycylalanine.

Many important hormones are peptides, such as **insulin**, **glucagon**, and **growth hormone**. **Growth factors** have specific, potent effects on tissues, and these are peptides. The hypothalamus, a key regulator of the overall function of our body, releases a number of peptide factors that control the secretions from other glands, particularly the pituitary gland. As we will see, even fat cells produce and secrete important peptide-signaling molecules.

STRUCTURE OF PROTEINS

Proteins are composed of one or more polypeptide chains. Many also contain other substances, such as metal ions (e.g., hemoglobin contains iron), carbohydrate (e.g., glycoproteins contain sugars attached to amino acids; these can be found on cell membranes, pointing into the fluid surrounding cells), and fat or lipid (e.g., blood lipoproteins transport fat such as cholesterol and triacylglycerol). The precise biological structure of a protein, and hence its function, is determined by the amino acids it contains.

Primary Structure of Proteins

As mentioned earlier, the primary structure of a protein (or peptide) is the sequence of amino acids starting from the N-(amino) terminus. With 20 different amino acids, the number of different possible sequences for even a small polypeptide is enormous. However, many proteins in your body rely on very exact amino acid sequences. Even replacing one amino acid with another could make a protein totally useless and possibly dangerous. For example, sickle-cell anemia, characterized by short-lived erythrocytes of unusual shape, results from the replacement of just one amino acid (valine) with another (glutamic acid) in one chain of the hemoglobin molecule. As shown in figure 1.4, valine has an uncharged hydrophobic side chain, whereas the side chain of glutamic acid carries a negative charge. Even this modest change can have a profound influence on the properties of the hemoglobin molecule. The cystic fibrosis transmembrane conductance regulator protein (CFTR) allows chloride ions to move across membranes. This protein contains 1,480 amino acids. The disease associated with this protein, cystic fibrosis, is due to the absence of a single phenylalanine residue (phe-508). However, in other cases, one amino acid may substitute for another amino acid with very similar properties (e.g., arginine for lysine, aspartate for glutamate), with little effect on the structure or function of the protein.

The amino acid sequences of many thousands of proteins in humans and other species are now known. The sequence can be learned through actual sequencing of the amino acids in each polypeptide or learned from the base sequence of the gene for that polypeptide. The sequence always begins at the N-terminus of each polypeptide chain if the protein has more than one. The N-terminus is designated number 1, with succeeding amino acids given higher numbers. Since we know the sequence of so many polypeptides, the specific amino acid is described along with its position in the polypeptide. For example, serine-283 (ser-283) is the 283rd amino acid residue in the polypeptide, and it is serine. Amino acid sequence determines the three-dimensional structure of the protein. Moreover, with computers, amino acid sequences, and known three-dimensional structures, one can make computer-guided predictions about the three-dimensional structure of a new protein, given its sequence. Scientists in biotech companies throughout the world are designing peptides to improve human function. The study of large compilations of proteins, as those expressed in a single organelle, cell, organ, or organism, is known as **proteomics**. Researchers in this area are proceeding at a rapid pace; in many cases, they are looking for specific differences in amino acid sequence that underlie particular diseases. For more on new developments in this field, see the Next Stage section at the end of this chapter.

Bonds and Interactions Responsible for Protein Structure

Proteins have a very specific structure in the cell that is essential for a protein to perform its unique function. As we have seen, the primary structure, or the sequence of amino acids in the polypeptide chain, is maintained by peptide bonds. These are strong, rigid bonds joining the individual amino acids together in the protein. However, the three-dimensional structure of proteins is very important to their function. The three-dimensional structure is maintained by a variety of strong and weak bonds. Figure 1.11 illustrates these bonds and how they are responsible for the overall three-dimensional shape of a protein.

Disulfide bonds are covalent bonds that join together the side chains of two cysteine residues ($-CH_2SH$) in the same polypeptide or different polypeptides, resulting in disulfide (S–S) bonds and the loss of two hydrogen atoms. Disulfide bonds in the

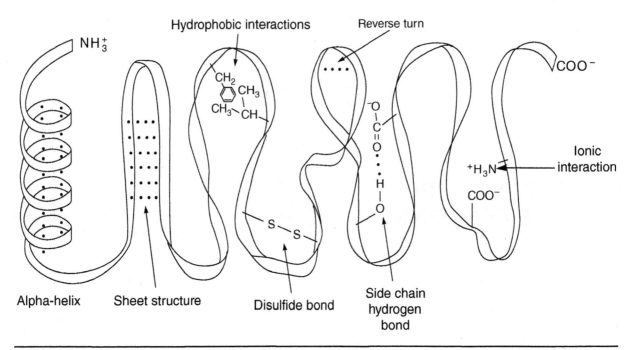

Figure 1.11 A schematic to show common bonds and interactions that maintain the secondary and tertiary structure of a protein. The ribbon shows the peptide backbone. The dotted lines represent hydrogen bonding.

same polypeptide chain are intramolecular bonds, whereas those joining different polypeptide chains are intermolecular bonds. For example, the polypeptide hormone insulin consists of two peptide chains held together by two intermolecular disulfide bonds.

Hydrogen bonds are weak electrostatic attractions between an electronegative element, such as oxygen or nitrogen, and a hydrogen atom that is covalently bonded to another electronegative element. Hydrogen bonds between individual water molecules, for example, are responsible for some of the unique properties of water. Figure 1.11 shows the hydrogen bond between the components of a peptide bond; such bonds are important in maintaining overall protein structure. Hydrogen bonds also form between polar parts of some amino-acid side chains and those in the protein molecule and water (see polar side chains of amino acids in figure 1.4). Although hydrogen bonds are weak, they are important because so many of them are involved in the structure of a protein. Figure 1.11 shows *ionic interactions* between oppositely charged groups, as found in proteins, such as the N- and C-termini and the charged carboxylate or ammonium groups in the side chains.

Amino acids are primarily found in a polar environment because they are in an aqueous medium with charged ions and polar molecules. Proteins spanning membranes are an exception because the interior of a membrane is hydrophobic. Water is a polar molecule and interacts well with polar and charged groups on amino acids. This means that inside cells and in the blood, polar and charged amino acids are most likely to be exposed to their aqueous environment. However, 9 of the 20 common amino acids are **hydrophobic**, or nonpolar, because their side chains are composed of carbon and hydrogen atoms and thus have no affinity for water (see figure 1.4). In fact, they can disrupt the relatively organized structure of liquid water. These side chains tend to cluster in the interior of the protein molecule, not because they have an affinity for each other but because they cannot interact with water. Fat globules also form tiny spheres in water to keep their exposed surface as small as possible. We call the clustering of hydrophobic groups on amino acids hydrophobic interactions (see figure 1.11).

Secondary Structure of Proteins

So far, a one-dimensional protein structure has been described, that is, the sequence of amino acids. As already mentioned, proteins actually have a three-dimensional structure. Imagine a peptide backbone with the side chains of each amino acid radiating out. The backbone consists of three elements—the α-carbon, the nitrogen group involved in the peptide bond, and the carbonyl group. The **secondary structure** is the spatial path taken by these three elements of the peptide backbone. Side chains (or R groups) of individual amino acids are not considered in secondary structure. Two common, recognizable structural features of the polypeptide backbone, the **alpha-helix** and the **beta-sheet**, are found in many proteins. These features result from the way the polypeptide backbone organizes itself. They are stabilized by hydrogen bonding between neighboring elements of the peptide bond, as shown in figure 1.11 with dotted lines. The reverse turn, also shown in figure 1.11, involves a sharp turn in the polypeptide backbone and is also maintained by hydrogen bonding.

> ☑ **KEY POINT**
>
> Proteins adopt stable, folded conformations mainly through noncovalent (hydrophobic, hydrogen, ionic) bonds. Covalent disulfide bonds between cysteines in different parts of the polypeptide chain also enhance the structural stability of proteins.

Tertiary Structure of Proteins

The tertiary structure is the overall three-dimensional arrangement of a polypeptide chain, including the secondary structure and any nonordered interactions involving amino acid residues that are far apart. The detailed three-dimensional structure of proteins is determined through X-ray crystallography, a technique that was perfected by Max Perutz and John Kendrew and for which they were awarded the Nobel Prize in chemistry in 1962. For a historical review on protein structure analysis using X-ray crystallography, interested readers should see Strandberg (2009). The very first high-resolution structure of a protein molecule was that of myoglobin, published by Kendrew and colleagues in 1960. Myoglobin is found in the muscles, and enables oxygen to be stored there.

Figure 1.12 shows the three-dimensional structure of the small oxygen-binding protein, myoglobin. This contains a single polypeptide chain and a single **heme** group. Most proteins are found in aqueous mediums where they assume a compact globular structure maintained by the forces already described (see figure 1.11). Examples include proteins in the blood and enzymes within cells. The hydrophobic side chains are buried in the interior of the protein, away from the aqueous medium, with **hydrophilic** side chains and the N- and C-termini exposed on the surface where they can interact with water or other polar molecules and ions. The tertiary structure also involves the spatial position of ions or organic groups that are part of the makeup of many proteins. The term **amphipathic** is often used to describe a part of

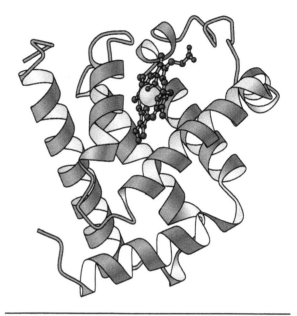

Figure 1.12 Three-dimensional view of the small oxygen-binding protein, myoglobin. In the upper middle portion, the heme group can be seen with its central iron ion (sphere). Much of the secondary structure of myoglobin is alpha-helices.

Reprinted, by permission, from J.M. Berg, J.L. Tymoczko, and J.L. Stryer, 2002, *Biochemistry*, 5th ed. (New York: W.H. Freeman), 31.

a protein that must interact with both a hydrophobic and a hydrophilic region. *Amphipathic* means that the section of the protein will have a polar region able to interact, for example, with the cytoplasm of the cell, and a nonpolar region that may interact with the hydrophobic part of the cell membrane.

Longer polypeptides are often folded into several globular units, called **domains**. For example, the tertiary structure of the Ca^{2+} transporting ATPase of the sarcoplasmic reticulum in muscle (known as the SERCA pump), which is a single polypeptide chain consisting of 1,001 amino acid residues, consists of a transmembrane (TM) domain with 10 TM helices, three well-separated cytoplasmic domains (A, actuator; N, nucleotide-binding; P, phosphorylation), and small luminal loops (see figure 1.13). Most domains in proteins contain between 50 and 300 amino acid residues. Generally, the residues forming a single domain are found within one continuous segment of the primary structure, but in some cases, a single protein domain may consist of sequentially noncontinuous segments because of the complex folded shape of the protein. For example, the P-domain shown in figure 1.13 is composed of two parts widely separated in the amino acid sequence, with the N-terminal part spanning residues 330-359 and the C-terminal part spanning residues 605-737 (Toyoshima et al. 2000).

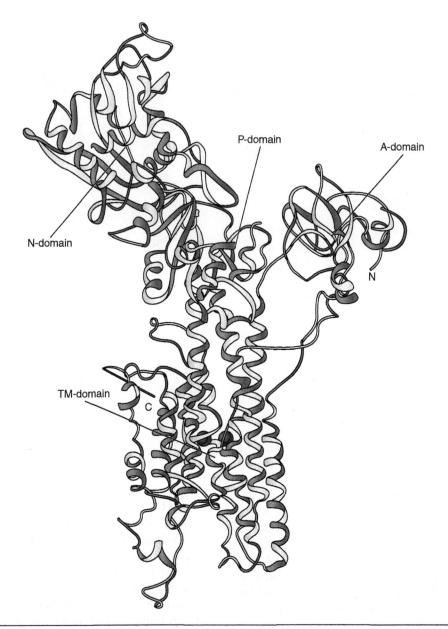

Figure 1.13 Ribbon diagram representing the crystal structure of the SERCA pump from rabbit skeletal muscle showing the transmembrane and cytoplasmic domains.

Adapted from *Journal of Molecular and Cellular Cardiology*, Vol. 34, D.H. MacLennan, M. Abu-Abed, and C.-H. Kang, "Structure-function relationships in Ca2+ cycling proteins," pgs. 897-918, Copyright 2002, with permission from Elsevier.

The secondary and tertiary structure of a polypeptide chain depends on the kind and sequence of its amino acids. In the cell, or wherever a protein is found, the overall structure or conformation of a protein must be maintained, because proteins must recognize and interact with other molecules. Biochemists describe a protein's three-dimensional form as its fold. However, the conformation of a protein in vivo is not fixed but changes in subtle ways as it carries out its particular function. For example, the crossbridges in the contractile protein myosin alter their conformation to generate force during muscle contraction. The SERCA pump protein undergoes large conformational changes as it pumps Ca^{2+} across

 **KEY POINT**

The sequence of bases in a gene specifies the amino acid sequence of a polypeptide, which determines secondary and tertiary structure through the interactions between the various amino acids. The biological function of the polypeptide is based on its overall structure.

the sarcoplasmic reticulum membrane from the cytosol to the lumen of the sarcoplasmic reticulum.

Quaternary Structure of Proteins

Many proteins consist of more than one polypeptide chain, each containing its own unique primary, secondary, and tertiary structures. We call these chains subunits. **Quaternary structure** refers to the arrangement of the individual subunits with respect to each other. The subunits in an **oligomeric** protein (i.e., one containing more than one subunit) are held together with noncovalent bonds (i.e., hydrogen bonds), electrostatic interactions, and hydrophobic interactions. Oligomeric proteins are common because several subunits allow subtle ways of altering the protein's function. For example, hemoglobin A, the adult form of hemoglobin, is a tetramer consisting of four subunits, two α (with 141 amino acids per subunit) and two β (with 146 amino acids per subunit). Each subunit contains one heme group that binds one oxygen molecule. The quaternary structure of hemoglobin refers to the way the two α and two β subunits interact with each other. The oligomeric nature of hemoglobin enhances its role in loading oxygen in the lung and releasing it at the cell level. In the lungs, where the oxygen concentration is high, binding of one oxygen molecule to one subunit facilitates the binding of oxygen molecules to the other subunits. This helps make hemoglobin saturated with oxygen in the lungs, maximizing its ability to transport oxygen. At the tissue level, where the oxygen concentration is much lower, the subunit interactions facilitate the unbinding of oxygen so that it is available for diffusion into adjacent cells.

Myosin, a contractile protein, is a hexamer (i.e., consists of six subunits). It has two **myosin heavy chains**, each with a molecular weight of about 220,000 (often expressed as 220,000 daltons or 220 kilodaltons, kD). The two heavy chains wind around each other, forming a single long, coiled tail for most of their length. At one end, they each form separate globular regions, known as heads. Each globular head contains two light chain polypeptides, which each have a molecular weight of about 18 to 25 kD. One is described as a regulatory light chain and the other is known as an essential light chain (see figure 1.14). The globular heads act as the crossbridges that attach to the protein **actin** to generate force when a muscle contracts. Each myosin head has an actin-binding domain and an ATP-binding domain. These domains act in concert to utilize the energy of ATP to generate a force and produce movement. Chapter 4 discusses the role of different heavy and light chains in terms of myosin function and fiber typing.

Denaturation of Proteins

The noncovalent forces that maintain the secondary, tertiary, and quaternary structures of proteins are generally weak. **Denaturation** refers to a disruption in the overall conformation of a protein (i.e., unfolding) with loss of biological activity. When denatured, proteins usually become less soluble and clump together, otherwise known as protein aggregation, due to attractions between hydrophobic groups of proteins that would normally be buried inside the

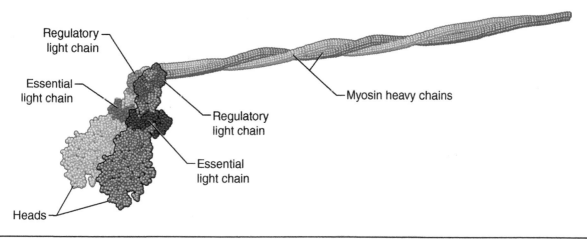

Figure 1.14 The contractile protein myosin. Myosin is a hexamer, composed of six subunits. Two heavy chains run the length of the molecule, terminating in two globular heads that act as crossbridges. Four myosin light chains are also part of the myosin molecule, two per crossbridge head; one is found in the neck region where each globular head begins and another is located further into each globular head.

Reprinted, by permission, from R. Vandenboom, 2004, "The myofibrillar complex and fatigue: A review," *Canadian Journal of Applied Physiology* 29(3): 330-356.

molecules. Denaturation can be caused by heat, acids, bases, organic solvents, detergents, agitation (e.g., beating egg whites to produce a meringue), or specific denaturing agents such as urea. We seldom encounter such harsh conditions that our body proteins are denatured. However, burning the skin or the reflux of acid from the stomach into the esophagus may cause some protein denaturation. Before proteins are broken down in their normal turnover process, they are denatured by enzyme-catalyzed chemical modification, and then broken down to their constituent amino acids. In high concentrations (8 M), urea, a normal product of amino acid breakdown in our cells, can be used as a laboratory denaturant.

NEXT STAGE

Proteomics—The New Systems Approach to Biomedical Health and Exercise Science

In humans, approximately 30,000 genes encode for nearly one million proteins (Humphery-Smith 2004). Proteomics is the study of the proteome, which is defined as the entire protein complement of a genome. Proteomics has expanded tremendously over the past 15 years as a research discipline and has evolved into many different subspecialties related to specific types of research questions about the proteome (for a review, see Kislinger and Gramolini 2010). The goal of *expression proteomics* is to identify all or many of the proteins expressed in a sample of interest. *Posttranslation modification proteomics* is aimed at the identification of an ever-growing body of enzymatic or nonenzymatic protein modifications. *Functional proteomics* involves mapping protein–protein interactions, with the premise being that much can be inferred about a protein's function by the proteins with which it associates, assuming that its protein partners already have well-defined functions. Other subspecialties include *chemical proteomics* (interaction of proteins with specific chemical structures), *structural proteomics* (the large-scale identification of 3-D protein structures), and *quantitative proteomics* (the relative or absolute quantification of proteins). Application of proteomics techniques, such as expression and posttranslation modification proteomics, can help exercise and health scientists answer long-standing questions, such as how endurance training can improve cardiac function and protect against heart disease, by identifying novel protein adaptations to exercise. In turn, this may reveal novel therapeutic targets for the treatment of heart disease.

The proteome varies under all physiological conditions, with pronounced changes observed in aging, disease, and in response to acute and chronic stressors such as exercise. Differences in protein expression, abundance, or structure between samples representing different biological (i.e., young vs. old, healthy vs. diseased) and physiological (i.e., sedentary vs. exercise trained) states may provide clues to the role of certain proteins in these different states. For example, proteomics research has identified a novel cardiac adaptation to endurance training involving an increase in phosphorylation of a protein called heat-shock protein 20, which previously was shown to be associated with improved heart-cell function and protection against cell death (Burniston 2009). Therefore, this adaptation could be important in mediating the beneficial effects of endurance training for the prevention of heart disease.

Proteomics can also be applied as a diagnostic tool in the clinic to predict or prevent diseases such as cardiovascular disease, diabetes, and cancer (Ouzounian et al. 2007). Pharmaceutical companies apply proteomics to identify and validate novel drug targets for a wide range of major human diseases, including cancers, neurodegenerative diseases, and cardiovascular disease (Blackstock and Weir 1999). To date, proteomics has only been applied to the exercise science field by a few researchers (Hittel, Hathout, and Hoffman 2007). However, enthusiasm is growing for the power of proteomics and for systems approaches in general to help us uncover the complexities of cellular and systemic adaptations to specific exercise-training programs or to chronic inactivity. These will ultimately result in improvements or decrements in strength, power, or endurance, respectively (Baldwin 2000; Brooks 2007; Keller et al. 2007).

SUMMARY

Maintenance of protein acid–base balance is critical to life. Acids are substances that can donate protons, and bases are proton acceptors. Buffers are chemicals that can help maintain pH in the face of added acid or base. Proteins are the action molecules of life, made from 20 different amino acids. Differences among amino acids are based on their characteristic side chains, known as R groups. Amino acids exist in all cells of the body and in the fluids outside the cells. Natural amino acids are in the L-configuration. Because each has at least one group that acts as an acid and one that acts as a base, they are said to be amphoteric. All amino acids have at least one charged group, and at the neutral pH associated with life, they have at least two charged groups. When an amino acid has no net charge, it is said to be a zwitterion. The pH where this occurs is known as the isoelectric point. Peptides are formed when amino acids are joined together by peptide bonds. In general, proteins are large peptides (polypeptides), usually containing 100 or more amino acids. The primary structure of a peptide is the sequence of amino acid residues starting from the end containing the free amino group, known as the N-terminus. Because of their large size, proteins have higher levels of three-dimensional structure, known as secondary and tertiary structures. If a protein consists of more than one polypeptide, the structural relationship of the polypeptide subunits is described as a quaternary structure. The forces such as hydrogen bonds and hydrophobic interactions maintaining the secondary, tertiary, and quaternary structures are generally weak.

REVIEW QUESTIONS

1. Using the information found in figure 1.4, answer the following questions about the peptide shown in shorthand: Gly-Arg-Cys-Glu-Asp-Lys-Phe-Val-Trp-Cys-Leu.
   a. How many amino acid residues are there?
   b. Identify the N- and C-terminal amino acids.
   c. What would be the net charge on this peptide at pH 7.0?
   d. Identify the amino acids that have a polar, but uncharged, side chain.
   e. Identify the nonpolar amino acids.
   f. What are the hydrophobic amino acids?
   g. Is it possible for this peptide to have an intramolecular disulfide bond? Explain.
   h. Write the primary sequence using single-letter amino acid abbreviations.
2. Using the Henderson–Hasselbalch equation, predict the pH of a solution in which cysteine carries a net negative charge.
3. Write the dominant molecular formula for the constituent of Pi that exists at pH of 6.0 and then 8.0.
4. Do all proteins have quaternary structure?

chapter 2

Enzymes

Enzymes are proteins that catalyze the thousands of different chemical reactions that constitute our metabolism. As proteins, they are large molecules, made from the 20 different amino acids. As catalysts, enzymes speed up chemical reactions without being destroyed in the process. As proteins, their lifetime is relatively short, so the concentration of most enzymes at any time reflects their relative use. This chapter begins with a review of catalysts, the role enzymes play as catalysts, and key factors that are critical to enzyme action in our cells. Many enzymes function with the aid of helpers or cofactors, especially in reactions where oxidation and reduction take place. We conclude with an overview of how enzymes and transport proteins are regulated and how we can measure enzyme action. The Next Stage section deals with insights on cardiovascular disease resulting from research on the sarcoplasmic reticulum Ca^{2+}-ATPase enzyme (SERCA pump) in vascular smooth muscle.

ENZYMES AS CATALYSTS

Enzymes speed up chemical reactions by lowering the energy barrier to a reaction—called the energy of activation—so that the reaction takes place at the low temperature of an organism (37 °C for a human). In the laboratory, chemical reactions may take place because we heat the reacting substances, overcoming the energy barrier that prevents the reaction from occurring at a lower temperature. Because our bodies function at a relatively constant, but low, temperature (37 °C), enzymes are essential to reduce the energy barrier or energy of activation. In this way, the thousands of chemical reactions that reflect our metabolism can take place speedily. In fact, enzymes are so effective that they can increase the speed of reactions by millions of times the rate of an uncatalyzed reaction.

The molecule (or molecules) that an enzyme acts on is known as its **substrate**, which is then converted into a product or products. For example, when we digest dietary protein (the substrate) with protein-digesting enzymes in our small intestine, we produce amino acids and small peptides as products. Enzymes are highly specific, catalyzing a single reaction or type of reaction. Left to themselves, many chemicals can interact over long periods of time to reach an equilibrium in which there is no net change in substrate or product concentrations. The enzyme allows the reaction to reach equilibrium much faster than it would if the enzyme were absent, but the position of equilibrium remains the same.

A part of the large enzyme molecule will reversibly bind to the substrate (or substrates), and then a specific part or parts of the enzyme will catalyze the detailed change necessary to convert the substrate into a product. The enzyme has amino acid residues that bind the substrate, as well as those that carry out the actual catalysis. There is thus a **binding site** and a **catalytic site**, although the term **active site** often represents both the binding and catalytic domains of the enzyme protein. Active sites are typically clefts in the three-dimensional enzyme structure, where amino acid residues from close or distant parts of the molecule can act on the substrate. The bonds and interactions discussed in the previous chapter maintain the conformation of the active site and interact with the specific substrate molecules.

We can write a general equation to describe a simple enzyme-catalyzed reaction in which a single substrate (S) is converted into a single product (P). The reaction could be irreversible, with all substrate molecules converted into product molecules and indicated by an arrow with only one head pointing toward the product. Irreversible reactions are typically described as nonequilibrium reactions. Alternatively, the reaction could be reversible, such that, given time, it establishes an equilibrium, with the ratio of product concentration to substrate concentration known as the **equilibrium constant** (K_{eq}). Reversible reactions are commonly called **equilibrium reactions**, and the two terms mean the same thing. An equilibrium reaction will be shown with a double-headed arrow, meaning that the product is also a substrate for the reverse reaction. Figure 2.1 illustrates these two types of reactions and outlines their properties.

- Irreversible reaction
- Large energy change
- Nonequilibrium reaction
- Less common

- Reversible reaction
- Small energy change
- Equilibrium reaction
- Product also a substrate for the reverse reaction
- More common

Figure 2.1 The properties of irreversible (left) and reversible (right) reactions, catalyzed by enzyme E, in which a substrate (S) is converted to product (P).

Most enzyme-catalyzed reactions are considered to take place in discrete steps or partial reactions (see figure 2.2). Enzyme E first combines reversibly with substrate S to form an initial enzyme–substrate complex ES. This complex is then converted to an enzyme–intermediate complex EI, which then changes to an enzyme–product complex EP. Enzyme–product complex then dissociates into free product, and the enzyme is released unchanged. The formation of an enzyme–intermediate complex often involves a conformational change in the enzyme protein structure that is reversed on dissociation of the product.

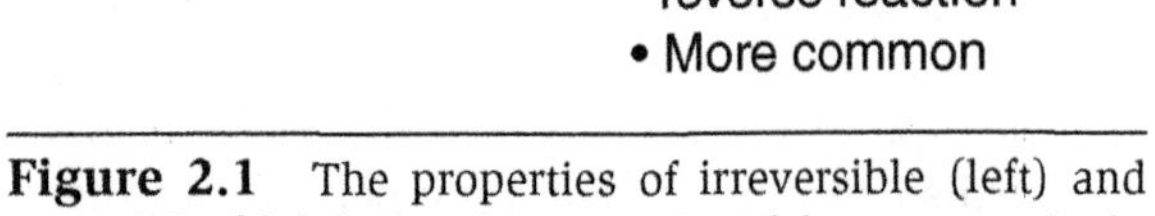

Binding Conversion Catalysis Release
E + S ⇌ ES ⇌ EI ⇌ EP ⇌ E + P

Figure 2.2 Enzyme-catalyzed reactions. These reactions are considered to take place in steps or partial reactions. Substrate (S) first binds with enzyme (E) to form an ES complex. This subsequently is converted into an enzyme–intermediate complex (EI), which is catalyzed to the enzyme–product complex (EP) until product is released. The release of P is so rapid that often the EP step is omitted from diagrams of the reaction.

RATES OF ENZYMATIC REACTIONS

The study of the rate at which an enzyme works is called *enzyme kinetics*. Consider the case of an enzyme-catalyzed reaction in which substrate S is converted to product P, catalyzed by enzyme E. This reaction is reversible, so that P is a substrate for the reverse reaction. Our goal is to measure the rate of this reaction in the direction S→P. If we begin with only S plus the enzyme, P will be formed. As the concentration of P increases and that of S decreases, the reverse reaction will take place, becoming more important as the concentration of P increases. Therefore, the reaction rate in the direction toward P decreases over time. Accordingly, the rate of the forward reaction must be measured quickly before any appreciable amount of P is formed and ideally when the amount of substrate is in substantial excess to the amount of enzyme. We call this quickly measured forward reaction rate the initial velocity. The initial velocity of an enzyme is influenced by several factors, which we will discuss in the next section.

Effect of Substrate Concentration

Let us carry out an experiment in which we set up 10 test tubes, each containing a solution at 25 °C, pH 7.0, and a fixed concentration of enzyme E. We add a specific amount of substrate S to each test tube, mix it, and quickly measure the rate of the reaction, either by measuring the rate of disappearance of S or the rate of appearance of P. Let us assume that the concentration of S in test tube 1 is 2 micromolar (2 μM), with higher and higher concentrations in the other tubes such that the concentration of S in tube 10 is 500 μM. After getting our initial velocities, expressed in units of micromoles of S disappearing per minute (or micromoles of P appearing per minute), we plot the initial velocity as a function of substrate concentration.

Figure 2.3 shows that initial velocity is higher as substrate concentration is increased. Note that the relationship is not linear but hyperbolic. That is, at low substrate concentration, the velocity increases linearly with increasing substrate concentration, but at a higher substrate concentration, the velocity flattens out, approaching a **maximum velocity** called Vmax. When this is reached, increasing substrate concentration will not produce any increase in the rate of the reaction. At this point, each enzyme molecule is working as fast as it can, converting S to P.

If we determine the value of the maximum velocity, divide this in half, then determine what substrate concentration will produce one-half of Vmax, we get a concentration known as the **Michaelis constant**, which is represented by the abbreviation K_m. The K_m is defined as the substrate concentration needed to produce one-half the maximal velocity of an enzyme-catalyzed reaction. The K_m has units of concentration because it represents an intercept on the X-axis. Vmax is the limiting rate for the velocity of an enzyme-catalyzed reaction at a fixed enzyme concentration. It occurs when enzyme active sites are so totally saturated with substrate that as a substrate molecule is converted to product and leaves the enzyme, another substrate molecule immediately binds.

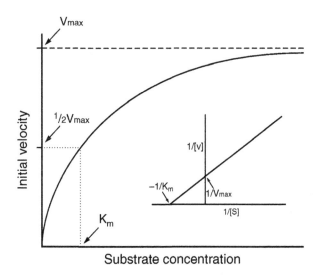

Figure 2.3 Relationship between the initial velocity of an enzyme-catalyzed reaction and the substrate concentration. Vmax is the maximum velocity that can be achieved using a fixed amount of enzyme. K_m is the Michaelis constant, the substrate concentration needed to generate one-half maximum velocity. The inset shows the Lineweaver–Burk plot generated by plotting the reciprocals of initial velocity versus the reciprocals of substrate concentration.

Determining reliable values for the kinetic parameters K_m and Vmax from the hyperbolic velocity–substrate concentration curve is difficult. A wide range of substrate concentrations must be tested to ensure a reasonable value of Vmax because we also need this value to determine K_m. An easier way, shown in the inset of figure 2.3, is a **Lineweaver–Burk plot**. Here, the reciprocals of velocity (1/velocity) and substrate concentration (1/[substrate]) are plotted. This allows us to determine accurately the kinetic parameters for an enzyme-catalyzed reaction because these values are determined from intercepts on the horizontal and vertical axes. When enzyme-catalytic activity follows Michaelis–Menten kinetics (i.e., hyperbolic) over the range of substrate concentrations tested, the Lineweaver–Burk plot is a straight line with:

$$\text{Y intercept} = 1/\text{Vmax}$$
$$\text{X intercept} = -1/K_m$$
$$\text{slope} = K_m/\text{Vmax}$$

We can thus obtain more accurate values for the kinetic parameters from fewer data points.

Vmax and K_m are known as **kinetic parameters** for an enzyme-catalyzed reaction. K_m is said to reflect the affinity of an enzyme for its substrate; the smaller the value of K_m, the greater the affinity an enzyme has for its substrate. In other words, if the enzyme has a high affinity for its substrate, it can convert the substrate to product even when the [S] is low. The reverse reaction will have its own kinetic parameters, because the product becomes the substrate for the backward reaction. The kinetic parameters for the reverse reaction are unlikely to be identical to those for the forward reaction. Enzymes that exhibit hyperbolic kinetics, as shown in figure 2.3, are said to exhibit Michaelis–Menten kinetics. Not all enzymes behave this way, as we shall discover later in this chapter.

☑ KEY POINT

Which direction a reaction actually takes is based on the relative concentrations of substrates and products compared to the individual K_m values. This direction is driven by the release of energy. A reaction taking place is analogous to a ball rolling down a hill. Balls roll down hills, not up. Reactions take place because energy is released. At equilibrium, nothing happens and no energy is released.

In the cell, the substrate concentration is generally equal to or less than the value for its K_m. This offers two advantages: (a) A substantial fraction of enzyme catalytic ability is being used, and (b) the substrate concentration is low enough that the enzyme can still respond to changes in substrate concentration because it is on the steep part of the curve (see figure 2.3). If the substrate concentration is much greater than K_m we will get efficient use of the enzyme, but it will respond less effectively to changes in substrate levels because it is on the flat part of the curve (see figure 2.3). Thus, K_m values for substrates generally reflect the concentrations of these substrates in the cell.

Some enzymes catalyze a given reaction in different tissues, but the enzymes have different kinetic parameters for the substrates. Often products of different genes, they are known as **isozymes** or isoenzymes. For example, when glucose enters a cell, the first thing that happens is that a phosphate group is attached to it. This reaction is catalyzed by four isozymes known as hexokinase I, II, III, and IV. Three of these isozymes have low K_m values for glucose (20-120 μM); the fourth (hexokinase IV, also known as glucokinase and found in the liver) has a high K_m for glucose (5-10 mM). The low-K_m isozymes can phosphorylate glucose even when the blood concentration is low. This is especially important to the brain, which depends solely on glucose as its fuel under normal circumstances. The high-K_m isozyme is found in the liver, where glucose is stored when blood concentration is elevated (for example, following a meal). The liver isozyme thus readily responds to glucose as a substrate only when blood glucose is elevated. Figure 2.4 shows the Michaelis–Menten kinetics for a low-K_m hexokinase and for glucokinase.

changing enzyme concentration increases Vmax alone; it has no influence on K_m. This is illustrated in figure 2.5, where a 67% increase in enzyme content produces a 67% increase in Vmax, but no change in K_m. We can use the relationship between enzyme concentration and maximum velocity to determine how much of a particular enzyme is present in a tissue or fluid (e.g., blood). When we undertake an exercise training program, the amount of some enzymes in the trained muscles may increase by a factor of 2 or more. This adaptation is in response to the increased demand placed on certain enzymes during training. The muscle cell responds by stimulating the synthesis of more enzyme proteins to reduce the stress of the training. The result is an improvement in metabolic homeostasis within the muscle cells, which decreases fatigue and improves exercise performance.

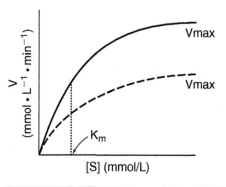

Figure 2.5 Enzyme velocity as a function of substrate concentration for two different concentrations of enzyme molecules. The solid curve is based on a 67% increase in enzyme concentration. Note that while Vmax is 67% higher when 67% more enzyme is present, the K_m is not changed.

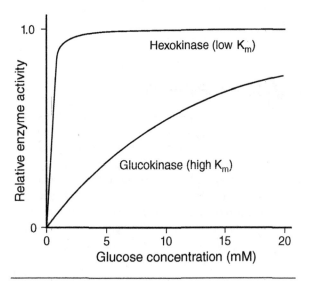

Figure 2.4 Relative velocities of a low-K_m and a high-K_m isozyme of hexokinase. The high-K_m hexokinase is known as glucokinase.

Effect of Enzyme Concentration

If an enzyme is saturated with substrate and is thus working as fast as possible, adding more enzyme increases the reaction velocity. Vmax is therefore proportional to enzyme concentration. However,

☑ KEY POINT

Tissues can adjust the amount of enzyme molecules by altering the rate of expression of the gene for specific enzymes, leading to higher or lower cellular content of the enzymes. For example, the muscle can vary the amount of enzymes that help metabolize carbohydrate and fat by adjusting the rate of expression of genes for key enzymes. This occurs in response to changes in the rate at which our muscles burn these fuels. The content of fat oxidation enzymes in muscle mitochondria can be increased by aerobic exercise training but will be decreased by bed rest.

Effect of pH

Most enzymes have a *pH optimum*—that is, a particular pH or narrow pH range where enzyme activity is maximal. As shown in figure 2.6, values of pH on either side of optimum produce lower reaction rates. The reason is that some of the forces holding an enzyme in its native conformation depend on charged groups. A change in pH can alter these charges, thus reducing enzyme function. The change may directly influence the active site or some other part of the enzyme that indirectly affects the active site. A change in pH may also alter the substrate for an enzyme, which could influence rate. For most enzymes, their pH optimum reflects where they are active in the body. As an example, the stomach enzyme pepsin has a pH optimum around 2 because the stomach is acidic. However, most biological enzymes in mammals have a pH optimum around 7.

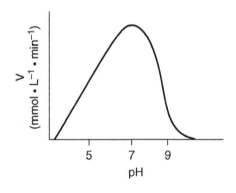

Figure 2.6 The effect of pH on enzyme-catalyzed reaction velocity for most biological enzymes.

Effect of Temperature on Enzyme Reactions

Like all chemical reactions, enzyme-catalyzed reactions increase in rate if the temperature is increased. However, since the forces holding the three-dimensional conformation of an enzyme are generally weak (see chapter 1), heating too much disrupts the conformation and decreases enzyme activity. Thus, if we plot the velocity of an enzyme-catalyzed reaction as a function of temperature, the curve rises with increasing temperature until about 50 °C, at which point most enzymes start to denature and the velocity drops quickly (figure 2.7). Biochemists describe the relationship between temperature and reaction rate by the quotient Q_{10}, which describes the fold increase in reaction rate for a 10 °C rise in temperature. For many biological processes, the rate of a reaction approximately doubles for each 10 °C increase in temperature. This means the Q_{10} is about 2. The resting temperature of human thigh muscles is approximately 33 to 35 °C, and it increases rapidly during moderate intensity exercise, possibly reaching steady state values of >40 °C (Saltin, Gagge, and Stolwijk 1968). Can you see how warming up muscles prior to competition can increase the rates at which the energy-yielding reactions take place by increasing the activities of the enzymes?

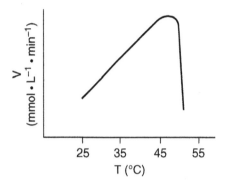

Figure 2.7 The effect of temperature on enzyme-catalyzed reaction velocity.

Turnover Number (k_{cat})

Turnover number, also known as the *catalytic constant* (k_{cat}), is defined as the maximum number of molecules of substrate converted to product per enzyme active site per unit of time (usually 1 s). It is a measure of how fast an enzyme can convert substrate to product. Since turnover number is considered a measure of the maximum catalytic activity for an enzyme, the enzyme must obviously be saturated with substrate for this to occur. Some enzymes are extremely fast as catalysts. For example, catalase, an antioxidant enzyme that breaks down hydrogen peroxide, has a k_{cat} of approximately 10 million. Carbonic anhydrase, which combines carbon dioxide with water to make carbonic acid, has a k_{cat} of approximately 1 million. However, most enzymes do not operate under conditions in which they are constantly saturated with substrate. A better way to express their catalytic efficiency in vivo is to use the expression k_{cat}/K_m, which gives a truer picture of enzyme function under physiological conditions. The k_{cat} directly reflects the speed with which an enzyme turns substrate into product, whereas the K_m is inversely related to the affinity of the enzyme for its substrate. Thus the ratio k_{cat}/K_m provides a measure for the catalytic efficiency of the enzyme—that is, how fast the enzyme can work given its physiological substrate concentration. While there may be large differences in values for k_{cat} among enzymes, the ratio k_{cat}/K_m provides far less diversity since most enzymes with large k_{cat} values generally have large K_m values.

Enzyme Inhibition

Enzymes can be inhibited by a variety of substances. We describe these inhibitors according to how they influence the enzyme. **Competitive inhibitors** resemble the normal substrate for an enzyme in that they bind to the active site, but they cannot be changed by the enzyme into product. In this sense, they act as mimics, but because of subtle differences between them and the normal substrate, they are not chemically altered by the enzyme. The inhibitor simply occupies the active site, binding, leaving, binding, and leaving, in a reversible fashion. Competitive inhibitors thus compete with the normal substrate for a place on the active site of the enzyme. This inhibition can be overcome by the addition of excess substrate. Accordingly, a competitive inhibitor will not affect the Vmax of the enzyme. It will, however, increase the K_m because more substrate will be needed to overcome the effects of the competitive inhibitor. Figure 2.8 shows the effects of a competitive inhibitor, using both a normal velocity–substrate curve and a Lineweaver–Burk plot.

A **noncompetitive inhibitor** does not resemble the normal enzyme substrate and does not bind to the active site. However, when bound to the enzyme, it interferes with its function, taking that enzyme molecule out of commission. Hence, noncompetitive inhibitors lower Vmax but do not alter K_m. Examples of noncompetitive inhibitors are heavy metal ions (e.g., Hg^{2+}, Pb^{2+}), known to cause many health problems because of their tight binding to reactive side chains of amino acids in proteins. Figure 2.9 shows the effects of a noncompetitive inhibitor using both a standard velocity–substrate relationship and a Lineweaver–Burk plot.

Many drugs are developed based on the principles of enzyme inhibition, with specific reactions in bacteria, viruses, or tumors targeted by antibiotics or antiviral drugs. With the sequences and three-

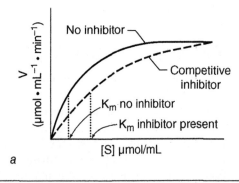
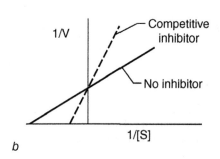

Figure 2.8 (a) The effects of a fixed concentration of a competitive inhibitor (dashed line) on the relationship between reaction velocity and substrate concentration for an enzyme-catalyzed reaction. In (b), the same data are shown using a Lineweaver–Burk plot.

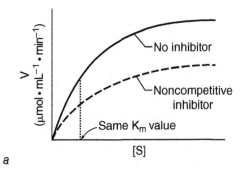
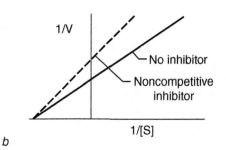

Figure 2.9 (a) The effects of a fixed concentration of a noncompetitive inhibitor (dashed line) on the relationship between reaction velocity and substrate concentration for an enzyme-catalyzed reaction. In (b), the same data are shown using a Lineweaver–Burk plot.

dimensional structure known for so many enzymes, researchers can design drugs to target specific active sites of enzymes by mimicking their substrates. Reducing the effects of inflammation and arthritis by targeting key enzymes such as cyclooxygenase-1 (COX-1) and cyclooxygenase-2 (COX-2) is a key example of how to use competitive inhibition of enzymes to improve health. COX-2 inhibitors such as nonsteroidal anti-inflammatory drugs (NSAIDs) are also often used in athletic settings to reduce pain and inflammation from injuries. However, care should be taken to prevent overuse of these enzyme-inhibiting drugs; longer term use will also inhibit muscle repair and muscle hypertrophy since inflammation is an important signal for activation of muscle **satellite cells** (Novak et al. 2009). Muscle satellite cells are essential for muscle repair and hypertrophy. Chapters 3 and 4 discuss this further.

ENZYME COFACTORS

Enzymes have reactive groups in the form of side chains of amino acids as well as the N- and C-termini. However, they may need other reactive groups not available on amino acids, called *cofactors*, in order to carry out their functions as catalysts. Cofactors may be metal ions, such as Zn^{2+}, Mg^{2+}, or Mn^{2+}, or they may be organic molecules called **coenzymes**. The polypeptide part of the enzyme is called the *apoenzyme*; when it is combined with the cofactor, we have the **holoenzyme**, illustrated as follows:

apoenzyme (*inactive*) + cofactor
= holoenzyme (*active*)

A cofactor tightly bound to the enzyme at all times is called a **prosthetic group**. An example of a prosthetic group is heme, bound tightly to hemoglobin and myoglobin (see figure 1.12 on p. 13). If the cofactor is not tightly bound but combines with the enzyme and the other substrate during the reaction, we can consider it a second substrate.

Eight B vitamins are necessary for humans because they form the basic components for coenzymes. Humans and many animals eating a plant-and-animal diet have lost the ability to synthesize the B vitamins, so these must come from the diet. Table 2.1 lists the eight B vitamins, the coenzymes they form, and their short-form names. The deficiency diseases associated with inadequate intake of specific B vitamins are due to insufficient catalytic power of enzymes because they lack their coenzymes.

People often stress the importance of vitamins but neglect the minerals. We need many more minerals in our diet than vitamins. Many minerals affect the function of enzymes, either as tightly bound components of prosthetic groups (copper, iron, manganese, selenium, and zinc) or because they are needed at the time of a reaction (calcium and magnesium). Zinc is an especially useful mineral in that it is an essential component of more than 300 different proteins, including enzymes for synthesizing RNA (ribonucleic

> ### ☑ KEY POINT
> The family of B vitamins performs various functions as coenzymes or critical components of coenzymes. Because they are so important to our overall metabolism and health, and because they are water soluble, we need B vitamins on a daily basis. Riboflavin, niacin, and pantothenic acid are particularly important in exercise since the coenzymes they form are critical components of the metabolic pathways that fuel our working muscles. Riboflavin is required for the synthesis of FAD, niacin is required for the synthesis of NAD, and pantothenic acid is required for the synthesis of coenzyme A.

Table 2.1 The B Vitamins, the Coenzymes They Form, and Their Common Abbreviations

B vitamin	Coenzyme	Abbreviation
Thiamine (B1)	Thiamine pyrophosphate	TPP
Riboflavin (B2)	Flavin adenine dinucleotide	FAD
Niacin (B3)	Nicotinamide adenine dinucleotide	NAD^+
Vitamin B6	Pyridoxal phosphate	PLP
Pantothenic acid	Coenzyme A	CoA
Folate (folacin)	Tetrahydrofolic acid	THFA
Biotin	Biotin	–
Vitamin B12	Methyl cobalamin	–

acid) and DNA, pancreatic digestive enzymes, and enzymes involved in carbohydrate, fat, protein, and alcohol metabolism. Magnesium (Mg^{2+}) is associated with the energy-rich ATP molecule, so it is involved in virtually all aspects of our metabolism. In addition to regulating the activity of many enzymes in a cofactor role, some minerals, such as calcium, act in signal transduction. Others, such as sodium and potassium, are critical to the chemical composition of intracellular and extracellular fluids.

CLASSIFICATION OF ENZYMES

Thousands of different enzymes are involved in our metabolism. The overwhelming preponderance of these enzymes is present in other organisms as well. Organizing these enzyme names in some form is a function of the International Union of Biochemistry and Molecular Biology (IUBMB). This classification system is shown in table 2.2. Six major classes of enzymes have been identified, and each of these contains subclasses and then sub-subclasses. Each enzyme is assigned two names: The recommended name is suitable for everyday use. The systematic name is formal and is used to remove any ambiguity. Each enzyme has a four-number designation as well. For example, lactate is reduced through loss of hydrogens and associated electrons (dehydrogenation) facilitated by a specific enzyme. The recommended name for this enzyme is *lactate dehydrogenase*; this is the name used by almost all biochemists. The systematic name is *lactate: NAD^+ oxidoreductase*. The classification is EC 1.1.1.27, where EC represents enzyme commission and the last number (27) refers to the specific enzyme.

In general, recommended names identify first the substrate, then the reaction type, with the ending *-ase* added. These are the enzymes we will encounter, for the most part. Some enzymes, however, have names that give no clue to either their substrate or the type of reaction they catalyze. For example, catalase, an enzyme we will encounter in chapter 5, is also a member of the oxidoreductase class, just like lactate dehydrogenase (LDH). One would not know this from the name.

PROTEIN TRANSPORTERS

The cell membrane and internal membranes (such as the mitochondrial membrane) act as barriers to the movement of molecules and ions. Since cells can function only if a two-way traffic of ions and molecules goes across their membranes, specific transport proteins, either embedded in or traversing the membrane, are needed to make this happen. Protein transporters are called *transmembrane proteins* because they span the entire biological membrane. They are also called **integral membrane proteins**, which refers to any protein or protein complex that is permanently attached to the biological membrane. All transmembrane proteins are integral membrane proteins, but not all integral membrane proteins are transmembrane proteins.

The membrane transport protein may go by such names as translocase, porter, carrier, pump, or transporter. The substances carried may be small ions (Na^+, Mg^{2+}, Ca^{2+}, and so on) or nutrient molecules, such as glucose, amino acids, or fat. The membrane protein may function as a simple conduit that allows an ion or molecule to flow down its concentration gradient, either into or out of the cell or organelle (mitochondrion). We call this **facilitated diffusion**. It is diffusion because the diffusing substance is moving from a higher to a lower concentration. The membrane transporter facilitates the process. Alternatively, the transport process requires the diffusing substance to move against its concentration

Table 2.2 Enzyme Classes, Including Subclasses We Will Encounter

Enzyme class: Number and name	Some common subclasses
1. Oxidoreductases	Dehydrogenases Reductases Oxidases Peroxidases Catalase
2. Transferases	Kinases Phosphomutases
3. Hydrolases	Phosphatases Thiolases Phospholipases Deaminases Ribonucleases Esterases
4. Lyases	Decarboxylases Aldolases Hydratases Dehydratases Synthases Lyases
5. Isomerases	Isomerases Mutases (but not all)
6. Ligases	Synthetases Carboxylases

gradient. Moving a molecule or ion from a lower to a higher concentration is similar to rolling a ball up a hill. This can take place only if driven by a release of energy. The energy release may come from the breakdown of ATP (adenosine triphosphate) or by the simultaneous movement of another molecule down a steeper concentration gradient. We call this energy-requiring membrane transport process **active transport**. For example, creatine is synthesized in the liver or taken in our diet from the meat we eat. Creatine circulates in the blood and enters cells that have a specific creatine transporter. Creatine is especially concentrated in muscle cells (fibers), at levels far above that in the blood. The reason muscle cells can take up creatine is that sodium ions (Na^+) are simultaneously transported into the muscle fiber down a steep gradient. The energy released in the flow of sodium allows the creatine to move up its gradient. In essence, the sodium drags the creatine into the fiber (figure 2.10).

Although membrane transporters are not technically enzymes, catalyzing reactions in which a substrate is converted into a product, they function in a way that is consistent with the kinetics of enzymes. They are similar to enzymes because the transporters recognize only specific transport substances and move them in a particular direction across the membrane. Their kinetics are similar to those of many enzymes in that we can describe the transport kinetics with Vmax and K_m. The transporters may also be subject to inhibition by other substances, just as enzymes can be inhibited.

> ### ✓ KEY POINT
> Transport of substances, such as fuel substrates (glucose, fatty acids), across cell membranes and intracellular membranes is critical to our metabolism. Transport proteins are selective in terms of the ions or molecules they recognize and the direction of transport of these. In all cases, transport across a membrane by either diffusion or active transport occurs with the release of energy.

OXIDATIONS AND REDUCTIONS

Oxidation–reduction reactions, or **redox reactions**, are extremely important for all organisms. In these reactions, something gets oxidized and something gets reduced. In earlier chemistry courses, you may have encountered the term **oxidation**, which means that something loses electrons, and **reduction**, which means that something gains electrons. The familiar memorization tools *LEO* (loses electrons oxidation) *says GER* (gains electrons reduction) and *OIL* (oxidation is losing) *RIG* (reduction is gaining) have been used by students for many years. Redox reactions are absolutely critical to life, and they underlie all aspects of metabolism. In all cases, the electron gain (reduction) is directly connected with electron loss (oxidation), as seen in the following reaction:

$$Fe^{3+} + Cu^+ \rightarrow Fe^{2+} + Cu^{2+}$$

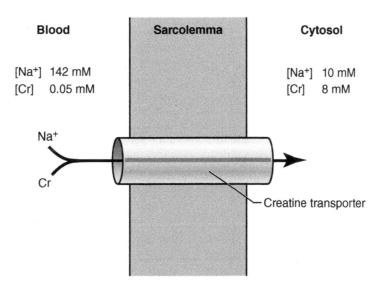

Figure 2.10 An example of active transport. Creatine (Cr) transport, across the sarcolemma from the blood and into skeletal muscle cells, takes place against a concentration gradient because of the simultaneous transport of sodium ions down their gradient.

In this example, the ferric iron (Fe^{3+}) is reduced because it gains an electron from the copper to form ferrous iron (Fe^{2+}), whereas the copper (Cu^+) is oxidized by losing an electron to the iron, thereby becoming cupric copper ion (Cu^{2+}).

In the cell, redox reactions occur in which ions are oxidized and reduced, as shown in the equation. However, in many important redox reactions, it is not easy to see that electrons are lost from one molecule and gained by another. For example, figure 2.11 shows two dehydrogenation reactions in which two hydrogens are lost from each of the two partial structures. The important thing is that when the hydrogens leave, they exit with electrons. In the first example (figure 2.11a), two hydrogen atoms are lost, and a carbon-to-carbon double bond is generated. In the second reaction (figure 2.11b), two hydrogens are lost from a secondary alcohol, generating a ketone group. In the former case, we consider that the hydrogen atoms each contain an electron. In the latter case, one hydrogen comes off as a hydride ion and the other as a proton. The **hydride ion** is simply a hydrogen atom (with one proton and one electron) plus an additional electron so that it has a negative charge. In summary, dehydrogenation reactions are oxidation reactions in which two electrons are lost as part of hydrogen atoms or a hydride ion.

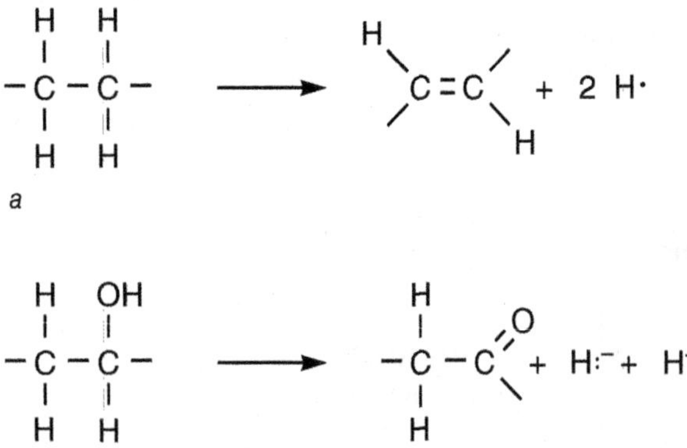

Figure 2.11 Dehydrogenation reactions. These reactions involve electron loss as (a) hydrogen atoms ($H\cdot$) or (b) a hydride ion ($H{:}^-$).

Figure 2.12 shows dehydrogenation, but these reactions are also accompanied by hydrogenation reactions, which means that a molecule is reduced because it accepts two hydrogen atoms (each with an electron) or a hydride ion (with two electrons). Two major coenzymes are involved in most cell redox reactions. The oxidized form of the coenzyme **FAD** (flavin adenine dinucleotide; see table 2.1) can accept two hydrogen atoms to become $FADH_2$ (the reduced form of FAD). The FAD coenzyme is involved in redox reactions in which two hydrogen atoms are removed from chemical structures, such as shown in figure 2.11a. Those shown in the lower part of figure 2.12 involve the **NAD$^+$** (nicotinamide adenine dinucleotide) coenzyme, which accepts a hydride ion, becoming NADH. NAD$^+$ is the oxidized form, whereas **NADH** is the reduced form of this coenzyme.

Most of the energy needed by our bodies to grow and survive arises when electrons on fuel molecules are transferred, first to coenzymes in dehydrogenation reactions and then eventually to the oxygen we breathe. In the top reaction of figure 2.12, part of the citric acid (tricarboxylic acid or Krebs) cycle, succinate, is oxidized to fumarate through the loss of two hydrogen atoms, forming the carbon-to-carbon double bond. The two hydrogen atoms, each carrying an electron, are picked up by the coenzyme FAD, forming $FADH_2$. Thus, succinate is oxidized to fumarate, and FAD is simultaneously reduced to $FADH_2$ in this redox reaction. The coenzyme (FAD or $FADH_2$) is a tightly bound coenzyme for the enzyme succinate dehydrogenase. In the lower example, lactate loses a hydride ion and a proton when oxidized to pyruvate. At the same time, NAD$^+$ accepts the hydride ion, becoming the reduced form of the coenzyme NADH. In this latter reaction, catalyzed by the enzyme lactate dehydrogenase, the coenzyme (NAD$^+$ or NADH) binds to the enzyme only at the time of the reaction. Therefore, as the reaction proceeds from left to right as shown in figure 2.12, NAD$^+$ is a substrate and NADH is a product. As shown in the two examples in figure 2.12, most redox reactions are reversible, so pyruvate can be reduced to lactate or fumarate can be reduced to succinate. Notice in figure 2.12 that we write L-lactate because the middle carbon is chiral, and the L refers to the absolute configuration of the lactate. Lactate comes from lactic acid, a modestly strong acid (pK_a about 3.8). Thus, at the neutral pH where this reaction occurs, lactic acid exists as the ion lactate because it has lost its proton. The same can be said for pyruvate, which comes from pyruvic acid.

The net direction of redox reactions depends on the relative concentrations of the oxidized and reduced forms of the substrates and coenzymes. During exercise, when the rate of pyruvate formation increases in muscle, the enzyme lactate dehydrogenase attempts to maintain equilibrium, producing lactate. Therefore, the lactate dehydrogenase reaction in glycolysis is normally drawn in the direction pyruvate → lactate, reflecting muscle lactate production during intense exercise. The lactate can travel from

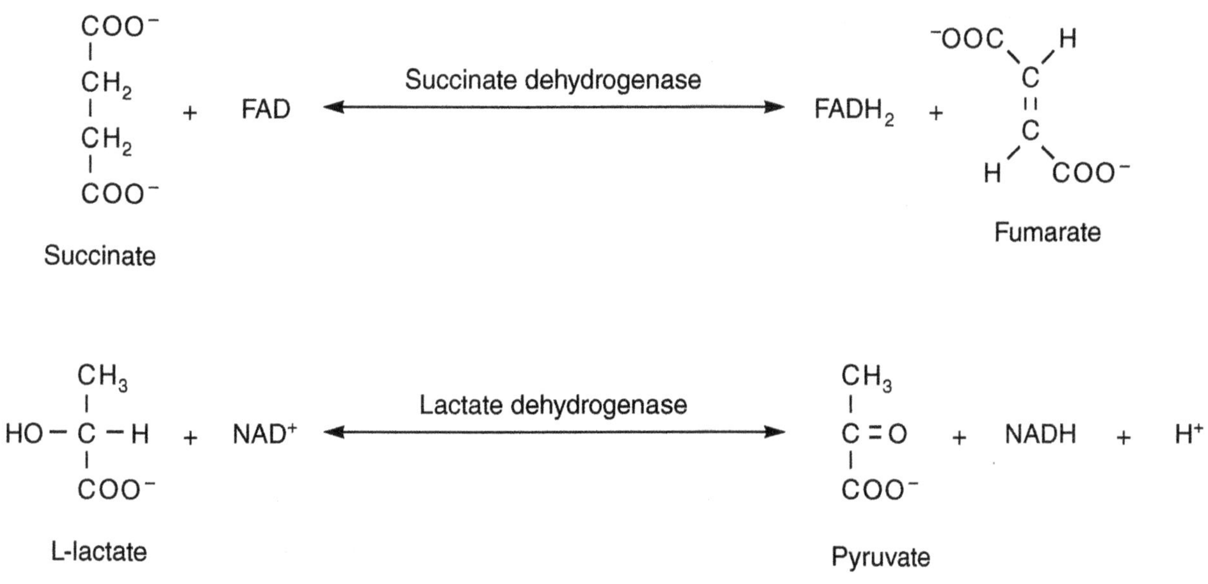

$$\text{Succinate} \; (\text{COO}^-\text{-CH}_2\text{-CH}_2\text{-COO}^-) + \text{FAD} \underset{\text{Succinate dehydrogenase}}{\longleftrightarrow} \text{FADH}_2 + \text{Fumarate}$$

$$\text{L-lactate} + \text{NAD}^+ \underset{\text{Lactate dehydrogenase}}{\longleftrightarrow} \text{Pyruvate} + \text{NADH} + \text{H}^+$$

Figure 2.12 Two biologically important redox reactions, showing substrates, products, and coenzymes.

the muscle cell to the blood, where its concentration is often used as an indicator of the exercise intensity. However, we will see in chapter 6 that in some highly aerobic muscle tissues, including the heart, lactate is an important fuel source for energy production during exercise, where the reverse reaction, lactate → pyruvate, is favored.

Another factor that influences the direction of the lactate dehydrogenase reaction is the type of lactate dehydrogenase isozyme that is expressed in different tissues. Lactate dehydrogenase enzymes are homo or hetero tetramers composed of muscle (M) and heart (H) protein subunits encoded by the *LDHA* and *LDHB* genes. The major isozyme of fast-twitch glycolytic skeletal muscle and liver, M_4, has four M subunits, while H_4 is the main isozyme for heart and slow-twitch oxidative muscle containing four H subunits. The other variants contain both types of subunits. The expression of the H subunit of the LDH increases in proportion to the oxidative capacity of the muscle. As shown in figure 2.13, the lactate dehydrogenase isozyme expressed in heart and slow-oxidative muscle fibers (H_4) favors pyruvate production. In contrast, lactate production is favored in fast-glycolytic muscle fibers that express different lactate dehydrogenase isozymes (M_4).

Redox reactions represent the molecular basis for energy generation in the cell. Electrons from substrates, generated from the foods we eat, are transferred to coenzymes NAD^+ and FAD, generating NADH and $FADH_2$. Subsequently, these electrons are transferred from the reduced coenzymes through a series of carriers embedded in the inner membrane

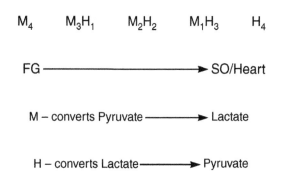

Figure 2.13 Five different lactate dehydrogenase (LDH) isozymes expressed in skeletal muscle and heart. Fast-glycolytic skeletal muscle fibers express LDH isozymes that favor conversion of pyruvate to lactate (i.e., M_4 and M_3H_1). Slow-oxidative skeletal and heart muscle fibers express LDH isozymes that favor conversion of lactate to pyruvate (i.e., H_4 and M_1H_3). The *M* and *H* represent muscle and heart isozymes, respectively.

of mitochondria until they are passed to oxygen (O_2). Reduction of oxygen through accepting these electrons and H^+ to form water molecules is the final step (see the following equation). During the passage of electrons from food-derived substrates to oxygen, a tremendous amount of energy is released and captured through formation of ATP. The ATP is then used to drive energy-requiring processes in the body. This is the basis of the process called **oxidative phosphorylation**, which is covered in detail in chapter 5.

$$O_2 + 4e^- + 4H^+ \rightarrow 2H_2O$$

> ### ✅ KEY POINT
>
> The harder we exercise, the more ATP we need. This means we must accelerate the rate of transfer of electrons from fuel molecules (dehydrogenation reactions) to oxygen. If we continue to increase the intensity of our exercise, a weak link in the electron transport process is reached. This limitation may be the availability of oxygen to accept electrons, the rate at which we can remove electrons from the fuels, or the actual kind and amount of fuel available.

REGULATION OF ENZYME ACTIVITY

Each chemical reaction in a cell is catalyzed by a specific enzyme. As discussed previously, the rate of an enzyme-catalyzed reaction depends on the concentration of substrates and products. The term *mass action* is used to describe the regulation of enzyme activity through changes in the concentration of substrates or products. This is an important controller of both equilibrium and nonequilibrium reactions. Refer back to figure 2.1. Increases in [S] or decreases in [P] will increase the forward rate of the reaction S→P. Reaction rates can also be controlled by the amount of enzyme protein, as well as by the enzyme's location within the cell. In subsequent chapters, we will encounter a variety of examples in which the activity of existing enzymes in the cell must be modified so that cellular metabolism is appropriate. Simple enzymes, obeying the Michaelis–Menten kinetics shown in figure 2.3, are very common but are rarely involved in controlling metabolism. For this, enzymes with more complex kinetics or properties are needed. The activity of existing enzymes can be controlled in two major ways: modification by effector molecules and covalent modification.

Allosteric Enzymes

The activity of one group of enzymes depends not only on substrate and product concentrations, but also on the presence of positive or negative effectors. Such enzymes are usually composed of multiple subunits with multiple active sites and are typically placed in metabolic pathways where they can control the overall pathway rate. We call these enzymes **allosteric enzymes**. The positive and negative effectors are called positive and negative *allosteric effectors*. Positive and negative allosteric effectors are also called *activators* and *inhibitors*, respectively. The term *allosteric*, derived from the Greek word *allo* (or other), is used because these enzymes have sites, other than the active site, to which effectors can bind. Such binding affects the overall rate of the enzyme-catalyzed reaction by facilitating or diminishing velocity for a given substrate concentration. Allosteric effectors (often just called *effectors*) can be substrates or products of the allosteric enzyme or other molecules whose concentration provides a message about how active the allosteric enzyme should be.

Molecules that bind to large molecules are known as **ligands**. The ligands may be substrates or products that bind to the active site, as well as allosteric effectors that bind to **allosteric sites**. Therefore, we could classify enzymes as (a) those whose ligands are only substrates and products that bind to the active site and (b) those that bind substrates and products, as well as one or more ligands that bind at allosteric sites. Ligands are also molecules, such as hormones, that bind to other proteins such as membrane receptors.

In general, allosteric enzymes are composed of subunits; that is, they have quaternary structure. The kinetics of allosteric enzymes are more complicated than the simple Michaelis–Menten kinetics shown in figure 2.3. Allosteric enzymes typically demonstrate sigmoidal (S-shaped) kinetics, as shown in figure 2.14. Here, the presence of positive or negative allosteric effectors greatly modifies the response of the enzyme to its typical substrate. For example, a much higher concentration of substrate would be needed to achieve one-half of Vmax in the presence of a negative effector, but a much smaller amount of substrate is needed to achieve one-half of Vmax with a positive effector bound to its allosteric site. Thus allosteric effectors markedly influence K_m values.

Enzymes displaying sigmoid kinetics do so because **cooperativity** in substrate binding exists at the active site in each subunit. This means that the binding of a substrate molecule to the active site of one of the subunits induces a conformational change in adjacent subunits that lowers the K_m for the binding of the substrate molecule at the active sites of the adjacent subunits. The degree of cooperativity can be quantified by determining the steepness of the enzyme's initial velocity–substrate concentration curve. The steepness of the curve is quantified by the **Hill slope**, also called the *Hill coefficient*. This is named for A.V. Hill, who first devised the coefficient to explain the cooperative

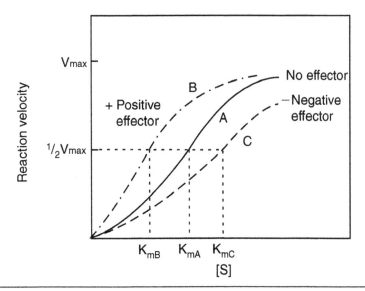

Figure 2.14 The effect of substrate concentration on reaction velocity for an allosteric enzyme (A) when neither positive nor negative allosteric effector is present, (B) when a positive allosteric effector binds at its allosteric site, and (C) when a negative allosteric effector binds to its allosteric site. In the example shown, Vmax is the same for the three cases, but the positive allosteric effector reduces K_m (i.e., K_{mB}), while the negative allosteric effector increases the value of K_m (i.e., K_{mC}).

binding of oxygen to hemoglobin. A coefficient of 1.0 indicates noncooperative substrate binding, whereas numbers greater than 1.0 indicate positive cooperativity. Numbers less than 1.0 indicate negative cooperativity.

Allosteric enzymes play an important role in the regulation of cell metabolism. For example, in a muscle fiber at rest, the breakdown of carbohydrate is very low, primarily due to negative effectors binding to the allosteric enzyme phosphofructokinase (PFK). Phosphofructokinase is located near the beginning of the pathway of glycolysis, and it acts to regulate the breakdown of carbohydrate. When the fiber becomes active in an exercise situation, the rise in the concentration of positive effectors and the decline in the concentration of negative effectors enormously increase the activity of PFK. On the other hand, under rest conditions, PFK activity is reduced by low concentrations of positive allosteric effectors and by higher levels of two negative allosteric effectors. This forces rested muscle to consume more fat as its fuel. We discuss the role of PFK in carbohydrate metabolism in chapter 6. Hexokinase, the enzyme that first interacts with glucose upon entering a cell, is a dimer; that is, it is composed of two polypeptide subunits. Its substrates are glucose and ATP. Its product, glucose 6-phosphate, can inhibit the activity of the enzyme if its concentration increases. This is an example of **feedback inhibition**, which helps to regulate metabolic pathways by sensing an oversupply of product.

✓ KEY POINT

Metabolic enzymes that are controlled by multiple allosteric effectors are usually located near the beginning of metabolic pathways and can often determine the rate of flux of an entire metabolic pathway. Therefore, these enzymes are described as **rate limiting** enzymes. PFK, an enzyme that regulates the breakdown of carbohydrate in muscle, is a good example of a rate-limiting metabolic enzyme because flux through glycolytic reactions that take place downstream of PFK is determined simply by mass action. Therefore, it is limited by PFK activity.

Covalent Modification of Enzymes

Allosteric regulation is a fine-tuning type of enzyme activity modulation. On the other hand, the activity of some enzymes can be rapidly turned on or off by the covalent modification of specific amino acid residues in the enzyme protein. One example of this is acetylation (addition of an acetyl group) to specific residues in a class of DNA-binding proteins known as histones. Acetylation of histones is associated with the enhancement of gene transcription, whereas removal of the acetyl groups (deacetylation) decreases transcription of closely associated genes.

We will now discuss two common covalent modifications of enzyme activity in more detail: (1) protein phosphorylation and dephosphorylation, which is the most prevalent reversible covalent modification of enzymes, and (2) thiol oxidation and reduction, which is emerging as an important mechanism for regulating the activity of several enzymes. The terms *thiol* and *sulfhydryl* are used interchangeably. We will use thiol from now on. Recall that the R group of the amino acid cysteine is a single thiol (—SH).

Protein Phosphorylation and Dephosphorylation

One of the most common ways to modify the activity of enzymes or receptors is to attach a phosphate group to the hydroxyl part of the side chains of the amino acids serine, threonine, and tyrosine (see figures 1.4 [p. 8] and 2.15). This process is known as **phosphorylation**, specifically protein phosphorylation, since the phosphate is added to part of a protein. Removal of the phosphate group is **dephosphorylation**. The addition of a phosphate group in place of a single hydrogen atom drastically alters a protein. The phosphate group is not only far larger, but it also contains two negative charges, thus making a major change in protein conformation. Of course, the drastic change in protein conformation following removal of the phosphate group reverses the effect. Protein phosphorylation–dephosphorylation is a common and effective mechanism to rapidly and reversibly alter the activity of key enzymes. The source of the phosphate group is ATP (adenosine triphosphate), a molecule we will encounter throughout all subsequent chapters. Protein phosphorylation is catalyzed by a class of enzymes known as **protein kinases**, which are specific for serine, threonine, or tyrosine side chains. For example, there are protein serine kinases, protein tyrosine kinases, and protein ser/thr kinases; the latter can phosphorylate both serine and threonine side chains. A class of enzymes known as **phosphoprotein phosphatases** catalyzes removal of the phosphate groups by hydrolysis. These phosphatases are specific to the amino-acid side chain, as in the case of protein tyrosine phosphatases. A summary of this process is shown in figure 2.15.

Control of enzyme activity by phosphorylation plays a critical role in controlling and integrating our metabolism. For example, when we begin to exercise, the hormone **epinephrine** (adrenaline) is released from the adrenal medulla, and it binds to a membrane receptor on muscle and fat cells. In active muscle cells, epinephrine binding leads to the phosphorylation and the resulting activation of glycogen phosphorylase, the enzyme that breaks down glycogen in muscle. In fat cells, adrenaline binding to its receptor leads to the phosphorylation and subsequent activation of hormone-sensitive lipase, an enzyme involved in catalyzing the breakdown of stored fat. Many other protein kinases are also activated when external molecules (e.g., a hormone) bind to specific sites on the membrane of the cell. For example, the hormone insulin, on binding to the insulin receptor on the cell membrane, results in the activation of several protein kinases, leading

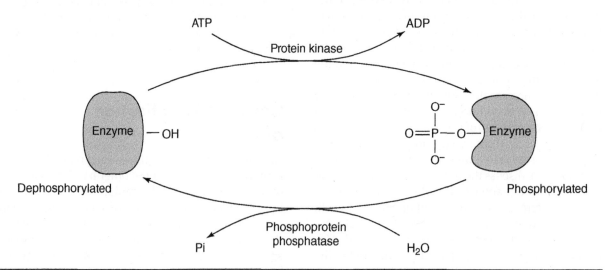

Figure 2.15 Modification of enzyme activity. Enzyme activity can be turned on or off by the covalent attachment of a phosphate group from ATP to the OH part of the side chain of the amino acids serine, threonine, and tyrosine. A class of enzymes known as protein kinases attaches the phosphate group to the side chain, transferring the terminal phosphate from ATP. Hydrolysis of the phosphate group by a class of enzymes known as phosphoprotein phosphatases reverses the effect of phosphate group addition.

to a cascade of phosphorylation reactions on multiple downstream proteins. In skeletal muscle, this cascade leads to glucose uptake.

For every phosphorylation reaction, there must be a dephosphorylation—just as every light that is switched on must eventually be turned off. We shall see in the next chapter that protein phosphorylation is a powerful strategy for controlling enzymes and other protein signaling molecules that regulate gene transcription and protein synthesis.

You might naturally think that protein phosphorylation activates enzymes or increases their catalytic activity. This is true for many enzymes, but in some cases, protein phosphorylation turns off or decreases the activity of an enzyme. A key example provided in chapter 7 is the enzyme acetyl CoA carboxylase β, which is involved in the regulation of fatty-acid oxidation in muscle. This enzyme becomes inactivated when it is phosphorylated by another enzyme called AMP-activated protein kinase (see figure 7.25 on p. 237). Another example we will encounter in chapter 6 is the enzyme pyruvate dehydrogenase, which catalyzes the oxidation of pyruvate in the mitochondria to form acetyl CoA. Just like acetyl CoA carboxylase β, the phosphorylated form of pyruvate dehydrogenase (PDH) is the inactive form (see figure 5.16 on p. 132). It is not unusual for multiple phosphorylation sites to exist on the same enzyme, allowing for graded modulations in the activity of the enzyme. The enzyme catalyzing the storage of glycogen in liver and muscle, glycogen synthase, has 10 sites where it can be phosphorylated. The more numerous the attached phosphates are, the less active the glycogen synthase will be.

Thiol Oxidation and Reduction

Another form of covalent modification, the reversible oxidation and reduction of reactive cysteine thiol groups of proteins, also plays an important role in regulating the activity of several enzymes (including metabolic enzymes), membrane channels, and transporters. We call this type of regulation *redox control of protein function*. As chapter 5 shows, certain reactive oxygen and nitrogen species, such as superoxide and nitric oxide, are produced constantly in the body. Under physiological conditions, reactive oxygen and nitrogen species are important signaling molecules that can oxidize protein thiols and alter protein function.

As seen in figure 2.16, thiols may form sulfenic (P-SOH), sulfinic (P-SO$_2$H), or sulfonic (P-SO$_3$H) acids; intra- or interprotein disulfides (P-S-S-P); or nitrosothiols (P-SNO) when oxidized. In the presence of reactive oxygen and nitrogen species, protein thiols may also form mixed disulfides with glutathione (P-S-SG), a nonenzymatic process called S-glutathionylation. Reversible oxidation or covalent modification of cysteine thiols found within the active sites of proteins provides an on-off mechanism for protein function, whereas reversible oxidation or covalent modification of nonactive-site cysteine thiols

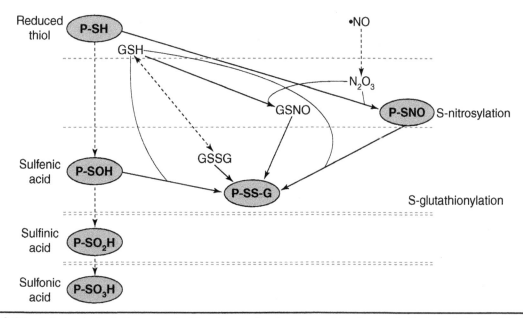

Figure 2.16 Schematic depicting several of the chemical pathways resulting in protein S-nitrosylation and S-glutathionylation, which are involved in redox control of enzyme activity.

A. Martínez-Ruiz and S. Lamas, "Signaling by NO-induced protein S-nitrosylation and S-glutathionylation: Convergences and divergences," *Cardiovascular Research*, 2007, 75: 220-228, by permission of Oxford University Press.

provides an allosteric-type mechanism to fine-tune enzyme activity (either up or down; see figure 2.17). Finally, protein function may be activated (recruited) or deactivated (inhibited) through macromolecular interactions involving disulfide cross-linking with other proteins (figure 2.17). Protein disulfides, nitrosothiols, sulfenic acid, and S-glutathionylation can be reversed in reactions catalyzed by specific protein-reducing enzymes called glutaredoxins, thioredoxins, and peroxiredoxins. The function of these enzymes in regulating protein function is analogous to phosphoprotein phosphatases, which reverse the effects of protein phosphorylation on protein function. Excessive accumulation of reactive oxygen and nitrogen species may result in the irreversible oxidation of protein thiols to sulfinic or sulfonic acid (see figure 2.16), which would result in the loss of ability to regulate protein function through redox control mechanisms. This problem contributes to many diseases (see the Next Stage section of this chapter).

> **KEY POINT**
>
> Enzyme activity may be regulated through allosteric regulation or by covalent modification of proteins. Allosteric regulation of enzyme activity involves the reversible binding of effectors or oxidation of nonactive-site amino acid residues that alter the conformation of the protein and affect enzyme activity. Allosteric modification of enzyme activity acts like a dimmer switch, grading the activity of the enzyme. Covalent modification of proteins occurs through two major types of signaling systems—one is phosphorylation and the other is redox based. Reversible phosphorylation or oxidation and subsequent S-nitrosylation or S-glutathionylation of specific amino-acid residues within the active site of proteins works like a light switch in that it has an all-or-nothing effect. The enzyme molecule is either active or inactive.

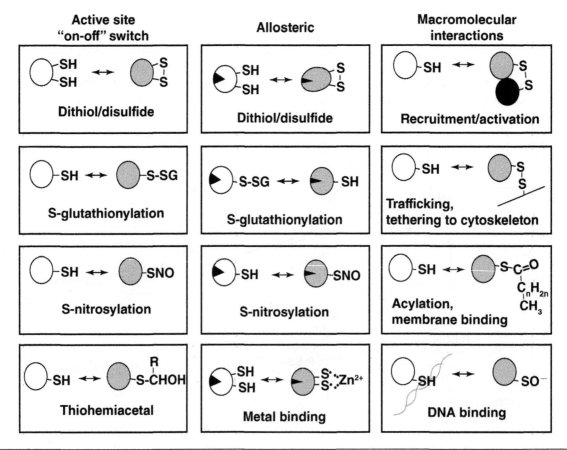

Figure 2.17 Schematic depicting different types of redox control of protein function. Cysteine (Cys) is found in the active site of many proteins, and reversible oxidation, S-nitrosylation, or S-glutathionylation provides an *on-off* mechanism for protein function (left). Reversible modification of nonactive-site Cys residues also provides an allosteric type mechanism to regulate activity (center). Cys residues are also important regulatory elements that control the macromolecular interaction of proteins (right).

Reprinted from D.P. Jones, 2008, "Radical-free biology of oxidative stress," *American Journal of Physiology-Cell Physiology* 295: C849-C868. Used by permission.

MEASUREMENT OF ENZYME ACTIVITY

We sometimes need to know how many functional enzyme molecules exist in a fluid (e.g., blood) or tissue. Because the number of molecules of functional enzyme is proportional to the Vmax, measurement results are in units of reaction velocity (i.e., change in substrate or product concentration per minute) per unit weight of tissue, per unit amount of protein, or per volume of fluid. Examples are micromoles of product formed per minute per gram of tissue, millimoles of substrate disappearing per minute per milligram of protein, or micromoles of product formed per minute per milliliter of blood, respectively. The expression of the Vmax or *enzyme activity*, as it is commonly called, is important in a variety of physiological and clinical conditions. For example, the activity of mitochondrial enzymes can almost double given the appropriate exercise training stimulus. This means that the maximal flux or traffic of substrate through the enzyme is twice what it was before training. We may want to determine if a particular exercise training program alters the metabolism of a muscle by measuring the activity of selected enzymes. We can also tell if a particular tissue is damaged by measuring the activity of isozymes specific for that tissue that are released to the blood due to cell-membrane damage.

When measuring the maximal activity of an enzyme, it is necessary to establish and rigidly follow certain principles. First, we need to make the measurements at a concentration of substrate (or substrates) high enough to generate a true Vmax. The pH of the reaction and the temperature at which it is measured should also be standardized so that meaningful comparisons can be made. Finally, we need to have a simple method for measuring the disappearance of substrate or appearance of product. This determination becomes possible if the substrate or product is colored or if it can be made to generate a colored complex. For example, phosphate appearance can be readily measured because it forms a colored complex with a number of agents. One useful technique takes advantage of two properties of the coenzyme NADH. First, NADH absorbs light at a wavelength of 340 nanometers (nm), so we can measure its rate of formation or disappearance with a spectrophotometer. The relationship is as follows: A 0.1 mmol solution of NADH has an absorbance of 0.627. Second, NADH fluoresces when bombarded with light of a specific wavelength; thus, we can measure its appearance or disappearance with a fluorometer. If the specific reaction does not actually involve NADH, it can be coupled with a reaction that does. The rate of the reaction in question then dictates the rate of a coupled reaction in which NADH is formed or lost.

Biochemists use the term **international unit** (IU) to express the activity of an enzyme. One international unit of enzyme activity is the amount of enzyme that converts one micromole of substrate to product in one minute. Thus, if an enzyme has an activity of 15 international units per gram, 15 μmol of product forms per minute per gram of tissue. Since enzyme activity is so sensitive to temperature, the composition of the medium, and pH, one must specify these conditions when describing enzyme activity in international units.

✓ KEY POINT

Measurements of the Vmax of enzymes under standardized conditions can provide useful information about the content of functional enzymes in a tissue because the number of molecules of functional enzyme is proportional to Vmax. However, it is also important to be aware that the Vmax of enzymes can be altered through allosteric and covalent modifications, which are independent of changes in enzyme concentration. For example, the Vmax of some muscle metabolic enzymes increases in response to acute exercise in the absence of any changes in enzyme concentration.

NEXT STAGE

Redox Regulation of the SERCA Pump in Vascular Physiology and Disease

Nitric oxide is an endothelium-derived relaxing factor that causes arterial smooth muscle relaxation, thus causing blood vessels to dilate (vasodilation). Its mechanism of action has been well characterized through 25 years of vascular biology research (see, for example, the review by Rush, Denniss, and Graham 2005). In response to a number of physical and chemical stimuli to the endothelium (a thin layer of cells that lines the interior surface of blood vessels), nitric oxide is generated in the cytosol from the amino acid L-arginine in a reaction catalyzed by the endothelial isoform of the enzyme, nitric oxide synthase (eNOS). Nitric oxide diffuses to underlying vascular smooth muscle cells, where it stimulates a signaling molecule called soluble guanylate cyclase (sGC), leading to accumulation of cyclic guanosine monophosphate (cGMP) and activation of protein kinase G (PKG). Once activated, PKG phosphorylates several ion channels and contractile proteins, resulting in lowered cytosolic Ca^{2+} concentration of vascular smooth muscle cells, relaxation of vascular smooth muscle, dilation of the blood vessel, decreased vascular resistance, and, ultimately, increased blood flow through the vessel.

Most undergraduate physiology courses probably do not cover more than this classic cGMP-dependent signaling pathway to explain nitric oxide–mediated vasodilation. However, nitric oxide relaxes vascular smooth muscle through both cGMP-dependent and cGMP-independent mechanisms. A number of elegant studies published by Richard Cohen and colleagues have shown that activation of the **SERCA pump**, which actively transports Ca^{2+} from the cytosol into the sarcoplasmic reticulum in vascular smooth muscle, occurs through redox signaling. In this process, nitric oxide and superoxide anion, through the formation of peroxynitrite, activate the SERCA pump directly by reversible S-glutathionylation of a specific cysteine residue (Cys674), resulting in arterial relaxation (see Tong, Evangelista, and Cohen 2010). This mechanism is predominantly cGMP-independent and is impaired in a variety of cardiovascular diseases, including diabetes, hypercholesterolemia, and atherosclerosis (Tong, Evangelista, and Cohen 2010). These cardiovascular diseases are all characterized by excessive oxidant production (i.e., oxidative stress), which leads to increased oxidative damage to proteins, including SERCA pumps. Recall that not all cysteine oxidation products are reversible (i.e., sulfinic and sulfonic acid) and that redox signaling is lost if key cysteine residues become irreversibly oxidized. Cohen and his colleagues have demonstrated that SERCA Cys674 from diseased arteries is irreversibly oxidized to sulfonic acid (Tong, Evangelista, and Cohen 2010). As a result, it cannot be S-glutathionylated in response to nitric oxide, resulting in reduced stimulation of Ca^{2+} uptake and impaired vasodilation (Tong, Evangelista, and Cohen 2010). One idea that is emerging from these findings is that preservation of SERCA function may be regarded as a new target for the treatment of impaired vascular function in cardiovascular disease.

SUMMARY

Enzymes are biological catalysts—specialized proteins that speed up reactions in cells enormously. Highly specific, they catalyze reactions involving single substrates or a closely related group of substrates. Enzymes have a Michaelis constant, K_m, which is the substrate concentration needed to produce one-half the maximal velocity (Vmax) of the enzyme reaction. The K_m, a characteristic constant for an enzyme–substrate pair, inversely reflects the affinity of the enzyme for its substrate. The maximal velocity of an enzyme-catalyzed reaction, or Vmax, is proportional to the amount of enzyme present. It can be determined only when the enzyme is saturated with its substrate. Measurement of the Vmax thus determines the amount of enzyme present. An international unit is defined as the amount of enzyme needed to convert one micromole of substrate to product in one minute. The action of enzymes can be hindered by the presence of inhibitors—specific substances that resemble the normal substrate and compete with it (competitive inhibitors) or that irreversibly alter the structure of the enzyme (noncompetitive inhibitors). In the cell, regulation and integration of the thousands of chemical reactions can be affected by modulation of the activity of a subset of key enzymes. The activity of allosteric enzymes can be changed by the binding of ligands, including substrates or products, to specific sites known as *allosteric sites*. Binding of these effector molecules may increase or decrease enzyme activity by changing the enzyme's response to a particular substrate concentration. A more profound change in enzyme activity accompanies the covalent attachment of a phosphate group to an enzyme, catalyzed by a class of enzymes known as *protein kinases*. Dephosphorylation by phosphoprotein phosphatases reverses the change in activity accompanying phosphorylation. Reversible oxidation of key active-site and nonactive-site protein thiols can alter enzyme activity in much the same way as phosphorylation.

Membrane transport is carried out by a class of protein molecules with properties similar to those of enzymes. Membrane transport can be described by kinetic constants, Vmax and K_m, and can be subject to competitive and noncompetitive inhibition. Many enzymes require the presence of nonprotein substances to function. These cofactors can be organic molecules (i.e., coenzymes) or they can be metal ions. Most coenzymes are derived from the B vitamins in our diet, while our need for many specific mineral nutrients relates to their role as enzyme cofactors. Isoenzymes (isozymes) are closely related. They are enzyme molecules that catalyze the same reaction but differ in certain properties, such as K_m. Some of the most important enzymes are the dehydrogenases, which add or remove electrons from their substrates. They play a major role in producing energy in the process of oxidative phosphorylation.

REVIEW QUESTIONS

1. Using the following data generated by measuring the rates of an enzyme-catalyzed reaction at constant enzyme concentration, determine the K_m and Vmax.

Substrate concentration	Reaction rate
1 mM	2 mmol · ml^{-1} · min^{-1}
2 mM	4 mmol · ml^{-1} · min^{-1}
4 mM	7 mmol · ml^{-1} · min^{-1}
10 mM	15 mmol · ml^{-1} · min^{-1}
50 mM	25 mmol · ml^{-1} · min^{-1}

2. If the data in question 1 were obtained at 25 °C, approximately what would be the Vmax at 35 °C?

3. One lab reports that the Vmax for the mitochondrial enzyme malate dehydrogenase in the quadriceps of a group of elite cyclists averages 40 μmol of substrate per gram wet weight of tissue per minute. Your lab, on the other hand, measures the same enzyme in a similar group of trained subjects, yet your average enzyme activity is double that of the other lab. What are some possibilities that could contribute to this discrepancy between two apparently similar groups of subjects?

4. Change the units for the activity of citrate synthase from 20 μmol per gram per 120 s to (a) millimoles per kilogram per minute and (b) international units per gram.

5. The experiment described in question 1 was repeated with a known inhibitor of the enzyme and using the same substrate concentrations. The data are paired as follows, using the same substrate and reaction-velocity units: S/V—1/1.7; 2/3.2; 4/5.8; 10/10.5; 40/21.1. What number did you get that was different? What kind of inhibitor was it?

6. What activates PDH: PDH kinase or PDH phosphatase? Explain.

chapter 3

Gene Transcription and Protein Synthesis

As mentioned in chapter 1, the information to make a protein is coded in the sequence of bases in DNA in cell nuclei. As chapter 5 discusses, mitochondria also contain double-stranded circular DNA that encodes 13 mitochondrial proteins necessary for the electron transport chain. The nuclear and mitochondrial DNA information is copied to make a sequence of bases in an RNA molecule, known as messenger RNA, in a process called transcription. The information in the base sequence of messenger RNA is translated into a precise sequence of amino acids in a protein. This chapter reviews DNA, the different kinds of RNA, and the genetic code. It then focuses on the processes of transcription and translation, introducing concepts regarding the regulation of these processes. Since proteins are continuously turned over, the processes that break down old proteins so that newly synthesized proteins can take their place are also discussed. The signaling mechanisms for gene regulation involved in exercise-induced adaptation in skeletal muscle are also covered in some detail. The Next Stage element highlights the role of microRNA molecules in muscle. This chapter begins with a review of DNA and RNA structure and the genetic code, topics many have already studied in biology courses.

DNA AND RNA

You are no doubt familiar with the double-helix structure of DNA and aware that DNA in each nucleus provided the information that allowed us to grow from a tiny embryo into adulthood. With the exception of human eggs and sperm, each nucleus in every cell contains 46 **chromosomes**. Each chromosome consists of a single, but very long, DNA molecule, plus many different kinds of proteins. Human males contain two copies of chromosomes 1 through 22, plus an X and a Y chromosome. Human females also have duplicate copies of chromosomes 1 through 22 but differ in that they have two X chromosomes and no Y chromosome. Our **genome** consists of 3.1647×10^9 bases, and the information contained in it codes for an estimated 30,000 genes. While we all think we are unique in terms of both our fingerprints and our DNA, and indeed we are, 99.9% of the sequence of bases in your DNA is exactly the same as everyone else's.

The three main forms of RNA are ribosomal, transfer, and messenger RNA. In general, RNA molecules are much smaller than DNA. However, a number of structural similarities exist between them. Ribosomal RNA represents more than 80% of the total mass of RNA in a typical cell, with transfer RNA and messenger RNA accounting for most of the rest.

DNA and RNA Structure

The components of DNA and RNA are a pentose (a 5-carbon sugar molecule), a phosphate, and four different bases. In DNA, or deoxyribonucleic acid, the pentose is deoxyribose. It also has two **purine** bases (**adenine** and **guanine**) and two **pyrimidine** bases (**cytosine** and **thymine**) (figure 3.1). Besides containing the sugar ribose instead of deoxyribose, RNA, or ribonucleic acid, has the pyrimidine base **uracil** instead of thymine (see figure 3.1). The difference between deoxyribose and ribose can be seen in figure 3.1c, where a hydrogen atom at the 2' position in deoxyribose replaces an OH group, as found in

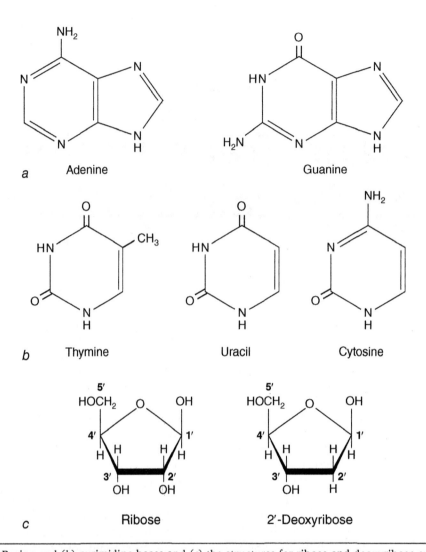

Figure 3.1 (a) Purine and (b) pyrimidine bases and (c) the structures for ribose and deoxyribose sugars.

ribose. In DNA and RNA, a base is attached to the 1' carbon of the sugar. The combination of a purine or pyrimidine base to ribose or deoxyribose generates a **nucleoside** or *deoxynucleoside*. The addition of one or more phosphate groups to the 5' carbon of the sugar component of a nucleoside by an ester bond produces a **nucleotide**. Figure 3.2 shows the structures for deoxyadenosine, a nucleoside, and adenosine 5'-monophosphate (AMP), a nucleotide.

When nucleotides polymerize to form nucleic acids, which are single-stranded polynucleotide chains, the OH group attached to the 3' carbon of the sugar of one nucleotide forms an ester bond to the phosphate of another nucleotide, eliminating a water. This is shown in a shorthand way in figure 3.3. If the sugar is deoxyribose and the base is thymine, we call it single-stranded DNA; if the sugar is ribose and the base uracil replaces thymine, it is a single strand of RNA.

Deoxyribonucleic acid is found as a double-stranded molecule in which the two strands are wound around a central axis, forming a right-handed double helix (figure 3.4). The two strands are held together by hydrogen bonding between bases, as shown in the middle of figure 3.4. We say the two strands are **complementary** in that a guanine (G) in one strand is always opposed by a cytosine (C) in the other, while an adenine (A) and a thymine (T) match each other in all cases, providing complementary **base pairs** (bp). Two hydrogen bonds are possible between each A and T base pair and three hydrogen bonds are possible between each G and C base pair. Uracil (U) in RNA similarly binds only to A, its complementary nucleotide, which is essential for the DNA sequence to be accurately coded into RNA. Note that the backbone of each strand in a DNA molecule is made up of deoxyribose and phosphate groups. Each strand in DNA has a polarity. One strand begins with

Deoxyadenosine   Adenosine 5′-monophosphate

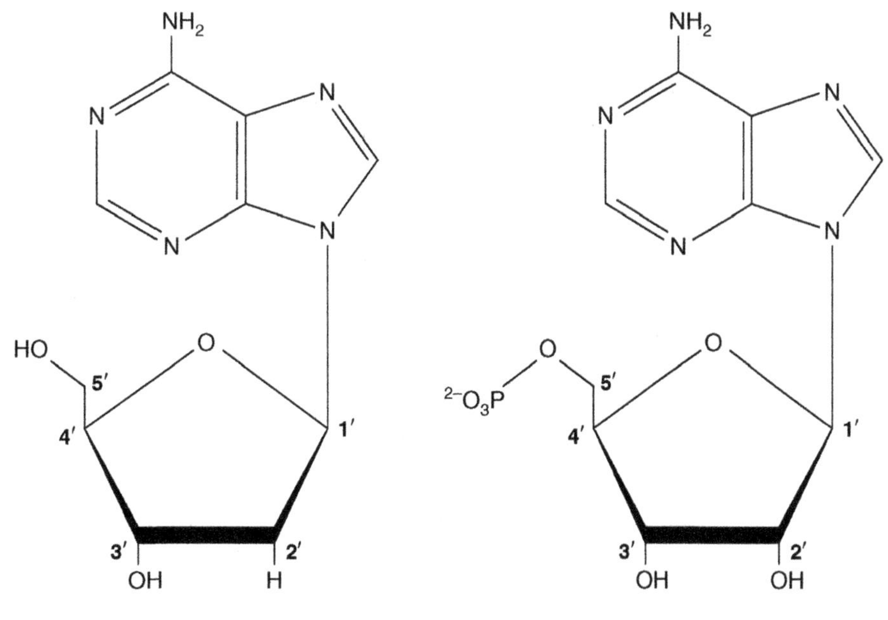

Figure 3.2 Structures for deoxyadenosine and adenosine 5'-monophosphate. The base adenine attached to the 1' carbon of deoxyribose creates the nucleoside deoxyadenosine. Addition of a phosphate group to the 5' carbon of the sugar ribose, itself attached to adenine, produces the nucleotide adenosine 5'-monophosphate (5'-AMP or simply AMP).

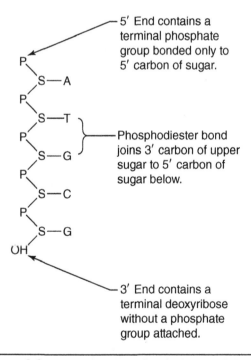

Figure 3.3 Representation of a single-stranded polynucleotide, shown with only five bases, identified by A, T, G, and C. S refers to the sugar deoxyribose, and P represents the phosphate group. The sugar-phosphate backbone is maintained by phosphodiester bonds that join the 3' carbon of one deoxyribose to the 5' carbon of the next deoxyribose.

a free phosphate group attached to the 5' carbon of deoxyribose. The end of the strand contains a free 3' OH group on the terminal deoxyribose. Note that the opposite strand has reverse polarity, with a free 3' OH at the top and the free 5' phosphate group at the bottom. Therefore, the two strands are said to be **antiparallel**. The polarity of DNA is very important. You should also understand that the content of G in a DNA molecule is equal to that of C; as well, A equals T.

KEY POINT

With a common sugar-phosphate backbone, the message carried by DNA lies in the order of the four bases. Therefore, we can say that the language of DNA is based on only four letters. The capital letters A, G, C, and T can represent the four bases in DNA. These letters are also used to designate the nucleotides (base + sugar + phosphate) that form the basic units of polynucleotides, such as DNA and RNA.

Chromatin and Nucleosomes

Deoxyribonucleic acid molecules are very large, containing millions of base pairs. The DNA in a single cell nucleus would reach about 2 m (6.6 ft)

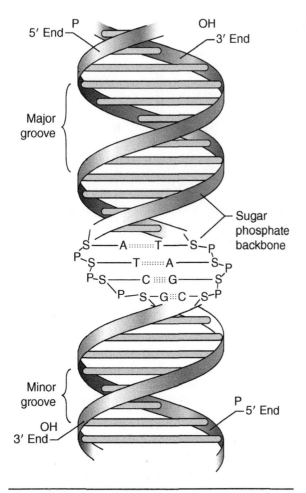

Figure 3.4 The double-helix structure of a portion of a DNA molecule, showing the two antiparallel polynucleotide strands. The two strands are held together by hydrogen bonding (dotted lines) between adjacent bases, with two hydrogen bonds between each A and T and three hydrogen bonds between each C and G. The arrangement of the helix creates both major and minor grooves. One complete turn of the helix spans a distance of 3.4 nm. Note that at the 5' end there is a phosphate (P) group and at the 3' end there is an OH group.

in length if extended. Nuclear DNA is compacted more than 100,000 times by folding and coiling (Chakravarthy et al. 2005). Each chromosome has roughly an equal mass of protein and DNA. Many of these proteins are needed to control the copying of sections of DNA. The DNA-protein complex known as **chromatin** is densely packed and distributed within the nucleus. The structural organization of chromatin is based on a class of proteins known as **histones**, which are intricately arranged with DNA to produce repeated structures known as **nucleosomes**. Each nucleosome consists of a disclike core containing two pairs each of the histones H2A, H2B, H3, and H4. DNA wraps tightly around each histone disc with 147 bp of DNA, making almost two complete turns (figure 3.5). Nucleosomes are joined by a linker DNA section that varies in length between 15 and 55 bp. A histone H1 also plays a role in linking adjacent histones. Histone proteins are rich in the basic amino acids lysine and arginine, so they carry a net positive charge that complements the negatively charged sugar-phosphate backbone of DNA. Because of nucleosomes, chromatin can look like a beaded necklace: Each nucleosome core particle represents a bead and the linker is the string that holds the beads together.

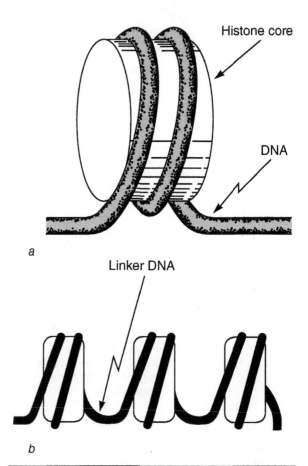

Figure 3.5 Nucleosomes. (*a*) A single nucleosome, showing 147 bp of DNA wrapped almost twice around a core of histone proteins, consisting of two of each of the histones H2A, H2B, H3, and H4. (*b*) Three nucleosomes joined together by a strand of linker DNA.

Chromatin has other levels of structural organization in which nucleosomes are packed together in even more condensed forms, generating higher levels of structure. Regions of chromatin with low levels of transcription activity are very tightly packed into a form referred to as *heterochromatin*. Such dense DNA packing in transcriptionally silent areas contrasts with more open chromatin packing in regions with higher rates of gene expression, known as *euchromatin*. The structure of the histone core shown in figure 3.5 does not reveal that parts of each histone protein in the core have tails that extend outside; these are described as *histone tails*. Lysine residues in these tails have positively charged side chains that allow interaction with neighboring DNA to produce more densely packed structures. We will see how modification of these histone tails (e.g., with exercise) can influence the transcription of genes.

Genetic Code

The sequence of four bases in a gene in DNA is transcribed to form a sequence of four bases in mRNA that must specify the sequence of the 20 different amino acids used to make proteins. We may ask ourselves how 20 different amino acids can be uniquely described by only four different bases. The only way that four different bases in mRNA (identified as A, G, C, and U) can specify 20 different amino acids in a polypeptide chain is for the bases to be read in groups of three, known as **codons**. Four different bases read three at a time gives rise to 4^3 different possibilities, or 64 codons. This process is called triplet coding; that is, three bases read together to give a message. Since mRNA has two ends, one of them—the 5' end—is considered to be at the beginning, and the message is read in groups of three toward the 3' end.

The **genetic code** is the relationship between the base sequence of DNA (A,T,G,C), transcribed to mRNA (U,A,C,G), and the sequence of amino acids in a polypeptide. It has a fixed starting position, an initiation codon, from which point the bases are then read in groups of three nonoverlapping bases (table 3.1). We usually consider the genetic code from the perspective of the codons of mRNA that spell out amino acids. Of the 64 possible codons, 61 specify amino acids, and the remaining 3 are stop or termination signals. Because there are 61 codons for only 20 amino acids, most amino acids have 2 or more codons. We say, therefore, that the genetic code is **degenerate**. Two amino acids (tryptophan and methionine) have a single codon, while some (leucine, serine, and arginine) have 6 codons. The codons for an amino acid with more than 1 codon are very similar. This similarity makes sense if minor errors are made. For example, the 4 codons for the amino acid glycine are GGA, GGG, GGC, and GGU. Notice that these all contain the same first two bases (letters). If amino acids are similar in structure, their codons are similar. For example, aspartic acid has the codons GAC and GAU, whereas glutamic acid, which is closely related in structure, has the codons GAA and GAG. If there is a base reading error, the same or a similar amino acid can still result.

Table 3.1 The Genetic Code Revealing the Three-Letter Codon for Each Amino Acid Shown by Its Three-Letter Abbreviation

First position (5' end)	SECOND POSITION				Third position (3' end)
	U	C	A	G	
U	UUU Phe UUC Phe UUA Leu UUG Leu	UCU Ser UCC Ser UCA Ser UCG Ser	UAU Tyr UAC Tyr UAA Stop* UAG Stop*	UGU Cys UGC Cys UGA Stop* UGG Trp	U C A G
C	CUU Leu CUC Leu CUA Leu CUG Leu	CCU Pro CCC Pro CCA Pro CCG Pro	CAU His CAC His CAA Gln CAG Gln	CGU Arg CGC Arg CGA Arg CGG Arg	U C A G
A	AUU Ile AUC Ile AUA Ile AUG Met**	ACU Thr ACC Thr ACA Thr ACG Thr	AAU Asn AAC Asn AAA Lys AAG Lys	AGU Ser AGC Ser AGA Arg AGG Arg	U C A G
G	GUU Val GUC Val GUA Val GUG Val	GCU Ala GCC Ala GCA Ala GCG Ala	GAU Asp GAC Asp GAA Glu GAG Glu	GGU Gly GGC Gly GGA Gly GGG Gly	U C A G

*Stop codons do not have an amino acid assigned to them.

**Codes for the amino acid methionine but also is the start or initiation codon.

The codon AUG is the initiation or **start codon** that signals the start of translation. This codon also represents the amino acid methionine; thus, the first amino acid used in protein synthesis is methionine. However, not all functional proteins have methionine as the first amino acid, since it can be removed after the polypeptide is completely formed. The codons UAG, UGA, and UAA are **stop (termination) codons**; they say that translation of the mRNA message is ended. The codon sequence beginning with a start codon and ending with a termination codon is known as a *reading frame*. The genetic code is universal for all organisms studied except in mitochondria, where some minor variations are noted. However, the amount of mitochondrial DNA is small (16,569 bp in humans)—just enough to code for 13 polypeptides, two kinds of rRNA, and 22 tRNA molecules.

> ### ☑ KEY POINT
>
> Each mitochondrion contains multiple copies of mitochondrial DNA. Our mitochondrial DNA is obtained from the egg, not the sperm, and the distinctive characteristics of our mitochondrial DNA can provide a history of the mothers in our families. The small circular mitochondrial DNA molecule is discussed in more detail in chapter 5.

TRANSCRIPTION

The process of transcription involves copying a gene on DNA to make an RNA molecule. A simple definition of a gene is that it is a section of DNA that provides the information to synthesize a single polypeptide or functional unit of RNA, such as a tRNA or rRNA molecule. For our purposes, we will focus specifically on the production of mRNA for protein synthesis. For genes coding for amino acid–sequence information for a polypeptide, the actual gene is considered to consist of a *coding region* and a region in front of this *(regulatory region)* that controls the transcription of the coding region. The precursors needed to make RNA are nucleoside triphosphates (NTP) such as CTP, GTP, ATP, and UTP. The formation of mRNA during transcription is catalyzed by a large oligomeric enzyme known as **RNA polymerase II**.

As shown in simplified form in figure 3.6, during transcription, only a portion of one strand of the DNA in the gene is copied. This is known as the **template DNA strand**, and it is read in the 3' to 5' direction. The RNA formed, called the **primary gene transcript**, will be complementary to the template DNA strand, with the exception that the RNA will contain the base uracil (U) rather than thymine (T). The primary transcript will be formed in the 5' to 3' direction, that is, antiparallel to the template strand. The DNA strand in the gene that is not copied is known as the *sense strand* because it will have the same base sequence as the RNA transcript, except that U will replace T. The polarity of the sense strand and the RNA are the same. In this context, directions are often described in terms of river flow. **Upstream** refers to the 5' direction, and, by convention, the frame of reference is the sense strand of DNA. Likewise, **downstream** refers to the 3' direction of flow—that is, the direction in which the template strand is transcribed. To facilitate understanding of the genetic code, figure 3.6 also shows the amino acids spelled out by codons on the mRNA molecule.

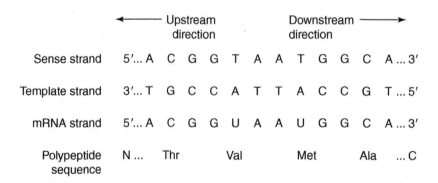

Figure 3.6 A section of a polypeptide-coding gene showing the two strands of DNA. The template strand is copied during transcription, generating a complementary copy of single-stranded mRNA. Note that the base (nucleotide) sequence of mRNA matches that of the sense strand of DNA, with U in RNA replacing T in the sense strand. To reveal the correspondence between base sequences on DNA (sense strand) and mRNA, three-base sequences on the latter are converted into an amino acid sequence determined from table 8.1. N and C refer to the amino and carboxy ends of the polypeptide, respectively.

Steps in Transcription

Figure 3.7 shows the three major phases of transcription. *Initiation* begins when general transcription factors (proteins, which we will discuss next) and RNA polymerase II bind to the double-stranded DNA just upstream of the **start site**, forming a preinitiation complex. The start site is where transcription actually begins, and the first DNA base copied is given the number +1 and is usually shown with a bent arrow pointing downstream. General transcription factors help the RNA polymerase II find the start site and facilitate initiation of the transcription process. Bases immediately upstream are located in a region called the **promoter** and are given negative numbers. The promoter region can also include some bases just downstream from the transcription start site.

When RNA polymerase II and the general transcription factors bind to the promoter region, the DNA at the start site is unwound (melted), exposing approximately 14 nucleotides in the template strand (figure 3.7b). The first nucleotide triphosphate that will become the first nucleotide on mRNA binds to the complementary base on nucleotide 1 on the template strand. Then the second NTP comes in, recognizing its complementary base, number 2 on the template strand. A phosphoester bond is formed between the 3' OH of ribonucleotide 1 and the 5' phosphate of ribonucleotide 2, and a pyrophosphate is released. During initiation, the RNA polymerase II does not move along the DNA.

The elongation phase begins when the RNA polymerase II dissociates from the promoter and general transcription factors and moves along the template strand of DNA, making a complementary RNA strand (figure 3.7c). As it moves, it unwinds the DNA double helix, catalyzing the formation of the RNA strand. Behind it, the DNA double helix reforms (figure 3.7d). For each new nucleotide added to the terminal-free 3' OH of the growing RNA polynucleotide, only one of

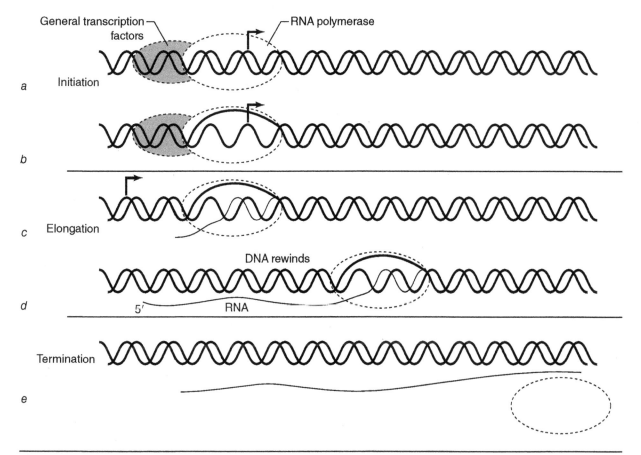

Figure 3.7 The three distinct phases of transcription. *Initiation (a, b)* occurs when RNA polymerase II binds along with general transcription factors to a region encompassing the transcription start site, shown with a bent arrow. Deoxyribonucleic acid is first unwound (b), exposing the template strand to be copied. During *elongation (c, d)*, RNA polymerase II catalyzes the addition of nucleotides, one at a time, to the 3' end of the growing RNA molecule. (d) As the polymerase moves along the template strand, the DNA behind it rewinds. (e) When a *termination* signal is reached, transcription ceases, and the RNA polymerase II and the primary RNA transcript dissociate from the DNA.

the three phosphate groups on the incoming nucleoside triphosphate is needed; the other two are released as **inorganic pyrophosphate**, PPi. Figure 3.8 shows the details of the addition of a new ribonucleotide to the growing RNA chain. The termination phase begins when the RNA polymerase II has moved along the template strand of DNA and reaches a sequence of bases indicating that the gene message is terminated. Some call this the terminator. At this point, the RNA polymerase II and the RNA strand dissociate from the DNA. The RNA polymerase II is now free to transcribe the same gene again or a different gene. It is estimated that RNA polymerase II catalyzes transcription at a rate of approximately 1,000 nucleotides per minute.

Regulation of Transcription

In complex organisms such as humans, regulation of transcription in a cell depends on a complex interaction of hormonal, metabolic, nutritional, and environmental signals. It is important to emphasize that DNA is found with proteins as chromatin. This means that the substrates for transcription are actually sections of chromatin. Control of gene expression occurs primarily through directing the initiation stage of transcription. Two major types of control decisions must be described for a cell. Irreversible decisions turn specific genes on or off completely. For example, during embryo development, certain genes are initially expressed, then turned off completely. Other genes are irreversibly turned on. But the second type of decision is adjustable in terms of transient increases or decreases in the rate of transcription of an already expressed gene in response to various environmental or metabolic conditions. Consider the analogy of a light switch with a dimmer. The irreversible decision is that the switch is either on or off. Once on, however, it can be adjusted to produce a low, moderate, or high level of light intensity. Some genes in a cell are permanently turned off. One mechanism whereby this occurs is the structural organization of DNA. As a subsequent section discusses, inactive genes are found in sections of chromatin that is condensed, precluding the binding of RNA polymerase II and other proteins at the promoter region to initiate transcription.

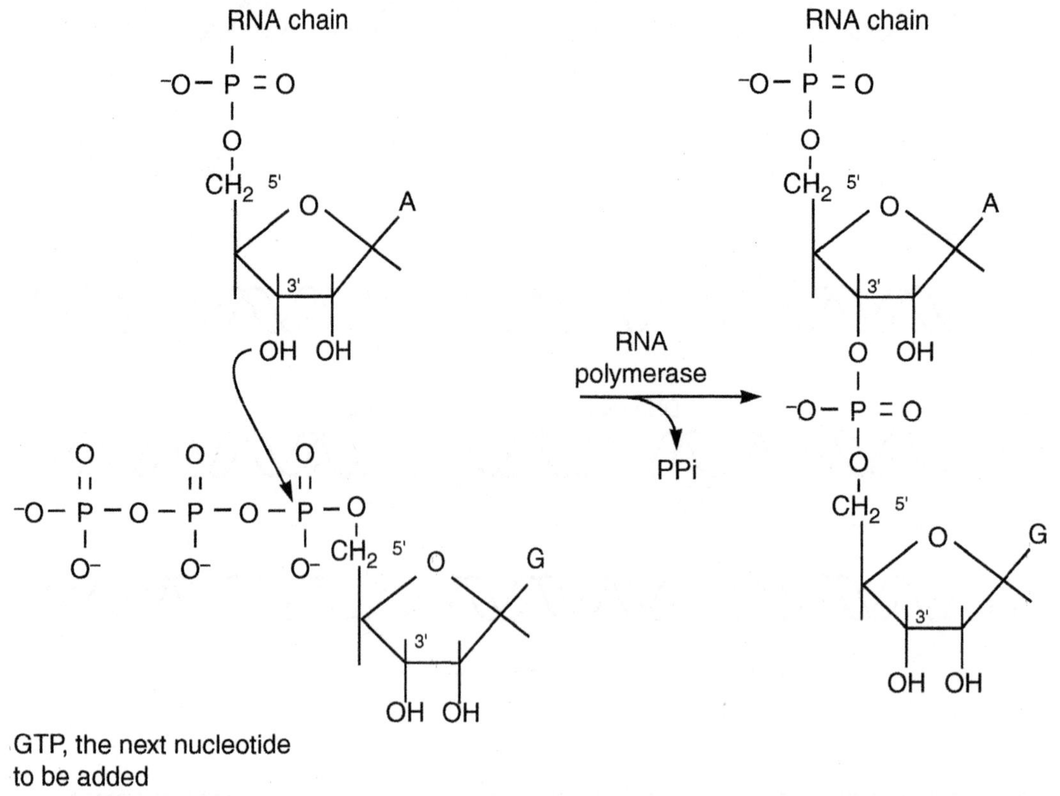

Figure 3.8 The RNA chain during transcription. The RNA chain grows by the addition of nucleoside triphosphates to the 3' end of the RNA chain. In the example shown, the new nucleoside triphosphate is guanosine triphosphate (GTP). On the left side of the figure, the OH group at the 3' end of the RNA chain attacks the α-phosphate of GTP, creating a new phosphoester bond and releasing inorganic pyrophosphate (PPi). The chain on the right side is now lengthened by one nucleotide.

KEY POINT

Proteins direct virtually all the events in a multicellular organism associated with a healthy life. The information for protein structure is based on the sequence of amino acids, dictated by the sequence of bases in a segment of DNA called a gene. Since multicellular organisms have highly specialized cells, we can assume that this specialization is due to the production and function of specific proteins. This means that specific genes must be properly transcribed in all cells.

Transcription Factors and Response Elements

Proteins are responsible for most of what happens in an organism, including the regulation of gene transcription. DNA-binding proteins that regulate transcription are known as **transcription factors** or gene regulatory proteins. They are products of their own genes, most likely on different DNA molecules from the gene or genes they regulate. If they promote transcription, they are **activators**; if they have a negative effect on transcription, we call them **repressors**. We can define two major classes of transcription factors: (1) general transcription factors needed for initiating transcription from most protein-coding genes and (2) those that modulate general transcription factors through activation or repression and provide specificity for different cell types (i.e., skeletal muscle) and differentiation. We will discuss these shortly. In their role as regulators, transcription factors must bind to specific DNA sequences, called *regulatory sequences* or *response elements*, as part of the *transcription-control region*; some sources describe this as the gene control region. Gene regulatory sequences are normally short, usually 6 to 10 bp, although some may be larger. These can be at the promoter region (or promoter), which is the sequence of bases at the site where RNA polymerase II and the general transcription factors bind to initiate transcription. A common feature of the promoter of many protein-coding genes is the **TATA box**, an A- and T-rich regulatory sequence located 25 to 35 bp upstream of the start site where the transcription initiation complex binds. Some sources also identify a *proximal promoter*, a region upstream of the promoter, typically 100 to 200 bp upstream of the transcription start site. The transcription-control region may also include **enhancers**, which are regulatory sequences typically far away from the promoter, either upstream or downstream, or even within the coding region of the gene. The response elements that make up a typical gene are shown in figure 3.9.

The gene regulatory proteins include general transcription factors that bind at the promoter, as well as several thousand other proteins that act to control the transcription of specific genes, since each gene or any closely related gene in a cell appears to have its own specific regulation arrangement. Most gene regulatory proteins have a *DNA-binding domain*. This is a specific portion of the protein molecule, often described as a motif, that recognizes a particular short stretch of DNA. While the details are beyond the scope of this book, you have probably learned about some common DNA-binding motifs found in **eukaryotic** species such as the *homeodomain*, *zinc finger*, *leucine zipper*, basic zipper (*bZip*), and *basic helix-loop-helix* (bHLH) motifs. Gene regulatory proteins often bind to their response elements in pairs (as dimers). The two transcription factors may be either the same protein or different proteins—that is, homo- and heterodimers, respectively. Therefore, we should expect that these transcription factors need dimerization domains that allow them to bind

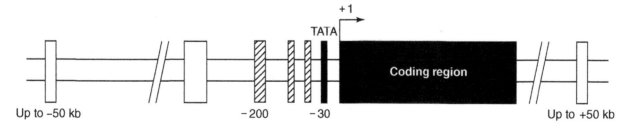

Figure 3.9 Transcribing a typical gene. Control over transcription is exerted by the binding of a variety of transcription factors to particular DNA control (response) elements, shown as boxes in this figure. Some control elements are located far upstream of the start site, shown as a bent arrow, while others may be far downstream of the coding region. The TATA box, located in the promoter, is where the transcription initiation complex binds. Immediately upstream of this, up to approximately −200 bp, are the promoter proximal elements, shown crosshatched. Other response elements are shown as open bars. Here, kb represents kilobase.

to the other protein forming the dimer complex at the response element.

Transcription factors also contain a region or regions that allow them to interact with other regulatory proteins, including general transcription factors. These regions are usually called activation domains because most of the interactions promote transcription, as opposed to repressing it. Activator proteins promote transcription by helping to attract and position other proteins at the promoter to facilitate transcription initiation. They may also act on the chromatin to expose the DNA to the transcription machinery. Given that several thousand different gene regulatory proteins exist, as well as the fact that a number of these are needed to initiate transcription, binding as both homo- and heterodimers, it is no wonder that there is likely a unique transcription initiation apparatus for almost every protein-coding gene.

Gene repression in eukaryotic species involves two kinds of repressor proteins. Some repressors act to keep large sections of chromatin unavailable for transcription because of the tight way it is packed in regions; this was described earlier as heterochromatin. Other repressors act locally to inhibit transcription initiation in much the same way that activators facilitate transcription. Whereas activators facilitate the assembly of the transcription initiation complex by interacting with other proteins, including those classified as general transcription factors, repressors inhibit this process. They may do so by masking the activation domain of an activator, by blocking access to a response element of an activator, or by altering the chromatin structure at the promoter to render the gene less accessible for transcription. Another class of gene regulatory proteins, the *coactivators* and *corepressors*, have activation and repression domains, respectively, but lack DNA-binding domains. Therefore, they are unable to bind to response elements for the gene. By binding to transcription factors, their role is primarily to facilitate or inhibit transcription initiation. The peroxisome proliferator γ coactivator-1 (PGC-1) family of coactivators is a very important family of coactivators that plays a major role in the regulation of carbohydrate and fat metabolism. It is also a major regulator of muscle fiber–type specialization and mitochondrial content and function. PGC-1 is discussed further in this chapter and in chapters 6 and 7.

Basal Transcription Apparatus

RNA polymerase II cannot initiate transcription by itself. Rather, it requires a number of general transcription factors to interact with it in the core promoter region and a large protein complex known as the *mediator*. The term *general transcription machinery* is used to describe RNA polymerase II, the general transcription factors, and the mediator that assembles at the promoter region of the gene. RNA polymerase II has 12 protein subunits and the mediator has 20 subunits. Additional proteins that will change the structure of the DNA to facilitate transcription are also necessary. Without the influence of other transcription factors bound to their related response elements in the promoter proximal region or of distant enhancers, the rate of transcription based only on general transcription factors is low or nil. In this section, we focus only on the basal transcription of a protein-coding gene. Figure 3.10 outlines the various protein factors found in the promoter region of a typical gene.

The general transcription factors are identified as such since they are found in the promoters of all genes transcribed by RNA polymerase II. They help to position RNA polymerase II and bind in the promoter region. They also facilitate the opening of the DNA to expose the template strand. After transcription is initiated, general transcription factors help to release RNA polymerase II so it may continue on during the elongation phase. A TATA-binding protein (TBP) is first to bind at the TATA box, followed by general transcription factors IID and IIB. Transcription factor IIF and RNA polymerase II binds next, so that RNA polymerase II is positioned over the transcription start site. Subsequent additions are at least transcription factors IIE and IIH. A large protein complex, containing at least 20 polypeptide subunits and known as the *mediator*, plays a key role in positioning RNA polymerase II and binding to activation domains of other transcription factors involved in the regulation of transcription. It is believed that to generate a basal level of transcription for most mammalian genes, at least five general transcription factors (IIB, IID, IIE, IIF, and IIH), along with RNA polymerase II and the mediator, are necessary. General transcription factor IIA is found in some promoters, but its role in all polypeptide-coding genes is unclear.

Some protein-coding genes lack a TATA box, having instead an alternate promoter element known as an *initiator*. Other protein-coding genes begin their transcription over a rather long region up to 200 bp in length. Their primary transcript can thus have multiple 5' beginnings. Genes lacking both TATA boxes and initiators may instead have *GC-rich regions* in their promoter, 100 or so bp upstream of their start site region. Such genes are generally coded at a consistent but low rate and are known as *housekeeping genes*. Their protein products are typically metabolic enzymes.

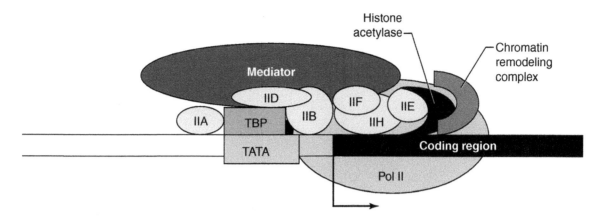

Figure 3.10 Protein factors in the promoter region of a typical gene. Transcription initiation at a typical protein-coding gene requires the presence of RNA polymerase II (Pol II), a large protein complex known as the mediator, and a number of general transcription factors, including the TATA-binding protein (TBP). Common general transcription factors found at the initiation of genes transcribed by Pol II are IIB, IID, IIE, IIF, and IIH. Many promoter regions also contain general transcription factor IIA. Transcription begins at the start site, shown as a bent arrow. Chromatin remodeling, essential for transcription, is facilitated by chromatin remodeling proteins, along with histone acetylase complex.

Higher Levels of Transcription Control

With thousands of protein-coding genes in our genome, regulation must be complex and specific. We have already discussed the fact that gene response elements may be located in positions far from the actual start site of the gene. In addition to the TATA box, which is so widely found in the promoter region of protein-coding genes, other base sequences are recognized by specific proteins that can lead to higher levels of transcriptional control. These response elements may be found in the proximal promoter region (see figure 3.9), regions immediately upstream of the proximal promoter, or thousands of base pairs removed from the transcription start site. The terms *enhancers* and *repressors* have been applied to distant response elements that act to enhance or inhibit transcription initiation by the general transcription machinery. The ability of distant enhancers or repressors to influence transcription initiation at the promoter is attributed to the way higher orders of DNA bending and looping can bring distant response elements close to the promoter. Figure 3.11 illustrates how distant response elements (enhancers) with their activators bound as homo- or heterodimers can interact with the general transcription machinery via the mediator complex at the site where transcription begins. Such interactions can play an enormous role in facilitating or repressing transcription initiation (Conaway et al. 2005).

Transcription factors that augment transcription at the point of initiation may be ubiquitous—that is, found in and active in a variety of cell types and species—or they may be highly specific, expressed in a single cell type. If they are ubiquitous, the response elements they bind to will be the same in a variety of species. We say that these are conserved sequences. It is important to emphasize that transcription factors are proteins, products of their own genes, indicating that whether the gene is expressed or to what extent it is expressed can dictate the expression of possibly many other genes. Figure 3.12 illustrates a number of common gene-regulating factors in the control region of a hypothetical gene. Transcription factors and the response elements to which they bind are shown in a linear way to simplify the picture.

The transcription factor *NFκB* (nuclear factor kappa B) activates a number of genes in response to conditions such as inflammation, infection, or another immune system–activating factor. Normally, NFκB is found in the cell cytosol attached to an inhibitor, *IκB*, which prevents NFκB from functioning in the nucleus as a transcription factor. With the appropriate stimulation, IκB becomes phosphorylated by an IκB kinase. Phosphorylation of IκB leads to its dissociation from NFκB, which allows two subunits of NFκB (p50 and p65) to be translocated to the nucleus; here, they bind to a response element located in more than 150 genes.

Activator protein-1 (AP-1) represents a family of transcription factors that bind as dimers. The best known of the AP-1 transcription factors are the proteins *Fos* and *Jun*, which act to regulate the transcription of many genes, especially in the immune system. Fos and Jun are the protein products of genes that

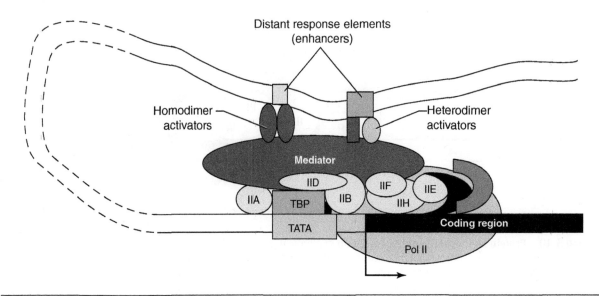

Figure 3.11 Interactions between distant response elements and the general transcription machinery. Distant response elements or enhancers interact with the general transcription apparatus, including the TATA-binding protein (TBP), through the large mediator complex. Looping and bending of DNA allow these distant response elements to be in the region of the promoter. The dashed lines are to illustrate that there are large distances along a DNA molecule between distant response element and promoter.

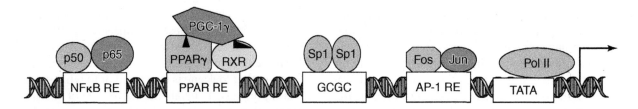

Figure 3.12 The control region of a hypothetical gene, often shown as a linear DNA molecule with specific base sequences (response elements, RE) where their cognate transcription factors bind as homo- or heterodimers. Nuclear factor κB (NFκB) has two main DNA binding subunits, shown as p50 and p65. Peroxisome proliferator-activated receptor γ (PPARγ) often binds to its response element as a heterodimer with the retinoid X receptor (RXR). Note that both PPARγ and RXR need to bind specific ligands (shown as small triangles) in order to function as transcription factors. PGC-1γ is a PPARγ coactivator. Sp1 is a common transcription factor, which binds as a dimer to a GC-rich response element. Fos and Jun are transcription factors often found together, known as AP-1. Ribonucleic acid polymerase II (Pol II) is found at the start site (shown as a bent arrow) near the TATA box.

are turned on very early following certain stimuli to a cell. Genes that are rapidly activated in response to some abrupt change are often called *immediate early genes* because their protein products are needed as transcription factors to turn on other genes.

Peroxisome proliferator–activated receptor γ (PPARγ) is one member of a small family of transcription factors (the others are PPARα and PPARβ/δ) that play very significant roles in the metabolism of carbohydrates and lipids. By themselves, PPARs are not active as transcription factors, but they can assume this role when they are activated by coactivators and a variety of natural ligands (e.g., unsaturated fatty acids). The PPARs are also targets of some drugs that can lead to decreases in blood lipid levels or increases in insulin sensitivity, depending on the specific drug that targets PPAR (Chinetti-Gbaguidi, Fruchart, and Staels 2005). PPARs function as transcription factors when they bind with the *retinoid X receptor* (RXR) as heterodimers at PPAR response elements. The RXR is itself activated as a transcription factor when it binds its ligand, a form of retinoic acid. Peroxisome proliferator γ coactivator 1 (PGC-1γ) binds to PPAR and RXR heterodimers and interacts with other transcription factors as a coactivator.

Four highly specific transcription factors that are found only in skeletal muscle and are essential for their development are the **myogenic regulatory factors**: MyoD, myogenin, MRF4, and myf5. These proteins contain secondary and tertiary structural components, known as the basic helix-loop-helix, that recognize a six-base response element termed an *E box*. The myogenic regulatory factors form heterodimers with another class of proteins called *E proteins* (e.g., MEF2) at the E box. The E box is found in the regulatory region of some specific muscle genes, including those expressing the fast myosin light and heavy chains (see figure 1.14, showing heavy and light chains of myosin). Artificial expression of the MyoD gene family in other cell types can cause those cells to express genes transcribed only in skeletal muscle. This emphasizes the significant role of the myogenic regulatory factors as transcription factors in the development of skeletal muscle.

Cell Signaling and Transcription

The integration of cells in complex organisms such as humans requires carefully regulated signaling throughout the body. You are familiar with this because you have learned about particular hormones in earlier studies. Besides hormones, there is a growing list of special signaling proteins known as **cytokines**, many of which are described as growth factors. These are secreted from a wide variety of cells. They may interact in different parts of the body by traveling in the blood in an endocrine manner. Alternatively, they may act locally, on cells in their immediate environment (**paracrine** effect), or even on themselves (**autocrine** effect). This sequence, in which an external molecule such as a hormone or growth factor binds to its cell surface receptor and thereby influences events within the cell, is known as **signal transduction** or signaling. As the following examples show, signaling can often lead to changes in gene transcription.

The **steroid hormones** (glucocorticoids, testosterone, estrogen, and progesterone) as well as thyroid hormone, the active form of vitamin D, and retinoids (from vitamin A) circulate in the blood and readily enter cells because their lipophilic nature permits them to diffuse across the cell membrane. They can enter the nucleus and bind to their specific protein **hormone receptors**, which are also products of specific genes. The receptors are known as the *nuclear receptor superfamily* because they can either permanently reside in the nucleus or enter the cell nucleus after binding with their specific hormone in the cytosol. Each nuclear receptor has a DNA-binding domain and a ligand-binding domain. Nuclear receptors bind as homo- or heterodimers in the control region of a number of genes. In the absence of their ligands, some members of the nuclear superfamily repress transcription. When the specific hormone or active vitamin form binds to its cognate receptor in the control region of a gene, drastic alterations take place in the conformation of the receptor to activate transcription.

Estrogens are steroid hormones that play major roles in both women and men. They are involved in reproduction, as most people are aware, but they also have effects on the cardiorespiratory system, the bones, and even our behavior. Because of its hydrophobic nature, *estradiol*, the dominant form of estrogen, can enter cells by diffusing through cell membranes. The dominant pathway involves entry of estradiol into the nucleus, where it binds to its specific estrogen receptor, forming an estradiol-estrogen receptor complex that binds as dimers at an estrogen-response element to facilitate transcription of specific genes in much the way we have shown for distant enhancers in figure 3.11.

As described in chapter 2, protein phosphorylation and dephosphorylation play major roles in modulating the activity of a host of proteins. Therefore, we should not be surprised to learn that protein phosphorylation–dephosphorylation can influence gene transcription through changes in transcription factors that accompany addition or removal of a phosphate group. Figure 3.13 outlines a signaling pathway in which an extracellular agonist, defined as a ligand for a receptor that activates that receptor (e.g., a hormone), eventually alters the conformation of a transcription factor to modulate transcription. The agonist in figure 3.13 binds to its receptor, generating a conformational change in the receptor that is communicated to **adenylyl cyclase** via a stimulatory

> ### ✓ KEY POINT
> Signal transduction, or signaling, is the process that provides the total network of communication among the various cells, tissues, and organs of the body. Cells cannot survive without communication, but with it, the normal pattern of cell growth, differentiation, and metabolism can be regulated and the rates of gene expression in individual cells can be appropriately matched to the overall needs of the organism. Exercise training provides an excellent example of this, whereby disruption to cellular homeostasis in response to individual training bouts activates specific signaling mechanisms that alter gene expression so that disruptions to cellular homeostasis during subsequent bouts of exercise are minimized.

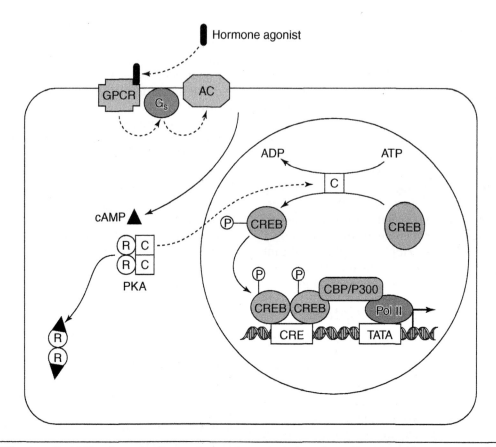

Figure 3.13 A cyclic AMP pathway leading to changes in transcription of specific genes. A hormone agonist, shown as a small oval, binds to a G protein–coupled membrane receptor (GPCR). This leads to activation of adenylate cyclase (AC) by way of a stimulatory G protein (G_s), creating cyclic AMP (cAMP, small triangle). The cAMP binds to the regulatory (R) subunits of protein kinase A (PKA), releasing the catalytic subunits (C), which translocate to the nucleus, where each subunit can catalyze the phosphorylation of cAMP response element binding protein (CREB). Phosphorylated CREB binds as dimers at a cAMP response element (CRE), attracting coactivator CBP/P300 to facilitate transcription of target genes.

G protein (G_s). Cell-membrane receptors that interact with G proteins following ligand binding are often described as G protein–coupled receptors (GPCR). When activated, adenylyl cyclase converts ATP into **cyclic AMP** (cAMP). The cyclic AMP binds to the regulatory subunits of protein kinase A, releasing the catalytic subunits, which are then free to transfer a phosphate group from ATP to substrates such as the cyclic AMP response element binding protein (CREB). Phosphorylated CREB transcription factors bind as dimers at the cyclic AMP response element (CRE), attracting the coactivator CBP/P300, which links CREB to the basal transcription machinery as represented by RNA polymerase II.

DNA Organization and Gene Transcription

The tight coiling of chromatin plus the nucleosome structure (see figure 3.5) suggests that initiating transcription must be rather difficult, considering the intricate arrangement of transcription factors and DNA response elements needed. Indeed, as they exist, nucleosomes block proper transcription initiation and make transcription elongation by the large RNA polymerase II complex difficult at best. In fact, nucleosomes are regarded as general repressors by some (Boeger et al. 2005). How then are genes transcribed? A number of findings, including the fact that nucleosome structure is temporarily broken down at the start site to expose the gene promoter region, suggest that the initiating event in the transcription of a particular gene is to expose the promoter region by remodeling local DNA structure.

Histone proteins in the core particles of nucleosomes have protruding short polypeptide segments, described as histone tails. Certain lysine residues in histone tails can be acetylated; that is, an acetyl group is transferred from acetyl CoA to the side-chain ammonium group of lysine residues. Lysine acetylation of histones involves the changing of a small, positively charged ammonium group on the lysine side chain to a larger N-acetyl group without a charge, as shown in figure 3.14. The tight binding of DNA to core histones is facilitated by attractions between the positive charge on the many lysine side chains and the negatively charged DNA backbone. Reduction in this binding facilitates the unraveling of DNA, exposing the promoter and other control regions and facilitating transcription elongation by RNA polymerase II. A class of enzymes found within the nucleus carries out this acetylation process. These are known as *histone acetylases* (also known as *histone acetyltransferases*, or HATs). Acetylation is not a permanent state, so a related enzyme known as *histone deacetylase* (HDAC) removes the acetyl group, creating a positive charge and facilitating nucleosome formation. Consistent with the idea that histone acetylation facilitates transcription, it has been discovered that transcriptionally active regions of the genome are highly acetylated. Regions with low or zero rates of gene transcription have little histone acetylation. Previously, we noted the role of coactivators and corepressors as proteins without DNA-binding domains that nevertheless facilitate or inhibit transcription, respectively. Some view HAT and HDAC as coactivators and corepressors, respectively (Roeder 2005).

Histone proteins also undergo other modifications that influence transcription rate. While a detailed description of these is beyond the scope of this book, a few should be mentioned. Serine and threonine residues in histone tails can be phosphorylated by kinases. Chapter 2 discusses the addition of negatively charged phosphate groups to amino acids with side chain hydroxyls and how this alters enzyme function. Lysine residues in histone tails can also be methylated. As might be expected, modification of histone tails by whatever mechanism (acetylation, phosphorylation, or methylation) changes both the charge and the conformation of the tails, altering their ability to interact with DNA and thereby facilitating or inhibiting promoter accessibility of neighboring genes (Chakravarthy et al. 2005).

Found in the nucleus are ATP-dependent chromatin remodeling enzymes that utilize the free energy of ATP hydrolysis to alter the conformation of chromatin, exposing repressed regions due to nucleosomes and tight higher levels of chromatin packaging (Flaus and Owen-Hughes 2004). It is likely that modified histones act as targets for the ATP-dependent chromatin remodeling enzymes in promoter regions of genes. As well, modified histone tails may serve to enhance or inhibit binding of the mediator, general transcription factors, and RNA polymerase II at the promoter. The complex protein–DNA and protein–protein interactions involved in transcription initiation, plus the ability of chromatin to be remodeled through histone acetylation and deacetylation, phosphorylation, and dephosphorylation, as well as by ATP-dependent remodeling enzymes, underscore the complexity of gene transcription.

POSTTRANSCRIPTIONAL MODIFICATIONS OF RNA

The process of DNA transcription yields RNA. Unlike DNA, RNA is composed of a single strand of nucleotides. The ends of RNA molecules are designated as 5' if there is a free phosphate on the 5' carbon of ribose, or 3' if the ribose sugar has a free 3' OH group. Thus, like each DNA strand, an RNA molecule has a direction. Ribonucleic acid molecules are created when a section of a strand of DNA (a gene) is copied in the transcription process. The length of the RNA molecule and the sequence of bases (nucleotides) dictate what role the RNA molecule will perform. As chapter 1 discusses in the context of the sequence of amino acids in a protein, the sequence of bases in an RNA molecule is also known as its primary structure. The three main types of RNA are as follows:

- **Messenger RNA** (mRNA) is the actual template for protein synthesis in the cytosol. This means that

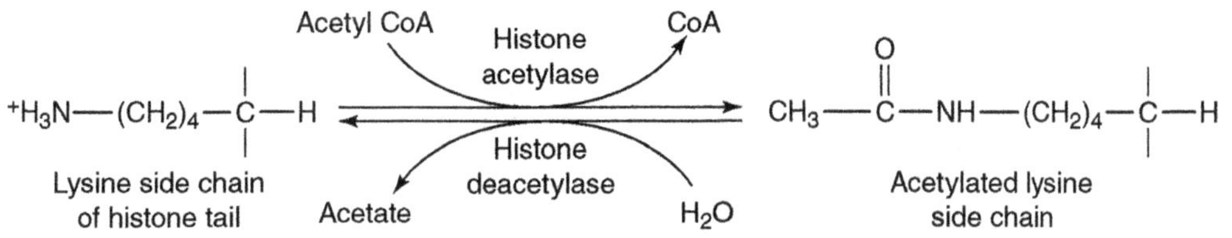

Figure 3.14 Lysine acetylation of histones. Lysine side chains found in the tails of core histone proteins can be acetylated by acceptance of an acetyl group donated by acetyl CoA in a reaction catalyzed by histone acetylase (also known as histone acetyltransferase). The reverse reaction is catalyzed by histone deacetylase. The addition of the acetyl group to the ammonium group in the side chain of lysine residues removes the positive charge and adds a bulky group, thus changing the conformation of the histone tail.

the base sequence of mRNA specifies the sequence of amino acids in a polypeptide chain. Most of our genes generate specific mRNA, which has a short lifetime—usually ranging from a few minutes to several hours. Messenger RNA is also the least abundant of the three types of RNA, representing about 2% of total cellular RNA.

• *Transfer RNA* (tRNA) is the smallest of the RNA molecules, usually between 73 and 93 nucleotides (Nt) in length. Transfer RNA attaches to specific amino acids and brings them to the complex of mRNA and ribosomes on which a polypeptide is formed. At least one tRNA molecule is present for each of the 20 amino acids involved in protein synthesis. All the tRNA molecules represent approximately 15% of the total cellular RNA.

• *Ribosomal RNA* (rRNA) is the most abundant RNA, representing more than 80% of all the RNA in a cell. A ribosome is a complex of protein and ribosomal RNA where proteins are synthesized.

One other type of RNA exists, known as **small nuclear RNA** (snRNA). The snRNAs are found in the nucleus associated with protein in particles described as small nuclear ribonucleoprotein particles or snRNPs. These are involved in the processing of the primary RNA transcripts during their conversion to mRNA molecules.

Proteins are synthesized on ribosomes in the cell cytosol. The mRNA message is translated into a sequence of amino acids. Before any of this can take place, however, the mRNA molecules made by transcription must be modified to generate the active form of mRNA.

Formation of mRNA Molecules

Following transcription of genes coding for polypeptides, the product is known as the primary transcript or *pre-mRNA*. Because in any nucleus, many hundreds of genes are being transcribed simultaneously, there will be a huge variety of pre-mRNA and other RNA molecules. Each pre-mRNA or other RNA molecule is associated with protein right from the time it is freed from its RNA polymerase II until when it is transported out of the nucleus. The collection of all of the transcripts in the nucleus is described as *heterogeneous nuclear RNA* or hnRNA. hnRNAs, when combined with the nuclear proteins with which they are associated, are known as *heterogeneous ribonucleoprotein particles*, or hnRNPs.

The pre-mRNA molecules undergo three kinds of changes before they become mature mRNA molecules ready to be exported from the nucleus. These alterations are referred to as RNA processing. First, each pre-mRNA has a cap added to its 5' end. Next, it has a tail added that consists of multiple copies of adenine nucleotides. Finally, it has its **introns** removed. This last step is very important because it means that the genes in eukaryotic organisms are interrupted—that these genes are in pieces. Parts of the gene, known as **exons**, contain information that will appear in the mature mRNA; these can be called coding regions. Other parts contain base sequences, known as *introns*, or intervening sequences that will not appear in the mature mRNA. As shown in figure 3.15, a gene can be illustrated as a segment of DNA with a transcribed section bounded at the 5' end by the start site and the poly A tail. Just upstream of the start site is the promoter. Within the transcribed portion are introns and exons. Figure 3.15*b* shows a typical way in which exons and introns can be illustrated. The primary transcript or pre-mRNA contains nucleotides corresponding to both introns and exons. The pre-mRNA in figure 3.15 also has a 5' cap and a poly A tail, which we will discuss next. The final processing of the pre-mRNA molecule involves splicing out the introns so that between the cap and poly A tail is a section composed only of exons.

Capping the 5' End of the RNA

As you will recall, transcription of genes proceeds as the template strand of DNA is read in the 3' to 5' direction. The RNA transcript is generated in the 5' to 3' direction. While the RNA transcript initially stays with the RNA polymerase II and loosely binds to the template strand of DNA by **hybridization** between complementary bases, the 5' end becomes free before transcription proceeds very far. Capping of the free 5' nucleotide of RNA involves a reaction between the 5'-phosphate group on the first ribonucleotide and a 7-methyl GTP (guanosine triphosphate). This creates a cap consisting of 7-methylguanine bound to sugar ribose, which in turn is attached to the first nucleotide of the RNA via three phosphate groups. This process is generally described as capping and is shown in figure 3.16*a*.

Forming the Poly A Tail

All mRNA molecules in eukaryotes, except those for the histone proteins, have a poly A tail. This consists of 200 to 250 adenine nucleotides added to the 3' end of the pre-mRNA. A schematic of the process is shown in figure 3.16. Transcription proceeds past a consensus sequence (AAUAAA) and the poly A site to a region that generates a G- and U-rich sequence in the pre-mRNA. Two major processes now take

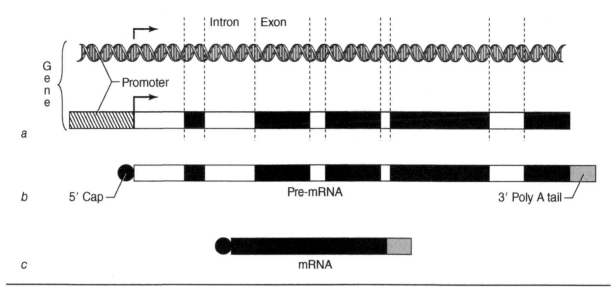

Figure 3.15 (a) A gene shown as the familiar double helix (top) and as a bar (bottom). Most eukaryotic genes have alternating sections containing coding information (exons—solid) and noncoding sequences (introns—open) downstream of the transcription start site (bent arrow) and the promoter region (hatched region). (b) The pre-mRNA molecule has introns and exons as well as a 5' cap and a 3' poly A tail (shaded region). (c) The final processing of the pre-mRNA involves removal of sections of RNA that represent the introns.

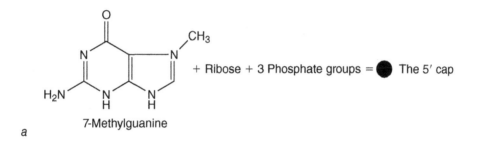

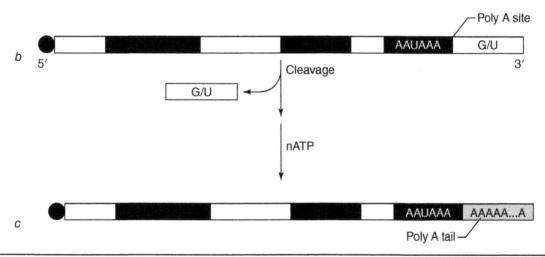

Figure 3.16 Processing of pre-mRNA. (a) A cap is added to the 5' end of the transcript. The cap consists of a guanine base (modified by the addition of a methyl group) that is attached by three phosphate groups to the 5' carbon of the ribose on the initial nucleotide of the transcript. (b) The transcript illustrates the introns (open) and exons (solid). (c) A site where a sequence of 200 to 250 adenine nucleotides will be added subsequent to removal of a section of the 3' end of the transcript, leading to the pre-mRNA with both a cap and a poly A tail (gray).

place. First, the pre-mRNA is cleaved at the poly A site, releasing a small RNA particle with the G- and U-rich sequences. Then an enzyme known as *poly A polymerase* adds about 200 to 250 adenine nucleotides to the poly A site, which is some 20 to 35 nucleotides downstream (in the 3' direction) from a key signal sequence (AAUAAA). This signal sequence is absolutely essential for polyadenylation to take place.

Splicing to Remove Introns

The next step is to remove the introns from the capped, polyadenylated, pre-mRNA through a splicing process in which the junction between introns and exons is cleaved at both the 5' and 3' ends. The introns are then removed and contiguous exons are joined up. Generally, the splicing out of introns occurs following the addition of the cap and poly A tail to the pre-mRNA. For very long transcripts, splicing can begin before the transcription of the long gene has been completed.

The locations of the intron–exon splice sites have been studied in detail, and we know that consensus nucleotides clearly delineate splice sites. The details of this process are complicated, involving small nuclear RNA molecules known as snRNAs. When combined with the pre-mRNA and associated nuclear proteins they form a *spliceosome*. The actual process of splicing out the intron is shown in figure 3.17. The hypothetical pre-mRNA molecule starts with a 5' cap and 3' poly A tail and contains a single intron. A phosphoester bond exchange occurs between the consensus nucleotides at either end of the intron, forming a lariat structure. This is removed from the second exon. Next, the two exons join together by their consensus terminal guanine nucleotides. The lariat is then attacked by RNA-digesting enzymes (RNases), releasing the nucleotides as nucleotide monophosphates.

As a result of the 5' capping, the 3' polyadenylation, and intron removal by cutting and splicing, we now have what has been called a mature mRNA that is ready to leave the nucleus for the cytosol, where it acts as a template on which a polypeptide (protein) will be made, employing ribosomes. Transport out of the nucleus takes place through *nuclear pores*, or openings in the nuclear membrane. Within the

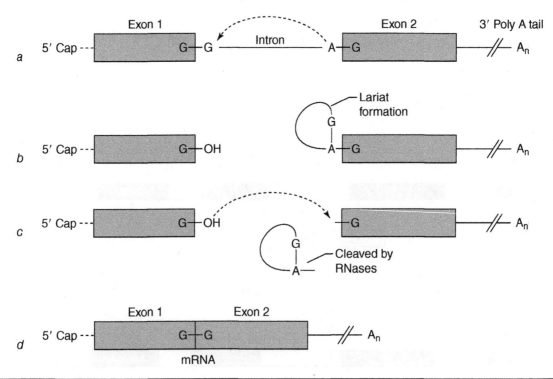

Figure 3.17 Splicing out of an intron in a simple pre-mRNA molecule containing a 5' cap and 3' poly A tail. (*a*) Exchange of the phosphoester bond of the A nucleotide residue at the 3' end of the intron with the G nucleotide residue at the 5' end of the same intron. (*b*) This forms a lariat structure in which the A nucleotide residue at the 3' end of the intron is still attached to the 5' G nucleotide residue of exon 2. (*c*) The intron, now a freed lariat, is attacked by RNA-cleaving RNases. (*d*) The 3' terminal G nucleotide residue of exon 1 forms a phosphoester bond with the 5' terminal G nucleotide residue of exon 2, linking the two exons together and forming the mRNA molecule.

nuclear pores is a complex of filaments (the *nuclear pore complex*) that are involved in allowing the transport of ions and proteins (e.g., transcription factors) into the nucleus and the export of molecules such as mRNA, rRNA, and tRNA from the nucleus. Newly formed mRNA molecules, associated with their nuclear proteins, are exported through the nuclear pore complex using a specific mRNA exporter protein.

For many genes, every intron is removed and every exon is incorporated into the mature mRNA. However, it is not uncommon for one gene to give rise to two or more mRNA molecules and, thus, two or more final proteins. An example of the alternate splicing of a primary gene transcript is shown in figure 3.18. The gene CGRP (calcitonin gene-related peptide) is expressed in the thyroid gland and in the brain. The primary transcript CGRP pre-mRNA contains six exons, but it is treated in different ways in the two tissues. In the thyroid, exons 5 and 6 are spliced out, creating an mRNA with just four exons. The protein product, *calcitonin*, is released as a hormone from the thyroid gland in response to elevated levels of blood-calcium ions. In brain tissue, exon 4 is spliced out of the CGRP primary transcript, producing a different mRNA that contains exons 1, 2, 3, 5, and 6. The CGRP has several functions, one of which is to act as a vasodilator. Alternate splicing of genes is common for some of the contractile proteins in skeletal, cardiac, and smooth muscle. It can occur in these instances:

1. When transcription is initiated at different promoters, resulting in different 5' exons in the mRNA.
2. When transcription terminates differently because of more than one site of polyadenylation, resulting in different 3' exons.
3. When different internal exons in the gene are included or not included in the splicing process, giving rise to different mRNA molecules (thus two or more different polypeptides) in a process known as alternate mRNA splicing. This alternative splicing can occur in the same cell at the same time, in the same cell at different times during cell differentiation, or in different cells. Control of alternative splicing occurs through the action of specialized proteins.

Formation of rRNA and tRNA Molecules

It is important to emphasize that, like mRNA, rRNAs and tRNAs are also encoded by genes. Modification of primary transcripts from genes for rRNA and tRNA also occurs. Two different genes are needed to produce the rRNA molecules that make up much of the ribosomes. Multiple copies of each rRNA gene exist; these are located in a part of the nucleus known as the *nucleolus*. The two different rRNA transcripts are so large that their mass is represented in Svedberg

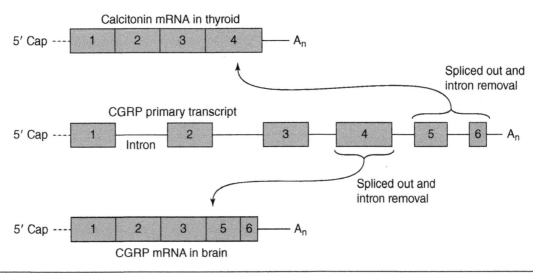

Figure 3.18 An example of alternate pre-mRNA splicing to generate two different mRNA products. The CGRP gene is expressed in the thyroid gland and brain as a primary transcript with six exons. In thyroid tissue, exons 5 and 6 are spliced, and the resulting mRNA provides the amino acid sequence information to make the hormone calcitonin. In brain tissue, the CGRP primary transcript is spliced a different way, removing exon 4. The resulting mRNA codes for the CGRP polypeptide.

units (S), which describe their velocity of sedimentation in an ultracentrifuge.

For very large molecules or molecular complexes, size is measured by how far they travel in an ultracentrifuge, expressed in Svedberg units, rather than by molecular weight. The larger the molecule, the faster it sediments in a centrifuge and the larger the value of its S becomes. Unlike units such as kilodaltons (kD) or base pairs (bp), which are typically used to describe the size of protein or DNA molecules, respectively, Svedberg units are not linearly related to size. Thus, a whole ribosome, with a mass of 80S, is made up of two subunits of size 60S and 40S because the sedimentation behavior of particles in the centrifuge is not linearly related to size.

The two primary rRNA transcripts are described as 45S and 5S. The smaller transcript, 5S (120 bases), remains unchanged, whereas the larger transcript, 45S, undergoes some splitting and modification, resulting in three new rRNA molecules: 18S, 28S, and 5.8S. The 18S rRNA molecule combines with 33 protein molecules to make the small ribosomal subunit called the 40S ribosomal subunit. The 5.8S, 28S, and 5S rRNA molecules combine 50 protein molecules to make the large ribosomal subunit known as the 60S ribosomal subunit. During protein synthesis, the 40S and 60S ribosomal subunits combine to make the complete ribosome, termed the 80S ribosome.

> ### ☑ KEY POINT
> Although there are only two different genes for rRNA, multiple copies of these exist because so many rRNA molecules are needed to make up the hundreds of thousands of ribosomes in a cell. Whereas each mRNA message gets amplified because it results in thousands of polypeptide chains, there is no such amplification of the four rRNA molecules, since each is the final product. The cell contains more rRNA (by weight) than all the mRNA and tRNA molecules combined.

Primary transcripts for genes for the various tRNAs undergo some modifications. For example, nucleotides are removed from the 5' and 3' ends, and, if present, an intron is removed. Then the 3' end has a nucleotide sequence CCA added. Finally, some of the bases are modified. Figure 3.19 shows the main features of a fully functional tRNA molecule. At one end is a 5' phosphate group. The 3' end has the sequence CCA with a free 3' hydroxyl group to which the amino acid becomes attached.

Notice the hydrogen bonding that helps maintain the rough cloverleaf structure, characteristic of tRNA molecules. Three bases at the bottom of tRNA represent the **anticodon** bases. These will correspond to the mRNA codon for the particular amino acid, except that they will be antiparallel and complementary. For example, if the codon is UCG (always specified in the 5' to 3' direction) as in figure 3.19, the anticodon will be AGC, in the 3' to 5' direction.

Because there are 61 codons in mRNA specifying the 20 amino acids, we would expect there to be 61 tRNA molecules, each with one of 61 different anticodons. However, the anticodon–codon binding has a bit of flexibility because the first base in the anticodon (the one at the 5' end that pairs with the third base at the 3' end in the codon) can often recognize two bases. For example, the codons CUA and CUG could be recognized by the anticodon GAU. This flexibility in codon–anticodon recognition is known as **wobble**. As a result of wobble, fewer than 61 different tRNA molecules are needed to ensure that all 20 amino acids are able to participate in protein synthesis. Indeed, on a theoretical basis, considering all the possibilities created by wobble, only 31 different tRNA molecules would be needed. The actual number of different tRNA molecules found in eukaryotic cells is 32.

TRANSLATION

In the process of translation, which takes place on ribosomes in the cytosol of the cell, the mRNA message is converted into a sequence of amino acids in a polypeptide chain. During translation, synonymous with protein synthesis, the mRNA is read according to the genetic code, which relates the DNA sequence to the amino acid sequence in proteins. The rate of protein synthesis in cells is influenced not only by the rate of gene transcription but also by the rate or efficiency of translation, which is controlled at many levels and is sensitive to changes in physical activity and diet.

Steps in Translation

The process of translation can be described as a series of steps, beginning with the attachment of each specific amino acid to its respective tRNA molecule. Translation then proceeds in three stages: initiation, elongation, and termination. Initiation of translation involves the formation of the initiation complex consisting of ribosomal subunits and initiator tRNA (always methionyl-tRNA, abbreviated Met-tRNA) at the start codon of mRNA. The elongation

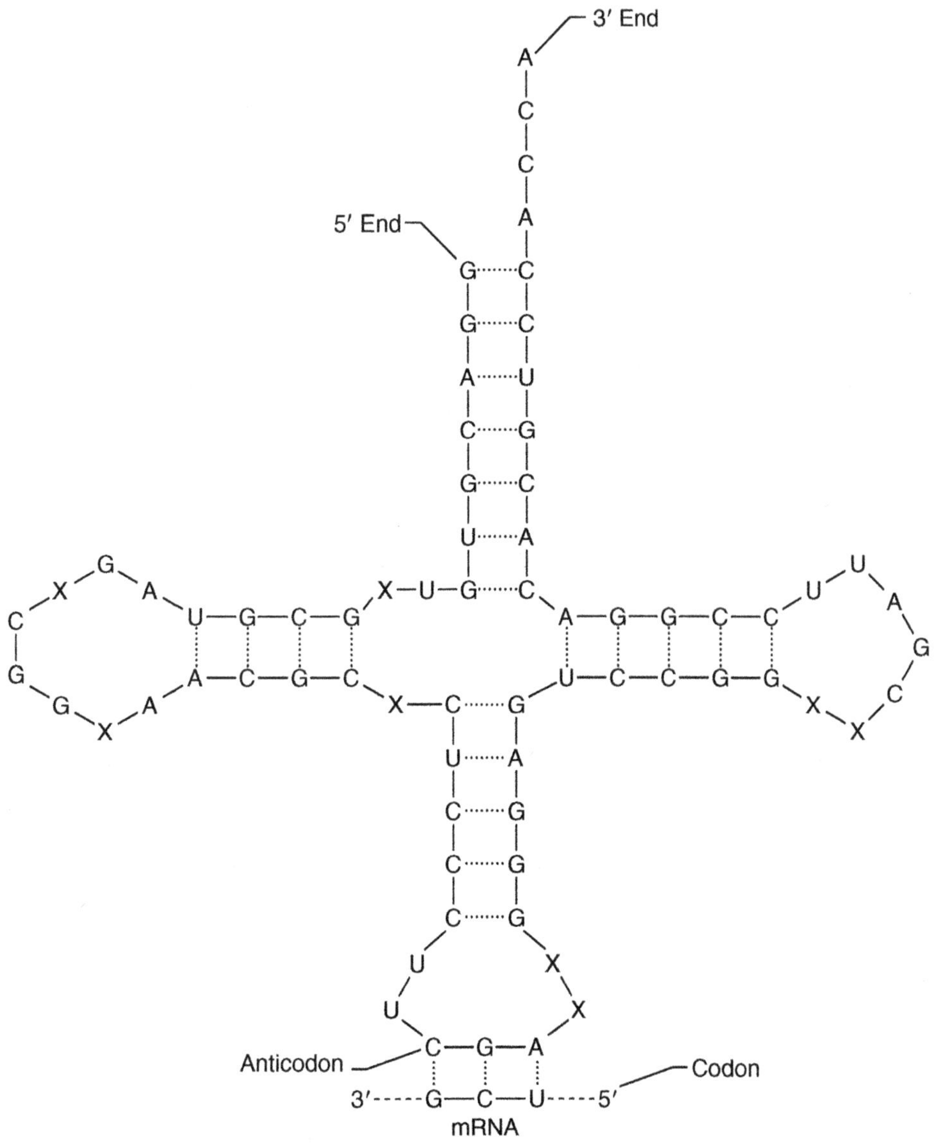

Figure 3.19 An example of a tRNA molecule that has been modified from its primary transcript. The finished product has a cloverleaf-type structure that is maintained by hydrogen bonding between complementary bases (dotted lines). Solid lines represent the covalent attachment of adjacent nucleotides shown by the letter representing their base. The X refers to one of several different types of bases that are generated after transcription. The 3' end contains a terminal adenosine (A) whose 3' ribose OH group binds a specific amino acid. Each tRNA has a three-base sequence at the bottom, known as the anticodon, that complements the codon on the mRNA that represents the particular amino acid. The anticodon and complementary codon are for the amino acid serine.

phase, as the name implies, involves elongation of the polypeptide chain through the addition of the proper amino acids, one at a time, to the carboxy terminal end of the existing polypeptide chain. Elongation continues until a stop codon in the mRNA is reached, at which point termination of translation occurs and the polypeptide is released from the last tRNA. We will now describe each stage of translation in more detail, starting with the formation of aminoacyl-tRNA.

Formation of Aminoacyl-tRNA

Each amino acid has at least one tRNA to which it will be attached. Each tRNA has an anticodon that matches one of the mRNA codons for that amino acid. Each amino acid also has a specific enzyme to catalyze the joining of its alpha carboxyl group to the 3' OH group of the terminal adenosine of the single tRNA (or duplicate tRNAs if there are more than one for that amino acid) by means of an ester bond. The enzyme that joins the amino acid to its tRNA is known as *aminoacyl-tRNA synthetase*. Each synthetase works to bond the correct amino acid and its tRNA, a reaction shown in the following equation:

$$\text{amino acid} + \text{tRNA} + \text{ATP} \rightarrow \text{aminoacyl-tRNA} + \text{AMP} + \text{PPi}$$

The energy needed for the formation of the ester bond between the amino acid and its tRNA is provided by the hydrolysis of ATP. The products of this hydrolysis reaction are AMP and inorganic pyrophosphate (PPi). Hydrolysis of PPi by inorganic pyrophosphatase ensures that the reaction is driven to the right. Amino acids attached to their respective tRNA molecules are often characterized as charged amino acids. Some aminoacyl-tRNA synthetases attach their amino acid directly to the 3' OH group of the terminal ribose on their tRNA, whereas others attach their amino acid to the 2' OH group of ribose. This latter attachment is only temporary, since the amino acid is subsequently moved to the 3' position. The role of the individual aminoacyl-tRNA synthetase enzymes, to recognize the correct amino acid and the correct tRNA, places these enzymes in a critical position to ensure that correct proteins are produced by the cell. Errors are extremely rare.

Role of mRNA

Figure 3.20 illustrates the essential components of a functional mRNA molecule. The coding region, headed by the AUG start codon, contains the information to indicate the amino-acid sequence for the polypeptide chain, which is followed by a stop codon (UAG, UGA, or UAA). The term **open reading frame** (ORF) is often used to describe a region of triplet coding, bordered on one end by the start codon and at the other end by a stop codon. We could use ORF to refer to a gene in DNA or, more likely, to refer to an mRNA molecule. Flanking the coding region are a 5' noncoding region and a 3' region that is also noncoding. Because the base sequence in these regions is not translated into an amino-acid sequence, they are also known as the 5' UTR (untranslated region) and 3' UTR, respectively. Base sequences in the 5' UTR or 3' UTR can play a role in the regulation of translation. Except for histone proteins, virtually all mRNA molecules contain the poly A tail.

Initiation of Translation

Like transcription, the translation process has three stages: initiation, elongation, and termination. This section does not provide great detail but instead focuses on the overall process. The major players in the initiation of translation are the two ribosomal subunits 40S and 60S, the mRNA molecule, a number of protein factors to control the initiation process (eukaryotic initiation factors, or eIFs, such as eIF1A and eIF2), and a source of energy from the hydrolysis of ATP and GTP. Initiation cannot take place without methionine, attached to its tRNA. However, while methionine has a single codon (AUG) and a single methionine tRNA synthetase, it has two different tRNAs. One of these, tRNA$_i$, is used only for the initiation of translation at the first start codon of the open reading frame. The other tRNA, which binds methionine, does so when it is incorporated into the peptide chain beyond the initial methionine.

Figure 3.21 illustrates the initiation part of translation, which can be considered to take place as five discrete steps. In the first step, ribosome subunit 40S binds to eIF3. In step 2, a preinitiation complex is formed with the methionyl-tRNA (Met-tRNA$_i$), eIF1A, and a complex of eIF2 bound to GTP. In step 3, eIF4 and the mRNA bind, with the cap on mRNA recognized by and located at eIF4. This creates the

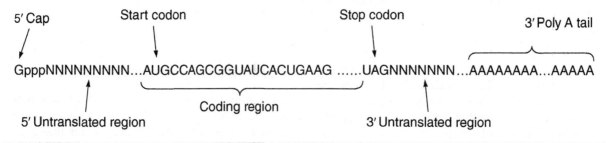

Figure 3.20 A model of a mature mRNA molecule, ready for translation to form a polypeptide chain. At the left end is the 5' cap and at the 3' end, the poly A tail. The region that contains the information for the sequence of amino acids in a polypeptide chain (coding region) begins with the start codon AUG and ends with a stop codon (UAG is shown, but UAA or UGA also signal stop). Flanking the coding region are the 5' and 3' untranslated regions. N refers to nucleotides whose designation is not shown.

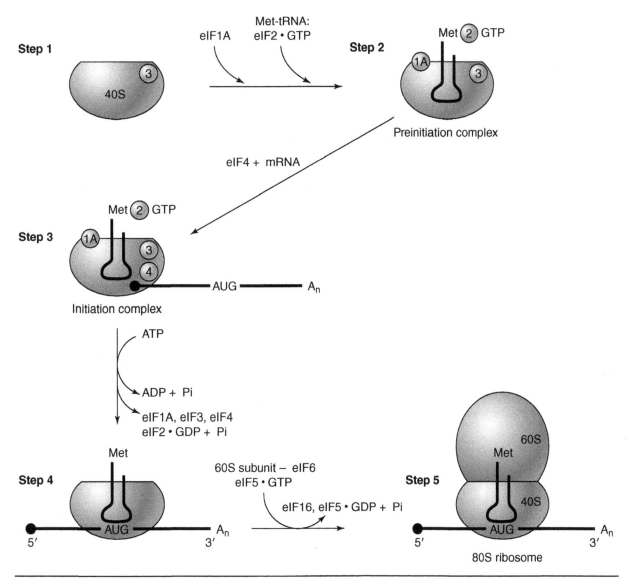

Figure 3.21 Stages in the formation of an 80S ribosome particle. A 40S ribosome subunit combines with eukaryotic initiation factor 3 (shown circled). Subsequent additions include the initiation tRNA (Met-tRNA$_i$) attached to GTP by eIF2 and eIF1A, forming a preinitiation complex (step 2). Messenger RNA is positioned at its 5' cap on the 40S subunit aided by eIF4. This generates the initiation complex (step 3). The mRNA is scanned for an initiation codon (AUG), and the ribosome moves along the mRNA in a process driven by the hydrolysis of ATP and GTP (step 4). Addition of the large ribosome subunit (60S) to the 40S subunit, forming an 80S ribosome, is facilitated by addition of initiation factors (eIF5 and eIF6) and the energy from GTP hydrolysis (step 5).

initiation complex. With the help of the initiation factors and energy released from ATP and GTP hydrolysis, the mRNA is scanned in the 3' direction to find the initiation codon (step 4). During step 4, the initiation factors and the hydrolysis products of ATP and GTP are released. Finally, the 80S ribosome complex is formed in step 5, when the 60S ribosome subunit bound to eIF6 joins with the 40S subunit. Energy is provided by hydrolysis of GTP. Subsequently eIF6, eIF5, GDP, and Pi are released. GTP hydrolysis during the formation of the 80S ribosome complex helps to maintain the attachment of the 40S and 60S subunits throughout translation of the mRNA molecule.

Elongation of Translation

The elongation phase involves the addition of amino acids, one at a time, to the carboxy terminal end of the existing polypeptide chain (see figure 3.22). Let us start with the 80S ribosome as shown in step 5 of figure 3.21. For clarification, anticodons are shown on the tRNAs, as well as relevant codons on mRNA. Recall that each amino acid is attached to the 3' end of its respective tRNA. In step 1, shown in figure 3.22a, the next aminoacyl-tRNA (Phe-tRNA) comes in, its anticodon (AAA) recognizing the mRNA codon (UUU) on the 3' side of the initiator (AUG) codon. In step 2 (figure 3.22b), a peptide bond is formed between the methionine carboxyl and the free amino group of the next amino acid (Phe), still attached to its tRNA. The formation of this peptide bond between Met and Phe means the methionine is released from its tRNA. The methionine tRNA leaves, and we now have a dipeptide attached to the second tRNA. The 80S ribosome complex then slides three bases along the mRNA molecule in the 3' direction. Then the next aminoacyl-tRNA (Val-tRNA) comes in, its anticodon (GUG) recognizing the codon CAC. Another peptide bond is formed, the Phe-tRNA is released, and we now have a tripeptide. Again, the 80S ribosomal complex slides along the mRNA. This process continues with the growing polypeptide chain still attached to the last incoming tRNA. Eukaryotic elongation factors (eEFs) participate in the elongation process creating the polypeptide chain. The energy needed to make these peptide bonds comes from the hydrolysis of GTP to GDP and Pi; the equivalent of four ATP is needed for the synthesis of each peptide bond.

Termination of Translation

The process of elongation continues, with the ribosome and growing polypeptide chain moving along the mRNA three bases at a time until a stop codon is reached (UAA, UGA, or UAG). At this point, a release factor causes the release of the completed polypeptide chain from the last tRNA. The whole complex dissociates, and the 80S ribosome dissociates into its 40S and 60S subunits. As with peptide chain elongation, energy for the termination of translation is provided by GTP hydrolysis.

Because most mRNA molecules are large, there is room for more than one ribosome on a single mRNA molecule—each containing a growing polypeptide chain. We call the mRNA and the two or more ribosomes, each with its growing polypeptide chain, a **polyribosome** or polysome (see figure 3.23). The overall process of translation occurs with remarkably few errors, roughly one error per 10,000 codons translated under normal conditions.

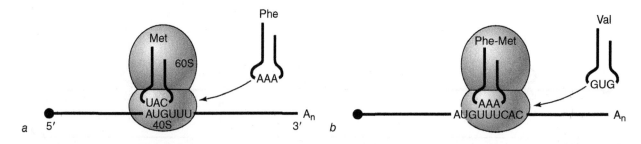

Figure 3.22 The elongation phase of translation. (a) The second aminoacyl-tRNA comes to the 80S ribosome, its anticodon AAA recognizing the second codon UUU, coding for the amino acid phenylalanine (Phe). (b) A peptide bond is formed between the carboxyl group of Met and the Phe, still attached to its tRNA. The 80S ribosome moves three bases along the mRNA toward the 3' end. Not shown are elongation factors and the hydrolysis of GTP, which provides energy for peptide bond formation and movement of the 80S ribosome along the mRNA.

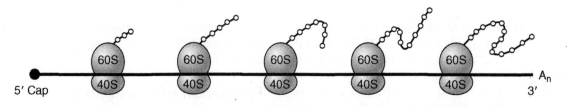

Figure 3.23 A polyribosome, composed of five ribosomes. Most mRNA molecules are long compared to the dimensions of an 80S ribosome complex, allowing for more than one 80S ribosome complex with its growing polypeptide chain. Five ribosomes appear here, each with its polypeptide chain (shown as small connected circles). The complex of a number of ribosomes on one mRNA molecule is known as a polyribosome or polysome. Note that the farther along the mRNA moves toward the 3' end, the longer the polypeptide chain becomes.

Regulation of Translation

When gene expression regulation is discussed, terms such as *transcription control* and *posttranscription control* are often used. We have already discussed transcription control, primarily from the perspective of regulation of the initiation of gene transcription by RNA polymerase II. Here, we look at posttranscriptional control because the term refers principally to the regulation of translation of mRNA. At the end of this chapter, we look more specifically at posttranscriptional control in muscle, where the rate of protein synthesis is exquisitely sensitive to a variety of nutritional and exercise states.

The amount or rate at which a specific polypeptide product is generated in a cell by translation depends on the number of molecules of its mRNA and the rate at which translation is initiated and terminated for each mRNA. Control regions in some mRNA molecules are found in both the 5' and 3' untranslated regions (UTR), which can regulate the ability of the mRNA message to be initiated and can influence the stability of each mRNA. Control regions in the 5' UTR can bind proteins, which may block the ability of the 40S ribosome subunit to bind to the 60S ribosome subunit such that an actively translating 80S ribosome cannot be created. On the other hand, binding of proteins at control regions in the 3' UTR may increase the stability of the mRNA, preventing it from being degraded by ribonucleases. This is important because if the rate of translation of mRNA molecules is fixed, then the amount of polypeptide product generated from each mRNA depends on how long it remains capable of being translated.

The number of mRNA molecules in a cell depends on how fast the gene template for the mRNA is transcribed and the mRNA is modified and transported to the cytosol, where translation takes place. We must also consider the average length of time during which each mRNA is capable of being translated. The lifetime of mRNA molecules, measured as a half-life of mRNA, can vary from a few minutes (for immune system–signaling molecules such as cytokines) to many hours. Obviously, then, the lifetime of mRNA must be considered as a level of control. How long the mRNA lasts in the cytosol is controlled by degradative enzymes known as **ribonucleases**. These enzymes cleave the sugar-phosphate backbone of RNA, releasing individual nucleoside monophosphates. The length of time an individual mRNA molecule lasts in the cytosol depends on its ability to prevent being degraded by the cytosolic ribonucleases. This is a very complicated area, since some proteins stabilize mRNA and others destabilize it. Regulation of these proteins can therefore indirectly influence the lifetime of the mRNA, as well as regulate the ability of the mRNA to be translated.

During its lifetime, the poly A tail of most mRNA molecules is gradually shortened by a *deadenylase*. Since most mRNA molecules are stabilized in the cytosol by a poly A–binding protein, shortening the poly A tail reduces the ability of the poly A–binding protein to stabilize the mRNA, including the 5' cap. The mRNA molecule then becomes susceptible to decapping and attack from the 5' end. Some mRNA molecules lose their cap without polyadenylation. Apparently nucleotide sequences in the 3' untranslated region render the cap more or less susceptible to removal and subsequent attack from the 5' end. Signals in the 5' and 3' UTRs of mRNA molecules can regulate not only the initiation of translation of the message but also the length of time the mRNA lasts.

☑ KEY POINT

Regulation of translation is important to many athletes who are attempting to increase muscle size. This means increasing the frequency of translation of each mRNA and increasing the lifetime of each mRNA. Simple strategies, such as adopting an appropriate exercise training stimulus (which is not the same for everyone) and adjusting the timing and content of meals, can play dominant roles.

POSTTRANSLATIONAL PROCESSING OF POLYPEPTIDES

Translation of the mRNA message generates a one-dimensional amino acid sequence in a polypeptide. As chapter 1 illustrates, higher structural levels of polypeptide chains must be created before the polypeptide can function as a single unit, combined with one or more cofactors, or combined with other polypeptides in a protein with quaternary structure. All the information needed to make a complete three-dimensional structure for a protein is found in its amino acid sequence. However, the cytosol, where proteins are synthesized, is a very crowded space. Improper associations could certainly be made. Even if a polypeptide could spontaneously assume its correct native shape after it is released from the translation apparatus, it would likely need to bind to other molecules (cofactors or other polypeptide chains) or move to a different location in the cell to function properly. A group of proteins known as *chaperones* bind and stabilize completed polypeptide chains or partially folded polypeptides, as well as participate in the process of complete folding of proteins. They are often characterized as *molecular chaperones* if they bind to and stabilize newly synthesized and partially folded polypeptides, or **chaperonins** if they function to directly fold proteins. Many of these proteins are **heat shock proteins** (HSP), a class of protein molecules initially found to be synthesized in response to a heat stress. HSP70 is particularly important as a molecular chaperone to stabilize partially folded proteins. The number following the designation HSP identifies the molecular weight in kilodaltons.

Chaperonins are a subgroup of chaperone proteins that function primarily in protein folding and assembly of polypeptide subunits into molecular aggregates. Remember that some polypeptides must also bind with other polypeptides to generate the active oligomeric protein. For example, myosin is a hexamer, so six polypeptides must combine in the proper way to generate a functional myosin molecule. Moreover, once formed, myosin hexamers must be inserted into thick filaments in order to be able to hydrolyze ATP and generate a force (see chapter 4). Some polypeptides need to have a prosthetic group attached to them in order to be effective. For example, myoglobin, the protein in muscle that aids in storing and moving oxygen in the cytosol of cardiac and skeletal muscle, needs to have a heme group attached to its polypeptide.

Other polypeptides must be packaged for export because they function in the blood; liver is actively involved in synthesizing and secreting a host of plasma proteins. Others must be incorporated into membranes because much of the mass of a membrane is made up of a variety of proteins. Mitochondria represent an important organelle that needs more than 1,000 different protein molecules to function. Since the mitochondrial genome codes for only 13 polypeptide subunits, at least 99.9% of all proteins in mitochondria are coded by nuclear genes and imported. Moreover, a functional mitochondrion has four separate areas where specific proteins must be inserted: the outer and inner membranes, the matrix, and the intermembrane space. To facilitate the proper positioning of proteins in the various mitochondrial locations, cytosolic-generated polypeptides contain N-terminal signal sequences that, much like shipping labels, target a given polypeptide to its proper location (Stojanovski et al. 2003). These signals are recognized by membrane protein translocases TOM and TIM (translocase of the outer membrane and inner membrane, respectively). The signals are typically 15 to 40 amino acid N-terminal sequences that are removed after the mitochondrial protein is properly located. Mitochondria represent one important example of the complexity of a cell in that protein synthesis on ribosomes is only one key step in a process to create a functional mitochondrial protein product.

PROTEIN DEGRADATION

Chapter 1 discusses the concept of protein turnover. Turnover relates to the balance between the opposing processes of protein synthesis and protein degradation. The content of any protein in a cell is based on the relative rates of protein synthesis and protein degradation. This can be described by the following equation:

cellular protein content = synthesis rate
− degradation rate

This tells us that the moment-to-moment change in the content of any protein is a reflection of the differences in rates of its synthesis and degradation. We have already seen how proteins are synthesized. The opposing process is carried out by three major systems of protein-degrading enzymes, known as **proteinases** (also termed *peptidases* or *proteases*).

For years, the content of a particular protein was thought to be regulated primarily by changes in the transcription of its gene and subsequent translation of this message. The degradation side of the preceding equation was considered to be relatively constant. We now know that protein degradation is extremely important, involved in regulating cellular processes such as growth and atrophy, transcription, rates of metabolic pathways, development, and some disease states such as cancer. To exercise scientists, the regulation of protein degradation in skeletal muscle is important to understand due to its potential role in the adaptive response to exercise training or physical inactivity. Since loss of muscle mass accompanies many chronic diseases and is associated with muscle weakness and exercise intolerance, clinicians are also interested in the mechanisms involved in regulating protein degradation in skeletal muscle. Three main protein-degrading systems operate in healthy skeletal muscle: the calpain (calcium-activated neutral protease) system, the lysosomal system, and the ubiquitin-proteasome pathway. Activation of these and other protein-degrading systems, particularly caspases (or cysteine-aspartic proteases), is increased under catabolic conditions associated with muscle atrophy, such as cancer,

AIDS, kidney disease, and diabetes mellitus (Du, Hu, and Mitch 2005). The calpain, caspase, and lysosomal pathways are likely involved in the initial fragmentation of structural proteins, contractile proteins, and membrane proteins. Further destruction of the protein fragments is primarily carried out by the ubiquitin-proteasome pathway. Finally, cytosolic peptidases degrade the remaining small peptides to free amino acids. As these mechanisms are complex and incompletely understood, we will briefly summarize how the three main protein-degrading systems function.

Ubiquitin-Proteasome Pathway

The ubiquitin-proteasome pathway plays a major role in targeting and degrading cellular proteins. This pathway must recognize and label a specific protein for degradation, then carry out the process. The recognition part is based on marking a protein by covalent attachment of a small ubiquitous protein known as **ubiquitin**. This is a protein containing 76 amino acids that is found in all eukaryotes, virtually unchanged. Ubiquitinated proteins are subsequently degraded to small polypeptides in a 26S **proteasome** complex (figure 3.24). The 26S proteasome complex is formed by the union of two 19S regulatory particles and the 20S core particle. The barrel-shaped core particle is arranged as four stacked rings, with proteolytic enzymes inside that can break polypeptides down into small peptide subunits.

Proteins are marked for degradation by the addition of ubiquitin units, termed polyubiquitination. Three conjugating enzymes are involved in this process, designated as E1, E2, and E3. Two very important muscle-specific E3 enzymes are called muscle-specific atrophy F box (MAFBx, also known as atrogin-1) and muscle-specific really-interesting-novel-gene finger protein 1 (MuRF1). Using the free energy released from ATP hydrolysis and the three conjugating enzymes, at least four ubiquitins are added to the target protein, generating a polyubiquitinated protein. Thus modified by ubiquitin addition, the protein binds to a 19S regulatory particle

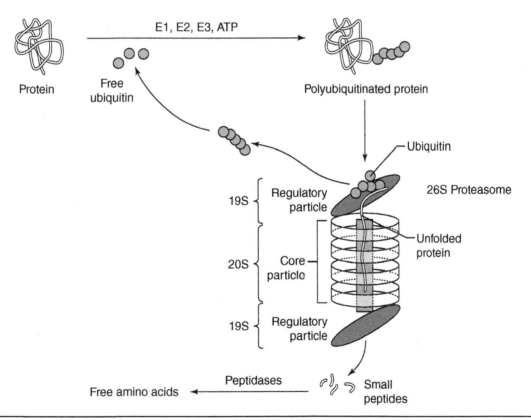

Figure 3.24 Summary of the steps involved in the degradation of an intracellular protein by the ubiquitin-proteasome system. Proteins are tagged for degradation by the attachment of multiple ubiquitin subunits to the protein using three enzymes, identified as E1, E2, and E3, and the energy from ATP hydrolysis. The polyubiquitin chain binds to the 19S regulatory particle and feeds the polypeptide chain into the 20S core particle, where the protein is hydrolyzed and small peptides are released. The polyubiquitin chain is not degraded by the proteasome but is released, and the individual ubiquitin subunits are freed. Peptidases cleave the small peptides released from the proteasome to free amino acids.

of the 26S proteasome complex. The target protein is unfolded, and its polypeptide chain is fed into the 20S core particle, where the polypeptide will be degraded by proteases generating and releasing small peptides. These are subsequently degraded to free amino acids by cytosolic peptidases. The polyubiquitin marker is freed by the regulatory particle, and free ubiquitin units are subsequently released by hydrolysis of the peptide bonds holding them together. Many forms of exercise result in decreased activity of the ubiquitin-proteasome system, whereas disuse as a result of illness or immobilization is associated with increased activity (Taillandier et al. 2004).

Lysosomal System

Lysosomes are organelles, found in most cells, that are responsible for degrading a variety of cell constituents. In particular, lysosomes contain a battery of protein-degrading enzymes known as *cathepsins* that break peptide bonds in the interior of the protein molecule. In addition, they contain proteinases that cleave amino acids, one at a time, from the amino and carboxy ends of the polypeptide. To be degraded by lysosomes, proteins must enter via a process known as endocytosis.

Calpain System

The calpains are a family of nonlysosomal calcium-activated proteases found in the cell cytosol. Skeletal muscle fibers contain both ubiquitous calpains, μ-calpain and m-calpain (also known as calpain-1 and calpain-2, respectively), and also a muscle-specific isoform, calpain-3 (or p94) (Lamb 2009). Both μ-calpain and calpain-3 are activated in physiological conditions over [Ca^{2+}] in the micromolar range, whereas m-calpain requires millimolar Ca^{2+} for activation. When activated by Ca^{2+}, calpains autolyze themselves, which makes them much more sensitive to Ca^{2+} and more proteolytically active. Even in rested muscle fibers, a small fraction of μ-calpain is autolyzed and remains proteolytically active, which is suggestive of a role in normal protein turnover in healthy muscle (Gailly et al. 2007). Importantly, increases in cytosolic Ca^{2+} associated with normal muscle activity (e.g., sprinting or endurance exercise) do not result in autolysis and activation of calpains in muscle (Lamb 2009). However, if exercise is excessively severe or if it involves an eccentric action where the muscles are stretched while contracting, such as occurs during downhill walking or running, then calpain autolysis and activation are increased due to disturbances in Ca^{2+} regulation and prolonged elevations in cytosolic [Ca^{2+}]. Because calpain acts on structural proteins in muscle, it is believed that its activation may be a first step in the response to exercise-induced injury (Belcastro, Shewchuk, and Raj 1998).

REGULATION OF GENE EXPRESSION IN EXERCISE AND TRAINING

The control of gene expression in skeletal muscle has been an area of active research for many years. We have long known that endurance training increases the content of mitochondrial proteins involved in oxidative phosphorylation, without any significant changes in contractile proteins or muscle hypertrophy. On the other hand, high-intensity resistance training increases the mean cross-sectional area of muscle fibers and induces muscle hypertrophy by increasing the content of contractile proteins, but without a notable effect on mitochondrial proteins. This tells us that the stimulus induced by the specific exercise activity must selectively modulate the transcription of some muscle genes, leading to increased levels of mRNA and thus proteins. Muscle hypertrophy and mitochondrial adaptations are the result of cumulative effects of repeated acute bouts of high-intensity resistance exercise and endurance exercise, respectively. The primary question is how the training stress is linked to the activation of key genes. Kristian Gundersen (2011) suggests that an "excitation-transcription coupling" must exist in skeletal muscle, whereby primary signals generated by muscle contractions are deciphered by intracellular molecules that act as sensors and are transmitted via signaling pathways that ultimately regulate transcription factors, coactivators, and corepressors of specific genes (i.e., contractile protein and mitochondrial genes). A simplistic flow chart for how muscle contraction signals could be processed by the muscle fiber to alter gene transcription is given in figure 3.25.

Endurance Training

Endurance exercise involves repeated low-intensity contractions that can be performed for prolonged periods of time at low frequencies. Here, the term *frequency* refers to the rate at which the motoneuron delivers action potentials to the muscle with each contraction that is low in endurance exercise (relative to high-intensity exercises like sprinting or

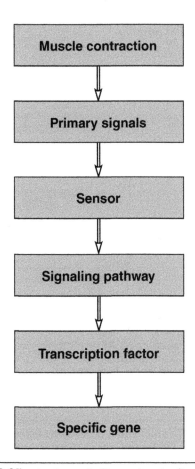

Figure 3.25 A schematic illustrating the flow of information in excitation-transcription coupling.

resistance exercise). Many muscle contraction signals associated with endurance exercise may serve as primary signals in excitation-transcription coupling, resulting in the adaptations to training noted previously. However, it is becoming increasingly apparent that oscillations in cytoplasmic [Ca^{2+}], [AMP]/[ATP], and reactive oxygen species (ROS) are key signals that correlate with the frequency, duration, and intensity of contractions. They activate signaling pathways in skeletal muscle that control the expression of mitochondrial proteins. See chapters 4 and 5 for more information on the importance of AMP/ATP, ROS, and calcium changes in muscle during exercise, as well as their regulation.

In skeletal muscle, two primary Ca^{2+}-dependent transcriptional pathways that are activated with endurance exercise are the Ca^{2+}-calmodulin-dependent serine/threonine protein phosphatase, calcineurin (CaN), and the Ca^{2+}-calmodulin-dependent kinase II and IV (CaMKII and CaMKIV) pathways (Chin 2010). One of the substrates for CaN is a transcription factor known as **NFAT** (nuclear factor of activated T cells). When activated by Ca^{2+}-calmodulin, it is possible for CaN to dephosphorylate NFAT, allowing it to bind to its response element in the control region of a number of its target genes in skeletal muscle. One of the substrates for CaMK is a class of HDAC enzymes that, when dephosphorylated, interact with and inhibit a transcription factor called myocyte-enhancer factor 2 (MEF2). CaMK induces nuclear export of these HDACs through phosphorylation, which increases MEF2-dependent transcription.

HDACs can also be phosphorylated by another kinase called AMP-activated protein kinase (AMPK) (McGee and Hargreaves 2011). AMPK is activated during metabolic stress by increased [AMP]/[ATP]. Several studies have shown that AMPK is activated in skeletal muscle in response to endurance exercise. Therefore, AMPK signaling may also contribute to MEF2-dependent transcription and altered gene expression with endurance training. Moreover, AMPK signaling promotes **mitochondrial biogenesis** in skeletal muscle through its effects on PGC-1α expression and activity. As mentioned previously, PGC-1α is a transcriptional coactivator that is considered to be a major regulator of mitochondrial biogenesis (reviewed in Uguccioni, D'souza, and Hood 2010). PGC-1α interacts with several transcription factors, such as nuclear respiratory factors (NRF)-1 and -2, the estrogen-related receptor, ERRα, and PPARs, that induce the expression of mitochondrial genes. When it is bound to a transcription factor, PGC-1α binds several HAT enzymes which, as discussed previously, remodel histones on chromatin, thereby allowing greater access of the transcriptional machinery to DNA for initiation of transcription.

Another important signaling pathway involved in the adaptive response of skeletal muscle to endurance exercise is the **mitogen-activated protein kinase** (MAP kinase, or MAPK) pathway. The word *mitogen* refers to something that stimulates cell proliferation (mitosis), but only some of the actions of the MAP kinases ultimately lead to the formation of new cells. Activation of the p38γ MAPK in skeletal muscle phosphorylates and activates PGC-1α but also activates other transcription factors, namely MEF2 and activating transcriptional factor 2 (ATF2), that increase transcription of the PGC-1α gene. Increased ROS production and CaMK activity appear to be important for activating p38γ MAPK during endurance exercise. The major signaling pathways underlying endurance exercise–induced adaptation in skeletal muscle (see figure 3.26) have been the subject of several reviews (Coffey and Hawley 2007; Gundersen 2011; McGee and Hargreaves 2011; Yan et al. 2011).

Booth and Neufer (2005) summarized studies that all monitored gene expression in subjects over

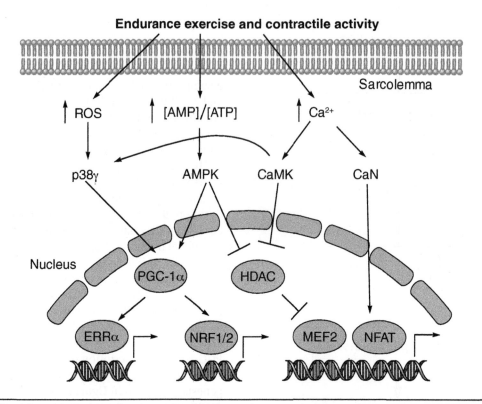

Figure 3.26 Schematic representing the major signaling pathways underlying endurance exercise–induced adaptation in skeletal muscle. AMPK is adenosine monophosphate kinase; CaMK is Ca^{2+}-calmodulin-dependent kinase; CaN is calcineurin; ERRα is estrogen-related receptor α; HDAC is histone deacetylase; MEF2 is myocyte-enhancer factor 2; NFAT is nuclear factor of activated T cells; NRF1/2 is nuclear respiratory factors 1 and 2; p38γ is p38γ mitogen-activated protein kinase; PGC-1α is peroxisome proliferator–activated receptor γ coactivator-1α.

the course of a short-term one-legged endurance training program. They described the response of genes in terms of how rapidly they responded, the duration of the response, and the peak of the response based on mRNA transcript levels. Genes that were turned on very quickly, but transiently, were described as "stress response genes." These coded for proteins, such as transcription factors, whose content was rapidly elevated in a number of models under high-stress conditions. A second category of genes, the "metabolic priority genes," demonstrated peak expression several hours after the exercise-training bout. Genes in this latter category coded for enzymes that played regulatory roles in carbohydrate metabolism. The third gene category, "metabolic/mitochondrial enzymes," was slower to respond and had a lower peak response, but the response persisted over a longer period of time. These genes coded for mitochondrial proteins, involved in oxidative phosphorylation. Studies such as those summarized by Booth and Neufer (2005) remind us that there are priorities in the transcription of genes, the ultimate effect of which is to enhance the ability to respond to further training stimuli.

Resistance Training

Compared with endurance exercise, resistance exercise consists of much higher intensity contractions (e.g., usually 70% to 80% of one repetition maximum) repeated at high frequencies (referring again to motoneuron firing frequency) that can only be sustained for short durations due to the fatiguing nature of the exercise. Therefore, the pattern of cytoplasmic $[Ca^{2+}]$ oscillations and changes in [AMP]/[ATP] and ROS production with skeletal muscle contractions is different between endurance and resistance exercise. For example, endurance exercise likely results in extended periods of moderately elevated $[Ca^{2+}]$, while resistance exercise would generate short cycles of significantly higher intracellular $[Ca^{2+}]$ in skeletal muscle (Chin 2010). Therefore, differences in the magnitude and pattern of these primary signals generated by endurance and resistance exercise will result in the activation of different signaling pathways and different gene-expression and protein-synthesis responses in skeletal muscle. A number of studies have shown that resistance exercise results in increased rates of muscle-protein synthesis, about two- to fivefold after exercise for periods up to 48 h before declining to baseline values. Although acute resistance and endurance-type exercise result in a similar global anabolic response in untrained skeletal muscle, increases in muscle mass (i.e., hypertrophy) and strength are significantly greater following chronic resistance training compared with chronic endurance training. This means that chronic resistance training, but not endurance training, increases the rate of muscle-protein synthesis to levels above the rate of protein degradation.

Activation and differentiation of muscle satellite cells into new muscle cells that fuse with existing muscle fibers can also contribute significantly to the hypertrophic response to resistance exercise. Muscle fibers are large multinucleated cells. They are also postmitotic cells in that they no longer have the ability to divide and reproduce themselves. Additionally, in order to significantly increase in size or hypertrophy, muscles have to add more nuclei, since a nucleus is only able to supply mRNA for new protein synthesis to a limited amount of cytoplasm in its vicinity. Hence, a specific ratio of nuclei to cytoplasm must be maintained. In order to do this, skeletal muscles have small specialized satellite cells, or *myogenic precursor cells*, located at the periphery of their outer membranes. When stimulated by resistance training or by exercise-induced muscle damage, these specialized cells will be activated and induced to create daughter cells or proliferate. These new satellite cells will then fuse with the existing muscle fibers and add their nuclei to the cells to support greater protein synthesis and increase muscle hypertrophy or repair.

It is well known that the primary factor determining the hypertrophic response to contractile activity is the load across the muscle or the mechanical stretch/strain imposed on the muscle fibers, which is higher in resistance exercise than in endurance exercise. Ultimately, these mechanical/force signals are transduced by signaling pathways that regulate transcriptional and translational processes and satellite-cell activation.

For several years, researchers have focused on the role of insulin-like growth factor (IGF)-1 as a primary signaling molecule that mediates skeletal-muscle growth in response to resistance exercise, given that resistance exercise stimulates the secretion of IGF-1. IGF-1 is known to induce muscle hypertrophy by binding to its receptor on the muscle-cell surface and activating the classical growth factor pathway (see figure 3.27). IGF-1 binding to the receptor activates phosphatidylinositol 3-kinase (PI3K), which

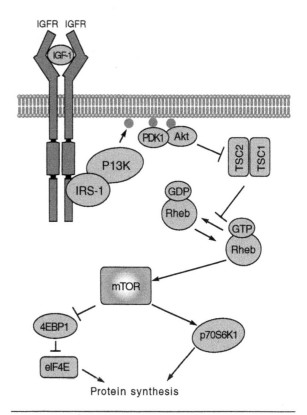

Figure 3.27 Classic growth factor–signaling pathway from IGF-1 to mTOR and protein synthesis. Arrows indicate that phosphorylation activates the signaling molecule and bars denote inhibition. 4E-BP1 is eukaryotic initiation factor 4E binding protein; eIF4E is eukaryotic initiation factor 4E; IGF-1 is insulin-like growth factor-1; IGFR is insulin-like growth factor-1 receptor; IRS-1 is insulin receptor substrate-1; mTOR is mammalian target of rapamycin; p70S6K1 is 70KDa ribosomal S6 protein kinase; PDK1 is 3'-phosphoinositide-dependent protein kinase 1; PI3K is phosphatidylinositol 3-kinase; Rheb is ras homologous protein enriched in brain; TSC1/2 is tuberosclerosis complex 1/2.

> **KEY POINT**
>
> The reaction to exercise-induced muscle-fiber damage and the stimulus for muscle hypertrophy are similar in that both activate muscle satellite cells to facilitate repair and to increase muscle-fiber size. They only differ in degree, since a small amount of damage can induce hypertrophy, while more damage will direct all the muscle response to repair it without allowing for hypertrophy to occur. Hence, appropriate amounts of overreaching in training that induce small degrees of muscle damage will also provide the stimulus for muscle hypertrophy and training adaptations facilitated by activation of muscle satellite cells.

leads to the activation of Akt, a serine/threonine protein kinase. Akt phosphorylates and inactivates tuberosclerosis complex (TSC2), resulting in the activation of the ras homologous protein enriched in brain (Rheb) and **mammalian target of rapamycin** (mTOR). mTOR phosphorylates and suppresses the eukaryotic initiation factor 4E binding protein (4E-BP1) to blunt 4E-BP1 inhibition of translation-initiation cap-binding protein eIF4E (see figure 3.21). mTOR also phosphorylates the 70KDa ribosomal S6 protein kinase (p70S6K1), resulting in an increase in protein synthesis.

It is well accepted that mTOR signaling plays a dominant role in the adaptive response to resistance training; however, researchers have questioned the importance of IGF-1 signaling in mediating this response, based on evidence from several studies employing pharmacological and knockout-mouse approaches to systematically manipulate the IGF-1-PI3K-Akt pathway (Philp, Hamilton, and Baar 2011). It appears that the IGF-1 signaling pathway is not required for mTOR activation or increased protein synthesis that is induced by resistance-type exercise. It is possible that mechanical signals working through stretch-activated membrane channels, for example, might be able to activate mTOR and its downstream targets. However, this hypothesis needs to be examined experimentally (Philp, Hamilton, and Baar 2011).

NEXT STAGE

MicroRNAs and the Adaptive Response to Exercise Training

MicroRNAs (miRNAs) are a class of short, noncoding RNA molecules that bind to mRNA molecules and play a central role in regulating gene expression through posttranscriptional gene silencing (reviewed in Bushati and Cohen 2007). Most miRNAs are encoded in introns of protein-coding genes and are transcribed by RNA polymerase II as long primary-miRNAs (pri-miRNA) that encode a single miRNA or a cluster of miRNA species. Processing of pri-miRNA species in the nucleus produces stem-loop structures of ~70 nucleotides, termed precursor-miRNA (pre-miRNA). These pre-miRNAs are transported to the cytoplasm where they are further processed, giving rise to the mature ~19 to 22 bp of miRNA. The mature miRNA is incorporated into a ribonucleoprotein complex known as the RNA-induced silencing complex (RISC). Generally, miRNAs inhibit protein synthesis by binding (base-pairing) in the 3' untranslated regions of target mRNAs, either repressing translation or bringing about deadenylation and subsequent degradation of mRNA targets. Individual miRNAs can target hundreds of genes, while individual mRNAs can be targeted by multiple miRNAs, making this one of the most complex gene-regulatory processes.

Studies have uncovered a cluster of muscle-specific miRNAs that regulate muscle differentiation and modulate diverse aspects of muscle function (reviewed in van Rooij, Liu, and Olson 2008). The most highly studied are miR-1, miR-133 and miR-206, which are induced during differentiation of myoblasts into myotubes (Callis et al. 2008) and play an important role in muscle mass maintenance. Other miRNA species (miR-23, miR-103, miR-107, and so on) are proposed to play an important role in regulating expression of genes encoding metabolic pathway enzymes in skeletal muscle and other tissues (Wilfred, Wang, and Nelson 2007). Interestingly, studies have shown the potential importance of miRNA regulation in skeletal-muscle adaptations to exercise (reviewed in Roth 2011). For example, miR-1 and miR-133a are downregulated in mouse skeletal muscle during functional overload–induced hypertrophy of the plantaris muscle (McCarthy and Esser 2007). In response to endurance exercise, miR-23, a putative negative regulator of the transcriptional coactivator peroxisome proliferator α coactivator 1 (PGC-1α), was downregulated in mouse skeletal muscle (Safdar et al. 2009). Importantly, downregulation of miR-23 was associated with increased expression of PGC-1α mRNA and protein, along with several downstream targets of PGC-1α signaling.

You are no doubt aware that the ability to increase muscle size in response to resistance and strength training is greater for some people than others. Have you ever wondered why strength training causes large gains in muscle mass in some people (i.e., *high responders*), whereas others gain very little muscle mass in response to the same training stimulus (i.e., *low responders)*? This is true even after accounting for differences in age, training status, exercise adherence, and diet. A study by Davidsen and colleagues (2011) helps to shed some light on this issue. In their study, vastus lateralis biopsies were taken from the top and bottom 49 responders, in terms of muscle mass gain, of 56 men who completed a 12-week strength-training program. The expression level of 21 abundant miRNAs was measured to determine whether variation in these miRNAs was able to explain the variation in resistance training–induced gains in muscle mass. They indentified 4 miRNAs that showed uniquely different responses between high responders and low responders. MiR-378, miR-29a, and miR-26a were downregulated in low responders and were unchanged in high responders, while miR-451 was upregulated only in low responders. Therefore, the regulation of protein synthesis by miRNAs may play an important role in explaining the variability in strength training adaptations. However, further research is required to uncover how these miRNAs themselves are regulated and whether they can be targeted for therapeutic interventions.

SUMMARY

A DNA molecule is composed of deoxynucleotides, joined together to make huge DNA molecules. Each deoxynucleotide consists of the sugar deoxyribose, phosphate, and one of four different bases: adenine, thymine, guanine, and cytosine. Deoxyribonucleic acid molecules are double stranded and are arranged to form the well-known double helix. The bases in one strand are complementary to the bases in the other strand because hydrogen bonding between adenine and thymine and between guanine and cytosine is an absolute necessity. The strands in DNA are antiparallel; one strand runs from the 5' phosphate at one end to a free 3' hydroxyl group of the last deoxyribose sugar, and the complementary strand runs in the 3' to 5' direction.

Transcription of genes produces an RNA molecule called the *primary RNA transcript* that will be a complementary copy of the DNA strand transcribed, with the exception that the RNA transcript will have the base uracil instead of thymine and the sugar will be ribose. The primary transcript, or pre-mRNA, is modified to produce a messenger RNA (mRNA). The base sequence of the mRNA, beginning with the 5' end, is read in groups of three called codons, which constitute the words of the genetic code. Of the possible 64 codon combinations, 61 spell out the 20 amino acids, so the genetic code is said to be degenerate. Three of the codons (UAA, UGA, and UAG) stop the message. Transcription of genes encoding polypeptides is carried out by RNA polymerase II. During transcription, the polymerase unwinds the two DNA strands and copies part of one, reading it in the 3' to 5' direction. The nontranscribed DNA strand, called the *sense strand*, has the same base sequence as the primary transcript, except that uracil replaces thymine in RNA.

The start site is the first nucleotide in the DNA strand that is copied during transcription. The region of DNA just before the transcription start site is called the promoter. The promoter, and other regions of the DNA molecule that may be far from the start site, contain sequences of bases known as response elements to which specific protein molecules called transcription factors can bind to regulate the process of transcription. General transcription factors are needed for the transcription of all genes. These bind upstream of the start site at a very common response element called the TATA box, help position RNA polymerase II, and start the transcription process. Virtually all genes require other tissue-specific or developmentally specific proteins to provide an additional level of transcription control so that genes are expressed when they should be and at appropriate rates to ensure the overall health of the body. These additional transcription factors may stimulate (activators) or inhibit (repressors) transcription of the gene. Circulating hormones and growth factors can influence transcription in a process termed *signaling* or *signal transduction*. Steroid hormones are hydrophobic; they can diffuse into cells and bind with their specific receptors in the nucleus, which subsequently act as transcription factors. Other hormones and growth factors bind to receptors on the cell membrane and generate signals within the cell that alter the rate of transcription. The protein-DNA complex in the nucleus known as *chromatin* is organized into basic structures called nucleosomes. The tight DNA loops within nucleosomes and higher orders of DNA coiling in the nucleus repress transcription. Before genes can be transcribed, the nucleosomes and other higher levels of DNA organization must be remodeled to allow access to transcription factors.

An example of DNA remodeling takes place when histone tails in nucleosomes are altered by the addition of acetyl, phosphate, or methyl groups, or by other ATP-dependent chromatin remodeling enzymes, adding a further level of complexity to transcription control.

The information for the sequence of amino acids in a polypeptide chain is based on the sequence of nucleotides (bases) in the messenger RNA molecule. Created during the transcription of a protein-coding gene, the pre-mRNA molecule undergoes extensive modifications, including the addition of a cap to the 5' end, the addition of multiple adenine nucleotides at the 3' end (poly A tail), and the removal of sequences not involved in the gene message (introns) from those that do provide amino acid–sequence information (exons). The sequence of nucleotides (bases) in the mRNA, read in groups of three, specifies amino acids. The coding region, or open reading frame, describes the area between the start codon (AUG), where protein synthesis will begin, and a stop codon indicating that the message is terminated.

Protein synthesis occurs in the cell cytosol using mRNA, two ribosome subunits (described on the basis of their mobility in a centrifugal field as 40S and 60S), amino acids, energy sources, and a large number of protein factors. A significant first step is the attachment of amino acids to specific tRNA molecules by a specific aminoacyl-tRNA synthetase. Each tRNA contains a three-base sequence called an anticodon; the complementary base pairing between the anticodon on tRNA (containing its attached amino acid) with the codon on mRNA provides fidelity in amino-acid sequence in a protein.

Translation initiation is the most complex step in protein synthesis. It involves the formation of a complete ribosome subunit (80S) at the start (AUG) codon on mRNA with a methionyl-tRNA. Energy provided from GTP and the assistance of a number of eukaryotic initiation factors (eIFs) is also necessary. Elongation is the step at which individual aminoacyl-tRNA molecules come to the ribosome complex and form a peptide bond between amino acids attached to their respective tRNAs. Individual ribosomes then move along the mRNA molecules three bases at a time. Each time the ribosome complex moves along the mRNA, a new amino acid is added to the growing polypeptide chain. Elongation continues the building of a polypeptide chain until a stop codon on mRNA is encountered. The ribosome complex then dissociates from the mRNA, and the completed polypeptide is released. Each mRNA molecule may have a number of ribosomes moving along independently, each with a growing polypeptide chain. Since the quantity of individual proteins in a cell can dictate rates of cell metabolism, control of the translation process must be carefully regulated. Translation, mainly initiation, is controlled at a number of different sites and may be based on nucleotide sequences in the 5' or 3' untranslated regions of the mRNA. The lifetime of its mRNA molecules also determines the amount of a cell's protein.

The amount of a specific protein in a cell is related to its rate of breakdown as well as its synthesis. Protein degradation is an important process that plays a significant role in the function of a cell. Three major systems are involved in normal protein degradation in muscle. The ubiquitin-proteasome pathway involves targeting proteins for degradation by attaching multiple units of a small protein known as ubiquitin. Subsequent degradation to small peptides is carried out by a proteasome complex known as the 26S proteasome. Lysosomes, which are internal organelles, degrade endocytosed proteins using a battery of proteinases. Calpains are calcium-activated proteinases that may play a significant role in the response of muscle to severe exercise stress.

Specific cell-signaling responses occur following endurance or strength training that ultimately signal adaptations in skeletal muscle, such as hypertrophy or mitochondrial biogenesis. These responses are specific to the type of training performed and will ultimately lead to enhanced exercise performance.

REVIEW QUESTIONS

1. The base sequence of one of the strands of a 21 bp duplex of DNA is 5'-AGTCCAGCGTTAGACCGAAGT-3'. What is the base sequence of its complementary strand?

2. If the complementary strand of the same 21 bp duplex of DNA is part of the coding region of a gene, what is the base sequence of the mRNA it generates?

3. Using the sequence of bases in the mRNA molecule from the previous question, determine the amino acid sequence.

4. Knowing that the sequence of amino acids in a polypeptide is Glu-Lys-Met-Ala-Gly-His-Thr, design a base sequence to represent the sense strand of the DNA that generated this amino acid sequence.

5. If a functional protein contains 200 amino acids, what is the minimum number of nucleotides in the coding region of the mature mRNA molecule for this protein? How many nucleotides would there be in the open reading frame for this protein?

6. Explain how autolysis increases the proteolytic activity of calpains.

7. What signaling pathways regulate mitochondrial biogenesis, and how might these pathways be influenced by endurance training?

part II

Metabolism

Regulation and Adaptation to Exercise and Training

Money is the central commodity that allows modern societies to function. People work for their money and use this money to buy things. If you operate like most folks, the amount of money that comes in as salary balances the money that goes out for housing, food, and other expenses. Living organisms operate in the same way, except that the currency is ATP, not money. We create ATP by breaking down fuels such as carbohydrate and fat obtained from our diet. We use this ATP to build proteins, use our muscles, transport things into cells and cell organelles, and generally grow and maintain our bodies. All the reactions that create and use ATP are collectively called metabolism. That branch of metabolism in which ATP is generated is known as catabolism, and the branch in which ATP is used to make things is known as **anabolism**. Part II of this text focuses on the detailed processes for the breakdown of the major fuel molecules (fat, carbohydrate, and amino acids from protein) to make ATP. It also looks at how ATP is used, especially in the formation and storage of fuels, and how the rate of ATP synthesis is tightly coupled with the rate of ATP use.

Throughout these chapters, we emphasize not only the details of the processes but also the mechanisms that regulate them, including the roles played by diet and various forms of exercise. We begin in chapter 4 with an overview of metabolism, outlining the three energy systems that provide ATP to contracting skeletal muscles. This is followed by a section that illustrates how we can quantify energy. Oxidative phosphorylation, the dominant process generating ATP in the body, is the theme of chapter 5. For this process to function, carbohydrate (chapter 6), lipids or fat (chapter 7), and amino acids (chapter 8) are degraded, and their chemical energy is converted into ATP.

chapter

4

Energy Systems and Bioenergetics

The ability to use energy is the hallmark of all living organisms. This energy, in the form of **ATP (adenosine triphosphate)**, is found in all cells but in small amounts, ranging from about 3 mmol per kilogram in most tissues to about 6 mmol per kilogram of skeletal muscle. The content of ATP remains the same even though it is used constantly, because it is being regenerated at a similar rate. This need to maintain almost constant ATP levels in cells even in the face of highly variable metabolic rates and ATP turnover is central to how our cells determine which metabolic pathways to utilize and which fuels to select for ATP regeneration during times of intense exercise or times of modest physical activity. Adenosine triphosphate is used to drive a number of processes in cells, such as making cell proteins, storing fuels, synthesizing RNA (ribonucleic acid) molecules, and transporting substances into cells and the organelles within cells. It also aids in signaling to regulate cellular processes. In addition, ATP is the central currency in converting energy stores into fuel for muscular contraction and physical activity. Skeletal muscles, especially in elite athletes, are incredible systems. Consider that a 70 kg marathon runner may cover 42 km (26.2 miles) in less than 130 min, but may do the following:

- Expend approximately 3,000 kilocalories (12,600 kilojoules)
- Oxidize approximately 600 to 700 g of carbohydrate
- Oxidize approximately 30 to 40 g of fat
- Utilize about 600 L of oxygen
- Break down and reform more than 150 mol of ATP, which weighs about 63 kg (139 lb)

This chapter looks at the energy requirements of skeletal muscle and the role of ATP, an energy-rich phosphate compound that directly drives muscle contraction. The ATP concentration of muscle is very small. We will discuss three energy systems responsible for maintaining the ATP concentration in cells. The last part of the chapter deals with quantitative aspects of cellular energy metabolism. The Next Stage section highlights developments in the new field of epigenetics as it relates to exercise and training.

ENERGY REQUIREMENTS OF SKELETAL MUSCLE

Skeletal muscle is a special tissue that can suddenly increase its rate of ATP use to more than 100 times that of rest, during very intense exercise such as sprinting. Skeletal muscle can maintain isometric contractions, as in holding a baby. It can be used at a high level for long periods of time, in the case of a marathon. We will start with the structure of skeletal muscle and how it contracts.

Structure of Skeletal Muscle

Skeletal muscle, the largest tissue in the body, consists of a number of elongated cells known as fibers, which contain a number of parallel *myofibrils* (see figure 4.1). Muscle fibers have a cell membrane, known as the **sarcolemma**, and mitochondria arranged under the sarcolemma and between myofibrils (figure 4.2). Muscle fibers differ from other cells because they are so large, running

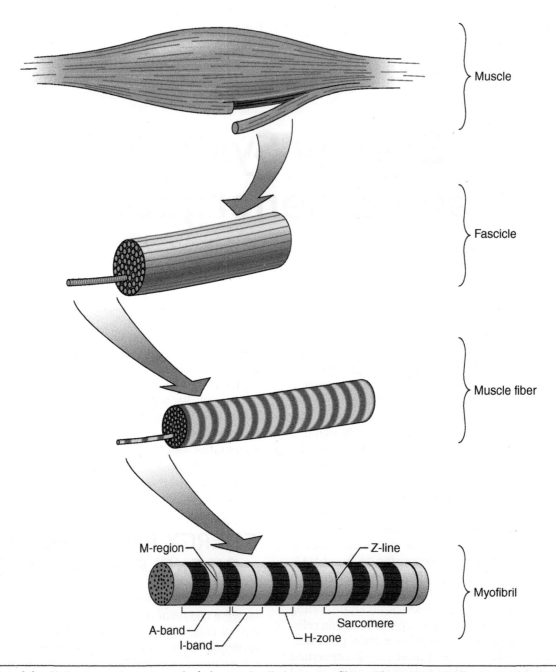

Figure 4.1 Skeletal muscle, composed of elongated cells known as fibers. These are typically arranged in bundles known as fascicles. Within a muscle fiber are myofibrils, aligned in parallel with the fiber's long axis. Striations in the fiber and myofibril are due to the arrangement of contractile proteins, creating bands in the sarcomere.

Adapted, by permission, from B. MacIntosh, P. Gardiner, and A.J. McComas, 2004, *Skeletal muscle: Form and function*, 2nd ed. (Champaign, IL: Human Kinetics), 3.

from one end of a muscle to the other. This large size explains the need for many nuclei in each skeletal muscle cell. Neurons, emerging from the spinal cord, innervate each fiber at the *neuromuscular junction*. Nerve impulses crossing from the neuron to the sarcolemma activate the fiber to make it contract. Contraction is initiated when calcium ions are released from intracellular storage vesicles known as **sarcoplasmic reticulum**.

The striations seen in muscle fibers are due to the arrangement of proteins aligned in parallel in two kinds of filaments in each myofibril. Figure 4.3 shows parts of three thin filaments and a section of a thick filament. The **thick filament** is composed of

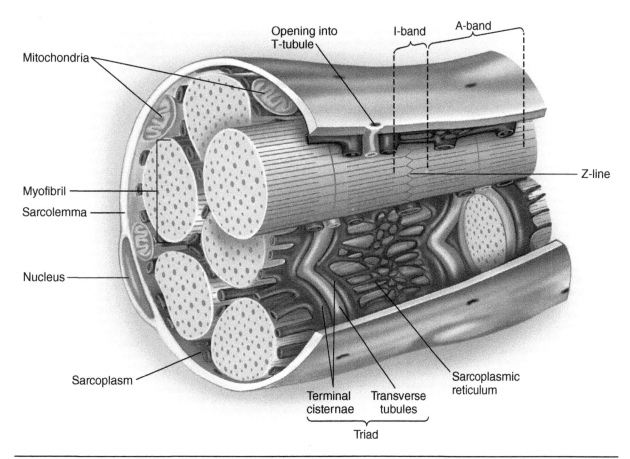

Figure 4.2 Schematic of a muscle fiber showing longitudinal and cross-sectional perspective. A myelinated motor neuron, originating from the spinal cord, attaches on the cell membrane (sarcolemma) at the neuromuscular junction. Muscle fibers have mitochondria located under the sarcolemma and between individual myofibrils. Sarcoplasmic reticulum is a network of filaments and sacs that store and release calcium ions. Because of their large size, muscle fibers have numerous nuclei.

Reprinted, by permission, from J.H. Wilmore and D.L. Costill, 2004, *Physiology of sport and exercise*, 3rd ed. (Champaign, IL: Human Kinetics), 36.

individual myosin molecules arranged in overlapping parallel arrays. Three myosin molecules, each with two protruding heads known as crossbridges, can be seen in the thick-filament section in figure 4.3. Calcium ions released when the fiber is activated bind to the thin-filament protein troponin, allowing strong, force-generating interactions between myosin and actin, driven by ATP hydrolysis, to take place. The sum of the forces from millions of myosin crossbridges, each attached to an actin molecule, leads to shortening of the overall muscle.

A **thin filament** contains three kinds of proteins. Individual molecules of actin (actin monomers), shown as spheres in figure 4.3, align themselves into two strands that form a helix. Arrayed along the actin fibers are the proteins *tropomyosin* and troponin. When **troponin** binds calcium, it can alter the position of the tropomyosin filaments, exposing sites on each actin monomer that can bind a myosin crossbridge.

> ### ☑ KEY POINT
> Troponin is the name given to a protein complex containing three subunits. Troponin C (named for its calcium-binding function) can bind reversibly with calcium when the calcium concentration increases, as it does when the fiber is activated by its nerve. Calcium binding by troponin C results in alterations in the two other subunits, troponin I (named for its inhibitory function) and troponin T (named for its tropomyosin-binding function), leading to a change in the position of tropomyosin on the thin filament and better access by myosin crossbridges to actin-binding sites. Since the binding and release of calcium are critical for actin–myosin interaction, troponin is a key regulator of muscular contraction and relaxation during physical activity.

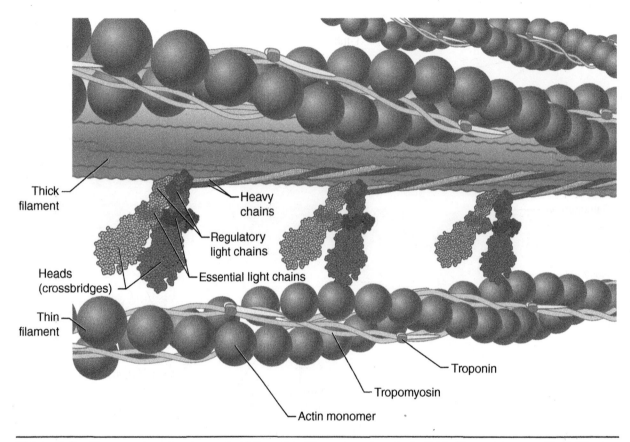

Figure 4.3 Close-up view showing parts of three thin filaments and part of a thick filament in a myofibril. The two crossbridges on three individual myosin molecules in a thick filament can each interact with individual actin monomers (shown as spheres) in the thin filament when the muscle fiber is activated.

Adapted, by permission, from R. Vandenboom, 2004, "The myofibrillar complex and fatigue: A review," *Canadian Journal of Applied Physiology* 29(3): 330-356.

Myosin and Muscle Contraction

A myosin molecule consists of six polypeptide chains. Two of these polypeptide chains are very large and are twisted together from one end (the tail) to the other end, where they form the heads or **crossbridges**. These are the *myosin heavy chains* (MHC). The four other polypeptides are called *myosin light chains* (MLC). Two of these are associated with each head or crossbridge. The light chains in the neck region are known as the *regulatory light chains* because each of these can be phosphorylated by accepting a phosphate group from ATP. Phosphorylation of myosin regulatory light chains causes subtle changes in the performance of the myosin crossbridge. The other two light chains, located farther into each crossbridge, are known as the *essential light chains*.

Myosin has two important properties. Each crossbridge can bind to a single actin molecule in the thin filament. The second property is that each head can act as an enzyme to hydrolyze ATP, as shown in the following equation. The word *hydrolysis* is used because it refers to breaking the covalent bonds of a molecule by the addition of water.

$$ATP + H_2O \rightarrow ADP + Pi$$

This ability to break down ATP into the products **ADP (adenosine diphosphate)** and inorganic phosphate (Pi) is called *myosin ATPase* activity, because it is myosin acting as the **ATPase** enzyme.

The characteristics of a myosin molecule are dominated by the heavy chain properties. Each crossbridge or head contains an actin-binding site and an ATP-binding site. The rate at which a myosin crossbridge can hydrolyze ATP is determined by the heavy chain. By itself, myosin ATPase activity is low, but it can be increased by about 100-fold when actin binds to a myosin crossbridge. Actin serves as a powerful activator for myosin ATPase activity, so we describe this as *actin-activated myosin ATPase* activity or *actomyosin ATPase* activity. Actomyosin ATPase activity is the rate at which it can hydrolyze or break down ATP, and it is strongly correlated with the maximum speed at which a muscle or muscle fiber can contract (shorten).

The four different kinds of skeletal-muscle myosin heavy chains are identified as *MHC I*, *MHC IIA*, *MHC IIX*, and *MHC IIB*. The Roman numeral I is used to designate myosin that has a lower actin-activated myosin ATPase activity; sometimes the word *slow* is used instead of the Roman I. The Roman II means fast. The addition of A, X, or B is intended to subtype the particular fast MHC. Human skeletal muscles have three main myosin heavy chains, MHC I, MHC IIA, and MHC IIX. Smaller animals have MHC IIB, which has the fastest actomyosin ATPase activity. Two additional MHCs are found in embryonic and neonatal muscles and in adult muscles undergoing regeneration.

KEY POINT

In the past many people described the fastest-contracting fiber type in human skeletal muscle as type IIB when, in fact, it is really IIX. One of the reasons for this is that the IIX myosin heavy chain was identified as distinct from the IIB myosin heavy chain much later. Accordingly, much of the early literature mistakenly identified IIX human fibers as IIB fibers. You can still read of IIB human muscle fibers in older textbooks.

We have focused on the MHCs because these dominate the contraction characteristics of the myosin molecule. Nonetheless, myosin is still a molecule with six subunits (two MHCs and four MLCs). Accordingly, exercise scientists talk about *myosin isozymes* (also known as myosin isoforms or isomyosins), different molecular forms of the same enzyme (myosin) that catalyze the same reaction but with different speeds of contraction and ATP hydrolysis rates. The properties of the muscle fibers that contain different myosin isozymes can be illustrated by how fast these fibers shorten or how fast they can undergo a twitch in response to a single electrical stimulus. Thus, slow-twitch (Type I) fibers are dominated by the MHC I, and fast-twitch (Type II) fibers are dominated by more of the fast MHCs. Table 4.1 summarizes the major MHCs and the corresponding fiber types for mammalian skeletal muscles. As an added note, skeletal muscle fibers have many nuclei, each with the ability to express MHC genes. Therefore, a single muscle fiber may have two or more different MHCs, although one tends to dominate, thus leading to a single fiber-type characterization.

As already mentioned, it is the action of myosin crossbridges exerting force on actin monomers in the thin filament that leads to muscle shortening and therefore movement. Each crossbridge undergoes a sequence of binding with an actin monomer, ATP hydrolysis, force generation, and then detachment from the actin. This crossbridge cycle is shown in figure 4.4. In figure 4.4a, one of the two crossbridges of a myosin molecule attaches loosely (weakly) to an actin monomer. This is depicted as the darker myosin head on the figure, with the lighter color representing the nonbinding (distal) myosin head. An ATP molecule is split into the products ADP and Pi, but these remain together so that the actual energy of this splitting is not released. This step can be pictured as a spring compressed between two fingers. Next, some of the energy of the ADP-Pi couple is released, and a strong binding conformation is attained. In figure 4.4b, the Pi is released as a consequence of the strong binding, and the power stroke occurs. The nature of the change in the head–neck region of the myosin molecule strains elastic elements in this region. The strain on the elastic elements exerts a force on the actin monomer. The actual attachment of the myosin head to actin does not change during the power stroke. In figure 4.4c, the myosin head, acting through the strain on the elastic

Table 4.1 Fiber Types in Mammalian Skeletal Muscle, Their Dominant Myosin Heavy Chain Type, and Relative Speed of Contraction

Fiber type*	Myosin heavy chain	Relative speed of contraction
Type I (ST)	MHC I	Slow
Type IIA (FTA)	MHC IIA	Fast
Type IIX (FTX)	MHC IIX	Faster
Type IIB (FTB)**	MHC IIB	Fastest

* Type I and ST (slow-twitch) and Type II and FT (fast-twitch) are used synonymously.

** Type IIB (FTB) fibers with MHC IIB are found in skeletal muscles of small mammals, but they do not appear in humans.

elements in the head–neck region, exerts a force on the actin monomer. At the end of this force-generating stroke, ADP is released, a new ATP binds, and, as figure 4.4d shows, the previously attached, force-generating crossbridge detaches from actin. In the detached state, ATP is hydrolyzed to ADP and Pi, and the crossbridge is available to bind weakly to actin. The sum of millions of crossbridges, each acting on its actin monomer, results in actual movement of the thick filament with respect to the thin filament and muscle fiber. Hence, overall muscle shortening occurs.

Figure 4.5 summarizes the overall process of muscle contraction, which includes the following steps:

1. Acetylcholine released by the neuron at the neuromuscular junction diffuses across the gap between the neuron and the sarcolemma, binds to its receptor, and causes depolarization of the sarcolemma.
2. A wave of depolarization passes over the sarcolemma and down into the interior of the fiber via surface invaginations known as **T-tubules**. The depolarization is associated with sodium ions (Na^+) moving into the fiber and some potassium ions (K^+) moving out of the fiber.
3. T-tubule depolarization is linked to the release of calcium ions (Ca^{2+}) from a specialized form of endoplasmic reticulum known as *sarcoplasmic reticulum* or SR.
4. The calcium-ion concentration in the fiber interior increases from about 10^{-7} M to about 10^{-5} M (a 100-fold increase).
5. The calcium ions bind to the protein troponin in the thin filament in the sarcomere.
6. This binding allows crossbridges to bind to individual actin monomers in the thin filament.
7. Using the energy released from ATP (by the enzyme actomyosin ATPase), the bound myosin crossbridges exert force on the actin in the thin filament.
8. The force exerted by millions of crossbridges in many fibers causes the overall muscle to shorten and do work.

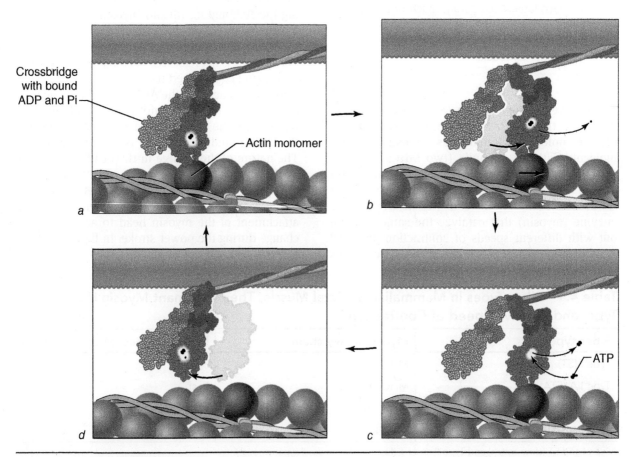

Figure 4.4 The crossbridge cycle.

Adapted, by permission, from R. Vandenboom, 2004, "The myofibrillar complex and fatigue: A review," *Canadian Journal of Applied Physiology* 29(3): 330-356.

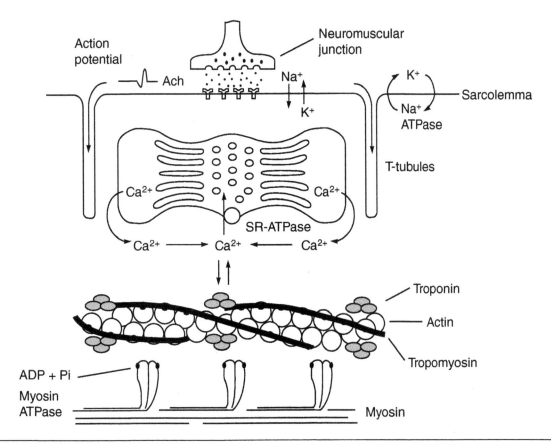

Figure 4.5 An overview of the steps in muscle contraction. A nerve impulse causes depolarization of the sarcolemma when acetylcholine (Ach) binds to its receptor. A wave of depolarization spreads over the surface of the muscle fiber and into the interior through the T-tubules, resulting in the release of Ca^{2+} ions from the sarcoplasmic reticulum (SR). Ca^{2+} binding to troponin on the thin filament allows myosin crossbridges to bind to actin, creating force when ATP is broken down. In relaxation, Ca^{2+} is transported into the SR by the SR-ATPase using energy from ATP. The Na^+-K^+ ATPase restores the balance of Na^+ and K^+ across the sarcolemma.

9. The Ca^{2+} is pumped back into the SR in a process requiring energy from ATP. A special membrane ATPase known as SERCA (sarcoplasmic–endoplasmic reticulum calcium ATPase) uses the energy from ATP hydrolysis to pump two calcium ions inside the reticulum of the SR or endoplasmic reticulum in other cells.

10. At the same time, a sarcolemma sodium–potassium ATPase (Na^+-K^+ ATPase) pumps three sodium ions from the inside of the muscle fiber to the outside while simultaneously moving two potassium ions back inside. This restores the normal polarized sarcolemma and also utilizes ATP.

Sites of ATP Use in Muscle

When we think of the energy requirements for muscular exercise, we correctly think of actomyosin ATPase activity as the major site of ATP utilization. While the majority of ATP hydrolysis during muscle activity (60% to 70%) does occur due to myosin ATPase, a significant portion of the total energy expended as ATP hydrolysis is due to two other ATPase enzymes that power calcium or sodium and potassium movements across membranes. As noted above, the SERCA ATPase hydrolyzes ATP to pump calcium ions back into the sarcoplasmic reticulum during the relaxation phase of muscular contraction. Since the calcium ions must be transported from areas of lower concentration in the cytoplasm to those of higher concentration in the sarcoplasmic reticulum, energy is required to facilitate this movement of calcium across the sarcoplasmic reticulum membrane. It has been suggested that at least 20%, and possibly 30% or more, of the ATP utilized during muscular activity may be hydrolyzed by the SERCA ATPase to power calcium uptake into the sarcoplasmic reticulum during the muscle-relaxation phase (Barclay, Woledge, and Curin 2007). The sodium–potassium ATPase also hydrolyzes ATP to restore or repolarize the muscle membrane during muscle relaxation. Its contribution to total ATP hydrolysis during muscular activity is relatively less than that of the myosin ATPase and the SERCA ATPase, since it accounts for less than 10% of the total ATP consumed (Barclay, Woledge, and Curin 2007). Figure 4.6 depicts the sites of ATP utilization in skeletal muscle.

> **KEY POINT**
>
> Muscle ATP breakdown by ATPase enzymes occurs at three main sites during muscular contraction. Although the actomyosin ATPase hydrolyzes the most ATP during contraction, the SERCA ATPase and the sodium–potassium ATPase also account for a significant amount of ATP utilization during exercise.

ENERGY-RICH PHOSPHATES

We have already discussed ATP as the energy currency in the cell and illustrated how it is used to drive the contraction of muscle. As we will learn in the next section, most of the ATP in the cell is produced by the breakdown of fuels, such as fat, carbohydrate, and amino acids (from protein). These fuel molecules are oxidized to simple products such as CO_2 and H_2O. The energy released drives the phosphorylation of ADP to make ATP, as shown in the following equation:

$$ADP + Pi \rightarrow ATP + H_2O$$

Note that the formation of ATP in this reaction is opposite to that of ATP hydrolysis shown earlier in the chapter. Adenosine triphosphate synthesis in this reaction is accomplished when energy released from the **catabolism** (breakdown) of fuels is harnessed through ADP phosphorylation. The energy released by the hydrolysis of ATP drives a process that would not ordinarily take place without an input of energy, such as the crossbridge cycle in muscle.

> **KEY POINT**
>
> To understand the fact that ATP hydrolysis releases a lot of energy, consider the analogy of a spring, compressed between your thumb and second finger. Keeping the coils of the spring together requires force. If you quickly release the pressure on the spring, it rapidly pops free, reaching a longer length and releasing energy.

Pool of Phosphates in the Cell

Adenosine triphosphate and ADP are examples of a class of molecules known as nucleotides. As chapter 3 discusses, nucleotides are molecules that consist of

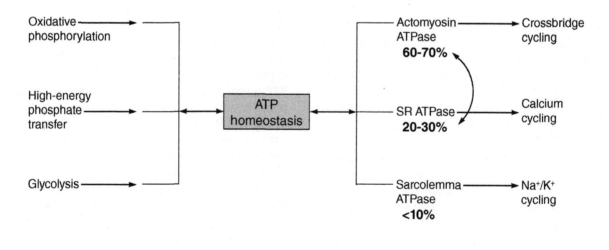

Figure 4.6 During exercise, ATP is utilized by three enzymes at three sites to power muscular contraction: the actomyosin ATPase, which powers crossbridge cycling, the SERCA ATPase (SR ATPase), which powers calcium cycling, and the sarcolemma ATPase, which powers sodium–potassium (Na^+/K^+) cycling. Their rate of ATP utilization drives and matches the rate of ATP resynthesis via oxidative and glycolytic metabolic pathways.

three kinds of components: either a pyrimidine or a purine base (adenine in the case of ATP), the sugar ribose, and one or more phosphate groups. When the base adenine is combined with the sugar ribose, we get a new molecule known as a nucleoside, specifically adenosine. If we add one phosphate group to the 5' position of the sugar ribose in **adenosine**, we get a nucleoside monophosphate (NMP), specifically adenosine 5'-monophosphate (abbreviated 5'-AMP). If we add another phosphate group to the one existing in 5'-AMP, we get adenosine 5'-diphosphate or 5'-ADP—a nucleoside diphosphate (NDP). Another phosphate added to 5'-ADP gives us the **nucleoside triphosphate** adenosine 5'-triphosphate or 5'-ATP (an NTP). Normally, we drop the 5' from these molecule names and simply call them **AMP (adenosine monophosphate)**, ADP, and ATP, respectively. The molecule ATP is shown in figure 4.7 as it would exist in the cell, associated with a magnesium ion (Mg^{2+}). Each of the three phosphate groups is identified by a Greek letter. The bonds between the α and β phosphates and the β and γ phosphates are **anhydride bonds**. Note that there are negative charges on the phosphate groups. The combination of the anhydride bonds plus the negative charges in close proximity means that hydrolysis of ATP, according to the two reactions shown in figure 4.7, yields a lot of energy. In reaction 2, one of the products is PPi, or inorganic pyrophosphate. The bond between the two phosphate groups in PPi is also an anhydride. Its splitting also yields additional energy, accomplished in the cell by a ubiquitous enzyme, *inorganic pyrophosphatase*.

In addition to the nucleoside mono-, di-, and triphosphates represented by AMP, ADP, and ATP, respectively, three other important kinds of nucleotides are shown as nucleoside triphosphates: *GTP* or guanosine triphosphate contains the purine base guanine, *UTP* or uridine triphosphate contains the pyrimidine base uracil, and *CTP* or cytidine triphosphate includes the pyrimidine base cytosine. Their corresponding nucleoside di- and monophosphates are GDP and GMP, UDP and UMP, and CDP and CMP, respectively. Hydrolysis of the anhydride bonds in GTP, UTP, and CTP generates large energy releases, just the same as with ATP. However, the hydrolysis of ATP drives most of the reactions or processes that cannot take place by themselves and need a source of energy. Guanosine triphosphate is specifically involved in the synthesis of proteins and signaling at the plasma membrane of cells. UTP is used to make glycogen, the storage form for carbohydrate in the cell. ATP, GTP, UTP, and CTP are also used to make RNA molecules, as

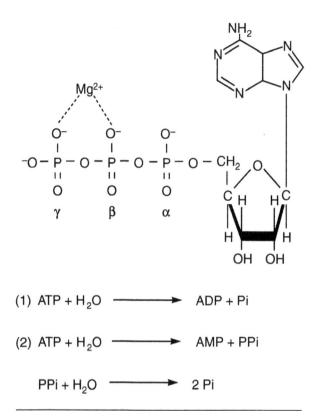

Figure 4.7 Adenosine triphosphate found in the cell in association with magnesium ions. The three phosphate groups in ATP are identified by Greek letters. Hydrolysis of ATP can take place between the β and γ phosphate groups (reaction 1) or between the α and β phosphate groups (reaction 2).

discussed in chapter 3. All four of these nucleoside triphosphates yield large quantities of energy when they are hydrolyzed. As a result, we say that they are energy-rich molecules.

As we will discuss in more detail later in this chapter, the hydrolysis of ATP or any other nucleoside triphosphate results in the release of a lot of energy. Reactions in which a great deal of energy is released tend to go to completion, so that the initial reactant (ATP) virtually disappears. In fact, if we were to put ATP in a test tube solution containing a buffer to keep the pH near 7, plus Mg^{2+} in quantity sufficient to form an ionic interaction with the phosphates on ATP (as shown in figure 4.7) and the ATP-hydrolyzing enzyme myosin, within a few minutes virtually no ATP would remain. We would have a solution with concentrations of ADP and Pi equivalent to that of ATP at the start. No equilibrium would be established, so we would have a mixture with roughly equal portions of ATP, ADP, and Pi. Reactions such as ATP hydrolysis release so much energy that they go almost completely to the right. In fact, if we performed the calculations to see what the concentration of ATP would be in our test tube, assuming that the end point could be called an equilibrium point, we would have a very difficult time finding any ATP molecules that are not hydrolyzed. This is because for every ATP, there would be more than 10 million ADP molecules.

Adenosine triphosphate would be useless as an energy currency in the cell if the ATP were allowed to be converted completely into its products, ADP and Pi. At equilibrium, there would be no net ATP hydrolysis and no energy release because chemical reactions at equilibrium release no energy. For ATP to be the important energy currency it is, the concentration in the cell is kept very far from equilibrium so that the concentration ratio of ATP divided by free ADP is very high—almost 500 in a skeletal muscle cell. Moreover, under most conditions, when ATP is used as an energy source, it is replenished at the same rate so that its concentration does not decrease. As we will see in the last part of this chapter, maintaining a high ATP-to-ADP ratio ensures that the energy released when ATP is hydrolyzed is very high. During very severe exercise, the concentration of ATP does decline. Studies using whole muscle have suggested that muscle ATP concentrations rarely drop below 60% of resting levels. However, during very intense exercise, which depletes PCr stores, fast muscle fibers in humans have been reported to have up to 80% ATP depletion (Allen, Lamb, and Westerblad 2008). In addition, localized sites of high ATP consumption and relatively low ATP diffusion within muscles, such as the sarcoplasmic reticulum-T-tubule site where concentrations of ATP-utilizing SERCA pumps are high, may also significantly deplete ATP levels (Allen, Lamb, and Westerblad 2008). However, this situation is very short lived. ATP levels are very rapidly restored at the onset of fatigue (i.e., when exercise stops), since the maintenance of ATP levels is a high priority in the regulation of muscle metabolism.

✓ KEY POINT

Because more than 90% of all of the ADP in muscle is bound with proteins, especially actin, we use the term *free ADP* to describe that pool of ADP that is not contained in a protein complex but is free to be involved in chemical reactions. The fact that most ADP does not exist as free ADP helps maintain [ADP] [Pi]/[ATP] ratios during exercise, which helps preserve the ΔG ATP and free energy available to do work.

The adenine nucleotides (i.e., ATP, ADP, and AMP) are primarily involved in coupling anabolic and catabolic reactions in the cell. Adenosine triphosphate is formed from ADP when fuel molecules are broken down, and ATP hydrolysis drives most energy-requiring processes. As mentioned earlier, GTP and UTP have their own special roles in metabolism, and all nucleoside triphosphates are necessary to make the various molecules of RNA. When the four nucleoside triphosphates are used as energy sources and to make RNA, the products are NDP molecules (ADP, GDP, UDP, and CDP), NMP molecules (AMP, GMP, UMP, and CMP), or both. Since synthesis and breakdown of RNA molecules in the cell is continuous, the need to maintain a balanced level of NTP molecules is evident. This means that there must be specific reactions to interconvert the various components of the nucleoside phosphate pool. For example, nucleoside diphosphates must convert to nucleoside triphosphates, and nucleoside monophosphates to nucleoside diphosphates. For the first interconversion, a nonspecific enzyme known as nucleoside diphosphate kinase transfers phosphate groups to and from ATP and ADP, and NTP and NDP, as follows:

$$NDP + ATP \leftrightarrow NTP + ADP$$

In this freely reversible reaction, an NDP (such as UDP, GDP, or CDP) is phosphorylated to NTP by accepting a phosphate group from ATP. Because the reaction is freely reversible, its net direction depends on the relative concentrations of ATP, ADP, and the other NTPs and NDPs. The overall effect of the nucleoside diphosphate kinase reaction is to maintain a balance in the ratio of NTP to NDP in a cell.

To convert NMP to NDP, nucleotide-specific enzymes transfer a phosphate group from ATP to the nucleoside monophosphate, making ADP and a nucleoside diphosphate. They are said to be specific because one exists for each nucleoside monophosphate. For example, *adenylate kinase* catalyzes the following reaction:

$$AMP + ATP \leftrightarrow ADP + ADP$$

Uridine monophosphate kinase catalyzes a similar reaction:

$$UMP + ATP \leftrightarrow UDP + ADP$$

These freely reversible reactions also need magnesium ions because these ions are bound to the ATP molecules. The net effect of the nonspecific nucleoside diphosphate kinase and the specific nucleoside monophosphate kinase (e.g., AMP kinase and UMP kinase) is to maintain a balance among the NTP, NDP, and NMP in the cell.

> **✓ KEY POINT**
>
> The need to maintain a high ATP-to-ADP ratio and the very small amount of ATP found in muscle necessitates a close coupling of the rate of energy expenditure, or ATP utilization, with the rate of ATP resynthesis. Metabolic pathways vary greatly in their rate and capacity for ATP resynthesis. The selection of which of those pathways is predominantly used for ATP resynthesis is highly dependent on the rate of ATP utilization or the intensity of the exercise. By tightly coupling ATP utilization rate to ATP resynthesis rate, the muscle ensures that during most kinds of physical activity, ATP levels are kept as high as possible.

Phosphagens

The ATP concentration in most tissues is fairly low, about 3 to 8 mmol per liter of cell water, or about 2 to 6 mmol of ATP per kilogram of tissue. Since ATP represents the immediate energy source to drive energy-requiring processes, problems could arise if ATP is needed at a rapid rate and is therefore used up quickly. In cells with a slow acceleration of ATP-consuming reactions, ATP concentration can be easily maintained by a gradual acceleration of ATP-producing reactions, such as fuel oxidation. However, in muscle, this could be a big problem because during the transition from rest to maximal exercise, the rate of energy expenditure in a human muscle can increase more than 100 times. The rate of energy turnover in a rested muscle is about 1 mmol of ATP per kilogram of muscle per minute. During sprinting, an elite athlete is able to turn over ATP at a rate of about 4 mmol of ATP per kilogram of muscle per second (240 mmol · kg^{-1} · min^{-1}). For normally active individuals, achieving 75% of the maximum rate of ATP turnover of an elite athlete is certainly feasible. During peak activity, all of the muscle cell's ATP could be consumed in about 2 s if it were not regenerated.

As we will see in the next section, ATP in cells is regenerated from ADP through the breakdown of fuel molecules (primarily carbohydrate and fat) using aerobic or anaerobic catabolic processes. However, ATP-regenerating processes cannot produce ATP at the same rate at which it is hydrolyzed to drive muscle contraction during sprinting. Moreover, these processes take time to gear up to maximum speed, whereas at the start of a sprint, the rate at which ATP is hydrolyzed is about maximal. To prevent muscle cells from using up their ATP at the start of maximal or near-maximal contractions, an alternate energy-rich molecule, known as a **phosphagen**, is capable of regenerating ATP at a very high rate. In vertebrate muscle, the phosphagen is **phosphocreatine** (abbreviated PCr), also called creatine phosphate (abbreviated CP). In some invertebrate muscles, the phosphagen is arginine phosphate. In humans, 92% to 96% of the body's total PCr is found in skeletal muscle; the remainder is in cardiac muscle, brain, and testes.

> **✓ KEY POINT**
>
> Relative to ATP levels, muscle has a great deal of creatine kinase. It is found near the contractile proteins, where ATP is hydrolyzed. It is found with the sarcoplasmic reticulum, at the level of the mitochondrial membranes, and free within the muscle cell's cytoplasm. In short, creatine kinase is found at the places where ATP is both consumed and produced.

Phosphocreatine (or creatine phosphate) has its phosphate group transferred to ADP to yield ATP and creatine (Cr), in a reaction catalyzed by an enzyme known as creatine kinase:

$$ADP + PCr + H^+ \leftrightarrow ATP + Cr$$

This reaction is freely reversible. During muscle contraction, the forward direction is favored in order to regenerate ATP. During recovery, when PCr is much reduced and free Cr is elevated, the backward reaction is favored to regenerate PCr. Because a little more energy is released when PCr is hydrolyzed, compared to when ATP is hydrolyzed, the reaction just presented is tilted a little more to ATP formation (and thus PCr disappearance) during rapid rates of ATP hydrolysis. Note that a proton (H^+) is consumed as a substrate when the creatine kinase reaction proceeds toward ATP formation. This is due to the acid–base characteristics of the phosphate groups. Consumption of a proton is advantageous, since it can partially reduce the acidification of muscle during very vigorous exercise. We note this later in this chapter. The actual concentration of PCr in muscle is about three or four times that of ATP (about 18-20 mmol per kilogram of muscle). This is not that much, considering how fast ATP can be used in very intense muscle activity. However, the extremely high activity of creatine kinase ensures that ATP is regenerated almost as fast as it is broken down near the beginning of sprint-type activities. Although limited in quantity, there is enough PCr to act as a temporary ATP buffer until other ATP-regenerating processes reach maximal rates and to allow for brief periods of exercise at intensities that may exceed the rates of the other ATP regeneration processes. In general, both chemical analysis and the use of phosphorus nuclear magnetic resonance spectroscopy reveal that the content of PCr and ATP is higher in muscle or muscle fibers with the highest rates of ATP hydrolysis. The ATP level in muscle must not be allowed to drop to critical levels. If it does, a condition known as rigor occurs, which can be damaging to muscle cells. In fact, rigor mortis is caused by the loss of muscle ATP some time after death, as a result of the inability of the muscle cells to regenerate ATP. This results in a lack of ATP to bind the myosin head and to allow it to release from actin, thereby preventing the thick and thin filaments from sliding past each other, and resulting in muscle rigor.

In skeletal muscle, the rate of activity of creatine kinase exceeds that of all other enzymes. This means that the creatine kinase reaction is very important in muscle for sprinting or burst-type activities, such as bounding or maximum-force weightlifting. Animals in which the gene for creatine kinase is knocked out demonstrate very poor performance during high-intensity exercise. Because skeletal muscle contains more creatine kinase than any other enzyme, damage to a muscle-cell membrane, typically through unaccustomed exercise or hard eccentric exercise, results in creatine kinase leaking from the muscle into the blood. Damage to the heart muscle can also result in the appearance of the cardiac isozyme of creatine kinase in blood. Determination of the activity and isozyme type in blood can reveal something of the nature and extent of damage to cardiac and skeletal muscle.

☑ KEY POINT

Cardiac muscle contains creatine kinase as a dimer that includes both an M subunit and a B subunit. Damage to cardiac muscle membranes can involve loss of the MB isozyme to the extracellular fluid, including blood. Measurement of the MB isozyme has been used to determine that damage to the heart has occurred. Damage to skeletal-muscle membranes results in elevated levels of the MM isozyme in blood. Muscle often sustains minor damage during intense or unaccustomed exercise. Measuring changes in blood levels of creatine kinase activity provides a crude estimate of the degree of muscle membrane disruption, as well as its rate of repair.

ENERGY SYSTEMS

Cells use ATP or related nucleoside triphosphates, such as GTP and UTP, to power everything that requires an input of energy. This can range from transporting substances into the cell or across internal membranes to signaling information to the interior of the cell, storing fuels, and synthesizing protein and RNA. Muscle cells use ATP for these roles, but during contractile activity, ATPase enzymes become dominant in the breakdown of ATP. As previously noted, the ATPases are actomyosin ATPase, the ATPase associated with the sarcolemma to pump sodium and potassium ions, and the ATPase that pumps calcium ions back into the SR (SERCA). The following equation shows the hydrolysis of ATP in complete detail:

$$MgATP^{2-} + H_2O \rightarrow MgADP^- + n\ H_2PO_4^- + 1\text{-}n\ HPO_4^{2-} + 1\text{-}n\ H^+$$

In the cell, all of the ATP is associated with magnesium ions. It is believed that much of the ADP is also in a complex with magnesium ions. The other product of ATP hydrolysis, which we have previously simply described as Pi, is actually a mixture of the dihydrogen ($H_2PO_4^-$) and monohydrogen phosphate (HPO_4^{2-}) ions. This is the case because the dihydrogen phosphate ion is a weak acid with an acid-dissociation constant (pK_a) just below 7. This means that at the pH of a typical cell (approximately 7), Pi is a mixture of the acid ($H_2PO_4^-$) and its conjugate base (HPO_4^{2-}) in almost equal proportions. It also means that for every monohydrogen phosphate produced by ATP hydrolysis, a proton (H^+) is generated so that ATP hydrolysis has an acidifying effect in a cell at neutral pH.

We have emphasized that there is not a lot of ATP in any cell and that, therefore, when ATP is used, it must be immediately regenerated at a rate that is as close as possible to the rate of its utilization. In skeletal and cardiac muscle, three energy systems are responsible for maintaining ATP concentrations. These are illustrated in figure 4.8 and can be identified as the phosphocreatine system, oxidative phosphorylation, and anaerobic glycolysis. In most cell types, oxidative phosphorylation is dominant in ATP regeneration. In erythrocytes or red blood cells, the only source for ATP regeneration is glycolysis. Skeletal and cardiac muscles use all three systems. In the next sections, we will briefly consider these energy systems. The determination of which systems to use and the extent to which their rates of ATP regeneration are utilized are intimately tied to the rate of ATP hydrolysis and, hence, the intensity of

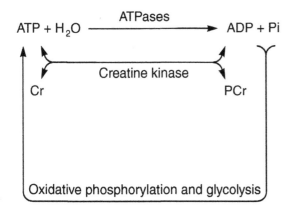

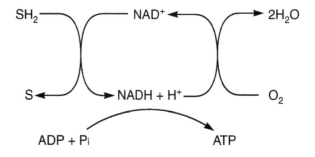

Figure 4.8 Three ATP-hydrolyzing enzymes (ATPases) in muscle, able to use ATP at a rapid rate. The ATP concentration is maintained, or diminishes only slightly, due to the rapid response of the creatine kinase reaction and the aerobic (oxidative phosphorylation) and anaerobic (glycolysis) breakdown of fuel molecules. Although the creatine-kinase reaction is freely reversible, during muscle activity, it is driven in the direction to regenerate ATP. During recovery, the reverse reaction is favored to make PCr.

Figure 4.9 A brief summary of oxidative phosphorylation. Electrons in the form of hydrogen atoms are removed from reduced substrates (SH_2) and transferred to the oxidized coenzyme NAD^+ to make it NADH. Through a series of carriers, electrons on NADH are transferred to oxygen, creating water. Energy released during the transfer of electrons from SH_2 is harnessed to phosphorylate ADP with Pi to make ATP.

exercise at any given time. Numerous controlling and signaling pathways are involved in this regulation and selection of ATP resynthesis pathways. Detailed discussions of oxidative phosphorylation and anaerobic glycolysis are presented in chapters 5 and 6, respectively.

Oxidative Phosphorylation

Oxidative phosphorylation is the fancy name for a process that many people know as the aerobic system or the fuel oxidation system. Biologists often call it cellular respiration, or simply respiration. The term *oxidative phosphorylation* is becoming more dominant because it describes the overall process better. We can define this process as the formation of ATP from ADP and Pi in association with the transfer of electrons from fuel molecules to coenzymes to oxygen. Products of oxidative phosphorylation are H_2O and CO_2, as well as ATP.

Figure 4.9 outlines the overall scheme of oxidative phosphorylation. In this scheme, SH_2 represents fuel molecules, such as carbohydrate, fat, amino acids, or alcohol. Electrons associated with the hydrogen atoms are transferred from SH_2 to coenzymes (represented by nicotinamide adenine dinucleotide, NAD^+), and then the electrons on the now-reduced coenzyme (NADH) are transferred on to oxygen, forming H_2O. During this process, enough energy is generated and captured to phosphorylate ADP with Pi to make ATP. In chapter 2, electron transfer using dehydrogenase enzymes was illustrated. We devote the next chapter to the steps involved in oxidative phosphorylation. For the present, we can summarize it using figure 4.9.

The following equation is based on the figure:

$$2\,SH_2 + O_2 + 5\,ADP + 5\,Pi \rightarrow 2\,S + 2\,H_2O + 5\,ATP$$

This very simple equation illustrates that the transfer of electrons to oxygen in the form of H atoms on fuels reduces the oxygen to water, and the energy is used to make ATP. As a summary equation, it does not show the role of the coenzymes NAD^+ or flavin adenine dinucleotide (FAD) during oxidation and reduction, which is discussed in chapter 2. Although it does reflect the true stoichiometry (the ratio in which molecules react with each other) of ATP production, it does not reveal that during oxidative phosphorylation, CO_2 is produced from the carbon atoms of fuel molecules, represented by SH_2.

Oxidative phosphorylation requires fuel molecules that we obtain from our diets, principally in the form of carbohydrate and fat. The oxygen comes from the air we breathe. Oxygen enters the lungs, diffuses to hemoglobin molecules in the blood, and is transported throughout the body using large and small arteries and arterioles, finally reaching every cell by way of capillaries. Oxygen diffuses from capillaries to cells to the mitochondria of cells, where it accepts electrons to form water and ATP.

It is fairly easy to measure the rate of oxidative phosphorylation by measuring the disappearance of the substrate oxygen. Indeed, this is the method employed by exercise physiologists when they talk about **oxygen consumption**, measured at the mouth with special breath-sampling techniques. The resulting $\dot{V}O_2$ is used as an index of whole-body metabolism. During exercise, $\dot{V}O_2$ increases in proportion to the rate of whole-body energy expenditure. With gradually increasing loads of exercise (for example, treadmill running or pedaling on a cycle ergometer), $\dot{V}O_2$ reaches a maximum or peak that is considered to be the highest rate of oxidative phosphorylation for the individual during that activity. If a large enough muscle mass is activated, this peak rate of oxidative phosphorylation is the maximal rate the individual can achieve, or the $\dot{V}O_2$max. It can be described in absolute units, liters of oxygen consumed per minute, or milliliters of oxygen per kilogram of body weight per minute.

For many activities, oxidative phosphorylation can provide all of the ATP needed by active skeletal muscles. When exercise level and $\dot{V}O_2$ are constant over time, oxidative phosphorylation is providing all of the ATP needed by the activity. Such a condition is known as a metabolic **steady state**. From a simple bioenergetics perspective, we can use the approximate relationship in which a $\dot{V}O_2$ of 1.0 L per minute is equivalent to an energy expenditure of about 5 kcal per minute (21 kJ/min). The precise relationship between $\dot{V}O_2$ in liters per minute and energy expenditure in kilocalories per minute depends primarily on the relative proportions of carbohydrate and fat that are being catabolized. The value is slightly higher for pure carbohydrate oxidation (about 5.047 kcal or 21.2 kJ) compared to pure fat (about 4.69 kcal or 19.7 kJ). See chapters 6 and 7 for details regarding fat and carbohydrate oxidation and respiratory quotient. Oxidative phosphorylation is considered a low-power process because it relies on oxygen from the atmosphere as the final acceptor of electrons in fuels. With the steps in oxygen transport from the air, to the lungs, to the blood, to the individual cells, and finally to the mitochondria within cells, oxidative phosphorylation cannot quickly reach a maximal rate. Indeed, the time needed to double the rate of oxidative phosphorylation is approximately 15 to 20 s. Although its power is limited in most cases by the oxygen transport system, oxidative phosphorylation is considered to have a high capacity (i.e., large fuel tank), because a major fuel for oxidative phosphorylation is fat. Even a very lean person has enough stored fat to provide fuel for several days of oxidative phosphorylation.

Glycolysis

Glycolysis is a **metabolic pathway**, a sequence of enzyme-catalyzed reactions in which a starting substrate is converted into a product after undergoing a number of steps. From the perspective of muscle, there are two starting substances for the glycolytic pathway, glucose and glycogen. **Glycogen** is simply a polymer consisting entirely of glucose units joined together. It is readily available in muscle and can quickly be used to fuel glycolysis. The products of glycolysis are the same whether the starting substrate is glucose or glycogen. The only difference is that a glucose unit from glycogen produces one more net ATP than a free glucose unit. The following are summary reactions for glycolysis, beginning with glucose and then continuing with glycogen.

$$\text{glucose} + 2 \text{ ADP} + 2 \text{ Pi} + 2 \text{ NAD}^+$$
$$\rightarrow 2 \text{ pyruvate} + 2 \text{ ATP} + 2 \text{ NADH} + 2 \text{ H}^+$$

$$\text{glycogen}_n + 3 \text{ ADP} + 3 \text{ Pi} + 2 \text{ NAD}^+$$
$$\rightarrow \text{glycogen}_{n-1} + 2 \text{ pyruvate} + 3 \text{ ATP} + 2 \text{ NADH} + \text{ H}^+$$

For the second reaction, we consider glycogen to consist of n glucose units. When one of these is removed from glycogen and converted to pyruvate, what is left is shown as glycogen$_{n-1}$. The ATP produced can be used to drive any energy-requiring process. The NADH is a source of electrons to feed into oxidative phosphorylation, as the next chapter discusses. The pyruvate formed in glycolysis has two major fates: (1) Pyruvate can be transported into a mitochondrion and be completely oxidized using the pathways of oxidative phosphorylation; (2) pyruvate can be converted to lactate as the next reaction illustrates:

$$\text{pyruvate} + \text{NADH} + \text{H}^+ \leftrightarrow \text{lactate} + \text{NAD}^+$$

This latter reaction is catalyzed by *lactate dehydrogenase*, abbreviated LDH. We describe the LDH reaction in chapter 2 as an illustration of a redox reaction. Going from left to right, pyruvate is being reduced and NADH is being oxidized. The double-

> ### ☑ KEY POINT
> Oxidation of carbohydrate produces more ATP per unit of oxygen consumed than does the oxidation of fat. In addition, the maximum rate of electron transport from fat to oxygen is approximately one-half the rate reached when carbohydrate is the sole fuel. These facts explain why carbohydrate is the preferred fuel for exercise at intensities beyond 50% of maximal oxygen consumption.

headed arrow reminds us that the reaction can also go from right to left. What actually happens—that is, the net direction—depends primarily on the relative concentrations of the various constituents on the left and right sides of the equation, but also on the type of LDH isozyme.

When the pyruvate from glycolysis is converted to lactate, the overall process from glycogen (or glucose) to lactate is referred to as *anaerobic glycolysis*. The word *anaerobic* is used because ATP is formed without the need for any oxygen. When the pyruvate in glycolysis enters mitochondria and gets completely broken down to carbon dioxide and water, the process is often termed *aerobic glycolysis* because the final breakdown of pyruvate does require oxygen. Oxidation of pyruvate is a rich source of ATP; two pyruvates can be obtained from each glucose unit, and the complete oxidation of two pyruvates can produce 25 ATP. We look at the oxidation of pyruvate in the next chapter.

The net production of ATP from glycolysis and PCr is known as **substrate-level phosphorylation**, in contrast to the production of ATP by oxidative phosphorylation. The enzymes catalyzing the reactions of glycolysis producing pyruvate from either glucose or glycogen have generally high activities in skeletal muscle. Thus, pyruvate can be generated at a considerably faster rate than it can be completely oxidized in mitochondria. Therefore, when glycolysis is proceeding at a rapid rate in a vigorously contracting muscle, much of the pyruvate formed will be reduced to lactate because it is formed at a rate exceeding the capacity of mitochondria to oxidize it, even if there is plenty of oxygen available. For this reason, we say that anaerobic glycolysis has a moderate power to generate ATP. As we will see, the power of glycolysis in fast-twitch muscle fibers is greater than in slow-twitch muscle fibers because there is a richer concentration of enzymes of glycolysis in the former. The capacity of glycolysis to produce ATP is moderate, much larger than the PCr system, but much less than that of oxidative phosphorylation. The moderate capacity may be related to the availability of glycogen to feed into the glycolytic pathway. A more likely explanation relates to the fact that skeletal muscle can become fairly acidic when high rates of ATP hydrolysis are accompanied by very rapid rates of anaerobic glycolysis. Unlike oxidative phosphorylation, which may take 2 to 3 min to reach its peak rate at the onset of exercise, glycolysis can be quickly started at the beginning of exercise. If the exercise is intense enough, the glycolytic pathway can reach a maximum rate in 5 to 10 s. The quick start for glycolysis is not as rapid as the onset of ATP regeneration from PCr, but it makes an important contribution to ATP formation at the beginning of intense muscle work.

As noted in chapter 2, lactic acid has a pK_a of approximately 3.8. This means that at physiological pH, it will dissociate a hydrogen ion. Lactic-acid production during intense exercise has been commonly associated with hydrogen-ion accumulation and a drop in pH, which has also been associated with muscle fatigue. This concept has been more recently challenged, since evidence is mounting that a significant portion of hydrogen-ion generation during intense exercise may be due to increased ATP breakdown to ADP and Pi, which also results in hydrogen-ion generation (Robergs, Ghiasvand, and Parker 2004). Nevertheless, lactic-acid generation may still contribute to ischemia-induced acidosis (Marcinek, Kushmerick, and Conley 2010). Lactate formation during exercise is related more to glycolysis and glycolytic production of pyruvate and $NADH^+ H^+$ (at rates that, during more intense exercise, exceed the ability of mitochondria to process these products of glycolysis) than to a lack of oxygen available to muscle. The formation of lactate by the donation of H^+ ions from $NADH^+H^+$ allows for the now reformed NAD to continue facilitating an intermediary step in glycolysis that could not occur without NAD to accept H^+ ions. The benefit of being able to produce lactate during intense exercise is that it allows for a continuation of high rates of glycolytic ATP resynthesis that otherwise could not occur. This allows us to exercise at a higher intensity for a longer period of time. Lactic-acid formation in combination with increased ATP turnover during intense exercise is associated with a drop in pH and muscle acidosis, which is a factor in muscle fatigue. However, lactate formation itself may not be the cause of this drop in pH. Nevertheless, the benefits of higher rates of ATP resynthesis outweigh the potential negative effects of acidification in the muscle in the short term. The mechanisms and implications of these events are discussed in more detail in chapter 6.

✓ KEY POINT

Since ADP is a substrate and allosteric activator of glycolytic enzymes, glycolysis will be turned on whenever the rate of ATP hydrolysis and, consequently, increases in ADP concentration in a fiber abruptly increase. For low-intensity exercise, little lactate formation takes place, but for higher rates of exercise, the concentration of lactate in muscle and blood increases rapidly.

Phosphocreatine System

Earlier in the chapter, we discussed the phosphagen PCr as a readily available, energy-rich molecule capable of regenerating ATP, as seen in the following equation:

$$ADP + PCr + H^+ \leftrightarrow ATP + Cr$$

Use of PCr to regenerate ATP is sometimes referred to by exercise physiologists as the *anaerobic alactic system* because it does not need oxygen and does not generate lactate. At the onset of very vigorous muscle activity, creatine kinase drives this reaction to the right, depleting PCr, maintaining ATP, and increasing creatine. The consumption of a proton (H^+) as PCr concentration decreases can be beneficial to the muscle during high-intensity exercise or games when high rates of ATP hydrolysis acidify the muscle.

Four genes for creatine kinase (CK) are present in mammals, coding for four protein monomers of about 40 kD each. Two of these monomers are designated B (for brain) and M (for muscle), and there are two mitochondrial subunit forms. Of the mitochondrial subunits, one is expressed primarily in heart and skeletal muscle and the other in brain and other tissues. The mitochondrial creatine kinase subunit is involved in the facilitation of oxidative phosphorylation, a topic discussed in chapter 5. The nonmitochondrial form of creatine kinase in skeletal muscle is active as a dimer, containing two M subunits that can be represented as MM. This is found in the cytosol attached to contractile filaments, at the inner side of the sarcolemma, and at the outer face of the SR membrane, as well as free in the cytosol during exercise. These locations ensure that creatine kinase is appropriately positioned in sites where ATP is hydrolyzed and regenerated.

Because creatine kinase activity is so high, it can maintain ATP levels remarkably well even during intense exercise. This ATP-buffering effect is substantial. For this reason, we can say that creatine kinase has a high power for regenerating ATP. The actual PCr concentration in resting human skeletal muscle, based on analysis of muscle biopsy samples, is approximately 18 to 20 mmol per kilogram wet weight of muscle (or 23-26 mM). No matter how it is expressed, supply of PCr is limited. Therefore, the CP system has a low capacity. During exercise, PCr levels fall in proportion to the relative intensity of the exercise, and the free creatine concentration rises in parallel. For all-out efforts to fatigue, PCr levels can decrease by 90% or more.

During recovery or rest periods, the reverse of the creatine kinase reaction dominates, and phosphate transfer to creatine from ATP produced by oxidative phosphorylation regenerates PCr. During recovery from intense exercise, PCr is rapidly resynthesized, such that muscle PCr stores can be more than 50% resynthesized following 30 s of recovery and more than 90% resynthesized within 2 min of recovery (see figure 4.10). Since resynthesis of PCr is a reversal of the breakdown reaction, the reaction now consumes ATP rather than generating it. Thus, PCr resynthesis during the initial stages of recovery from intense exercise must be linked to ATP resynthesis. During recovery from exercise, ATP resynthesis to power Cr phosphorylation comes from aerobic metabolism. Hence, some of the excess oxygen consumed during recovery is used to provide ATP for PCr resynthesis. This elevated oxygen consumption following exercise is called excess postexercise oxygen consumption, or **EPOC** (see figure 4.11). Oxygen consumption at the end of a bout of exercise does not immediately return to resting levels but instead can remain elevated for several minutes, or even as long as several hours, following exercise termination. The excess oxygen required to generate the ATP for PCr resynthesis accounts for much of the elevated oxygen consumption in the first few minutes after exercise cessation. The longer term elevation of oxygen consumption is accounted for by other factors related to elevated postexercise metabolism, which are due to hormonal stimulation and elevated body temperature and to lesser lactate use for glucose synthesis. Lactate metabolism is discussed further in chapter 6.

> ### ✓ KEY POINT
> The rapid rate of PCr resynthesis during recovery from intense exercise allows for rapid recovery of this high energy store in muscle. This ability to rapidly restore PCr during recovery can facilitate intense exercise training, which uses short rest intervals, followed by short, intense exercise. These short, intense training bouts (as seen, for example, in sprint-interval training in swimming or running) are highly dependent on repeated PCr use during the exercise and rapid PCr resynthesis during recovery by muscles. Repeated high-intensity training also leads to adaptations that increase the rate of PCr resynthesis during recovery.

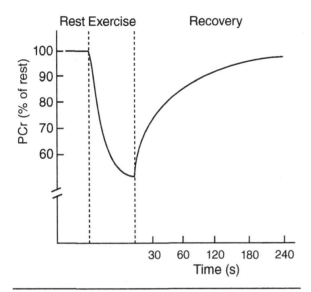

Figure 4.10 Following partial depletion of phosphocreatine (PCr) stores by intense exercise, PCr stores are rapidly restored during recovery. The restoration of PCr levels is facilitated by ATP generated aerobically during recovery. This elevates the need for oxygen consumption during the first few minutes of postexercise recovery. PCr stores are restored rapidly such that 50% of PCr is restored within the first 30 s of recovery.

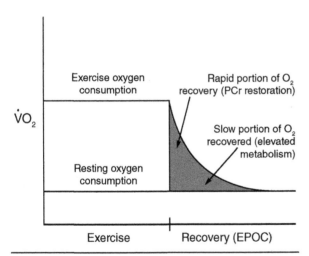

Figure 4.11 Excess postexercise oxygen consumption (EPOC) is utilized primarily to generate ATP for restoration of muscle phosphocreatine levels during the initial few minutes following intense exercise. Elevated metabolism, due partly to elevated epinephrine levels, can factor into prolonging EPOC beyond the first few minutes of recovery.

Properties of Creatine

PCr doubtlessly plays a very important role at the start of exercise and during sprint-type activities. Since the early 1900s, it has been known that feeding people extra creatine leads to retention of some of this creatine, presumably in the muscles. Indeed, Chanutin (1926) suggested that the retained creatine could have an effect on increasing body-protein content. The issue of creatine as a supplement to improve exercise performance and muscle mass did not approach the significance it now has until the early 1990s. Researchers in Sweden and the United Kingdom initially published a number of articles suggesting that supplemental creatine could increase muscle PCr levels and improve performance in exercise activities. Additional research followed in many other laboratories and countries, using a variety of human and animal feeding and exercise models. Today research on creatine may not be the hot topic it once was, but many serious exercise trainers consider supplementing with creatine to be a necessity.

To understand this interesting topic, it is necessary to learn something about creatine. Humans get much of their creatine in a two-step sequence of reactions that takes place primarily in the liver, using the amino acids glycine and arginine. We also know that creatine can be obtained in significant quantities by those who consume creatine-containing tissues—that is, meat. As a simple, water-soluble molecule, creatine is easily absorbed. In the blood, absorbed creatine would be indistinguishable from the creatine made by the liver. The average person needs only about 2 g of creatine a day to replace losses, an amount that can be obtained through biosynthesis or a combination of biosynthesis and dietary sources. The usual supplemental regimen is 20 to 30 g per day for about six days, followed by maintenance doses of 3 to 5 g per day. Some studies have demonstrated that carbohydrate ingestion in addition to the creatine supplement enhances extra creatine retention. Retention of supplemental creatine is also increased with exercise training, compared to inactivity.

An overall snapshot of creatine is shown in figure 4.12. Creatine in the blood, whether absorbed from food or synthesized by the liver, is generally in the concentration range of 25 to 100 μM. The lower end would likely be seen in those who are vegetarians and the upper range in those who consume a diet with plenty of meat. With large supplemental doses, blood creatine concentrations can be increased by more than 10-fold. Creatine concentration inside muscle fibers at rest is generally in the 7 to 8 mM range, so transport of creatine into a muscle fiber is an uphill process, going against a significant concentration gradient. A specific *creatine transporter* allows creatine to enter a cell, driven by the simultaneous transport of sodium and chloride ions down their significant concentration gradients. Once inside the muscle fiber, much of the creatine is phosphorylated using ATP and the creatine kinase enzyme. Loss of creatine from cells occurs when either free creatine or CP undergoes nonenzymatic loss of a water molecule, producing a new molecule, *creatinine*. Creatinine is released from cells, filtered from the blood by the kidneys, and excreted in the urine. The daily loss of body creatine due to formation of creatinine is roughly 1.7% of the total body amount, or about 2 g per day.

The total concentration of creatine in a muscle fiber (TCr) is the sum of the concentrations of PCr and Cr. Mean values for creatine, PCr, and total creatine in muscle are shown in table 4.2. These values are based on chemical analysis of skeletal muscle biopsy samples in human subjects at rest and before and after a program of creatine supplementation. On the basis of data from table 4.2, supplementa-

> ## ✓ KEY POINT
>
> Since the preponderance of all creatine is found in muscle, the excretion of creatinine each day should reflect skeletal muscle mass. Indeed, the ratio of 1 g of creatinine excreted per 20 kg of muscle is commonly accepted. Creatinine excretion rates can be used as a marker of kidney function. It should be noted that the rate of creatinine excretion can be elevated in people taking dietary creatine supplementation.

tion with creatine leads to a 16% increase in total creatine and an 11% increase in PCr. However, data from a variety of studies reveal that the amount of extra TCr retained can range from 10% to 30%. Although there are significant variations in TCr and the ratios of PCr to TCr in the literature, generally, we can say the following:

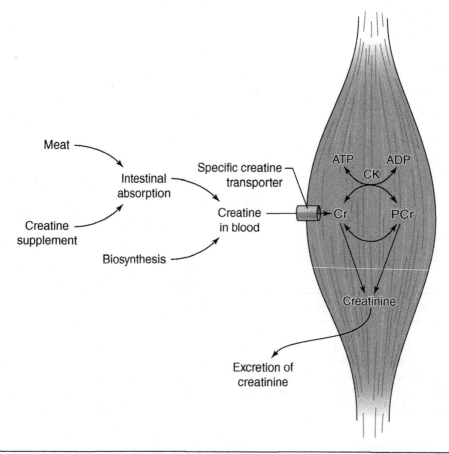

Figure 4.12 A summary of the sources of creatine and creatine uptake and metabolism in skeletal muscle. Creatine is synthesized in the body, obtained from dietary meat or through supplements. Creatine from all sources is taken up from the blood against a concentration gradient through a specific creatine transporter that also lets in Na^+ and Cl^- ions that move down their concentration gradients. In the cell, about 60% to 80% of creatine is phosphorylated from ATP to make creatine phosphate. About 2 g per day (1.7%) of cellular creatine or creatine phosphate undergoes nonenzymatic conversion to creatinine, which is excreted in the urine.

Table 4.2 Creatine (Cr), Phosphocreatine (PCr), and Total Creatine (TCr) Concentrations in Muscle

Condition	Cr (mmol/kg dw)	PCr (mmol/kg dw)	TCr (mmol/kg dw)	PCr/TCr
Presupplementation	46.4 ± 8.1	76.3 ± 7.1	122.7 ± 7.8	0.62
Postsupplementation	57.1 ± 7.9	84.7 ± 6.2	141.8 ± 6.4	0.60

Values are based on data from 10 selected studies published between 1994 and 2003 and are reported as means ± SD. The PCr/TCr ratio was calculated from the mean values for PCr and TCr. dw = dry weight.

1. The PCr/TCr ratio in rested muscle is generally in the range of 0.6 to 0.7.
2. The differences in total TCr and the ratio of PCr to TCr in human skeletal muscles are less pronounced compared to the values in animal muscles.
3. Lower TCr and PCr values tend to be found in those eating a meat-free diet.
4. Concentrations of TCr and PCr in individual Type II muscle fibers that are isolated from human skeletal muscle biopsy samples tend be about 20% higher than in Type I muscle fibers.

Effect of Creatine Supplementation on Exercise

Research data show that creatine supplementation does increase PCr and total creatine concentrations in skeletal muscles of most subjects, but does this lead to an improvement in exercise and sport performance? Hundreds of published studies have looked at the effects of creatine supplementation on human exercise and sport performance, and some conclusions can be made on the basis of these reports. Creatine supplementation appears to be most effective in short-term, high-intensity exercise lasting up to 3 min. Creatine appears to be especially helpful if the high-intensity activity is repeated with only a brief recovery period. We would expect results such as this because as previously noted, the half-time for restoration of PCr levels after an intense bout of exercise is approximately 30 s (see figure 4.10). Therefore, starting out with a higher PCr concentration before exercise and a higher Cr concentration after the first exercise bout would seem to be advantageous to performance on the next bout. Many athletes report that they can train harder in the gym or on the track with creatine because they recover faster after each set or repeat.

Many advocates of resistance training also believe that creatine increases body weight and strength gains. Such beliefs might suggest that creatine increases rates of muscle protein synthesis, either as an independent effect of creatine or because, as just noted, athletes can train harder. Studies by Tarnopolsky and his group demonstrated that creatine supplementation can upregulate the expression of a variety of skeletal muscle genes, especially those involved in intracellular signaling (Safdar et al. 2004). Additionally, creatine supplementation has been demonstrated to enhance activation and proliferation of muscle satellite cells and, consequently, increase muscle myonuclei in weight-training humans (Olsen et al. 2006). This study demonstrated that greater muscle-mass accumulation occurred in conjunction with creatine supplementation, even when the intensity or quantity of weight training was not different between the supplemented and placebo groups. It was concluded that creatine supplementation stimulated additional muscle hypertrophy consequent to weight training by inducing early stimulation of muscle satellite cells and by possibly upregulating certain myogenic regulatory factors (MRFs) such as myogenin and MRF-4, which can stimulate expression of myosin heavy chains at the transcriptional level.

Some of these effects of creatine may be relevant to maintaining strength and function in the elderly. When people age, they lose muscle mass. Specifically, this age-related loss of muscle mass, called **sarcopenia**, differs from muscle atrophy in that muscle fibers (or muscle cells) are actually lost through cell death as people age. Sarcopenia can result in a significant decline in functioning ability and mobility as old age progresses. Some data are now emerging that dietary creatine supplementation in conjunction with resistance training in the elderly can partly counteract this age-related loss of muscle mass.

Another side of creatine supplementation exists, but its long-term effects are not well known. Taking 10 times or more of the daily creatine allotment has been shown to decrease both the biosynthetic pathway for creatine in the liver and the expression of creatine transporters required for creatine uptake into cells (Snow and Murphy 2003). However, to date, no negative side effects of creatine supplementation as it is typically done have been demonstrated.

> ## ✓ KEY POINT
>
> Creatine supplementation can increase high-intensity exercise performance and training performance in many individuals; however, it will only do so in conjunction with an intense training regimen. Taking creatine supplements without training will not result in significant improvements in athletic performance. Creatine supplementation may also stimulate greater muscle hypertrophy in response to weight training by stimulating greater activation and proliferation of muscle satellite cells. This can be very important for athletic training and performance. Even more important is the ability of dietary creatine supplements to augment training-induced muscle hypertrophy, which may have significant implications for helping older people retain muscle mass and counteract sarcopenia during aging.

Role of Adenylate Kinase and AMP Deaminase

When we consider muscle metabolism during exercise, two other reactions must be taken into account. The first is one we have seen previously. It is written here in the direction opposite to the way it was shown earlier in this chapter, but this does not matter because the reaction is freely reversible with an equilibrium constant near 1.

$$2 \text{ ADP} \leftrightarrow \text{ATP} + \text{AMP}$$

The enzyme for this reaction is named *adenylate kinase*, but you will also see it identified as adenylyl kinase. It has also been called *myokinase* (*myo* for muscle) because its activity is so high in skeletal muscle. The significance of this reaction is as follows. During hard muscle activity, the rate of ATP hydrolysis is high. Although muscle efficiently regenerates ATP from ADP by the three energy systems previously described, an increase in ADP concentration occurs. What the adenylate kinase does is to cause two ADP molecules to interact, transferring a phosphate from one to the other, creating one AMP and one ATP. This interaction keeps the ADP concentration from building up to the extent it would otherwise. While this reaction also results in the production of ATP, the amount of ATP produced is minimal relative to other sources of ATP regeneration during exercise. In addition, forming an ATP from two ADP is advantageous during exercise, but the AMP also has an important role in stimulating glycolysis, as chapter 6 shows.

In the second reaction, *AMP deaminase* (also called adenylate deaminase) irreversibly deaminates (removes the amino group from) the base adenine of AMP, producing ammonia, NH_3. Deaminated AMP becomes *IMP* (inosine monophosphate).

$$\text{AMP} + H_2O \rightarrow \text{IMP} + NH_3$$

Because NH_3 is a weak base, it takes a proton (H^+) from the medium, becoming ammonium ion (NH_4^+). The ammonium ion formed during the AMP deaminase reaction can leave the muscle cell and travel in the blood to the liver and kidney for further metabolism. Adenosine monophosphate deaminase is found in higher levels in the fast-twitch (Type II) muscle fibers. These fibers are more active during intense exercise activity. The activity of AMP deaminase is low at rest but increases due to changes within a muscle fiber under vigorous exercise conditions, such as the fall in muscle pH and rise in ADP concentration. In a rested muscle cell, most of the AMP deaminase is free in the sarcoplasm, but increasing contractile activity facilitates AMP deaminase binding to myosin in the thin filaments. This myosin binding increases AMP deaminase activity.

Adenylate kinase and AMP deaminase can work in concert during vigorous muscle activity. In effect, two ADP molecules are converted into an ATP and an AMP by the AMP kinase reaction. Then in the AMP deaminase reaction, the AMP is irreversibly converted into IMP. These related reactions play a major role in maintaining optimal energy status in the muscle fiber, especially during intense muscle work. The irreversible AMP deaminase reaction drives the reversible adenylate kinase reaction to the right, to diminish the increase in the concentration of ADP ([ADP]). This reaction also maintains a high [ATP]/[ADP] ratio, which is absolutely important to maximize the free energy release of ATP hydrolysis, and is therefore essential for driving the ATPase reactions associated with muscle contraction. Since the AMP deaminase reaction is increased by a decrease in muscle pH, we would expect this reaction to be dominant in fibers with a high rate of ATP hydrolysis, a high capacity for glycolysis, and a lower capacity for oxidative phosphorylation. Not surprisingly, Type II fibers are the primary source of blood lactate and ammonia during exercise. The two reactions involving AMP are very important in muscle under certain exercise conditions, as evidenced by the fact that people with a genetic AMP-deaminase deficiency have a poor ability to perform intense exercise, experiencing muscle pain, cramping, and early fatigue during exercise attempts.

The product, IMP, which accumulates in muscle during severe exercise, has several fates in the postexercise period. Most of the IMP is converted back to AMP during a two-step process, part of the *purine nucleotide cycle* that is examined in chapter 8. Some of the IMP can undergo a loss of its phosphate group to form *inosine*. Inosine can lose its ribose group to form *hypoxanthine*. Without phosphate groups, inosine and hypoxanthine can leave the muscle fiber. Following hard exercise, both inosine and hypoxanthine can be found in blood, draining active muscle.

Muscle Metabolism in Exercise

When we ask our muscles to do something for us, whether it is getting out of a chair, walking to the store, doing a set of resistance exercises, or running a 10 K race, we are placing a specific energy demand on the muscle. Activated by our nervous system, individual fibers contract at certain rates (or perhaps not at all), hydrolyzing ATP. During low-level activity, the fibers that are active tend to be Type I, with a high capacity for oxidative phosphorylation. As one increases the demand on a muscle, more muscle fibers are brought into play, including Type II. At the highest levels of force demand, almost all muscle fibers are active. High levels of force can be maintained only for brief periods, since some large Type II fibers, with a high actomyosin ATPase activity and glycolytic capacity but a low ability for oxidative phosphorylation, fatigue easily and thus can only participate for brief bursts.

Through the energy systems already described, the ATP used is regenerated as fast as possible. Nonetheless, changes occur within the muscle fiber that reflect its ability to regenerate ATP, as well as its metabolic displacement from the rest condition. As previously emphasized, the rate of ATP utilization drives the rate of ATP regeneration as the muscle tries to the best of its ability to match ATP demand with ATP regeneration rate during various exercise intensities. The concentrations of ADP and AMP generally increase, anaerobic glycolysis can produce lactate, and PCr is used to make ATP. We may actually lose some adenine nucleotide (ATP, ADP, and AMP) if IMP is formed. We call these molecules *metabolites*, and how their concentrations differ from those at rest can tell us a lot about the previous contractile activity. For example, glycogen is an important fuel for anaerobic glycolysis. It is also a primary source of pyruvate that is oxidized in mitochondria. We would expect glycogen concentration to decrease in proportion to how much it is used and how it is used (anaerobic vs. aerobic glycolysis) because the anaerobic pathway uses glycogen at 10 times the rate of the aerobic pathway to get the same amount of ATP. Inorganic phosphate (Pi) will increase if there is a net decrease in PCr, and, of course, Cr will increase in proportion to the breakdown of total PCr.

☑ KEY POINT

Activation of muscle fibers during any type of physical activity is controlled by neurons that originate in the spinal cord. The amount of force that a muscle is required to generate determines the number of active muscle fibers, while the actual speed of movement is of lesser importance in muscle-fiber recruitment. Thus, slowly lifting a very heavy weight will recruit most Type I and Type II muscle fibers to contract, while quickly tossing a light table-tennis ball may only recruit a few Type I fibers to contract.

Researchers have various techniques to measure changes in muscle metabolites during exercise. One of these is to sample the muscle with a biopsy needle, which can extract a small (about 50 mg) piece of muscle. Chemical analysis of this muscle sample can give us a picture of the muscle at the time of the biopsy if the sample is immediately frozen in liquid nitrogen (–196 °C) after being removed from the muscle. This is considered an invasive technique because the biopsy needle must be inserted into the muscle each time a sample is needed. An alternative technique based on the properties of specific atomic nuclei can be used. For example, the isotope of phosphorus with mass number 31 (^{31}P) can give rise to a specific signal, depending on the type of molecule in which the phosphorus atom is found. Thus, we can get a particular signal from each of the three phosphate groups in ATP, the phosphate in PCr, or the two forms of Pi (HPO_4^{2-} and $H_2PO_4^{1-}$). This technique, known as *^{31}P NMR* (nuclear magnetic resonance) spectroscopy, has been used for more than 25 years to study human skeletal muscle. The term *magnetic resonance imaging* or *MRI* is often used now instead of NMR. This technique is noninvasive, and sampling of muscle can be done repeatedly. Some drawbacks exist, not the least of which is the expensive equipment needed. Moreover, the person has to exercise inside the bore of a very large magnet, which limits the type of exercise that can be used. In addition, the sensitivity of the technique is limited, so metabolites such as ADP and AMP cannot be measured. However, these can be estimated based on the assumption of equilibrium for the creatine-kinase and adenylate-kinase reactions, respectively.

The concentrations of metabolites in muscle are generally expressed in one of three ways. If metabolites from the frozen sample are extracted out and measured, we can express the quantity on the basis of the number of millimoles of particular metabolite per kilogram of wet tissue weight. If the muscle sample is first freeze-dried to remove all the water, we can express the concentration as millimoles per kilogram of dry weight of muscle. The water content of fresh muscle is roughly 77% of the total weight. Thus, a wet sample weighing 100 mg would be approximately 23 mg if completely dried. It is quite common now to express the concentrations of metabolites in muscle in mM units (i.e., millimoles of metabolite per liter of intracellular water). It is accepted that for every kilogram of muscle, 70% will be represented by intracellular water, or 0.7 L. The difference between intracellular and total water is the extracellular water—that is, the fluid around cells and in capillaries, small arteries, and veins.

Table 4.3 outlines some representative values for human muscle metabolites, expressed in mM units for conditions representing rest; mild fatigue; fatigue from severe, repeated contractions; and the state following a marathon run. These values do not reflect any specific person or study, but they give an indication of how metabolites change under different contractile conditions. Notice that for mild fatigue and following a marathon race, concentrations of many metabolites are not very different from those at rest. In particular, notice that ATP concentration is well defended by the energy systems operating in muscle. The severe fatigue is not something many people experience, but there are very large differences in PCr, Cr, Pi, IMP, ADP, lactate, and pH, suggesting the stress on the ATP-regenerating systems. For the marathon, the major difference is the huge drop in glycogen concentration, suggesting that this must be a very important fuel for a long race. Subsequent chapters look at exercise metabolism in greater detail.

As chapter 5 discusses, the calculation of energy derived from aerobic metabolism can be performed relatively easily by determining the amount of oxygen

Table 4.3 Hypothetical Metabolite Concentrations in Human Skeletal Muscle at Rest and After Different Kinds of Physical Activity

Metabolite	CELLULAR CONCENTRATION (MM)			
	Rest	Mild fatigue[a]	Severe fatigue[b]	Prolonged fatigue[c]
ATP	8.2[d]	8.1	6.2	8.0
ADP (free)[e]	0.016	0.022	0.13	0.020
AMP (free)[e]	0.00008	0.00012	0.002	0.0003
TAN[f]	8.217	8.123	6.337	8.032
IMP	<0.02	0.09	1.8	0.1
Pi	3.0	12	28	14
PCr	32	22	5	23
Cr	10	20	37	19
Lactate	2.2	5.9	26.3	2.5
pH	7.1	7.0	6.3	7.1
Glycogen[g]	175	140	80	10

[a]Mild fatigue refers to a state associated with moderate-speed cycling or after 30 min of an easy game of pickup basketball.
[b]Severe fatigue could be induced by an intense training session for a 400 m running race—a metabolic state that can be generated only in the trained muscles of well-motivated subjects.
[c]Prolonged fatigue is best represented by the state of muscle after a major marathon race.
[d]Metabolite concentrations are expressed as millimoles of metabolite per liter of intracellular water. It is assumed that each kilogram of skeletal muscle contains 0.7 kg (L) of water.
[e]The word *free* is used for the concentrations of ADP and AMP, meaning the fraction of each that is not bound to proteins in the cell.
[f]TAN (total adenine nucleotide) concentration is the sum of the concentrations of ATP, ADP, and AMP.
[g]The number refers to the millimoles of glucose units in glycogen, assuming they were free in the cell.

consumed. It is relatively more difficult to calculate the amount of energy derived from anaerobic metabolism, since this calculation requires an accurate assessment of changes in muscle PCr, as well as muscle-lactate production and changes in muscle ATP concentration. The assessment of such changes requires muscle-biopsy samples and samples of venous blood draining the exercising muscles (since some lactate produced in muscles will be released into the blood). These can be subject to methodological difficulties. Nevertheless, if accurate assessment of such metabolite changes can be obtained, a reasonable estimate of total anaerobic energy production can be determined. Thus, total anaerobic energy provision can be determined using the following formula:

$$\text{Total ATP provision} = \Delta \text{PCr} + 1.5 \Delta \text{lactate} + 2\Delta [\text{ATP}]$$

Here, Δ PCr represents the ATP produced that results from the drop in muscle PCr content from rest; 1.5 Δ lactate represents the ATP produced from the glycolytic production of lactate with a net gain in 1.5 ATP per lactate molecule generated, representing 3 ATP produced per glucose molecule derived from muscle glycogen; and 2Δ [ATP] represents the total ATP utilization from the stored ATP pool in resting muscle. Note that the reduction in muscle ATP with exercise is always equal to the increase in IMP. This means that 2 ATP are used for a net change in 1 ATP to occur, since IMP is produced from AMP in the AMP deaminase reaction, AMP is produced from 2 ADP in the adenylate kinase reaction, and 2 ADP are produced from 2 ATP in an ATPase reaction (i.e., actomyosin ATPase). However, the net decrease is only 1 ATP, since the adenylate kinase reaction also produces 1 ATP.

> ### ☑ KEY POINT
> The major change in metabolite concentrations after a marathon run is the reduction in muscle and liver glycogen. The lack of glycogen to support aerobic glycolysis is one of a number of reasons that most runners are slower at the end of a marathon race.

BIOENERGETICS

So far, our discussion has been very general in terms of energy. Energy in the body is produced by the hydrolysis of ATP and PCr, glycolysis, and oxidative phosphorylation. This energy is used for the anabolic part of metabolism (synthesis of protein, RNA, cell constituents, and so on), contraction (skeletal, cardiac, and smooth muscles), transport processes, and control mechanisms. We have neither described the energy in any detail nor shown how it can be quantified. These are the aims of this section.

Concept of Free Energy

Before we begin, however, let us review a few simple concepts, using figure 4.13 as our model. A loose cable is strung between two poles and fastened at positions A and B in figure 4.13*a* and C and D in figure 4.13*b*. You can slide down the cable holding a handle attached to a ring that fits over the cable. Intuitively, you know that if you start from position A, you will slide along the cable, stopping not partway down, but at B. If you start from C, you will slide toward D, but you will not reach it. You will stop, as shown, closer to D than to C. You could also start from D and slide along the cable. You know, based on the relative heights of the attachments of C and D, that you will slide along the cable toward C, but that you will not go far. The examples of energy involving a mechanical apparatus are simple. Potential energy based on the weight of your body and its elevation can be converted into kinetic energy as you slide along the cable. The kinetic energy of the ring sliding over the cable causes the molecules in the cable to become warmer; that is, they gain heat, meaning that the molecules in the cable and even some in the air around the cable increase their motion. You could use your kinetic energy of movement sliding down the cable to knock over a stack of blocks. Of course, when you are suspended between C and D and you are not moving, you will not be able to transfer your energy to anything else. You may also realize that some of the potential energy based on your elevation cannot be converted entirely into kinetic energy of motion. There will be losses of energy in friction with the cable, generating heat and resistance by the air. The lessons we can learn from a simple analogy such as this help us to understand the following statements about chemical reactions in cells.

- All chemical reactions involve energy changes. If there is no energy change, nothing happens.
- Chemical reactions in living organisms are catalyzed by enzymes.
- Enzymes attempt to drive the reaction they catalyze toward equilibrium.
- As enzyme-catalyzed reactions proceed toward equilibrium, they release energy.

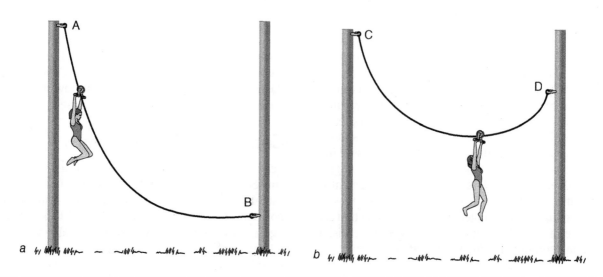

Figure 4.13 Example of energy involving a mechanical apparatus. A person sliding along a cable loosely hung between two posts from a higher to a lower position will stop (reach equilibrium) farther from the starting position the greater the difference in relative heights of the cable attachments. As in this mechanical example, chemical reactions proceed toward an equilibrium. The farther the reaction is from equilibrium, the more energy it can release. At equilibrium, no energy is released.

- The farther a reaction is from equilibrium, the more energy it can release.
- At equilibrium, no energy is released.
- Some energy released in a chemical reaction can be used to do work; the remainder is unavailable. In our muscles, the energy that is not captured to do work is released as heat.

In **thermodynamics**, we focus only on differences between initial and final states; in this sense, we talk about changes. The following thermodynamic terms are defined as applied to biological processes:

• **Enthalpy change** (ΔH): In any chemical reaction or sequence of reactions, the change in energy of the reactants when they are turned into products is called the enthalpy change. This can be measured as the total heat energy change, described by the abbreviation ΔH. When heat energy is given off or released in a reaction or process, we have a negative value for ΔH; such a reaction is an **exothermic reaction**. Reactions or processes with a positive value for ΔH are said to be *endothermic*. These can only occur with an input of energy.

• **Entropy change** (ΔS): Entropy is a measure of energy dispersal. The concept of entropy tells us that energy wants to move from where it is concentrated to where it is dispersed or spread out. Consider a hot frying pan after the heat is turned off. It is a concentrated source of heat energy, but we know that it will cool and that the air around it will become warmer. A ball placed on the side of a steep hill will roll down the hill; this is a spontaneous process. The potential energy of the ball on the hill will be dispersed first into kinetic energy as it begins to move, then to heat energy as it proceeds to the bottom of the hill. If we place a rock below the ball, we simply prevent it from rolling, but we don't change the fact that the ball's tendency is to roll down the hill if given a chance. This is entropy at work. Adenosine triphosphate is a concentrated form of energy, as we have discussed. If given a chance, say when a myosin crossbridge interacts with an actin monomer, the chemical energy stored in ATP is dispersed and turned into a bit of work and mostly heat. Processes that can occur by themselves if given a chance have a positive value for entropy change (positive ΔS). The reverse of the examples just given cannot occur along the same path. That is, a cool frying pan does not become hot by accepting heat energy from the cooler air around it, and a ball does not spontaneously roll up a hill.

• **Free energy change** (ΔG): Of the total energy released in a reaction or process, not all is available to do something useful. The free energy change, or ΔG, is the maximum energy available from a reaction or process that can be harnessed to do something useful. From a biological perspective, something useful would be a muscle contracting and lifting a load, moving ions across a membrane against their concentration gradient, or synthesizing a protein from amino acids. When free energy is released, the ΔG is negative; the reaction is an **exergonic reaction**. A reaction with a positive value for ΔG cannot occur by itself—as a ball cannot spontaneously roll up a hill—and is said to be an **endergonic reaction**.

The relationship among changes in enthalpy, entropy, and free energy for a reaction or process is given by the following expression:

$$\Delta G = \Delta H - T\Delta S \qquad (4.1)$$

This equation tells us that not all of the energy that is released in a reaction or process can be used to do something useful. The unavailable portion is given by the TΔS term. This term reflects loss of energy to the surroundings in the form of increased motion of molecules, perhaps in sound or light. We know from the law of the conservation of energy that we can neither create nor destroy energy. Thermodynamics tells us that we cannot capture 100% of the energy released in a reaction or process; some must be lost to the surroundings.

> ### ✓ KEY POINT
> In exercising muscle, the energy that is liberated from stored fuels and is not captured to do work is released as heat. Hence, our bodies grow warmer as we exercise, and we need to sweat to cool ourselves down. Humans are relatively efficient at capturing energy from metabolic reactions to perform work with about 20% to 25% of energy being captured in activities such as running or swimming. In contrast, the internal combustion engine in an automobile may only capture 2% to 3% of the energy released from gasoline to produce forward motion when driving.

From the earlier parts of this chapter, you should understand that reactions or processes with a positive value for free energy change can take place only if a simultaneous reaction with a negative ΔG occurs. As you have seen, the simultaneous reaction that drives most endergonic processes in living organisms is ATP hydrolysis. Thus, a crossbridge in muscle can only interact with an actin subunit in the thin filament and exert a force (a process with a +ΔG per se) when an ATP is hydrolyzed. The key point is that the energy required must be less than the energy provided by ATP hydrolysis so that overall, the combination has a negative value for ΔG.

Quantitative Values for Free Energy

In thermodynamics, we cannot measure exact values associated with an initial and final state of a reaction in terms of enthalpy, entropy, and free energy. However, we can measure energy changes over the course of a reaction as changes in enthalpy, changes in free energy, and changes in entropy. For the reversible chemical reaction given by the equation

$$A + B \leftrightarrow C + D$$

the free energy change in the direction from left to right is given by the expression

$$\Delta G = \Delta G^{\circ\prime} + RT \ln \frac{[C][D]}{[A][B]} \qquad (4.2)$$

where ΔG is the free energy change for the reaction.

ΔG°' is the **standard free energy change** for the reaction under what are considered to be standard conditions, when each of the reactants and products is at a concentration of 1.0 M. A prime is added to indicate a pH of 7.0; the traditional thermodynamic standard free energy is defined at a most nonbiological pH of zero. ΔG°' reflects the energy-generating potential of the reaction, and a value for ΔG°' can be determined from the equilibrium constant at a pH of 7.0 (K'_{eq}, discussed later).

R is the *gas constant*, which has a value of 8.314 joules per mole per degree K (1.987 cal per mole per degree K). Remember, 1 calorie (cal) is equivalent to 4.18 joules (J), or 1 kilocalorie (kcal) = 4.18 kilojoules (kJ). T is the absolute temperature expressed in degrees K. This can be determined by adding 273 to the temperature in Celsius units. [A] is the concentration of species A, [B] is the concentration of species B, and so forth.

In equation 4.2, ΔG°' and R are constants. At a particular temperature T, you should see that ΔG will depend only on the relative concentrations of A, B, C, and D. Suppose we start the reaction by adding A and B together. The reaction will quickly proceed to the right; as C and D accumulate, the net reaction (toward C and D) will gradually slow down. Eventually, the forward (toward C and D) and backward (toward A and B) reactions will occur at the same rate, and we will have reached **equilibrium**. From an energy perspective and using the analogies in figure 4.13, we know the following:

1. The farther a reaction is from equilibrium, the larger the value of ΔG (as a negative value). In equation 4.2, when the reaction begins, [A]

and [B] will be large, whereas [C] and [D] will be small. Accordingly, the natural logarithm (ln) term will be large and negative. Thus, at the start of the reaction, the ΔG will be large and negative.

2. As the reaction proceeds toward equilibrium, the numerical values for [A] and [B] will decrease, whereas the values for [C] and [D] will increase. Accordingly, ΔG will decrease, but will remain negative.

3. At equilibrium, there is no net reaction and no net energy change, so the value for ΔG is zero. Therefore, a reaction at equilibrium releases no free energy.

Rewriting equation 4.2 at equilibrium, where $\Delta G = 0$, we get

$$0 = \Delta G°' + RT \ln \frac{[C][D]}{[A][B]}$$

However, at equilibrium (and a pH of 7), the concentration ratio [C][D]/[A][B] is given by the equilibrium constant K'_{eq}. We use this fact and rearrange the equation:

$$\Delta G°' = -RT \ln K'_{eq} \quad (4.3)$$

This equation allows us to determine the value of the standard free energy change for a reaction at pH 7 if we know the temperature and the value of the equilibrium constant. We can say the following about the $\Delta G°'$ value for a reaction:

1. It is determined by the value for the equilibrium constant.
2. It reflects the energy-generating potential for the reaction.
3. It does not define the actual energy change for the reaction inside a cell, since this depends on the relative concentrations of reactants and products.

The actual free energy change for a reaction inside a tissue is not difficult to determine. Take a sample of tissue and quickly freeze it; this immediately stops all chemical reactions. Chemically analyze the amount of reactants and products for the reaction in question. From this, we can determine the **mass action ratio** (many people use the Greek symbol Γ for this) for the reaction. This is the ratio of product concentrations to reactant concentrations as they existed at the moment the tissue was frozen. We can determine the K'_{eq} for the reaction by allowing it to come to equilibrium in a test tube under conditions of temperature, pH, and ion concentration identical to those of the living tissue, and then analyzing the equilibrium concentrations of reactants and products. Once we have determined an equilibrium constant, we can plug it into equation 4.3 to determine $\Delta G°'$. Now we can determine the actual free energy change for the reaction in question at the exact time it was frozen by employing a variation of equation 4.2.

$$\Delta G = \Delta G°' + RT \ln \Gamma \quad (4.4)$$

We can rearrange this equation, taking advantage of the relationship between $\Delta G°'$ and K'_{eq}.

$$\Delta G = -RT \ln K'_{eq} \, \Gamma \quad (4.5)$$

This equation tells us that if K'_{eq} is $> \Gamma$, the free energy of the reaction (ΔG) will be negative; this means the reaction can proceed on its own, releasing energy. The larger the ratio K'_{eq}/Γ (that is, the more the reaction is displaced from equilibrium), the more free energy will be released by the reaction. If K'_{eq} is exactly the same as Γ, then ΔG is zero.

Free Energy in the Cell

We have emphasized that the free energy of ATP hydrolysis (or GTP or UTP) drives just about every reaction or process in our bodies. Therefore, it is essential to ensure that the free energy released during the hydrolysis of each ATP molecule is as high as possible. If the ATP hydrolysis reaction is allowed to come to equilibrium, the concentration of ADP is about 10 million times greater than that of ATP. In the cell, the complete opposite prevails; the concentration of ATP is hundreds of times larger than that of ADP under all but the most severe conditions of fatigue. The $\Delta G°'$ for ATP hydrolysis is -30.5 kJ/mol (-7.3 kcal/mol), but the real ΔG is far larger in magnitude with a negative sign. We could determine the actual free energy for ATP in the cell by using equation 4.4 and using actual values for [ADP][Pi]/[ATP] to get the mass action ratio (Γ).

$$\Delta G = -30.5 \text{ kJ/mol} + RT \ln [ADP][Pi]/[ATP]$$

We exclude the role of water in this equation, focusing on the key metabolites whose concentrations can change in the cell.

If you use the concentrations for ATP, ADP, and Pi from table 4.3, use 37 °C (310 K) for temperature, and do the calculation for ΔG using equation 4.4, you will see that this number is well defended (large and negative) except under conditions of severe stress. The smaller release of free energy under severe fatigue conditions can affect the force generated by each crossbridge or the ability of the two ATPases to pump sodium, potassium, and calcium ions.

The strategy of the cell to maximize the ΔG for ATP hydrolysis is based on keeping the ratio of [ATP] to [ADP] as high as possible. This is assisted by the location of several key enzymes. For example, creatine kinase activity is high in the region of the myofibrils, the SR, and the sarcolemma so that as rapidly as ATP is hydrolyzed to ADP and Pi by the actomyosin, SR, and sodium–potassium ATPases, the ADP will be immediately converted back to ATP at the expense of PCr. In addition, adenylate kinase plays an important role in reducing ADP concentration by reacting two ADP molecules together to get an AMP and an ATP. Besides helping to keep [ADP] low, the AMP produced is a potent stimulator of glycolysis.

☑ KEY POINT

Exercise training is not thought to greatly affect the efficiency of energy transfer from metabolized fuels to ATP synthesis or ATP utilization by specific ATPase enzymes. Similar amounts of energy are thought to be captured or released as heat by these enzyme-catalyzed reactions in both trained and untrained people. Where trained athletes do have an advantage is in the efficiency with which they perform the movements associated with physical activity and the efficiency with which they recruit appropriate muscle and muscle-fiber contractions. This allows the trained to expend relatively less energy while performing the equivalent amount of external work as untrained people of similar weight in activities such as running, swimming, or cycling.

Muscle Fatigue and Muscle Energy Balance

Muscle fatigue, or the loss of ability to maintain the rate and force of muscle contractions during exercise, is a complex and multifaceted phenomenon, which is briefly highlighted in the Next Stage section of chapter 6. It should be noted that some aspects of energy balance do contribute significantly to muscle fatigue in certain types of exercise. In their review, Allen, Lamb, and Westerblad (2008) note that increased concentrations of muscle Pi, as well as reduced ATP levels coincident with increased ADP concentration in muscle, result in a reduced ability to maintain muscular-contraction intensity and rate. In part, this muscular fatigue results in a slower rate of ATP utilization. Therefore, it can be viewed as a protective mechanism that prevents us from exercising to such an extent that we risk depleting muscle-ATP levels and consequently disrupting basic cellular functions, bringing on muscle rigor and possibly resulting in cell death.

Increases in muscle Pi that are seen primarily with PCr breakdown during intense exercise coincident with increased breakdown of ATP to ADP and Pi can interfere with muscle-force production and induce symptoms of fatigue. Increases in cytoplasmic Pi concentration can inhibit force production by direct action on muscle crossbridge function, primarily by reducing calcium sensitivity. In other words, the same amount of SR calcium release will result in fewer actin–myosin crossbridges generating force. In addition, when muscle ATP levels drop, calcium release from the SR will also be inhibited (Allen, Lamb, and Westerblad 2008). Different types of exercise may affect the ability of muscle SERCA to uptake calcium or that of the sarcoplasmic reticular calcium channels to release calcium (Tupling 2004). Among the possible causes of these effects are hydrogen-ion accumulation, accumulation of reactive oxygen species (ROS), or the accumulation of ATP hydrolysis products, such as ADP and inorganic phosphate (Pi), as well as a number of other possible factors. Other factors associated with muscle fatigue during longer term exercise, such as muscle-glycogen depletion, may also inhibit SR calcium release. These issues are discussed further in chapter 6.

Despite its common association with fatigue, muscle-lactate production per se may not be as critical to loss of muscle force during intense exercise as once thought. The complex issues of lactate production, clearance, and effect on muscle pH and fatigue are discussed further in chapter 6. Chapter 6 also discusses the potential fatigue-related mechanisms associated with muscle-glycogen depletion.

NEXT STAGE

Epigenetic Control of Skeletal Muscle Fiber Type and Metabolic Potential

The distribution of Type I and Type II muscle fibers in any specific skeletal muscle and their consequent distinct metabolic profiles have been seen to be genetically determined. In other words, the genes present in the nuclei of muscles will express proteins as coded in their

DNA along predetermined genetic lines. Inherited traits are passed on from generation to generation unless a mutation occurs in the genome that conveys an advantage to the ability of the species to reproduce. In this case, it will become the trait that is eventually passed on to future generations. This is classic Darwinian genetics. More recently, scientists have discovered that we can inherit changes in gene expression in the absence of changes to the genome. In other words, sometimes interactions between people and their environment, perhaps including exercise or training, alter the expression of certain genes. We know from animal models that very long and intense endurance training is able to alter gene expression in muscle fibers such that some of the Type II muscle fibers start to express genes that will produce proteins found in Type I fibers and suppress expression of genes found in Type II fibers, essentially transforming themselves from Type II to Type I muscle fibers. What is now becoming clearer is that even in the absence of the signaling factors associated with such changes, some changes in muscle-fiber types may be retained by cells derived from these muscles and grown in tissue cultures. The study of "heritable change in gene expression in the absence of changes to the sequence of the genome" (Baar 2010, p. 477) is known as **epigenetics**.

The epigenetic control of skeletal muscle fiber type was reviewed by Baar (2010). An interesting question raised by this study of epigenetics is whether environmental factors, such as diet and exercise training, that are known to alter gene expression in muscles and that affect muscle-fiber type or metabolic pathways and metabolism can be passed on to future generations in humans. For example, muscles cultured from endurance athletes had significantly higher glucose uptake (a training-induced adaptation) than muscles cultured from untrained subjects (Berggren et al. 2005).

Epigenetic changes in gene expression can be regulated in cells by adding methyl molecules (methylation to appropriate sites on DNA, adding acetyl or methyl molecules to histones that regulate DNA folding and DNA exposure to RNA synthesis), as seen in chapter 3. These types of mechanisms are known to affect muscle development, fiber type, and adaptation through regulators such as MyoD and myocyte-enhancer factor (MEF) enzymes, which affect histone acetylation and expression of Type I and Type II fibers. Endurance exercise has been shown to influence factors that regulate histone acetylation and methylation, which in turn regulates mitochondrial biogenesis and glucose transport into muscle (McGee et al. 2008).

These findings suggest that factors that influence epigenetic regulation of muscle-gene expression can be affected by exercise training. We do not yet know how heritable these types of changes are, particularly in humans. Animal muscles that were trained by electrical stimulation for long periods and that, as a consequence, significantly increased their Type I muscle fiber and mitochondrial content did not continue to express these changes when grown in tissue culture. This suggests that inherited fiber-type expression is highly conserved. Nevertheless, factors involved with aging that make our muscles less adaptable or plastic as we age show that effects of our environment, exercise history, and diets do influence factors that regulate epigenetic gene expression. An interesting question for future research is to what extent these factors may affect our rate of muscle loss or sarcopenia associated with aging over a lifespan. How epigenetic changes occur and how they may control things such as muscle-fiber type, sarcopenia, and heritability of muscle metabolic characteristics will continue to be researched in years to come (Baar 2010).

SUMMARY

The thousands of chemical reactions in our bodies that constitute our metabolism involve energy changes. Adenosine triphosphate, a nucleoside triphosphate, is the key mediator of metabolism because its hydrolysis generates a great deal of free energy that can be used to drive processes unable to occur by themselves. Adenosine triphosphate is generated when the energy stored in fuel molecules is released during catabolism. Although ATP ensures that a host of energy-requiring reactions and processes in all cells can take place, its rate of use can be extremely high in skeletal muscle. Each tiny crossbridge on a myosin molecule in a thick filament in a muscle fiber can bind to actin in a thin filament, generating force. Each of these interactions uses an ATP, and millions of these interactions take place each second as a muscle fiber shortens. Adenosine triphosphate is also needed to regulate the concentration of calcium, which activates myosin and actin interactions, and to control the concentrations of sodium and potassium ions that can change so much as the muscle fiber is activated by its neuron.

The ATP concentration in a muscle fiber is sufficient to drive only about 2 s of maximal work. Its concentration is maintained remarkably well by three key energy systems. Transfer of a phosphate group from PCr to ADP is a mechanism to renew ATP rapidly, but supplies of PCr are limited. Glycolysis, an anaerobic process in which glycogen or glucose is broken down to lactate, can provide ATP at a fairly rapid rate, but capacity for this process is limited. Oxidative phosphorylation, however, provides ATP at a low rate for a prolonged period of time. It is the dominant mechanism for producing ATP in almost all cells in our body, and it provides all the ATP during steady-state exercise. The selection of which of these processes is used to regenerate ATP in muscle depends primarily on the rate of ATP use or the intensity of the exercise. For example, those systems with high ATP regeneration rates but lower capacities, such as PCr, are generally only used at times of intense exercise. Exercise at a high rate to fatigue can lead to severe metabolic displacement in muscle, since the contracting fibers attempt to generate ATP to match its high rate of breakdown by the ATPases. Severe decreases in PCr and pH and large increases in ADP, lactate, ammonium, and phosphate ions characterize the fatigued state as a result of maximal exercise lasting 1 to 5 min. Some of these factors, particularly increased concentrations of Pi and ADP, as well as decreases in ATP, affect specific aspects of processes related to muscular contraction. In turn, these reduce our capacity to generate muscle force, thereby slowing down the rate of ATP utilization. This is important in preventing exercise from depleting ATP to below critical levels.

Differences in energy states between reactants and products in a chemical reaction contribute to the overall energy changes of the reaction or process designated as the enthalpy change (ΔH). Not all of the energy released in a chemical reaction can be captured to do something useful because there is an unavoidable dispersal of energy based on changes in entropy (ΔS) over the course of the reaction. What we can use is known as the free energy change (ΔG). Free energy release during hydrolysis of ATP is critical to cell function. This is ensured, since the concentration of ATP is kept high relative to its hydrolysis products, ADP and Pi. The three energy systems play critical roles in this, aided under more severe contractile conditions by the enzymes adenylate kinase and AMP deaminase.

REVIEW QUESTIONS

1. Using the data in table 4.3 and equation 4.5, as rearranged in the Free Energy in the Cell section, calculate the mass action (Γ) ratios for the hydrolysis of ATP for each of the four conditions shown. Be sure to use the free concentration of ADP.

2. Convert a metabolite concentration of 4.0 mmol per kilogram wet weight to the following:
    a. Micromoles per gram wet weight
    b. Millimoles per kilogram dry weight
    c. Millimoles per liter of intracellular water

3. Assuming that the adenylate kinase reaction is at equilibrium at rest in skeletal muscle, determine the equilibrium constant in the direction of ATP formation using the data in table 4.3.

4. Using equation 4.4 and the standard free energy of ATP hydrolysis for ATP as 7.3 kcal/mol, calculate the free energy for ATP hydrolysis in a skeletal muscle cell at rest at 37 °C using the data for ATP, ADP (free), and Pi concentrations found in table 4.3.

5. If the free concentration of magnesium ions in a resting muscle cell is 2 mM, what is the approximate total concentration of magnesium in the cell?

6. Answer question 4 again, but use the values for severe fatigue and the same temperature.

7. Algebraically sum the ATPase reaction and the creatine kinase reaction to show that the overwhelming source of increased Pi in an actively working muscle fiber is from PCr.

8. Why would it be necessary during intense exercise for the exercising muscle to augment ATP resynthesis via oxidative phosphorylation with ATP resynthesis via upregulation of glycolysis and PCr breakdown?

9. Why would muscles choose not to use PCr stores for ATP resynthesis during rest or modest intensity exercise?
10. What are the potential benefits of dietary creatine supplementation to exercise and skeletal muscle size? Can these benefits be realized without concurrent weight training?
11. Calculate the total ATP provided by anaerobic metabolism based on the following information:

Source	Rest	Exercise
ATP	30	26
PCr	80	30
Muscle lactate	5	20
Blood lactate	1	12

chapter 5

Oxidative Phosphorylation

Humans are open thermodynamic systems. This means we take in and release matter and energy from and to our environment. On the intake side, we take in oxygen, food, and water from the world around us. We release heat, carbon dioxide, urine, and fecal waste back to the environment. Our source of energy to carry out all the actions associated with life is the food we eat. In this chapter, we go through the steps in which oxygen is involved in the generation of more than 90% of all the ATP we need. Because this chapter on oxidative phosphorylation represents our first detailed study of a metabolic pathway, we begin with a general overview of cellular energy metabolism. Oxidative phosphorylation is a complex process taking place in mitochondria, where we will start. The citric acid cycle is the first major pathway that generates electrons, which are eventually transferred to the oxygen we breathe. The energy released is used to make ATP. The rate of oxidative phosphorylation must be coupled to and regulated by our energy needs, so more detailed accounts of regulation of oxidative phosphorylation and how this is quantified are included. We finish with a basic section on oxidants and antioxidants and a Next Stage section on aging, training, and mitochondrial function.

OVERVIEW OF METABOLISM

The food we eat is broken down in the digestive tract to simplest components, absorbed into the blood, and distributed to various tissues. There, the simple, absorbed molecules may be broken down to yield energy in the form of ATP, stored as energy molecules for future needs (glycogen and triglyceride), or used to create large molecules (e.g., proteins) specific to our needs or to maintain a proper internal milieu and support metabolism (minerals and vitamins).

We have already briefly covered the ATP-generating systems important to muscle: phosphocreatine, glycolysis, and oxidative phosphorylation. The only fuel source for glycolysis is carbohydrate, but oxidative phosphorylation can use carbohydrate, fat, and amino acids. The amino acids that are oxidized are obtained from the normal breakdown of proteins or from excess proteins in the diet that are not immediately converted into fat or glucose. Most fat is stored in specialized cells known as fat cells or **adipocytes**, primarily located subcutaneously (beneath the skin). Small amounts of fat are also stored in other cells, such as muscle. Fat stores (e.g., visceral fat) also accumulate in other locations, such as between organs in the body cavity. This accumulation of visceral fat, primarily associated with obesity, can contribute to some of the health problems connected to severely overweight conditions. When metabolism needs to be increased, nerve and hormonal signals cause triglyceride molecules to be broken down. Fatty acids are also released to be used as fuel.

The body's carbohydrate stores are not large; most carbohydrate is stored in liver and muscle as glycogen—a polymer made of individual glucose molecules joined together. Approximately 80% of all the body's carbohydrate is found in muscle as glycogen. Although liver has a higher concentration of glycogen than muscle, the liver is so much smaller in mass compared to the skeletal muscles that it holds only 10% to 15% of total body carbohydrate.

The remainder of the carbohydrate in the body is the glucose in blood and extracellular fluids. When needed, glycogen in liver is broken down to maintain a proper glucose concentration in the blood. In muscle, glycogen is broken down and fed into the glycolytic reactions. Glucose is normally the sole fuel for the brain, the tissue that the body treats as most important from an energy perspective. In addition to glycogen stores in muscle, blood glucose (from liver glycogen) can also be used by muscle for ATP generation. Intensity and duration of exercise will affect the source of carbohydrate for use in muscle oxidative phosphorylation and glycolysis, with more intense exercise preferentially utilizing muscle glycogen. As exercise duration increases, particularly if it exceeds 1 h, our muscle glycogen stores begin to get used up. Therefore, muscle will rely more on blood glucose as a fuel source. These issues are discussed in greater detail in chapter 6.

Figure 5.1 shows a section of a muscle fiber, representing the tissue that is or has the potential to be the main energy consumer. It also illustrates the major energy pathways in the cell, which are located in the cytosol and mitochondria. The fuels illustrated for these pathways are glucose from blood, either used directly in glycolysis or temporarily stored as glycogen, and fatty acids (FA), which can be taken up from the blood and used immediately in mito-

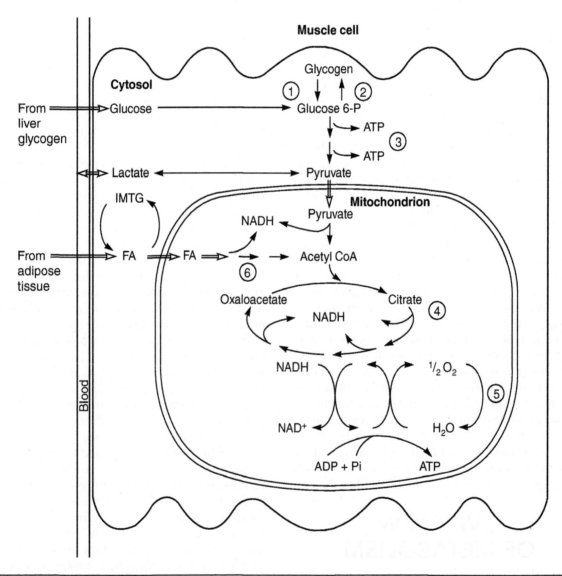

Figure 5.1 Overview of energy metabolism in a cross section of a muscle cell. Illustrated are the cytosol and mitochondrial compartments and the major pathways discussed in this chapter: (1) glycogen breakdown, (2) glycogen synthesis, (3) glycolysis, (4) citric acid cycle, (5) electron transport chain, and (6) beta-oxidation. A double-headed arrow means that a reaction or sequence of reactions is reversible. An open arrow means that a substance is crossing a membrane. IMTG represents intramuscular triglyceride, and FA represents fatty acids.

chondria or temporarily stored as *intramuscular triacylglycerol* (IMTG). Although our focus is on muscle, these same pathways operate in other cells as well.

In chapter 4, we defined oxidative phosphorylation as the formation of ATP from ADP and Pi, in association with the transfer of electrons from fuel molecules to coenzymes to oxygen. Although probably the best term to use, oxidative phosphorylation is also known by other names. The biologist may call it cellular respiration, whereas the exercise physiologist may refer to it as oxidative metabolism. As shown in figure 5.1, oxidative phosphorylation takes place in mitochondria. It encompasses the **citric acid cycle** (shown by the circle of arrows with oxaloacetate and citrate) and the **electron transport chain** (also known as the respiratory chain), which is shown below the citric acid cycle (CAC). The citric acid cycle starts with the acetyl group (CH_3CO) being attached to coenzyme A (CoA). The acetyl group can be derived from any kind of fuel molecule. The citric acid cycle produces electrons in the form of the reduced coenzymes (mostly NADH). These electrons feed into the electron transport chain and are passed through a series of carriers to the oxygen we breathe. Addition of electrons to oxygen allows it to combine with two protons (H^+) to produce water. Free energy produced in the electron transport chain is used to phosphorylate ADP with Pi to make ATP.

As shown in figure 5.1, acetyl is the substrate for the citric acid cycle, while CoA is the ubiquitous coenzyme needed to facilitate its entry into the citric acid cycle. Acetyl can arise from the breakdown of fatty acids using **beta-oxidation** (number 6 in figure 5.1). Acetyl is also formed in mitochondria from pyruvate, which is produced during glycolysis in the cytosol. In exercise, the proportion of acetyl coming from pyruvate in the breakdown of glycogen and glucose increases with exercise intensity. We discuss this further in chapter 6.

✓ KEY POINT

When we exercise, we place demands for ATP use on our muscles. This must be matched by ATP synthesis. For most activities, the bulk of the ATP is generated in mitochondria, when electrons on fuel molecules are transferred to coenzymes, through the electron transport chain, to oxygen. Free energy released during this process is captured via the formation of ATP from ADP and Pi. In this process, fuels and oxygen are consumed; the harder we exercise, the greater and more rapid the requirement for fuels and oxygen.

MITOCHONDRIA

Mitochondria are often called the powerhouses of the cell because so much of our ATP is regenerated through oxidative phosphorylation in these organelles. This simple view of mitochondria has been updated, for we now know that they are involved in cell calcium-ion control, cell signaling, and apoptosis (programmed cell death). We also know that mitochondria are not separate organelles but instead exist as a highly regulated reticular structure. Problems with and within mitochondria give rise to a number of pathological conditions, including tissue damage when blood flow is sharply reduced (ischemia) and then resumed (reperfusion), cardiomyopathy, diabetes, and many neurodegenerative diseases such as Parkinson's, Alzheimer's, and Lou Gehrig's (ALS) diseases (Duchen 2004). Mitochondria also start to malfunction as we age; this contributes to declining muscle size and function associated with the aging process. See the Next Stage section in this chapter for more information on aging-related issues and muscle mitochondria. As shown in figure 5.2, many mitochondria have an elongated shape. However, mitochondria are generally found as very complex structures, interacting with each other to form a reticulum and with other cell organelles, such as the sarcoplasmic reticulum in skeletal muscle and endoplasmic reticulum in other tissues (Dirksen 2009). The structural links between the muscle sarcoplasmic reticulum and mitochondria facilitate important two-way signaling involving oxygen radicals and ATP-exchange functions during muscular contraction and exercise. This linkage is discussed in more detail later in this chapter. In skeletal muscle, mitochondria are found just beneath the sarcolemma (**subsarcolemmal mitochondria**) and between myofibrils (**intermyofibrillar mitochondria**).

Mitochondria have two membranes. The *outer membrane*, consisting of roughly equal proportions of proteins and lipid, contains *porins* (special pores) that allow entry to most ions and molecules up to a molecular weight of 1,000 daltons (D). The *inner membrane* is even higher in protein content (nearly 80%) and is impermeable to most ions and polar molecules unless they have specific transporters or carriers. The two membranes come together at intervals to form junctional complexes. Between the two membranes is the **intermembrane space**. Within the intermembrane space are some key proteins, including mitochondrial creatine kinase and cytochrome c. The inner membrane is formed into bulges called **cristae**, which greatly increase its surface area. The density of cristae in mitochondria is generally much higher than shown in figure 5.2, especially in

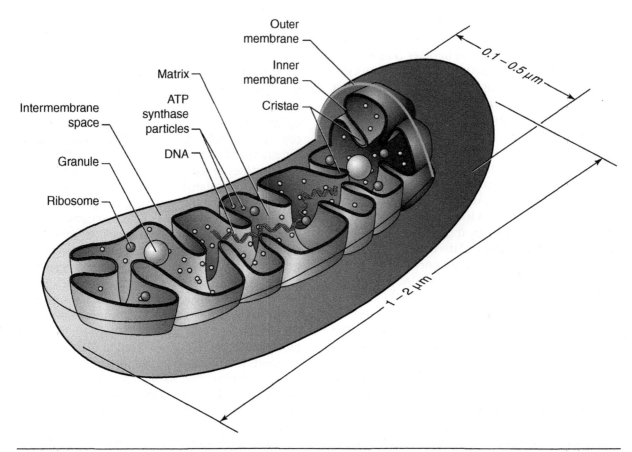

Figure 5.2 Cross section of a simple mitochondrion, showing the various components and approximate dimensions. Mitochondria have an outer membrane that is permeable to small molecules and an inner membrane that is folded into cristae studded with small knobs, identified as the ATP synthase. Between the outer and inner membrane is the intermembrane space. The interior space, known as the matrix, contains DNA, granules, and ribosomes. In the cell, mitochondria can assume quite complex shapes, but almost all have more densely packed cristae than shown here.

tissues where the rate of oxidative phosphorylation is high, such as the heart. The inner membrane is loaded with enzymes and proteins for transferring electrons to oxygen, as well as the enzyme **ATP synthase**, which converts ADP and Pi to ATP. In figure 5.2, ATP synthase is shown as small knobs in the mitochondrial inner membrane.

In the center of a mitochondrion is the **matrix**, a viscous medium containing all the enzymes of the citric acid cycle (except succinate dehydrogenase), three enzymes of beta-oxidation of fatty acids, other enzymes, and *mitochondrial DNA*. This small, circular DNA molecule codes for 13 of the thousand or so mitochondrial proteins (found in the outer and inner membranes, the intermembrane space, and the matrix); the remainder are coded by nuclear genes. Following transcription and translation, proteins coded by nuclear genes are imported into the mitochondrion, specifically targeted because of presequences of amino acids. Mitochondrial DNA also codes for two rRNA (ribosomal ribonucleic acid) molecules and 22 tRNA (transfer ribonucleic acid) molecules—in sum, a total of 37 genes. In each mitochondrion, there are multiple copies of mitochondrial DNA. Researchers have discovered that mitochondrial DNA is more sensitive to damage by a variety of substances than is nuclear DNA. Lacking histone proteins and with fewer repair enzymes than found in nuclear DNA, mitochondrial DNA deteriorates as we age, compromising our ability to generate ATP. This is manifest in part by a reduced $\dot{V}O_2$max as we age, although many systems participate in this loss of aerobic function. As well, a variety of neurological and neurodegenerative disorders owe their origin to **mutations** in mitochondrial DNA or damaged mitochondrial proteins. Virtually all mitochondrial DNA is maternally contributed—less than 0.1% arises from sperm. Mitochondria also play a major role in programmed cell death, a process known as **apoptosis**. The Next Stage section of this chapter discusses some of these issues in more detail.

The actual amount of mitochondria in various muscle types can be expressed by the percent volume of the total muscle cell occupied by mitochondria. This can range from about 1% in glycolytic fibers with a low oxidative capacity to 50% in cardiac muscle. Each milliliter of muscle mitochondrial volume has the remarkable ability to use up to 3 to 5 mL of oxygen per minute when maximally functioning (Moyes and Hood 2003). This is due primarily to the enormous ability to pack cristae inside the mitochondrion, creating crista areas up to 40 m^2 per milliliter of mitochondrial volume (Leary et al. 2003). Exercise training has been shown to increase mitochondrial volume by more than twofold, whereas bed rest, space travel, and immobilization can lead to a reduced volume of mitochondria (Hood 2001). The more energy demands that are placed on muscle, the more mitochondria it will tend to have. For example, the very high energy demands placed on flight muscle of insects, which may contract hundreds of times per minute, have much higher mitochondrial concentrations than seen in even the best trained human muscles. Recent studies have also highlighted the importance of repeated short, intense exercise bouts in stimulating mitochondrial biogenesis in muscles. As few as six training sessions, consisting of 8 to 12 cycling intervals at $\dot{V}O_2$max for 60 s, separated by 75 s of rest, over two weeks elicited significant increases in muscle mitochondria, aerobic enzymes, and endurance performance comparable to much longer traditional aerobic training (Little et al. 2010). The signaling mechanisms controlling gene expression that result in the adaptations we see from different types of training are outlined in chapters 3. The mechanisms and signaling for these mitochondrial adaptations resulting from repeated short, intense exercise training bouts are also discussed in the Next Stage section in this chapter.

✓ KEY POINT

The content of mitochondria in muscle parallels the need for oxidative phosphorylation by that muscle. It is exceptionally high in cardiac muscle and high in muscles involved in breathing. Signals generated by regular endurance or short, high-intensity interval training increase the expression of a variety of genes found in nuclear and mitochondrial DNA, leading to the formation of proteins that are imported into existing mitochondria, and thus expanding the total mitochondrial volume to better handle the exercise stress.

General Mechanism of Oxidative Phosphorylation

The chemical equations that follow illustrate the complete oxidation of two types of fuels: glucose, a representative carbohydrate, and palmitic acid, a very common fatty acid.

$$C_6H_{12}O_6 + 6\ O_2$$
$$\rightarrow 6\ CO_2 + 6\ H_2O \quad \text{(glucose)}$$

$$C_{16}H_{32}O_2 + 23\ O_2$$
$$\rightarrow 16\ CO_2 + 16\ H_2O \quad \text{(palmitic acid)}$$

These balanced equations summarize the complete oxidation of these fuels as would occur in the cell by oxidative phosphorylation. They also describe the oxidation of these fuels if they were completely broken down by burning in a very hot flame.

The **respiratory quotient (RQ)** is the molar ratio of CO_2 produced divided by the O_2 consumed during fuel oxidation. Using the previous equations, RQ is $6 / 6 = 1.0$ for glucose and $16 / 23 = 0.7$ for palmitic acid. These differences are due to the lower amount of oxygen atoms relative to carbon atoms in fat molecules compared to carbohydrate molecules. This can be seen in our examples of glucose, where equal amounts of carbon (6) and oxygen (6) molecules are present, and the fatty acid, palmitic acid, where 8 carbon atoms are present for each oxygen atom. Hence, when CO_2 and H_2O are formed via oxidative phosphorylation of carbohydrate and fat, relatively less oxygen needs to be delivered to the muscle and consumed with carbohydrate oxidation since a relatively greater amount of oxygen atoms are already available in the structure of the carbohydrate molecule for the purpose of CO_2 formation. When the oxygen consumed and carbon dioxide produced are measured at the mouth of an animal or human as $\dot{V}O_2$ and $\dot{V}O_2$, respectively, we use the term **respiratory exchange ratio** or **RER** (i.e., $\dot{V}O_2 / \dot{V}O_2$) instead of RQ.

In the cell, oxidation of glucose and fatty acids is tightly coupled to ADP phosphorylation—that is, phosphorylation of ADP with Pi to make ATP. Let's rewrite these two equations, now including the phosphorylation part; that is, we will include the number of moles of ATP that can be formed for each mole of fuel oxidized in the cell.

$$C_6H_{12}O_6 + 6\ O_2 + 32\ (ADP + Pi)$$
$$\rightarrow 6\ CO_2 + 32\ ATP$$
$$+ 38\ H_2O \quad \text{(glucose)}$$

$$C_{16}H_{32}O_2 + 23\ O_2 + 106\ (ADP + Pi)$$
$$\rightarrow 16\ CO_2 + 106\ ATP$$
$$+ 122\ H_2O \quad \text{(palmitic acid)}$$

The number of water molecules generated is increased by 32 for glucose and 106 for palmitic acid in relation to the equations that showed only oxidation, because when ATP is formed from ADP and Pi, a water molecule results. In contrast, with ATP hydrolysis, a molecule of water is needed to hydrolyze the ATP.

In these latter two equations, the numbers of molecules of ATP that can be formed during the oxidation of a molecule of glucose or palmitic acid are considered the maximal amounts; they may be lower, as we will discuss later in this chapter. These numbers may also be lower than those you have seen in early textbooks, but revisions to the ATP yield from specific fuels are based on a better understanding of the overall mechanism of oxidative phosphorylation. The **P/O ratio** (sometimes called the ATP/O ratio) is the number of ATP formed for each atom of oxygen consumed. For palmitic acid, the maximum value for the P/O is as follows:

$$106 / (23 \times 2) = 2.3$$

For glucose, the maximum value for P/O is as follows:

$$32 / (6 \times 2) = 2.7$$

Comparing these two numbers reveals that, for the same amount of oxygen consumed, you get more ATP from glucose than you do from a fatty acid. Because of the revisions in the number of ATP generated from glucose or palmitic-acid oxidation, the P/O ratios just presented are lower than the numbers 3.0 and 2.8, respectively, for glucose and fatty acids as found in some earlier biochemistry textbooks.

As we are defining it here, two major metabolic pathways are involved in oxidative phosphorylation. The citric acid cycle breaks down acetyl units derived from fuel molecules and generates the reduced coenzymes NADH and $FADH_2$ as well as CO_2. In the second pathway, the electron transport chain, the free energy released when electrons are transferred from reduced coenzymes (NADH and $FADH_2$) to oxygen is channeled into the phosphorylation of ADP with Pi to make ATP; that is, it drives the following reaction:

$$ADP + Pi \rightarrow ATP + H_2O$$

The mechanism behind oxidative phosphorylation parallels the way electricity is generated by falling water. A dam creates potential energy by raising water to a high level. When the water falls down through special channels, the kinetic energy of the falling water rotates turbine blades in a magnetic field, producing an electric current. During electron transfer from NADH and $FADH_2$ to oxygen, free energy is released. At three complexes of the electron transport chain, the free energy release from electron transport is employed to pump protons (H^+) from the matrix side of the inner membrane of the mitochondria to the outside or cytosolic side—that is, the intermembrane space. As a result, an **electrochemical gradient** is created in which the cytosolic side of the inner membrane is more positive in charge (the *electro* part of the gradient) and has a higher concentration of H^+ (the *chemical* part of the gradient). When protons (H^+) return down the gradient through a special protein complex, the free energy released is used to make ATP from ADP and Pi. In other words, return of protons down their gradient is harnessed into driving ADP phosphorylation, much as the energy of falling water is harnessed to make electricity. Figure 5.3 summarizes this process.

✓ KEY POINT

We describe the regions on either side of the inner mitochondrial membrane as the matrix and cytosolic sides. Yet as figure 5.2 illustrates, what we call the cytosolic side is in reality the intermembrane space. However, the term *cytosolic side* is used in this text because the outer membrane is so porous that the ion composition of the inner membrane space is essentially cytosolic in nature.

The term *proton pumping* refers to the forced movement of H^+ ions across the inner membrane using the free energy released when electrons are passed through the inner membrane protein complexes. The outer (cytosolic) side of the inner membrane (the intermembrane space) has a lower pH (higher [H^+]) by about 1.0 pH unit. This is the chemical part of the electrochemical gradient. In addition, the intermembrane space is more positively charged by about 0.14 to 0.18 volts (140-180 mV); this represents the electrical gradient, often referred to as the *transmembrane electric potential*, given by the symbol $\Delta\psi$ (ψ is the Greek letter psi). Thus, protons are pumped against both a concentration and a charge gradient, which means that this is an example of active transport. The transmembrane electrochemical potential gradient is also described as a **proton motive force** (pmf). Some books and articles use the symbol $\Delta\mu_{H+}$ rather than pmf.

$$\text{pmf } (\Delta\mu_{H+}) = \Delta\psi + \Delta pH$$

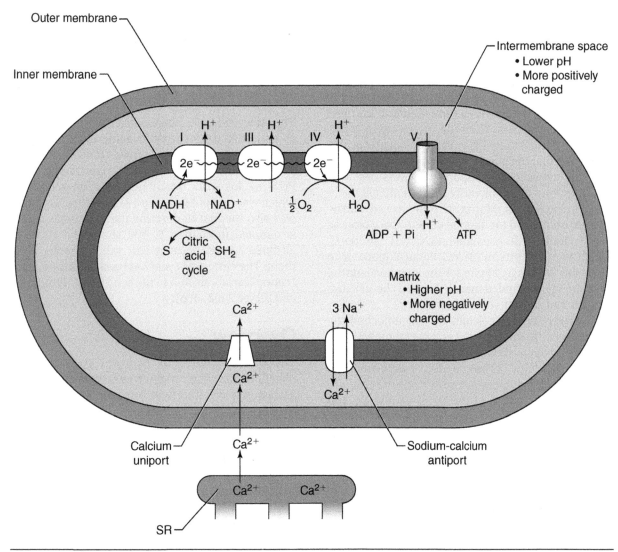

Figure 5.3 An overview of oxidative phosphorylation in mitochondria. Electrons on substrates (SH_2) are transferred to NAD^+, making NADH. The citric acid cycle is the principal pathway for this. Two electrons ($2e^-$) on NADH are then transferred through protein complexes in the inner mitochondrial membrane (I, III, IV) to oxygen, reducing it to H_2O. Energy released during the electron transfer is captured as protons are pumped from the matrix to the intermembrane (cytosolic) side of the inner membrane, creating an electrochemical gradient with a lower pH (higher $[H^+]$) and a more positive charge compared to the matrix. Protons flowing back down this gradient through a special complex V, the ATP synthase, are coupled to ADP phosphorylation to make ATP. Because of the electrochemical gradient, mitochondria can also take up Ca^{2+} ions using a calcium uniport. The source of the Ca^{2+} in muscle is primarily the sarcoplasmic reticulum. Removal of matrix Ca^{2+} occurs through a sodium–calcium antiport, which again takes advantage of the electrochemical gradient.

Calcium and Mitochondria

The electrochemical gradient across the inner membrane favors the uptake of calcium into the matrix from the cytosol whenever the cytosolic calcium concentration is elevated through release of calcium from the sarcoplasmic reticulum of muscle or the endoplasmic reticulum in other cells. The movement of Ca^{2+} ions is driven by both electrical and concentration gradients into the matrix through a special protein called a calcium uniport. Remember that the matrix is more negatively charged than the cytosol and typically has a low calcium-ion concentration. A membrane protein **uniport** allows one-way transport of a molecule or ion down its concentration gradient; this is an example of facilitated diffusion. Mitochondrial calcium uptake is facilitated in skeletal muscle because the calcium release areas on the sarcoplasmic reticulum can be very close to mitochondria. When calcium release channels on the sarcoplasmic reticulum are suddenly opened in response to depolarization of the T-tubule of skeletal or cardiac muscle, the Ca^{2+} concentration in the region next to the release channel is transiently elevated by more than 100-fold. If a calcium uniport is in the immediate area, calcium can flow into the mitochondrial matrix. A **sodium–calcium antiport** removes the calcium from the matrix, exchanging three Na^+ ions for one Ca^{2+}. This exchange is driven by the fact that the matrix is more negatively charged than the cytosol. An **antiport** transports two molecules or ions simultaneously across a membrane in opposite directions. Matrix sodium ions are subsequently exchanged for intermembrane space protons via a sodium–hydrogen antiport (Na^+–H^+ antiport), which uses the proton electrochemical gradient to make this exchange. Figure 5.3 shows mitochondrial calcium uptake and removal through the calcium uniport and sodium–calcium antiport.

The movement of calcium into the mitochondria is facilitated by the anatomical association between the sarcoplasmic reticulum and the mitochondria. This is important because a rise in mitochondrial calcium levels is important for signaling the calcium-dependent activation of the citric acid cycle and electron transport chain enzymes. This helps couple the signal for muscle contraction and the resultant increased ATP utilization tightly to rapid activation of ATP resynthesis via oxidative phosphorylation (Rossi, Boncompagni, and Dirksen 2009). Although calcium is only one of several activators of aerobic metabolism, it is an important factor in the upregulation of important aerobic enzymes, such as isocitrate dehydrogenase and α-ketoglutarate dehydrogenase, as well as the phosphatase enzyme that converts pyruvate dehydrogenase to its active form (Rossi, Boncompagni, and Dirksen 2009). In addition, mitochondrial signaling to the sarcoplasmic reticulum, which is related to its ability to control oxygen radical levels, also affects aspects of calcium release (Dirksen 2009).

CITRIC ACID CYCLE

The citric acid cycle, abbreviated CAC, is also known as the **TCA (tricarboxylic acid) cycle** or **Krebs cycle**. The latter name derives from Sir Hans A. Krebs, the biochemist who was coawarded a Nobel Prize for his work describing the citric acid cycle, which was first published in 1937. In 1932, Dr. Krebs was also the first to describe the urea cycle, which is outlined in chapter 8. You should be aware of all three names, because you will encounter all of them. The citric acid cycle has been described as the "central metabolic hub of the cell" (Berg, Tymoczko, and Stryer 2002, 466).

Overview

The prime function of the citric acid cycle is to completely oxidize (i.e., remove electrons) from acetyl groups in a way that will result in ATP formation. These acetyl groups are formed from all the oxidizable fuels in the body, including carbohydrate, fat, and amino acids from protein. The citric acid cycle removes electrons from acetyl groups and attaches them to nicotinamide adenine dinucleotide (NAD^+) and flavin adenine dinucleotide (FAD), forming NADH and $FADH_2$, respectively. In the electron transport chain, the electrons on the reduced coenzymes NADH and $FADH_2$ will be transferred through a series of carriers to oxygen. In two reactions in the citric acid cycle, the carbon atoms in the acetyl group are released as carbon dioxide. Each kind of fuel is converted to acetyl groups and attached to **coenzyme A** (CoA) (see figure 5.4). Our definition of oxidative phosphorylation is broader than that of most biochemistry books in that we include the role of the citric acid cycle, the pathway where most of the electrons on fuels are transferred to coenzymes.

> ☑ **KEY POINT**
>
> Calcium can be a dangerous substance to the cell because it can initiate cell death or apoptosis. Accordingly, its cytosolic concentration is generally kept very low and is carefully regulated. On the other hand, calcium has many positive effects for the cell, since changes in its concentration can signal alterations in many reactions and processes, such as initiating muscle contraction and oxidative phosphorylation. One example is the ability of mitochondrial calcium to stimulate some key reactions that regulate oxidative phosphorylation, such as when muscle contraction begins at the start of exercise.

> ☑ **KEY POINT**
>
> The citric acid cycle consumes acetyl groups. One limit to oxidative phosphorylation is the rate at which acetyl groups are provided to the citric acid cycle. The maximum rate of oxidative phosphorylation is markedly reduced if available carbohydrate stores are depleted. This suggests that the ability to produce acetyl CoA from fat can be a limiting factor in exercise.

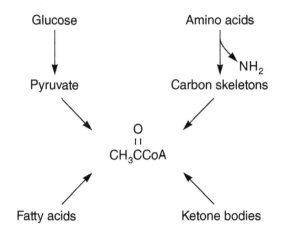

Figure 5.4 Oxidation of fuels. Glucose, fatty acids, ketone bodies (derivatives of fatty acid metabolism), and amino acids are broken down into two-carbon acetyl groups attached to coenzyme A (CoA). The resulting acetyl CoA feeds into the citric acid cycle. The amino groups of amino acids, which cannot be oxidized, are removed and excreted in the urine.

Coenzyme A is derived from **pantothenic acid**, a B vitamin. It acts as a handle to attach to a number of acyl groups, some of which we will see later. Coenzyme A has a terminal SH (sulfhydryl) group to which the acetyl group is attached, forming an energy-rich thioester bond. Coenzyme A and acetyl CoA are often written as CoASH and $CH_3COSCoA$, respectively. However, for simplicity, we use CoA and acetyl CoA.

Oxidation of acetyl CoA accounts for about two-thirds of the ATP formation and oxygen consumption in mammals. Acetyl groups enter the citric acid cycle, where their two carbon atoms are eliminated as CO_2, while the hydrogens and their associated electrons are removed. Figure 5.5 summarizes and simplifies this process. This picture also reinforces the concept that removal of electrons from fuels occurs using hydrogen. The $:H^- + H^+$ represents the hydride ion and proton, respectively, that are removed from many fuel substrates. As we saw in chapter 2, electrons accompany the hydrogen removed from substrates. The hydride with its two electrons, shown as dots, is attached to NAD^+, forming NADH; the negative charge on the hydride ion balances the positive charge with NAD^+. The two $H\cdot$ represent two hydrogen atoms removed from succinate, although succinate is not identified in the figure. The two hydrogen atoms, each containing one electron, become attached to FAD, forming $FADH_2$. Figure 5.5 shows that one turn of the citric acid cycle consumes one acetyl group and produces four pairs of electrons, one guanosine triphosphate (GTP), and two CO_2. Two water molecules are also consumed,

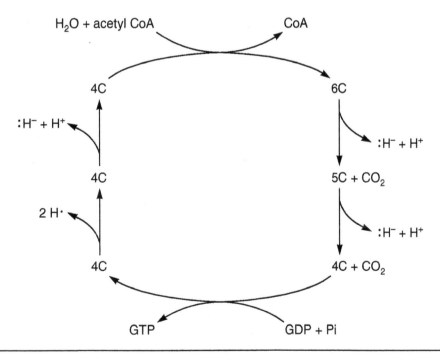

Figure 5.5 An overview of the citric acid cycle showing the path of carbon atoms and electrons that are recovered in association with hydrogen. The $:H^-$ is recovered in the form of NADH, while the two $H\cdot$ are recovered in the form of $FADH_2$. The two-carbon acetyl unit feeds into the cycle attached to coenzyme A (CoA). The carbon atoms are recovered as carbon dioxide (CO_2). During one step in the cycle, enough free energy is released to phosphorylate guanosine diphosphate (GDP) to make guanosine triphosphate (GTP).

although in the simplified diagram in figure 5.5, only one H_2O is shown. The GTP produced is an example of substrate-level phosphorylation—that is, formation of an energy-rich phosphate without the use of oxidative phosphorylation. The ATP produced in glycolysis also occurs via substrate-level phosphorylation.

> ### ✓ KEY POINT
> FAD is reduced to $FADH_2$ and electrons are transferred to oxygen in only a few reactions. Reduction of NAD^+ to form NADH occurs in many reactions—three in the citric acid cycle and one in glycolysis. For this reason, we simplify our thinking by focusing on NADH and not $FADH_2$.

Reactions of the Citric Acid Cycle

Figure 5.6 shows the complete citric acid cycle, including the structures of intermediates. Since many readers are not comfortable with organic chemistry, figure 5.7 also summarizes the citric acid cycle without the use of chemical structures. As is the custom for this book, double-headed arrows are used to show reactions that, if isolated, would reach equilibrium. Arrows showing one direction only define reactions in which there is a large free energy change; if studied in isolation, these reactions would use up virtually all substrate. The overall direction of the tricarboxylic acid cycle is to consume acetyl groups.

Hydrolysis of acetyl CoA to acetate and CoA has a $\Delta G°'$ of –32 kJ/mol (–7.7 kcal/mol). Thus, the reaction combining acetyl CoA, oxaloacetate, and water to form citrate, catalyzed by *citrate synthase*, is virtually irreversible. In the next step, the tertiary alcohol group on citrate is converted to a secondary alcohol group by two steps in the reaction catalyzed by the enzyme *aconitase*. The first step is a dehydration reaction in which the tertiary OH group is removed, producing cis-aconitate, which remains attached to the aconitase enzyme. A hydration reaction follows, resulting in the formation of isocitrate. Notice that the only difference between citrate and isocitrate is the position and type of OH group—tertiary alcohol on citrate and secondary alcohol on isocitrate.

Isocitrate undergoes oxidative decarboxylation, catalyzed by *isocitrate dehydrogenase*, to form α-ketoglutarate. This is the first of four oxidation–reduction reactions in the citric acid cycle. First, oxidation generates NADH and H^+; then decarboxylation forms CO_2. At this point in the cycle, one of

> ### ✓ KEY POINT
> The CO_2 produced during the citric acid cycle also serves to eliminate the carbon molecules from the carbohydrate, fat, and amino acids consumed in metabolism, which enter the citric acid cycle as acetyl. This occurs in much the same way that burning wood is reduced to ashes as smoke and heat are given off—the CO_2 produced by the citric acid cycle serves to eliminate the mass or weight of the body's fuel stores. Thus, for example, much of the actual weight of body fat lost resulting from diet and exercise is lost as the CO_2 we breathe out. In addition, as an excellent example of the interconnections and regulation of whole-body responses to exercise, CO_2 is also a primary driver of ventilation during exercise. As exercise intensity goes up, the increased CO_2 derived from the citric acid cycle serves to stimulate its removal from the circulation by signaling increases in ventilatory drive, which increases the rate and depth of breathing.

the two carbon atoms on the acetyl group is removed as carbon dioxide.

In the next step, an α-ketoglutarate (some call this 2-oxoglutarate) undergoes oxidative decarboxylation to succinyl CoA. This step begins with decarboxylation, followed by oxidation, generating NADH and H^+. In this process, the second of the two acetyl carbon atoms is lost as CO_2. The enzyme *α-ketoglutarate dehydrogenase* (2-oxoglutarate dehydrogenase) catalyzes the same kind of reaction as pyruvate dehydrogenase, a reaction we will encounter soon, and contains the same kinds of subunits and coenzymes. Both enzymes contain three types of polypeptide subunits and five coenzymes. The coenzymes are NAD^+ and CoA, which are loosely bound, and thiamine pyrophosphate, **lipoic acid**, and FAD, which are tightly bound and are not seen in the simple way the reaction is presented in figures 5.6 and 5.7. Recall from table 2.1 (p. 25) that **thiamine pyrophosphate** (TPP) is derived from the B vitamin, thiamine. During the α-ketoglutarate dehydrogenase reaction, enough free energy is released to generate the energy-rich succinyl CoA. Like the reactions catalyzed by citrate synthase and isocitrate dehydrogenase, the α-ketoglutarate dehydrogenase reaction is virtually irreversible, providing a one-way direction to the citric acid cycle.

In the next step, succinyl CoA is broken down to succinate and CoA in a reaction catalyzed by *succinyl CoA synthetase* (SCS). When this occurs, the free energy released when energy-rich succinyl CoA

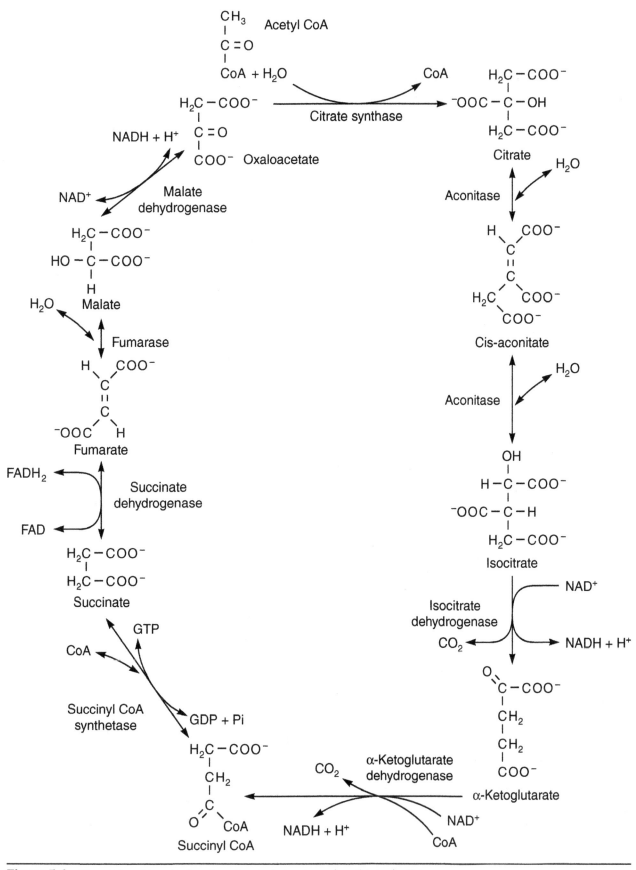

Figure 5.6 The citric acid cycle showing chemical structures for all intermediates.

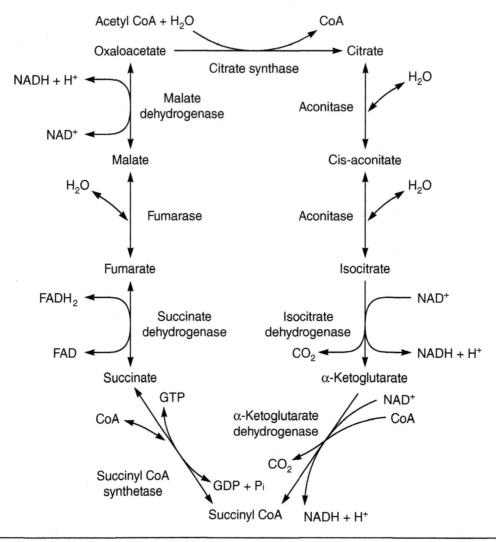

Figure 5.7 The citric acid cycle shown without chemical structures.

is hydrolyzed drives the substrate-level phosphorylation of GDP to make GTP. The SCS reaction is freely reversible, and the enzyme's name denotes the backward reaction, in keeping with the naming of similar reactions in biochemistry. The GTP produced may be used (a) for peptide bond formation during the process of translation, (b) to phosphorylate ADP to make ATP using the enzyme nucleoside diphosphate kinase, described previously, or (c) in certain types of cell signaling processes. SCS consists of a highly conserved α subunit and a β subunit that determines whether GTP or ATP will be produced during this reaction. Although GTP generation in the SCS step is commonly depicted in biochemistry textbooks, including this one, in highly oxidative tissues such as heart and skeletal muscle, it is actually ATP generation that predominates (Phillips et al. 2009). SCS can be activated by, among other things, increases in Pi concentration, which will occur when ATP is

hydrolyzed to ADP and Pi during exercise (Phillips et al. 2009). These issues are noted in chapter 4.

Once succinate is formed, the remaining three reactions of the citric acid cycle regenerate oxaloacetate (one of the starting substances of the citric acid cycle, along with acetyl CoA) and generate electrons for the electron transport chain. *Succinate dehydrogenase* (SDH) is an oxidation–reduction enzyme that contains a tightly bound FAD. Unlike the other enzymes of the citric acid cycle that are found in the mitochondrial matrix, SDH is a component of the inner mitochondrial membrane. In the SDH reaction, electrons are transferred from succinate to FAD and then to **coenzyme Q**. We will see later in this chapter that SDH and coenzyme Q are part of the electron transport chain. The product from the SDH reaction, fumarate, is hydrated (water molecule added) to malate by the enzyme *fumarase*. Malate contains a secondary alcohol group that is oxidized

✓ KEY POINT

The citric acid cycle is literally a cycle in that its start and end points are oxaloacetate. The cycle serves to take the carbon, hydrogen, and oxygen molecules derived from acetyl (and originally from carbohydrate, fat, and protein), which are initially attached to oxaloacetate to create citrate and to eliminate them via the generation of H^+ ions (attached to NAD and FAD), CO_2, and H_2O.

in the *malate dehydrogenase* reaction, generating NADH + H^+ and oxaloacetate, a starting substrate for a new round of the cycle.

The last three enzymes of the citric acid cycle carry out a three-step reaction sequence in which a methylene group (CH_2) in the first molecule (succinate) is converted to a carbonyl group (C=O) in the last molecule, oxaloacetate. This is accomplished by a dehydrogenation of succinate, generating $FADH_2$ and fumarate; a hydration reaction to make malate; followed by another dehydrogenation, forming NADH and oxaloacetate. This sequence of dehydrogenation, hydration, and dehydrogenation is common in life; we will next see this in the sequence of reactions breaking down fatty-acid molecules, a process known as beta-oxidation.

If we add all the reactions of the citric acid cycle algebraically, we get the following:

$$\text{acetyl CoA} + 3\ NAD^+ + FAD + GDP + P_i + 2\ H_2O \rightarrow 2\ CO_2 + GTP + 3\ NADH + 3\ H^+ + FADH_2 + CoA$$

As shown in this summary equation, the citric acid cycle does not involve the net production or consumption of oxaloacetate or any other constituent of the cycle. The only things consumed are an acetyl group and two water molecules. The summary equation also shows that oxygen is not directly involved in the citric acid cycle. As shown next, oxygen plays an obligatory secondary role in that without it, the citric acid cycle would quickly cease because there would be insufficient NAD^+ and FAD for the cycle to continue. Evidence is accumulating to show that the enzymes of the citric acid cycle are physically located in the matrix to enhance the flux of this pathway by transferring the product of one reaction directly to the next enzyme. The word **metabolon** has been used to denote a multienzyme complex in which products and substrates are physically channeled to enzymes, as opposed to cycle intermediates diffusing randomly in the matrix.

The reduced coenzymes produced in the citric acid cycle (NADH and $FADH_2$) are oxidized in the electron transport chain, and their electrons are transferred to oxygen. We can show this electron transfer as follows:

$$3\ NADH + 3\ H^+ + FADH_2 + 2\ O_2 \rightarrow 3\ NAD^+ + FAD + 4\ H_2O$$

Associated with the transfer of electrons from the reduced coenzymes to oxygen is the tightly coupled ADP phosphorylation reaction, producing ATP. On the basis of conservative estimates, the transfer of electrons from three NADH to oxygen will yield 7.5 ATP. The transfer of electrons from one $FADH_2$ to O_2 will generate 1.5 ATP. In counting the energy-rich phosphates, one must include the GTP formed during the succinyl CoA synthetase reaction. In summary, the complete oxidation of one acetyl group is associated with the formation of 10 ATP. Figure 5.8 illustrates the close coupling of the citric acid cycle, the electron transfer chain, and ADP phosphorylation.

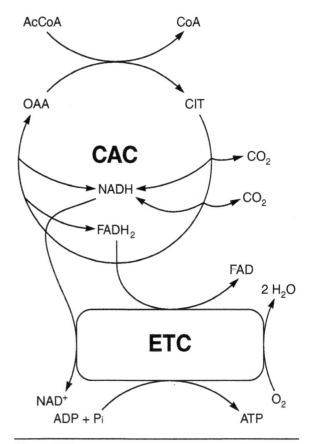

Figure 5.8 The citric acid cycle (CAC), electron transport chain (ETC), and ADP phosphorylation to ATP. All these are tightly coupled in mitochondria. OAA is oxaloacetate and AcCoA is acetyl CoA.

KEY POINT

For the citric acid cycle to operate at a high rate, there must be sufficient oxaloacetate to accept acetyl groups from acetyl CoA in the first reaction of the cycle. If oxaloacetate is used for any other purpose, the maximal power of the citric acid cycle may be reduced. In fact, a small amount of Krebs cycle intermediaries are constantly lost to other cellular reactions. Oxaloacetate needs to be continually recreated from other sources during metabolism. Carbohydrate serves as the primary source of this recreation of oxaloacetate via metabolic pathways that are distinct from the citric acid cycle. Hence, our cells must constantly utilize a small amount of carbohydrate to continually replenish the oxaloacetate levels in the mitochondria. This is one of a number of reasons why, as we start to run out of carbohydrate stores during long-endurance exercise, we may start to experience fatigue, even if we have plenty of fuel stores still available as fat. This occurs because we are not capable of oxidizing only fats or lipids for fuel without a small or obligatory breakdown of carbohydrate. Some biochemistry textbooks colloquially note this relationship and suggest that fat is "burned in a flame of carbohydrate."

ELECTRON TRANSPORT CHAIN

From a functional perspective, the electron transport chain (ETC), commonly called the respiratory chain, consists of four protein–lipid complexes located in the inner membrane of the mitochondrion. The role of these complexes is to participate in the transfer of electrons to oxygen. It is worth emphasizing again that we talk about electrons being transferred to oxygen, and this is true. However, as we have seen repeatedly, the electrons are most commonly transferred in association with hydrogen atoms or hydride (H^-) ions. In three of the four complexes that make up the electron transport chain, the free energy released during electron transport is associated with proton pumping from the matrix to the cytosolic side of the inner mitochondrial membrane. As already indicated, the term *pumping* means that the protons are transported across the inner membrane against both an electrical and a concentration gradient. Figure 5.9 shows the four complexes (each identified with a Roman numeral and a name) involved in electron transport and provides a perspective on the energy released during electron transfer, based on the concept that moving down releases free energy. The standard free-energy axis in figure 5.9 helps to put the free

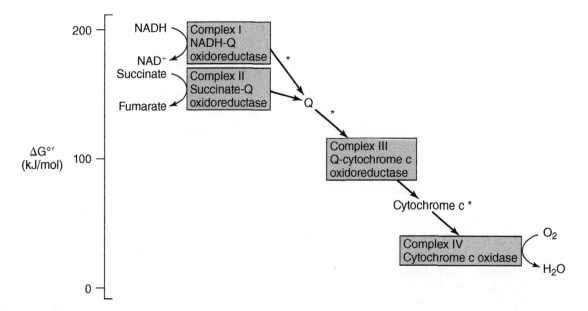

Figure 5.9 The path of electrons from NADH to oxygen uses complexes I, III, and IV. Electron transfer from succinate to oxygen goes through complex II with $FADH_2$, complex III, and complex IV. A perspective on the free energy released during the electron transfers can be seen by the relative position of the complexes in the chain, given as drops in standard free energy ($\Delta G°'$). Asterisks indicate where the free energy released can be captured by the pumping of protons across the inner mitochondrial membrane. The free energy release in electron transfer from succinate to Q, utilizing complex II, is insufficient to be captured as proton pumping.

energy release into numbers, as discussed in chapter 4. Electrons transferred from NADH through complex I to coenzyme Q (shown as Q) result in proton pumping. The free energy released when electrons are transferred from Q to cytochrome c, utilizing complex III, also results in enough energy to pump protons. When electrons on cytochrome c are transferred through complex IV to oxygen, more energy is captured in pumping protons. On the other hand, the free energy release in transferring electrons from succinate through complex II to coenzyme Q is too little to be captured as protons transported from the matrix to the intermembrane side on the inner membrane.

Two electrons are transferred via NAD^+, requiring dehydrogenases from fuel substrates (many in the citric acid cycle), through a series of electron carriers to oxygen. The sequences of electron flow are nothing more than a series of oxidation–reduction reactions, each of which can be shown as follows:

$$\text{reduced A} + \text{oxidized B} \rightarrow \text{oxidized A} + \text{reduced B}$$

As we will show near the end of this chapter, energy release during electron transfer can be measured in units of volts. When reduced substrate A in the preceding equation transfers electrons to oxidized substrate B, this is a downhill-type reaction that can be measured in volt units or displayed with energy units, as seen in figure 5.9. B has a greater affinity for electrons than does A. We can show this more clearly as the reduction of B, when it accepts two electrons associated with hydrogen in the equation:

$$AH_2 + B \rightarrow A + 2e^- + 2H^+ + B$$
$$\rightarrow A + BH_2$$

This is exactly what takes place in the electron transfer or respiratory chain. As shown in figure 5.9, electrons on NADH are transferred to Q using complex I. Q has a higher affinity for electrons than does NADH, so this reaction is associated with a release of energy. Similarly, cytochrome c has a higher affinity for electrons than Q, and so, utilizing complex III, electrons on reduced Q are transferred to cytochrome c. Oxygen has a very high affinity for electrons, so electrons on cytochrome c are passed to oxygen in a reaction catalyzed by cytochrome c oxidase. When oxygen accepts electrons and hydrogen, it becomes water. In each of the three electron transfer reactions, the energy released is captured through proton pumping that creates the electrochemical gradient across the inner mitochondrial membrane. Return of these protons through the ATP synthase is coupled to ADP phosphorylation with Pi to make ATP.

Figure 5.10 illustrates electron transfer in the inner membrane of a mitochondrion, showing the spatial relationships among the four electron transfer protein complexes and the two other intermediates of the electron transfer (respiratory) chain, coenzyme Q (Q) and cytochrome c (Cyt c). The two sources of electrons are NADH and succinate through $FADH_2$. Electrons from NADH are transferred to oxygen using complexes I, III, and IV. During the transfer of two electrons from NADH to oxygen, the free energy release is captured as a total of 10 protons are pumped across the inner membrane, creating

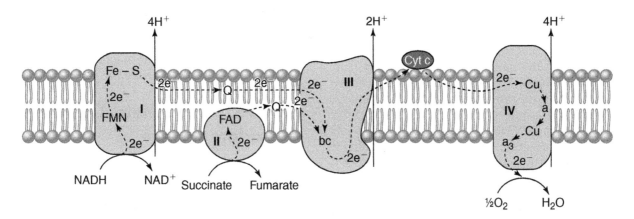

Figure 5.10 The flow of electrons in the inner mitochondrial membrane, showing spatially how the four complexes of the electron transfer (respiratory) chain are located with respect to each other and how the two other electron carriers, coenzyme Q (Q) and cytochrome c (Cytc), are arranged. Electrons are transferred from NADH through complexes I, III, and IV to oxygen. Electrons on succinate are transferred from FAD to Q and then through complexes III and IV to oxygen. The path of electron transfer is shown with the dotted lines. The number of protons pumped at each complex is illustrated. Complex I has an intermediate flavin mononucleotide (FMN). Complex III contains two other cytochromes, b and c1, as intermediates.

the electrochemical gradient. Four protons are pumped at complexes I and IV and two at complex III. The second source of electrons in figure 5.10 is succinate. Here, electrons are first transferred to an FAD in complex II, then through Q, to complex III, cytochrome c, and finally to oxygen using complex IV. The free energy released when electrons on succinate are passed to FAD in complex II and then to coenzyme Q is insufficient to pump protons across the inner membrane. Thus, a total of six electrons are pumped per pair of electrons passed from succinate to oxygen. In the following sections, we will look at each complex in more detail.

> **KEY POINT**
>
> It has been common to describe a pair of electrons as **reducing equivalents** or agents. Reducing equivalents or agents could be the electrons that are transferred from NADH or $FADH_2$. Since we talk about oxidation and reduction in terms of losing and gaining electrons, respectively, the term *reducing equivalents* is appropriate.

Complex I: NADH–Coenzyme Q Oxidoreductase

Complex I, which is also known as NADH dehydrogenase, is a huge complex consisting of at least 42 different polypeptide subunits, 7 of which are coded by genes in mitochondrial DNA; the remainder are coded by genes in nuclear DNA. The role of this largest complex is to transfer a pair of electrons from NADH in the matrix to the oxidized form of coenzyme Q (abbreviated Q; also known as **ubiquinone**) in the inner membrane, reducing it to QH_2 (**ubiquinol**). We have already seen a variety of substrates (lactate, isocitrate, α-ketoglutarate, and malate) whose oxidation is associated with the reduction of NAD^+, generating NADH. We will see more of these substrates and their reduction by NAD-dependent dehydrogenases later. Complex I also contains another cofactor, *flavin mononucleotide* (FMN), that is related structurally to FAD and is formed from the B vitamin riboflavin (see table 2.1, p. 25). Proteins containing tightly bound cofactors, such as FMN and FAD, are known as *flavoproteins*. Clusters of iron linked with sulfur also appear in this complex. Researchers who study the electron transport chain learn details about the individual complexes by blocking certain steps with inhibitors. The plant toxin rotenone specifically blocks complex I. When this is done, the remainder of the electron transport chain can be activated through the addition of succinate, the specific substrate for complex II.

Reduction of coenzyme Q (ubiquinone) by NADH to make NAD^+ and QH_2 (ubiquinol) is catalyzed by complex I. Details of this reaction are shown in figure 5.11. Overall, the reaction involving complex I causes the fully oxidized coenzyme Q (ubiquinone) to accept two electrons and two protons to form the fully reduced coenzyme QH_2 (ubiquinol). The reaction proceeds in two steps. The first step involves the transfer of a single electron to coenzyme Q, forming a short-lived species known as ubisemiquinone, which is a free radical. Normally, a second electron is quickly added to the **ubisemiquinone**, along with two protons, to produce the fully reduced ubiquinol form. However, in resting cells, the overall flow of electrons from NADH to oxygen through complexes I, III, and IV is slower, so the lifetime of the free radical ubisemiquinone is fairly long. As we will see later, ubisemiquinone can pass its one free electron to oxygen to produce **superoxide**, a free radical of molecular oxygen containing one extra electron. It is written as O_2^-. Coenzyme Q has also been touted as a dietary antioxidant since it can be obtained from diet supplements and can act as an antioxidant when present in the circulation (Tauler et al. 2008). We will discuss this and other issues related to oxygen radicals and antioxidants during exercise later in this chapter. Coenzyme Q is a very small molecule compared to proteins. It contains a large hydrophobic side chain, represented by R in figure 5.11. Thus coenzyme Q is at home in the hydrophobic interior of the mitochondrial inner membrane where it shuttles electrons back and forth. A considerable amount of free energy is released when two electrons are transferred from NADH to coenzyme Q using complex I. This is captured via pumping of four protons across the inner mitochondrial membrane.

Complex II: Succinate–Coenzyme Q Oxidoreductase

Succinate–coenzyme Q oxidoreductase is commonly known as succinate dehydrogenase (SDH); we have seen this as a component of the citric acid cycle. The role of SDH is to transfer electrons on succinate to coenzyme Q (ubiquinone), forming fumarate and coenzyme QH_2 (ubiquinol). The actual enzyme complex contains four polypeptide subunits, none of which are coded by mitochondrial DNA. Flavin adenine dinucleotide, a tightly bound cofactor in SDH, accepts electrons from succinate to become the reduced $FADH_2$ form. Subsequently, the elec-

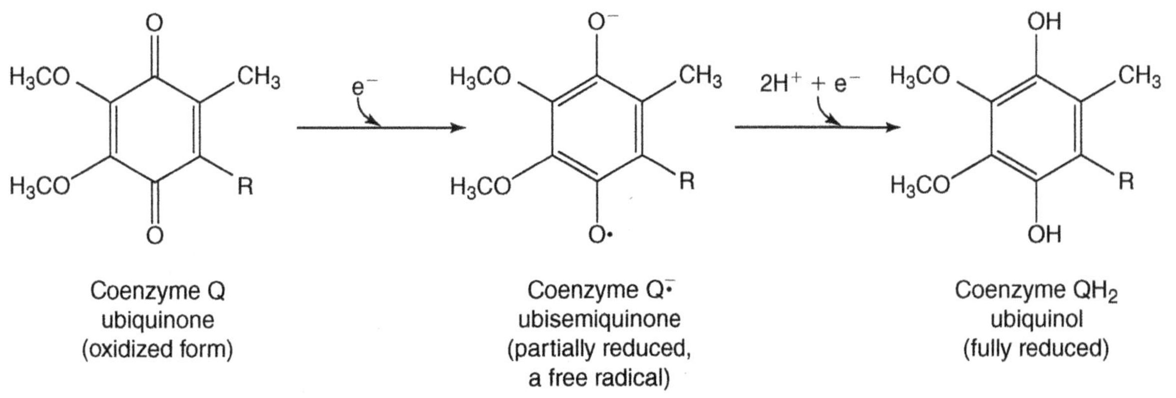

Figure 5.11 Reduction of coenzyme Q, also widely known as ubiquinone. The reduction is accomplished in two stages. In the first stage, a single electron is added, creating a charged, free radical species known as ubisemiquinone. Subsequently, a second electron and two protons are added to make the final product coenzyme QH_2, also known as ubiquinol.

trons are transferred to coenzyme Q, reducing it to coenzyme QH_2. Although we show the citric acid cycle enzyme SDH as producing $FADH_2$ as a product, strictly speaking, it is QH_2 that is the reduced product of this reaction. We acknowledge the true final product, QH_2, but show $FADH_2$ in keeping with a tradition passed down through a succession of biochemistry textbooks.

Other flavoprotein enzymes are also part of the inner mitochondrial membrane; using the tightly bound cofactor FAD, they transfer electrons from substrates to coenzyme Q. Strictly speaking, these other FAD-containing enzymes are also part of complex II. Unlike what happens with complexes I, III, and IV, the free energy change associated with complex II is not large; therefore, no protons are pumped across the inner membrane during electron transfer in this type of dehydrogenation.

Complex III: Coenzyme Q–Cytochrome c Oxidoreductase

Complex III or coenzyme Q–cytochrome c oxidoreductase transfers electrons from reduced coenzyme Q (QH_2) to cytochrome c. Complex III is a dimer, with each monomer containing 11 different polypeptide subunits; all but 1 are coded by nuclear DNA. The **cytochromes** are a class of heme electron transport proteins located in or on the inner membrane of the mitochondrion. In the center of the heme group is an iron ion that can exist in an oxidized (Fe^{3+}) or reduced (Fe^{2+}) state. In complex III, electrons on reduced coenzyme Q (i.e., QH_2) are transferred to cytochrome c, changing the iron from Fe^{3+} to Fe^{2+}. Since reduction of oxidized iron involves accepting only one electron, two cytochrome c molecules must be reduced to accept the electrons from each QH_2. Although complex III involves electron transfer from coenzyme Q to cytochrome c, two other types of cytochromes (known as cytochrome b and cytochrome c1) are intermediates in the stage from Q to cytochrome c. Electron transfer from ubiquinol (QH_2) to cytochrome c using complex III is not a simple process. More comprehensive sources provide details on the actual path of electron transport and proton pumping (described as the *Q cycle*) (Trumpower 1990).

Cytochrome c is a small protein that is not a part of any of the four electron transfer complexes. It is attached to the intermembrane side of the inner mitochondrial membrane. In its location, it can accept electrons from QH_2 and transfer them to oxygen using complex IV. During electron transfer from QH_2 to cytochrome c, enough free energy is released so that two protons are pumped across the inner mitochondrial membrane. In the following summary equation, note that only the electrons are transferred from QH_2 to cytochrome c. As a result, two protons are left over. Complex III can be specifically inhibited by the antibiotic antimycin A.

$$QH_2 + 2 \text{ cytochrome c-}Fe^{3+} \rightarrow Q + 2H^+ + 2 \text{ cytochrome c-}Fe^{2+}$$

Complex IV: Cytochrome c Oxidase

Cytochrome c oxidase contains 13 different polypeptide subunits coded by 10 nuclear and 3 mitochondrial genes. Two of these are cytochrome a and cytochrome a_3; two copper ions also appear in another polypeptide chain. This complex accepts electrons from reduced cytochrome c and passes them to an oxygen molecule, reducing it (after combination with four protons) to two water molecules. The sequence is as follows. A single electron on reduced cytochrome c is transferred to a protein-bound Cu^{2+}, reducing it to Cu^+. The electron is then transferred to cytochrome a, then to another protein-bound copper ion, to cytochrome a_3, and finally to oxygen. Because the oxygen molecule (O_2) contains two atoms of oxygen, we need four electrons to reduce it. We cannot break up the oxygen molecule as suggested earlier by the use of 1/2 O_2; this form of expression is for convenience. We can summarize in the following way the cytochrome oxidase reaction as it really occurs:

$$4 \text{ cytochrome c-Fe}^{2+} + O_2 + 4 H^+$$
$$\rightarrow 4 \text{ cytochrome c-Fe}^{3+} + 2 H_2O$$

The free energy released during electron transfer in cytochrome oxidase (complex IV) results in the pumping of four protons across the inner membrane. Complex IV can be strongly inhibited by cyanide, which accounts for the deadly nature of this substance.

In nature (outside the human body), oxygen radicals are formed as intermediaries when electrons and protons are naturally added in sequence to oxygen, resulting in the ultimate formation of water molecules. Cytochrome c oxidase largely prevents the formation of oxygen radicals in the mitochondria during this process by tightly regulating these intermediary steps and only allowing the finished product of H_2O to accumulate. Billions of years ago, when Earth's oxygen levels rose, the evolutionary development of cytochrome c oxidase allowed anaerobic life forms to adapt to a more oxygen-rich environment without greatly increasing their metabolic exposure to oxygen radicals. These issues will be discussed further later in this chapter.

Summary of the Electron Transfers

The steps associated with the electron transport chain can be summarized as follows:

- Two electrons in the form of a negatively charged hydrogen ion (the hydride ion, :H^-) are transferred from most substrates to NAD^+, creating NADH and a free proton (H^+). These reactions are catalyzed by NAD-dependent dehydrogenase enzymes, three of which we have encountered in the citric acid cycle.

- The electrons on NADH are transferred to oxygen utilizing three large protein complexes in the inner mitochondrial membrane: complexes I, III, and IV. Two other intermediates play key roles in this electron transfer. Coenzyme Q, often called ubiquinone, is a relatively small, hydrophobic molecule in the inner mitochondrial membrane. Coenzyme Q can accept two electrons and two protons from complex I to become the reduced form coenzyme QH_2, or ubiquinol.

- Electrons on ubiquinol are transferred to a small heme-containing protein, cytochrome c, on the cytosolic side of the inner mitochondrial membrane. Reduction of cytochrome c is catalyzed by complex III.

- Finally, electrons on reduced cytochrome c are transferred to oxygen utilizing complex IV.

During the transfer of electrons from NADH to oxygen, a great deal of free energy is released. This is captured by the creation of a proton and charge gradient across the inner membrane through a process of proton pumping at complexes I, III, and IV. For each pair of electrons transferred from NADH to oxygen, 10 protons are pumped across the inner membrane. The stoichiometry for this could be described as $10H^+/2e^-$.

Other reduced substrates (e.g., succinate) are oxidized by another class of dehydrogenase enzymes containing FAD. Two electrons in the form of two hydrogen atoms are transferred to FAD, reducing it to $FADH_2$. The electrons on $FADH_2$ are transferred to coenzyme Q, complex II, and then to oxygen using complexes III and IV. Less free energy is released in electron transfers from succinate to oxygen, and a total of only six protons are pumped across the inner membrane. The proton pumping to electron transfer stoichiometry for this would be $6H^+/2e^-$. Now that we have described the processes by which the citric acid cycle and electron transfer function to liberate the energy to form ATP, we move to considering how the ATP is actually generated by capturing a part of this energy.

☑ KEY POINT

The free energy released during electron transfers from NADH to oxygen is very large. Under standard conditions, it is −220 kJ/mol (−52.5 kcal/mol). Less free energy is released from electron transfer from substrates, such as succinate, that use FAD-dependent dehydrogenases. Under standard conditions, −200 kJ/mol (−47.8 kcal/mol) of free energy is released.

COUPLED PHOSPHORYLATION

We intuitively know that water flows downhill but must be pumped uphill. Therefore, the significance of the term *proton pumping* should be clear. The pump is driven by the energy released when the electrons flow from one complex to another that is more easily reduced. The cytosolic (intermembrane) side of the membrane has a lower pH (a higher H^+ concentration) and would have a higher positive electrical charge. Thus, protons must be pumped across the inner membrane against both an electrical and a chemical gradient, creating a proton motive force, based on a difference in electric potential ($\Delta\psi$) and a difference in pH (ΔpH). However, just as when water is raised above a dam, the electrochemical gradient can be exploited, since the energy released when protons flow back into the matrix is allowed to drive the phosphorylation of ADP, making ATP.

Coupled phosphorylation is based on the concept that electron transport is linked to ATP synthesis by way of proton pumping. This concept originated with the English biochemist Dr. Peter Mitchell in 1966. His concept is known as the **chemiosmotic hypothesis**, for which Dr. Mitchell received a Nobel Prize in 1978. As already mentioned, 10 protons (H^+) are pumped per pair of electrons transferred from NADH to oxygen. Similarly, for each pair of electrons transferred from $FADH_2$ to Q to oxygen, 6 protons are pumped.

ATP Synthase

The ATP synthase, or complex V, couples energy released during proton flow down the gradient into the matrix to phosphorylation of ADP with Pi to make ATP. This concept is similar to that of using energy from the flow of water to make electricity, as previously described.

A schematic of the ATP synthase is shown in figure 5.12. Adenosine triphosphate synthase consists of two parts. The F_o subunit, embedded within the inner membrane of the mitochondria, acts as a pore to allow protons to pass into the matrix. F_o consists of a disc of 10 c subunits plus an a and b subunit. Tightly associated with F_o is the F_1 subunit, which bulges into the matrix. It is composed of a central stalk (γ subunit) that can rotate and an immobile head complex composed of three α and three β subunits, involved in ADP and Pi binding and ATP formation. The current model, as shown in figure 5.12, has the protons passing through F_o. Energy released during this passage creates a rota-

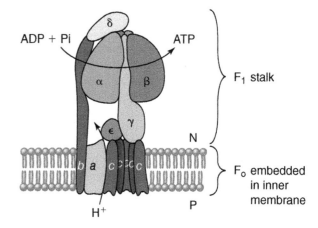

Figure 5.12 The synthesis of ATP accomplished by complex V, the ATP synthase. The ATP synthase consists of two major components. The F_o part is composed of a disc of c subunits and an a and a b subunit that make up the core within the inner membrane. Attached to this is the F_1 complex, which extends into the matrix side (N for negative) of the inner membrane and consists of a number of other polypeptide subunits, given by Greek letters. Protons passing through F_o create a rotation of the c subunits of F_o and the γ subunit of F_1. The rotational energy of the γ stalk of F_1, inside the fixed α and β subunits, drives ADP phosphorylation to make ATP. Three protons are thought to pass through the membrane from the cytosolic (P for positive) to the matrix side for each molecule of ATP manufactured by the complex.

Copyright © Dr. Boris Feniouk. Adapted from www.biologie.uni-osnabrueck.de/biophysick/Feniouk/home.html

tional motion in F_o that is transmitted to the F_1 part. The rotational energy in F_1 is thought to drive ADP phosphorylation to make ATP. In essence, the ATP synthase is acting like a rotatory engine, utilizing the rotational energy created by the flow of protons down their gradient to combine ADP and Pi to make ATP. The ATP is released to the matrix. The symbols N (for negative) and P (for positive) are often used to identify the matrix and intermembrane spaces, respectively, reminding us that the matrix is more negatively charged than the cytosolic side.

Three protons are thought to pass through the inner membrane from the cytosolic to the matrix side for each molecule of ATP manufactured by the complex. We have already confirmed that the proton pumping to electron transport stoichiometry is $10H^+/2e^-$ for electron transfer from NADH to oxygen and $6H^+/2e^-$ for electron transport from succinate through FAD to oxygen. Putting together this information, we could say that the ATP/O ratio for NADH is 3.3 ($10H^+/3$ protons for each ATP) and for $FADH_2$ is 2 ($6H^+/3$ protons for each ATP). Indeed, these are the numbers that have been used for many years as the P/O ratios for NADH and FADH$_2$. However, we will see that this is a gross oversimplification, and it is important to be mindful of this fact.

Complex V, ATP synthase, is also known as the F_o-F_1 ATPase because it can run backward. In this case, the free energy released from ATP hydrolysis (the ATPase part) drives proton translocation from the matrix to the cytosolic side—the opposite of what occurs normally during the synthesis of ATP. Such a reverse movement of protons driven by ATP hydrolysis would increase the electrochemical potential across the inner membrane. While reversal of the ATP synthase can artificially be made to occur in solution, to the best of our knowledge, there are no conditions under which this would normally occur in the body.

Uncoupled Oxidative Phosphorylation

Normally, electron transport from substrates to coenzymes to oxygen is tightly coupled to ADP phosphorylation, making ATP. However, protons can also leak across the inner membrane from the cytosolic side to the matrix side without accompanying ATP formation. This is known as *uncoupled respiration*, because oxygen would be consumed as the electrons are accepted from substrates, but ATP would not be made. This is shown in figure 5.13. As you may suspect, uncoupled respiration actually decreases the ATP/O (P/O) ratio. In this way, the free energy of the electrochemical gradient is immediately released as heat, warming the animal. Such leakage appears to be most prominent in the liver. It is more important in small mammals, where it can account for as much as 35% to 45% of mitochondrial oxygen consumption. Uncoupling of oxidation from phosphorylation allows the proton flow to be dissipated as heat only, helping small mammals maintain body temperature. This uncoupled respiration can also be caused by specific chemicals that make the inner membrane permeable to protons.

One type of adipose tissue (fat) is rich in mitochondria, giving it a brownish color. This **brown adipose tissue** (BAT) has **uncoupling protein** 1 (UCP1) in its mitochondrial inner membranes. The protein allows the electrochemical gradient across the membrane to be dissipated by letting protons flow down their gradient into the matrix without being coupled to ATP formation, using the ATP synthase. Many hibernating animals have significant amounts of BAT, since it allows them to maintain core temperature as they hibernate through the winter. Infant humans have also been reported to have small amounts of BAT to help them maintain body temperature. Typically, BAT levels diminish as children mature. However, studies have demonstrated that adult humans do have small but significant amounts of BAT and UCP1 in various locations throughout their body. In addition, humans also have uncoupling protein 2 (UCP2) in many tissues and UCP3 in skeletal muscle (Harper, Green, and Brand 2008).

Researchers have theorized that UCPs could be related to obesity and that those individuals with

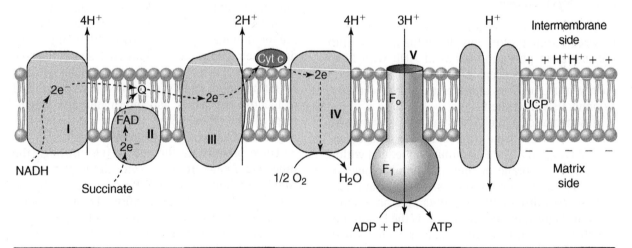

Figure 5.13 The four mitochondrial inner membrane protein complexes involved in electron transfer from NADH and succinate to oxygen. Energy is captured at complexes I, III, and IV as protons are transported across an electrical and concentration gradient. Free energy released on return of the protons through the F_o and F_1 components of complex V, the ATP synthase, is coupled to ATP formation from ADP and Pi. However, protons can also return from the intermembrane to the matrix side of the inner membrane through uncoupling proteins (UCP). Such flow dissipates the electrochemical gradient across the inner membrane but does not result in ATP formation.

higher levels of UCPs and, therefore, less efficient mitochondrial coupling could be less prone to obesity, since a greater amount of the energy released from metabolism would not be coupled with ATP synthesis but would instead be released as heat. UCP1 is found primarily in brown fat, and it is thought to be the UCP that is most closely related to increased thermogenesis. Genetically altered rodents lacking UCP1 have a greater tendency for obesity and are more sensitive to cold. In addition, transgenic overexpression of UCP3 in muscles of obese rodents does tend to reduce their rates of obesity. Less clear evidence exists for a role of UCPs in human obesity; however, studies in humans suggest an inverse association between brown fat and UCP1 activity and obesity (Tseng, Cypress, and Khan 2010). Evidence also exists that brown fat in humans is involved in cold-induced thermogenesis. Ongoing research is evaluating the potential effects of genetic variability in humans for brown fat and UCP1 activity in possibly explaining some of the variability in human metabolic rates and susceptibility to obesity (Tseng, Cypress, and Khan 2010). Evidence also suggests that UCPs in humans may be important in lowering and mitigating oxygen radical production in mitochondria and perhaps in regulating mitochondrial fatty-acid metabolism (Peterson et al. 2008). Exercise training reduces expression of UCP3 in muscle mitochondria of rodents and humans, possibly due to training-induced enhancement of resistance to oxidative stress, which then mitigates the need for higher UCP3 levels to perform antioxidant functions. This, coupled with the known upregulating effects of superoxide on UCP activation, strongly suggests that a function of UCP3 in skeletal muscle is related to antioxidant protection.

Mitochondrial Transport of ATP, ADP, and Pi

Most synthesis of ATP occurs in the mitochondria, but most ATP in muscle is hydrolyzed in the cytosol by ATPase enzymes associated with muscle contractile proteins, where it is used for transmembrane ion transport. Thus, we must have a way of getting ADP and Pi into the mitochondrion and ATP out of the mitochondrion. Adenosine diphosphate and Pi must enter the mitochondrial matrix by crossing the inner membrane, whereas ATP must exit the matrix in the opposite direction. Recall that the inner membrane is impermeable to most substances and that polar or charged molecules can cross only if they are transported (translocated) by a specialized carrier protein. Figure 5.14 illustrates how ADP and Pi enter the matrix and how ATP crosses to the cytosolic or intermembrane side. Remember that we ignore the outer mitochondrial membrane because it is so permeable to small molecules.

The **ATP–ADP antiport** (also commonly called the adenine nucleotide translocase [ANT] or ATP–ADP translocase) transports ADP and ATP. Similarly, the **phosphate transporter** allows Pi in the form of dihydrogen phosphate ($H_2PO_4^-$) to be exchanged for hydroxide ion (OH^-). This is an example of an antiport mechanism. The driving force for the ATP–ADP antiport comes from the charge difference between the ADP and ATP. Thus, the movement of ATP^{4-} from the more negative matrix region to the more positive region on the cytosolic side of the inner membrane allows the translocation to proceed with a free energy release. The movement of ATP^{4-} out of the matrix, coupled to movement of an ADP^{3-} into it, reduces the charge gradient by 1 but does not alter the proton concentration gradient. The phosphate transporter is electrically neutral, but it acts to dissipate a proton on the cytosolic side of the membrane because OH^- can combine with a proton (H^+) to form water. Thus, it is driven by discharge of part of the chemical imbalance of H^+ across the membrane. In summary, moving the constituents to take an ATP molecule into the matrix (e.g., ADP and Pi) and transferring the ATP out of the matrix costs the equivalent of one of the protons pumped during electron transport from reduced coenzymes to oxygen.

Figure 5.14 also shows transport across the inner mitochondrial membrane of calcium, as already discussed, and pyruvate, which we will discuss later in

☑ KEY POINT

The human body contains small amounts of brown fat and the uncoupling proteins UCP1, UCP2, and UCP3, but there is still some disagreement about their roles. Because these mitochondrial proteins can uncouple electron transport from ATP formation, it has been suggested that people who are especially endowed with these proteins, particularly UCP1 in brown fat, could have a higher resting metabolic rate and, thus, would be less likely to develop obesity. Data supporting such an assertion in humans, while limited, are still building. On the other hand, there seems to be support for the theory that uncoupling proteins can reduce the rate of formation of the free radical superoxide in mitochondria of resting cells. This may be an important role for these proteins, particularly UCP3, which is found exclusively in skeletal muscle.

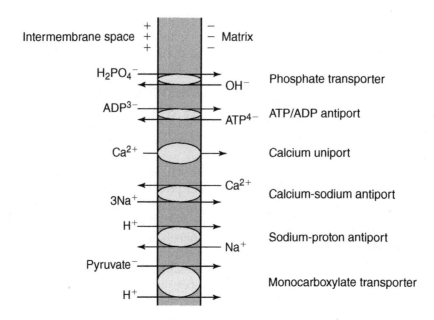

Figure 5.14 Transport of charged substances across the inner membrane of the mitochondrion, based on two principles. First, there is an electrochemical gradient (proton motive force) across the membrane, based on the fact that there is a higher concentration of protons (ΔpH) and a more positive charge ($\Delta\psi$) in the intermembrane space, or cytosolic side, than on the matrix side. Discharge of this gradient provides the energy to drive movement of charged particles. Second, all transport involves specific transporters or carriers that recognize certain charged species. In the examples shown, action of each transporter leads to a reduction in either ΔpH or $\Delta\psi$, or in both.

this chapter. Many other substances must be transported across the inner mitochondrial membrane, each requiring a specific carrier and being driven by energy release, based on charge or chemical differences across the membrane.

In trying to account for the ATP produced during transfer of a pair of electrons (reducing equivalents), we are assuming that 10 protons are translocated across the inner membrane when NADH is the source of electrons and that 6 protons are pumped when $FADH_2$ is the source of electrons. We also assume that only 3 protons are needed to flow down their gradient through the ATP synthase to provide enough energy to phosphorylate 1 ADP with Pi, making ATP. Further, transporting 1 ADP and 1 Pi into the mitochondrial matrix, across the inner membrane, is equivalent to using up 1 proton. Therefore, we should be able to generate 2.5 ATP from each pair of electrons transported in the respiratory chain from NADH, and 1.5 ATP for each pair of electrons transported from $FADH_2$. Nonintegral P/O ratios (e.g., 2.5 and 1.5) may be intuitively disturbing to some. As already mentioned, many earlier sources cite P/O ratios of 3 and 2 for NADH and $FADH_2$, but this is changing as the real stoichiometry of proton pumping and ATP synthesis is increasingly appreciated. We have been careful to refer to 2.5 and 1.5 as maximum P/O ratios for NADH and $FADH_2$, respectively, to account for the leakage of protons across the inner mitochondrial membrane through uncoupling proteins. We also discharge some of the proton electrochemical gradient in moving molecules and ions (besides ATP, ADP, and Pi), as figure 5.14 indicates. Therefore, true P/O ratios are difficult to determine, given species and individual differences, but the values will be less than the 2.5 and 1.5 we are using.

REGULATION OF OXIDATIVE PHOSPHORYLATION

Oxidative phosphorylation involves the oxidation of fuels to make ATP. From a whole-body perspective, we need systems to digest, absorb, and store the fuels that must come from our diets. We need systems to deliver oxygen from the air to the mitochondria, where it can accept electrons from fuels and use the energy released to generate ATP and to remove the CO_2 that is produced in the citric acid cycle. This is the whole-body picture. For the most part, however, oxidative phosphorylation is controlled at the level of the cell. It is here that ATP is hydrolyzed to generate the energy to drive endergonic reactions, and it is in each cell that this ATP is regenerated. Because oxidative phosphorylation generates most of the energy for the cell in the form of ATP, its rate should be precisely connected to the rate of ATP hydrolysis. The citric acid cycle is one component of oxidative phosphorylation. As shown in figure 5.8, the citric acid cycle and electron transport chain are tightly linked together because the citric acid cycle is the major producer of reduced coenzymes needed to funnel electrons into the respiratory chain. Therefore, what regulates the citric acid cycle can influence the electron transport chain, and vice versa.

If we neglect transfer of electrons from $FADH_2$ to oxygen, a single equation can represent oxidative phosphorylation, in which four electrons reduce a complete oxygen molecule and ADP is phosphorylated with Pi to make ATP:

$$5\ ADP + 5\ Pi + 2\ NADH + 2\ H^+ + O_2$$
$$\rightarrow 2\ NAD^+ + 5\ ATP + 7\ H_2O$$

Two of the seven molecules of water are generated by reduction of a molecule of oxygen, and five come about when five ATP are formed. A simple way to understand the regulation of oxidative phosphorylation is to consider which of the substrates on the left side of the equation (i.e., ADP, Pi, NADH, and O_2) actually limit the overall process. Our discussion focuses on muscle because it has an enormous range of metabolic rate, from complete rest to the vigorous contractions of sprinting. Figure 5.15 aids in this discussion. First, we must ask, where do the four potentially limiting substrates for oxidative phosphorylation come from?

Adenosine diphosphate and Pi are formed mainly in the cytosol when ATP is hydrolyzed to drive endergonic reactions. The increase in Pi is directly proportional to the decline in the concentration of phosphocreatine (PCr); that is, PCr decreases, and free creatine (Cr) and Pi increase.

The NADH comes from the citric acid cycle, beta-oxidation of fatty acids in the matrix of the mitochondrion, and the cytosol when pyruvate formed during glycolysis is oxidized in the mitochondrion. NADH is a potential limiting factor because adequate NADH for oxidative phosphorylation requires available substrates for the dehydrogenase reactions that generate it, as well as sufficient activity of the dehydrogenase enzymes to drive the NADH-forming reactions at a sufficient rate.

Oxygen is taken into the lungs when you breathe, where it diffuses to hemoglobin molecules in red blood cells (i.e., erythrocytes) in the capillaries, is pumped throughout the body from the heart, and is unloaded from hemoglobin molecules in the capillaries that reach all parts of the body. Oxygen is delivered by way of diffusion from the small capillaries, across the cell membrane and the cytosol, and then to mitochondria.

Figure 5.15 shows that ATP is hydrolyzed to ADP and Pi by the functional ATPases we discussed in chapter 4, whereas ATP is primarily regenerated via oxidative phosphorylation in the mitochondrion. For tightly coupled ATP hydrolysis and ATP regeneration to happen, ADP and Pi must cross from the cytosol into the matrix, and ATP must cross from the matrix to the cytosol to be used again. Transport of ADP, Pi, and ATP across the inner mitochondrial membrane is aided by specific carriers that we have already described. It has been generally assumed that ADP diffuses from sites of ATPase activity to mitochondria and that ATP diffuses back to the ATPase sites. However, it is now widely believed that in skeletal and cardiac muscle, a significant portion of the cytoplasmic transport of ADP and ATP occurs via Cr and PCr, respectively, as is shown in figure 5.15.

Adenosine triphosphate is hydrolyzed in the cytosol of muscle at three major sites: (a) where myosin interacts with actin (the actin-activated ATPase discussed in chapter 4), (b) when calcium ions are pumped back into the sarcoplasmic reticulum (the sarcoplasmic reticulum–Ca^{2+} ATPase, or SERCA), and (c) when sodium ions are pumped out of the cell and potassium ions are pumped back in (the sodium–potassium ATPase). At the site of these ATPases are the creatine kinase (CK) enzyme and PCr. Creatine kinase catalyzes the phosphorylation of ADP to make ATP, utilizing PCr and producing Cr. At this level, the net direction of the CK reaction is toward ATP formation. Next, Cr can diffuse through pores in the outer membrane to the inner membrane, where it is phosphorylated by an ATP, producing PCr and ADP. A mitochondrial creatine kinase (mtCK) catalyzes this reaction. The resulting PCr can then diffuse back to the site of the ATPases to rephosphorylate ADP. This process is called the *phosphocreatine shuttle* (or creatine phosphate shuttle), and it is illustrated in figure 5.15. The location of CK at sites of ATP hydrolysis and mtCK in the intermembrane space, adjacent to the adenine nucleotide translocase, promotes efficient diffusion of PCr and Cr.

Use of this shuttle does not preclude ADP from diffusing to the mitochondrion or ATP from diffusing back, as shown in figure 5.15. It just means that Cr and PCr can carry out the same process. Since diffusion depends on a concentration gradient, the fact that there are significantly larger changes in PCr/Cr concentrations than in ATP/ADP concentrations during muscle work makes the former molecules more likely candidates in this energy transport process. Moreover,

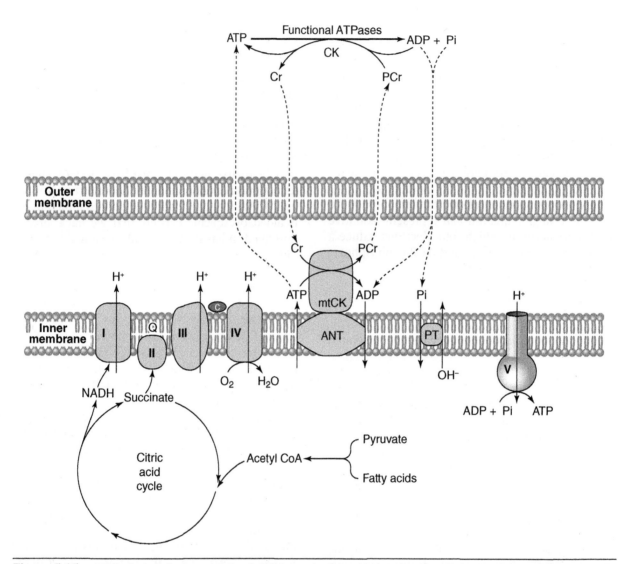

Figure 5.15 ATP hydrolysis in the cytosol and ATP formation by oxidative phosphorylation in mitochondria. Adenosine diphosphate and Pi can diffuse (dotted arrows) from the site of ATPases; when formed, ATP can diffuse back. Alternatively, Cr and phosphocreatine (PCr) can act as energy transfer agents, diffusing toward the mitochondrion and back to the ATPases, respectively. Electrons are transferred to oxygen using the electron transfer chain (complexes, I, II, III, and IV); the free energy released is coupled to proton (H+) translocation across the inner membrane, creating an electrochemical gradient within the intermembrane space. When protons flow down their gradient through the ATP synthase (V), the energy released is used to phosphorylate ADP to make ATP. ANT is adenine nucleotide transporter and PT is the phosphate transporter.

both Cr and PCr are smaller molecules than ADP and ATP, making diffusion easier. In addition, the normal cellular concentrations of the former are larger than those of ADP and ATP. The fact that the active form of mtCK (an octamer with eight subunits) is located on the intermembrane space side of the inner membrane, adjacent to the ATP–ADP antiport (adenine nucleotide translocase), lends further support to the phosphocreatine shuttle mechanism.

Figure 5.15 shows the role of the major players in oxidative phosphorylation (NADH, ADP, Pi, and O_2). Now, let us take a more detailed look at what limits the rate of oxidative phosphorylation.

☑ KEY POINT

For simplicity, we talk about limiting factors as if we have a chain with links that are not equal. It is important to point out that what limits oxidative phosphorylation in one metabolic state may not be the weak link in another condition. For example, oxygen availability at high altitude may be a limiting factor for oxidative phosphorylation during exercise, while other limiting factors, such as NADH availability, may prevail in carbohydrate depleted conditions.

Regulation of the Citric Acid Cycle

It has been estimated that in the transition from rest to very intense exercise, the flux through the citric acid cycle may increase 100-fold in contracting muscles of well-trained humans. Because the citric acid cycle is so tightly coupled to the electron transport chain, anything that limits the activity of electron transport to oxygen and ADP phosphorylation will stop the citric acid cycle. In other words, the citric acid cycle is the primary source of electrons (three-fourths as NADH, the remainder as succinate) for the electron transport chain. If the flow of electrons from NADH to oxygen in the electron transfer chain is slowed or blocked, NADH will accumulate and NAD^+ will be sharply reduced. Because NAD^+ is a substrate for three of the dehydrogenase reactions in the citric acid cycle, its concentration is critical for cycle operation. Other things must be considered, since three of the citric acid cycle enzymes catalyze reactions that are essentially irreversible. If uncontrolled, these reactions could convert all available substrate into product, thus compromising mitochondrial metabolism.

Table 5.1 summarizes the major loci of control for the citric cycle. At the level of the whole cycle, acetyl units attached to CoA are absolutely essential for the cycle to operate. The acetyl units come from fuel molecules, such as carbohydrate, following breakdown to pyruvate, fatty acids, ketone bodies, and carbon skeletons of amino acids. How fast these fuel molecules can be converted into acetyl CoA is critical to regulation of the citric acid cycle. As we will see later, the inability to generate acetyl CoA at a sufficient rate is a major mechanism accounting for the phenomenon of hitting the wall that many endurance athletes experience near the end of a prolonged event.

Citrate synthase (CS) should be controlled, because if it were not, it could consume the available acetyl CoA and oxaloacetate. It is inhibited by the allosteric effector NADH, which binds to an allosteric site on CS, increasing the K_m of CS for its substrate acetyl CoA (see the discussion on allosteric enzymes in chapter 2). In addition, citrate is a competitive inhibitor for oxaloacetate at the active site of CS. This means that a rise in citrate can competitively inhibit the binding of oxaloacetate (see chapter 2).

Isocitrate dehydrogenase (ICDH) is likewise inhibited by NADH at a negative allosteric site. Thus, at rest, when NADH concentration is high, ICDH is inhibited. In addition, ICDH is activated by Ca^{2+} ions. Flow of calcium into the matrix through its uniport increases the matrix $[Ca^{2+}]$, as we have already discussed. Calcium ions lower the K_m for the substrate isocitrate, thus increasing enzyme activity for the same isocitrate concentration. The more active a muscle fiber is, the higher and more sustained is the rise in $[Ca^{2+}]$ in both the cytosol and matrix, and the more ICDH is activated.

α-Ketoglutarate dehydrogenase (α-KGDH) is inhibited allosterically by NADH (just like CS and ICDH). Moreover, like ICDH, α-KGDH is activated by a rise in the concentration of Ca^{2+}, which lowers the K_m for its substrate, α-ketoglutarate. In addition, succinyl CoA is a competitive inhibitor for CoA. This means that a rise in the product succinyl CoA will competitively inhibit the binding of the normal substrate CoA. Thus, α-KGDH cannot tie up the available CoA in the matrix.

Control of the ICDH and α-KGDH by Ca^{2+} in muscle is such an elegant mechanism because the Ca^{2+} that initiates contraction, stimulating ATP breakdown, also signals the need for ATP synthesis. Moreover, the increase in mitochondrial Ca^{2+} that activates the citric acid cycle also activates the breakdown of pyruvate to acetyl CoA catalyzed by pyruvate dehydrogenase, as well as enhancing complex V activity in the inner membrane, the ATP synthase. Such coordinated regulation of the citric acid cycle, pyruvate dehydrogenase, and ATP synthesis reminds us of the extremely tight relationship between the primary mitochondrial reactions generating NADH and the electron transport system that consumes electrons.

Table 5.1 Regulation of the Citric Acid Cycle

Site of regulation	Substrates	Negative effectors	Positive effectors
Cycle as a whole	Acetyl CoA, NAD^+	–	–
Citrate synthase	Acetyl CoA, oxaloacetate	NADH, citrate	–
Isocitrate dehydrogenase	Isocitrate, NAD^+	NADH	Ca^{2+}
α-Ketoglutarate dehydrogenase	α-Ketoglutarate, NAD^+, CoA	NADH, succinyl CoA	Ca^{2+}

Regulation of Pyruvate Oxidation

Carbohydrate metabolism is covered in the next chapter. However, the control of pyruvate oxidation is closely linked with the regulation of oxidative phosphorylation. As shown in figure 5.1, pyruvate is formed during the glycolytic reactions. It may be reduced to lactate in the cytosol or may enter mitochondria to be oxidized to acetyl CoA. The latter then enters the citric acid cycle. Pyruvate is an important source of acetyl CoA. As we will see shortly, pyruvate availability and its oxidation to acetyl CoA can significantly influence peak rates of oxidative phosphorylation. The reaction for pyruvate oxidation, catalyzed by pyruvate dehydrogenase (PDH) in the mitochondrial matrix, is as follows:

$$\text{pyruvate} + NAD^+ + CoA \rightarrow \text{acetyl CoA} + NADH + H^+ + CO_2$$

The PDH reaction as shown in the equation is catalyzed by an enzyme complex with multiple copies of three enzyme subunits, identified as E1, E2, and E3. Although the reaction seems simple, in reality, much takes place that is not shown and is beyond the scope of this book. Three coenzymes involved in this reaction are not visible in the equation presented. *Thiamine pyrophosphate* is involved in the decarboxylation of pyruvate, and *lipoic acid* and FAD are involved in the oxidation part of the reaction, in which NADH is formed as a final product. As mentioned earlier, the PDH reaction is similar to that of α-KGDH in that both use the same five coenzymes (NAD^+, CoA, TPP, lipoic acid, and FAD), although the subunits of the enzyme complexes are different. We will use PDH to represent pyruvate dehydrogenase, but some sources use the abbreviation PDC for pyruvate dehydrogenase complex.

The PDH reaction must be carefully regulated because the irreversible conversion of pyruvate to acetyl CoA means that a potential precursor to make glucose is lost. That is, pyruvate can be converted to glucose in the liver, but acetyl CoA cannot. Because the brain needs glucose, being treated biochemically as the most important tissue in the body, the PDH reaction must be regulated to spare pyruvate from being irreversibly lost. Regulation of PDH occurs via phosphorylation, dephosphorylation, and allosteric mechanisms, discussed in chapter 2. A summary of the control of PDH is shown in figure 5.16. To prevent the unnecessary oxidation of pyruvate to acetyl CoA when other fuels, such as fat, can provide the acetyl CoA, PDHa is phosphorylated by *PDH kinase* into the inactive form, PDHb. Removal of the phosphate group, catalyzed by *PDH phosphatase*, leads to the activated form, PDHa. The relative activities of the two enzymes, PDH kinase and PDH phosphatase, determine the extent of phosphorylation and, hence, activity of PDH. Both the phosphate and kinase are part of the overall PDH complex. Different isoenzyme forms exist for pyruvate dehydrogenase kinase (PDKs 1 through 4) and pyruvate dehydrogenase phosphatase (PDPs 1 and 2). PDK4 and PDP1 are the predominant isoforms in cardiac and skeletal muscle. PDK4 is the isozyme in skeletal muscle that is most sensitive to changes in exercise and carbohydrate content of the diet (Horowitz et al. 2005).

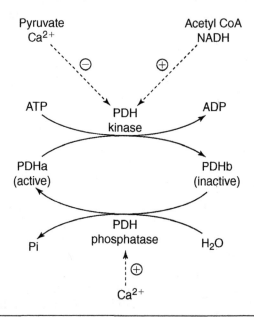

Figure 5.16 The activity of pyruvate dehydrogenase (PDH), regulated mainly by its state of phosphorylation. When the enzyme is phosphorylated by PDH kinase, it becomes inactive (PDHb). When the phosphate group is removed by PDH phosphatase, it is active (PDHa). The balance between the active (PDHa) and inactive (PDHb) forms determines the rate of formation of pyruvate to acetyl CoA. Pyruvate dehydrogenase kinase is activated allosterically by NADH and acetyl CoA. It is inhibited by Ca^{2+} ions and pyruvate. Pyruvate dehydrogenase phosphatase is powerfully activated by Ca^{2+} ions. Dotted arrows with + or – are used to show activation and inhibition, respectively.

In resting skeletal muscle, where fat oxidation provides the majority of fuel oxidized to make ATP, the PDH is predominantly in the inactive (PDHb) form. Under these conditions, the ratios NADH/NAD^+, ATP/ADP, and acetyl CoA/CoA are elevated, and intramitochondrial calcium concentration is low. At the onset of exercise, when there is a demand for acetyl CoA for the citric acid cycle, dephosphorylation of PDH is stimulated and the kinase is inhibited as the intramitochondrial concentration of Ca^{2+} increases and the concentrations of acetyl CoA and NADH decline. Research on the activation of PDH at the onset of exercise shows that changes in its activity can take place quickly (in < 1 min), and that activation is graded to the intensity of the exercise. It is suggested that the rise in intramitochondrial $[Ca^{2+}]$ provides the immediate increase in PDH activation and that the relative changes in $NADH/NAD^+$, ATP/ADP, and acetyl CoA/CoA are responsible for fine-tuning the activation of PDH to match the cell's need for ATP. Indeed, following a seven-week endurance training program, subjects had lower levels of activation of PDH, accompanied by lower pyruvate concentrations and increased ratios of acetyl CoA/CoA and ATP/ADP when performing an identical exercise task as before training (LeBlanc et al. 2004). Such results point to a training-induced adaptation that can spare the use of carbohydrate during exercise at the same absolute workload.

One interesting research topic relates to whether the relative activity of PDH can regulate oxidative phosphorylation at the onset of exercise. One theory holds that, at the transition from rest to exercise, a lag in PDH activation minimizes the availability of acetyl CoA for the citric acid cycle, thereby slowing $\dot{V}O_2$ (oxygen uptake) response. This theory has been tested by prior administration of a drug, *dichloroacetate* (DCA), into a vein. Dichloroacetate acts as an analog to pyruvate, which, as you can see from figure 5.16, inhibits the activity of PDH kinase. With DCA administered before exercise, PDH is primarily in the active PDHa form. Although many studies have employed DCA to enhance activation of PDH before, rather than during, exercise, the effects have generally been minor; in most cases, DCA has not significantly affected the rate of increase of $\dot{V}O_2$ during the first few minutes of exercise. Thus, it appears that the PDH reaction does not have a significant regulatory effect on oxygen consumption at the beginning of exercise.

It has been shown that a single prolonged bout of endurance exercise leads to a rapid increase in the activity of PDK4 enzymes, without an increase in their concentration in the exercised muscle (Watt, Heigenhauser, et al. 2004). It is believed that this increased activation is due to enhanced binding of PDK4 to the core of the PDH complex, which may allow the PDK4 enzyme to better turn down pyruvate oxidation during and following the prolonged exercise. This would lead to a decrease in glucose oxidation in the exercised muscle, so that the glucose utilized by the exercising muscle would be reduced in order to preserve the remaining liver glycogen and blood glucose for the brain and to minimize utilization of the now diminishing muscle glycogen stores in favor of increased oxidation of fatty acids (Watt, Heigenhauser, et al. 2004). This would also ensure that the blood glucose taken up by the fatigued muscle during recovery would be directed primarily to the restoration of glycogen used up by the exercise (Booth and Neufer 2005). Research has also demonstrated that rapid up- and downregulation of PDH activity in muscle can be induced by as little as 24 h of a diet either high or low in fat or carbohydrate, such that relative carbohydrate usage by the muscle would be responsive to the relative availability of fat or carbohydrate at any given time. These issues are discussed further in chapters 6 and 7.

> ☑ **KEY POINT**
>
> Although results from exercise studies in humans do not favor a major effect for PDH activation in terms of regulating oxygen kinetics, some studies show that rapid activation of pyruvate dehydrogenase at the onset of exercise plays a role in limiting the extent of lactate buildup. This likely occurs because the glycolytic pathway can produce pyruvate at a much greater rate than it can be oxidized in the mitochondria, even if PDH is maximally activated. However, the rapid activation of PDH may enhance aerobic metabolism of pyruvate at the start of exercise when aerobic metabolism is still in the process of becoming fully activated.

Regulation of Oxygen Delivery

Since we will be spending considerable time on metabolic responses to exercise and how exercise training modifies metabolism during exercise, it is worthwhile to summarize some basic material on the delivery of oxygen from the air to mitochondria. The overall rate of oxidative phosphorylation by the body is given by the Fick equation:

$$\dot{V}O_2 = \dot{Q}\,(a\text{-}\bar{v})O_2 \text{ difference}$$

$\dot{Q}$ is the *cardiac output*, a measure of the amount of blood pumped (liters) from the left ventricle into the aorta each minute. This is based on the product of **stroke volume** and heart rate. Stroke volume is the volume of blood ejected during each beat. Both stroke volume and heart rate can increase with exercise; the changes in heart rate during exercise are more pronounced than that of stroke volume.

The (a-v̄)O₂ **difference** is the average difference in oxygen content between the arterial blood and the mixed venous blood, usually expressed as milliliters of oxygen contained in 100 mL of arterial and venous blood. This increases during exercise. Table 5.2 shows representative values for $\dot{V}O_2$, heart rate, stroke volume, $\dot{Q}$, and (a-v̄)O_2 difference for an average young man before and after a comprehensive three-month program designed to improve his endurance capacity. For comparison purposes, cardiorespiratory values are given for rest, during submaximal exercise at 60% of maximal oxygen uptake, and during maximal exercise. It should be noted that these figures represent average responses and that a wide range of higher or lower changes in $\dot{V}O_2$max can be expected when training a large population.

As the values in table 5.2 reveal, the body responds to an increase in exercise intensity by increasing heart rate, stroke volume, and the (a-v̄)O_2 difference. With a relatively short-term exercise training program, an adaptation takes place in terms of stroke volume, which seems to be a major factor in dictating the maximal aerobic capacity ($\dot{V}O_2$max).

It is also important to understand how oxygen is transferred from the air to the mitochondria, where it will accept electrons to become water in the reaction catalyzed by cytochrome c oxidase (complex IV). Although cytochrome oxidase has a high affinity for oxygen, it is still debated whether the actual concentration of oxygen at the level of cytochrome oxidase limits overall $\dot{V}O_2$ or whether some other limiting factor exists. Nevertheless, the weight of evidence seems to suggest that, in most cases, it is oxygen delivery to the skeletal muscle that limits $\dot{V}O_2$max, and not necessarily the ability of muscle to utilize the oxygen that is made available.

We know that the fraction of oxygen in air is 0.209; thus, oxygen should contribute 20.9% to the total pressure of air. At sea level, with an air pressure of 760 millimeters of mercury (760 mmHg), the fraction of total pressure of the air attributable to oxygen (PO_2) can be expressed as 0.209 × 760 mmHg, or 159 mmHg. This is the *partial pressure of oxygen in air*. This value would be smaller at altitudes above sea level where air pressure is less, but the fraction of total pressure attributed to oxygen would remain the same.

> ### ✓ KEY POINT
> It is common in physiology to express the partial pressure of oxygen in units of millimeters of mercury. Some people use *torr* instead of mmHg, where 1 torr is 1 mmHg. Expression of pressure using kilopascals (kPa) is also common. The relationship between the two is that a pressure of 760 mmHg is equivalent to 101.3 kPa.

Figure 5.17 summarizes the path of oxygen from the air to the mitochondria. When we breathe, oxygen is drawn into tiny alveoli sacs in our lungs by the action of respiratory muscles, which reduces the lung pressure relative to the atmospheric pressure by expanding the ribcage and contracting the diaphragm. In alveoli, the partial pressure of oxygen is less than in the air, being diluted by water vapor and carbon dioxide returning in venous blood. Based on sea level values, the alveolar oxygen concentration, or P_AO_2, is now about 102 mmHg. As blood in capillaries passes alveoli in the lung, oxygen diffuses to capillaries, where it binds to the protein *hemoglobin* inside red blood cells (erythrocytes). Hemoglobin contains four polypeptide subunits, each subunit including a heme group with a central Fe^{2+} ion to which a single oxygen molecule can bind. Each hemoglobin molecule is thus capable of maximally transferring four oxygen molecules. Numerically,

Table 5.2 Representative Values for Cardiorespiratory Function

Cardiorespiratory function	REST		SUBMAXIMAL EXERCISE		MAXIMAL EXERCISE	
	Before training	After training	Before training	After training	Before training	After training
$\dot{V}O_2$ (L/min)	0.30	0.30	1.90	2.20	3.15	3.65
Heart rate (beats/min)	70	60	145	145	185	182
Stroke volume (mL)	75	87	92	107	94	110
Cardiac output (L/min)	5.25	5.20	13.34	15.46	17.40	20.00
(a-v̄)O_2 difference	6.0	6.0	14	14	18	18

each gram of hemoglobin can transport 1.34 mL of oxygen. The more hemoglobin in the blood, the more oxygen can be transported. For a male with a blood hemoglobin concentration of 150 g per liter of blood (15 g/100 mL), the oxygen content of the blood is 201 mL/L (150 g/L × 1.34 mL/g). For a young woman with 130 g of hemoglobin per liter of blood, the oxygen content in blood is 130g/L × 1.34 mL/g, or 174 mL of oxygen per liter. These blood oxygen concentrations are based on hemoglobin's being 100% saturated with oxygen. Normally, in blood leaving the lungs, the saturation of hemoglobin with oxygen is about 98%. Although oxygen can dissolve in blood plasma, the amount that can be transported this way is tiny, so we will omit this aspect from consideration.

Oxygen-rich blood leaves the lungs and is transported back to the heart, where it is pumped from the left ventricle into the aorta, to smaller arteries and arterioles and then capillaries that surround all cells in the body, in a process described as convective transport. The partial pressure of oxygen in the arterial blood leaving the heart (P_AO_2) is about 100 mmHg, although the actual content of oxygen can vary depending on the concentration of hemoglobin in the blood. At the capillary level, where blood and individual cells interact, the partial pressure of oxygen is much lower. Oxygen can leave the hemoglobin because the ability of hemoglobin to bind oxygen is critically dependent on the partial pressure of oxygen. Released from hemoglobin, oxygen diffuses down a concentration gradient; through the erythrocyte cell membrane, the endothelial cells of the capillary wall, the interstitial space surrounding cells, and the cell membrane of the cell; and into the cell cytoplasm. During maximal exercise, the PO_2 gradient between the capillary and the cytosol of the adjacent cell is very high (~ 30 mmHg), but the PO_2 gradient from the cell membrane to mitochondria is very shallow, approximately 2 mmHg (Richardson et al. 1999). Cardiac and skeletal muscle cells have the oxygen-binding protein *myoglobin*. The content of myoglobin parallels the mitochondrial content of the cell and is higher in the muscles of endurance-trained athletes (Duteil et al. 2004). Because oxygen is not very soluble in aqueous solutions, myoglobin can facilitate oxygen diffusion in the cytosol. The extent of oxygen bound to myoglobin can be used to determine the oxygen content in cells. In cell cytosol, the partial pressure of oxygen can be in the range of 3 to 18 mmHg.

Our discussion to this point has been from a physiological perspective. Now let us switch to how a biochemist views the situation. Molecules interact

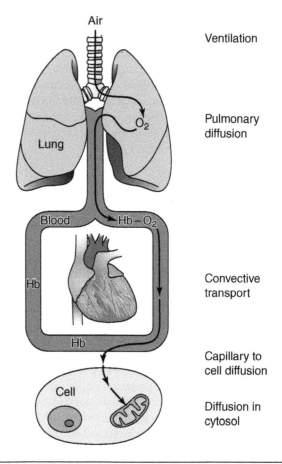

Figure 5.17 The path of oxygen from the air to mitochondria in cells. Air is drawn into the lungs by ventilation. Oxygen in the air diffuses from alveoli in the lung to the blood, where it is conveyed by convective transport bound to hemoglobin. At the cell level, oxygen is released from hemoglobin and diffuses from capillaries to cells, then diffuses in the cytosol to mitochondria.

with each other on a molecule-to-molecule basis (mole to mole), so the biochemist is interested in the concentration of oxygen expressed in molar units. In typical cells, the concentration of oxygen is in the 2 to 13 μM range (PO_2 3-18 mmHg). Oxygen is a substrate for the enzyme cytochrome c oxidase, whose affinity for oxygen is quite high. The concentration of oxygen needed to generate 50% of the maximal rate of cell respiration (K_m) is in the range of 1.5 to 1.7 μM for isolated liver and cardiac muscle mitochondria (Costa, Mendez, and Boveris 1997). This is well below the typical oxygen concentration in the cell cytosol, suggesting that oxygen is unlikely to limit oxidative phosphorylation under normal circumstances. However, the concentration of oxygen needed to support maximal cytochrome c oxidase activity (often described as the critical oxygen concentration) is estimated to be in the 2 to 5 μM range (Boveris et al. 1999). Alternatively, we could say that the critical mitochondrial partial pressure to support maximal respiration ($P_{mito}O_2$) is in the range of 3 to 7 mmHg. Although there is an apparent overlap between the critical oxygen concentration and the normal cytosolic oxygen concentration range in cells (i.e., 3-18 mmHg), it is likely that oxidative phosphorylation can be limited by oxygen supply under some exercise conditions.

✓ KEY POINT

Because more hemoglobin in the blood means a greater oxygen transport capacity, it is not surprising that athletes will go to great lengths to increase their hemoglobin concentration. Such practices as blood doping (adding extra red blood cells to the blood) or injection of erythropoietin, a factor that stimulates the formation of red blood cells, are unfortunately too common in endurance sporting activities. These measures will increase oxygen delivery to skeletal muscle in athletes, thereby enhancing oxidative phosphorylation, which is normally limited during intense exercise by the ability of the circulatory system to deliver oxygen. A legal method of achieving increased red blood cell concentration in athletes is exposure to high altitude or to low atmospheric oxygen tension, which can also stimulate natural erythropoietin release.

Regulation of Oxidative Phosphorylation in Rested Muscle

In a muscle at rest, the rate of energy expenditure is quite low, based mainly on maintaining protein synthesis, homeostasis, and normal cell function. Now consider the substrates in the simple equation to describe oxidative phosphorylation (shown earlier in this chapter). Which of these is likely to limit the rate of oxidative phosphorylation at rest?

$$5\ ADP + 5\ Pi + 2\ NADH + 2\ H^+ + O_2 \rightarrow 2\ NAD^+ + 5\ ATP + 7\ H_2O$$

As just discussed, oxygen is readily available in rested muscle; therefore, it cannot be limiting the rate. The concentration of Pi is also high enough to sustain a modestly high rate of oxidative phosphorylation (approximately 3 mM; see table 4.3, p. 98) and is therefore not limiting. Reducing power in the form of NADH in rested muscle is typically sufficient. This leaves the availability of ADP to the respiratory chain as the weak link limiting oxidative phosphorylation, including the citric acid cycle. Of course, the availability of ADP depends on the rate of ATP hydrolysis in the cytosol plus the entry of ADP into the mitochondria via the ADP–ATP antiport. On the basis of studies using isolated mitochondria, some authorities use the term **state 4 respiration** to describe the situation of mitochondria in a rested cell where the rate of oxidative phosphorylation is limited by availability of ADP. Actual consumption of oxygen occurs under state 4 conditions, but much of this is due to uncoupled oxidative phosphorylation, as we have discussed previously.

Many studies support the fact that availability of ADP limits oxidative phosphorylation for a variety of situations in muscle as well as in other tissues. For example, if mitochondria are isolated and placed in a well-oxygenated medium, the rate of oxygen consumption (used as an index of the rate of oxidative phosphorylation) is low. If either ADP or Cr is added, the rate of oxygen utilization greatly increases. We have already discussed how addition of ADP stimulates oxidative phosphorylation. However, addition of Cr also effectively increases oxidative phosphorylation by stimulating the mitochondrial creatine kinase enzyme, generating ADP and PCr (see figure 5.15). The ADP then enters the matrix using the ADP–ATP antiport to stimulate oxidative phosphorylation. When the oxygen utilization of isolated mitochondria is sharply increased by addition of an excess of ADP or Cr (the latter generates ADP), we say it is **state 3 respiration**.

Overall, the rate of oxidative metabolism for a muscle at rest, and for many other tissues, fits neatly into the concept that a limitation is due primarily to ADP within the mitochondria. This view of the regulation of oxidative phosphorylation is called the kinetic model, or **acceptor control** model.

Regulation of Oxidative Phosphorylation at the Onset of Exercise

When muscle undergoes a transition from rest to moderate exercise or when the intensity of moderate exercise is modestly increased, the rate of ATP hydrolysis undergoes an abrupt stepwise increase to match the new exercise intensity. The rate of oxidative phosphorylation increases, measured at the mouth as $\dot{V}O_2$. This follows an exponential time course, with a half-time of approximately 20 s (see figure 5.18). This is described as the $\dot{V}O_2$ kinetics or the kinetic response of $\dot{V}O_2$. It simply refers to the shape of the curve when breath-to-breath $\dot{V}O_2$ measures are plotted against time. Why is there a lag in oxygen utilization, a measure of the rate of oxidative phosphorylation, before it reaches a new level corresponding to the increased exercise intensity? Two hypotheses have been put forth. The metabolic inertia hypothesis suggests that a lag in electron transport and ATP synthesis occurs because of substrate limitations other than oxygen (Grassi 2001). This could arise from a lack of acetyl CoA for the citric acid cycle or from the partial inhibition of key enzymes in the cycle because mitochondrial Ca^{2+} is insufficient to fully activate them. The alternate view is that there is an oxygen limitation—an inability to supply oxygen to the electron transport chain due to a limitation in oxygen transport (Hughson, Tschakovsky, and Houston 2001).

Whether metabolic inertia, oxygen supply limitations, or their combination can account for the $\dot{V}O_2$ kinetics associated with stepwise increases in exercise intensity has been the subject of considerable research. The simple kinetic or acceptor control concept explains quite well what keeps oxidative phosphorylation at a low rate in rested muscle, but it does not adequately account for the responses that accompany changes in exercise intensity, since it is difficult to demonstrate a change in the concentration of any one of the potential limiting substrates that corresponds to the increase in oxidative phosphorylation. Use of ^{31}P nuclear magnetic resonance spectroscopy (NMR; see chapter 4) can provide concentrations of ATP, PCr, and Pi. From this, we can make calculations to determine ADP concentration. Values for intracellular PO_2 during exercise can be determined for exercising humans using the magnetic resonance signal from myoglobin, which is based on the extent of oxygen binding to myoglobin (Richardson et al. 2002). From these and other techniques, including studies with lower (hypoxia) or elevated (hyperoxia) concentrations of inspired oxygen, we learn that muscle is effective in maintaining adequate oxidative ATP formation under most situations. Despite different intracellular concentrations of oxygen, ATP formation by substrate oxidation can be maintained through the manipulation of concentrations of energy-rich phosphates (especially PCr and ATP), substrates (as in NADH), and controlling molecules (ADP, AMP, and Pi) (Hughson 2005).

Exercise biochemists have been more accepting of the idea that regulation is based on a combination of factors. For submaximal exercise, across a range of intensities, intracellular PO_2 remains consistently in the 4 to 5 mmHg range during the breathing of room air (normoxia) (Richardson et al. 2002). With a lower oxygen content in the air, exercise $\dot{V}O_2$ can still be maintained, although intracellular PO_2 is lower. With hyperoxia, in which inspired oxygen content far exceeds the normal 21%, intracellular PO_2 is elevated beyond that in normoxia, but $\dot{V}O_2$ for the same exercise intensity remains the same. How is this achieved? During hypoxic exercise, oxidative phosphorylation can be maintained with lower concentrations of PCr and ATP and elevated levels of ADP, Pi, and NADH. The increase in substrates for oxidative phosphorylation (ADP, Pi, and

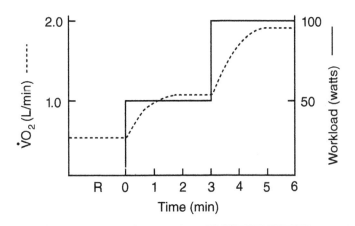

Figure 5.18 The relationship between the rate of oxidative phosphorylation, measured as the $\dot{V}O_2$, and the rate of ATP hydrolysis that tracks exercise workload for a hypothetical subject. $\dot{V}O_2$ is determined while a subject sits resting on a cycle ergometer (R). The clock starts, and the subject immediately begins pedaling at a workload of 50 W (watts). After 3 min, the workload is increased to 100 W. A lag of approximately 2 min transpires before the $\dot{V}O_2$ reaches a new steady state for both workloads.

NADH) would help to maintain ATP formation to balance the reduced cellular oxygen concentration. With elevated cellular oxygen levels, oxidative phosphorylation could be maintained with higher PCr and ATP and lower ADP, Pi, and NADH concentrations (Hughson 2005).

Biochemists have used the terms **phosphorylation potential** (that is, [ATP]/[ADP] × [Pi]) and the *mitochondrial redox state* (the ratio [NADH]/[NAD$^+$]) to define the energetic and metabolic potential of a cell. Combined with the cellular oxygen content as a measure of electron-accepting potential, these variables describe the effects of energy demand and potential for ATP supply by oxidative phosphorylation. For example, the phosphorylation potential ([ATP]/[ADP] × [Pi]) will decrease when any increase in exercise intensity occurs. While ATP concentration remains remarkably constant in most exercise conditions, it can be transiently reduced by more than 50% in parts of the muscle during very intense exercise. Couple this to a roughly parallel increase in ADP and a much larger increase in Pi due to a decline in PCr, and you can see how the phosphorylation potential can vary more widely than any of its constituents. Note also that the phosphorylation potential reflects PCr concentration because the increase in Pi reflects the decline in PCr. The mitochondrial **redox potential** ([NADH]/[NAD$^+$]) changes when the rate of electron transfer from NADH to oxygen is not matched by the rate of formation of NADH through dehydrogenase enzymes. Recall that the activities of three of these enzymes (pyruvate, isocitrate, and α-ketoglutarate dehydrogenases) respond to a complex pattern of substrate and allosteric effectors, including an increase in matrix [Ca^{2+}]. Finally, the oxygen availability to the respiratory chains in each mitochondrion (P$_{mito}$O$_2$) depends on a complicated combination of gas exchange, blood flow, and diffusion in getting from the outside air, through the lungs, into the arterial blood flow, and to the cytosol of each muscle fiber. Thus, the exponential response of V̇O$_2$ measured at the mouth at the onset of exercise, or during a step increase in exercise intensity, is the result of a complex mix of whole-body and cellular events providing the substrates for ATP synthesis.

Any step increase in exercise intensity will cause phosphorylation potential to decline. The extent of the decrease will reflect the increase in exercise intensity. This will stimulate the entry of ADP and Pi into the mitochondrion and electron transport from NADH to oxygen. As NADH is oxidized to NAD$^+$, the inhibitory effect of NADH on the three irreversible citric acid cycle enzymes is reduced and the citric acid cycle will speed up. In addition, the gradual rise in matrix calcium as a result of an increased cytosolic concentration in the more active fibers will further stimulate isocitrate and α-ketoglutarate dehydrogenase and activate PDH. Because of the initial mismatch between oxygen utilization by the cytochrome c oxidase complex and oxygen delivery from the air to the fiber, the oxygen tension within the fiber will decline. However, the rate of oxidative phosphorylation can be increased by a combination of a decrease in phosphorylation potential, an increase in mitochondrial redox potential, and a gradual increase in oxygen transport to the muscle mitochondria. The key point is that adjustments in phosphorylation and mitochondrial redox potentials can help to maintain oxidative metabolism in the face of declining oxygen availability to the respiratory chain (i.e., P$_{mito}$O$_2$). Any mismatch between ATP demand and ATP supplied by oxidative phosphorylation must be compensated with ATP provided by PCr and glycolysis. For small step increases in exercise intensity, the former will predominate. This has been demonstrated in a variety of models, showing that any slowing in the delivery of oxygen-rich blood to exercising muscle is compensated by a steeper decline in PCr. Of course, this means that phosphorylation potential would be lower compared to that in the exercise condition with intact blood flow. Any such mismatch could only be compensated for a short period of time, since muscle stores of PCr are limited. Once the ability to compensate ends, fatigue, or the inability to maintain the rate or intensity of muscle contractions, would manifest.

Oxidative Phosphorylation During Other Exercise Conditions

Much of the work and exercise we do is performed at a fairly constant rate of energy expenditure. For low to moderate sustained exercise, a steady rate of oxidative phosphorylation can supply virtually all the ATP needs. We call this steady-state exercise because the rate of oxidative phosphorylation is precisely matched to the ATP demands by a combination of phosphorylation potential and redox potential adjusted to the steady state content of oxygen in the muscle fibers. The phosphorylation potential would be inversely related to the intensity of exercise, while the redox potential should parallel exercise intensity.

So far, we have looked at control of oxidative phosphorylation only when the step increases in muscle activity are relatively small. The situation is in part similar, but more complicated, if we consider transitions to maximum exercise. For a huge and sudden increase in muscle activity, phosphorylation potential will decline even more. Although PCr can partially buffer significant declines in ATP and increases in ADP, both of these will change to a greater extent than in the situation discussed previously. Further, PCr concentration will decline faster and more precipitously, and Pi will rise in parallel. Oxidative phosphorylation will be stimulated even more than before, with larger changes in mitochondrial redox and matrix calcium concentration. However, we are talking about exercise intensities far beyond what could possibly be supported by even maximal rates of oxidative phosphorylation. Moreover, these supramaximal exercise intensities can be sustained for much less time than it takes to reach peak rates of oxidative phosphorylation. For these situations, the huge changes in cytoplasmic phosphorylation potential (and other factors discussed in chapter 6) rapidly turn on glycolysis. With its higher power for ATP formation, glycolysis plays a prominent role in providing the ATP needs.

The oxygen available to muscle mitochondria can limit oxidative phosphorylation under certain circumstances. For example, at higher altitude, where the oxygen content of the air is low, the time course in the response of $\dot{V}O_2$ (reflecting active muscle oxygen consumption) to a modest step increase in muscle activity is delayed. This would suggest that the oxygen content available to mitochondria is lower for any absolute work rate when a person exercises at higher altitude. Therefore, compared to the same exercise intensity at sea level, we would expect a lower phosphorylation potential and higher redox potential to offset the lower oxygen content of altitude. Oxygen available to the electron transport chain could also be limiting during isometric contractions when intramuscular pressure builds up sufficiently to reduce or completely cut off the blood flow. In situations of reduced mitochondrial oxygen availability due to compression on blood vessel walls, glycolysis becomes extremely important as a source of ATP.

Athletes engaged in prolonged exercise such as marathons or triathlons can experience a situation in which performance falls off near the end. This has been called *hitting the wall* but its cause is easy to understand. As we will see, oxidation of carbohydrate and fat provides the ATP to support submaximal exercise. The higher the exercise intensity, the more carbohydrate is oxidized and the less fat is used. Near the end of a marathon, for example, carbohydrate stores can become severely reduced. This means that the source of acetyl groups for the citric acid cycle must come increasingly from the beta-oxidation of fatty acids (discussed in chapter 7). However, numerous studies have revealed that provision of acetyl CoA to the citric acid cycle from beta-oxidation of fatty acids alone cannot match that of when pyruvate or a mixture of pyruvate and fatty acids is used. This means that the primary supply of reducing equivalents (electrons on NADH) to the electron transport chain is compromised. Accordingly, the athlete must reduce running, cycling, or skiing pace to a level where the rate of ATP demand by the exercise can be met by the new, lower level of oxidative phosphorylation, using fatty acids as the primary fuel.

☑ KEY POINT

A keenly debated topic is what limits the maximum rate of oxidative phosphorylation. Is it limited by oxygen transport from the air to cytochrome c oxidase (i.e., transport limited)? Alternatively, is the limitation based on the production of reducing equivalents for the electron transport chain (i.e., a metabolic limitation)? A program of endurance training generally increases the activities of citric acid cycle enzymes and electron transport chain protein components about three to five times more than it increases the ability to transport oxygen, as measured by $\dot{V}O_2$max. This is taken to mean that in fit people, oxygen transport places the upper limit on the maximum rate of oxidative phosphorylation. In addition, artificially increasing the ability of the circulatory system to deliver oxygen (e.g., via increasing red blood cells or blood doping) will increase $\dot{V}O_2$max in trained people, again pointing to oxygen delivery as the most likely limiting factor to $\dot{V}O_2$max. It is more likely that people who are unfit cannot generate electrons at a rate sufficient to match their ability to deliver oxygen to the mitochondria of exercising muscle (Mourtzakis et al. 2004).

QUANTIFICATION OF REDOX REACTIONS

We have discussed oxidation and reduction reactions throughout this chapter, but we have not put into numbers how easy it is for some molecules to be oxidized or reduced. Figure 5.9 was the closest we came to defining relative oxidation and reduction abilities based on a standard free energy scale. From this, we could derive the interpretation that NADH is easily oxidized and that oxygen is the easiest to reduce.

General Principles of Biological Redox Reactions

Biological oxidation–reduction (redox) reactions are catalyzed by enzymes. In a redox reaction, one substrate is reduced and one substrate oxidized. The names of the enzymes catalyzing biological redox reactions, besides describing a substrate, end with such words as dehydrogenase (pyruvate dehydrogenase), oxidoreductase (NADH–coenzyme Q oxidoreductase), oxidase, peroxidase, oxygenase, or reductase. Regardless of the name, the reactions catalyzed are redox reactions; one substrate is oxidized and one is reduced, as shown here:

$$A_{red} + B_{ox} \leftrightarrow A_{ox} + B_{red}$$

Not all redox reactions are reversible or will reach equilibrium. As we discussed in chapter 4, the energy change dictates whether it is an equilibrium or a nonequilibrium reaction. Redox reactions involve the addition or removal of electrons. Students can be reminded of what addition or removal of electrons means by using acronyms such as *LEO* (losing electrons oxidation) *says GER* (gaining electrons reduction) or *OIL* (oxidation is losing) *RIG* (reduction is gaining). In the general example just presented, A_{red} becomes oxidized as it loses electrons, whereas B_{ox} is reduced by gaining electrons. What is oxidized is the reducing agent, or reductant, and what gets reduced is the oxidizing agent, or oxidant. From a biological perspective, the term *pro-oxidant* refers to the oxidizing agent or oxidant, and *antioxidant* refers to the reductant or reducing agent. Table 5.3 explains these terms.

Redox Potentials

In oxidative phosphorylation, the electron transfer potential of NADH or $FADH_2$ to oxygen is converted into phosphoryl transfer potential of ATP, which we can measure as free energy change. Under defined conditions, in which concentrations of reactants and products are kept at 1 M concentration and pH is 7.0, we use standard free energy change ($\Delta G°'$). For electron transfer reactions, we use $E°'$ as the **standard redox potential**, under defined conditions in which the reactants and products are at 1 M concentration. The value of $E°'$ reflects the ability of a substance to act as an electron acceptor—that is, to be reduced—under standard conditions. Unlike $\Delta G°'$, where a negative sign preceding a large number reflects a reaction that is energetically favorable, the actual value of $E°'$ provides a direct measure of the ability to accept electrons under standard conditions. That is, a positive value shows strong ability as an oxidizing agent (to gain electrons). On the other hand, a negative value for $E°'$ shows a weak ability to be reduced (i.e., as an oxidizing agent or oxidant), but a strong ability to be oxidized, since oxidation and reduction are opposite events. Therefore, if a substance has a negative value for $E°'$, it will be a poor oxidizing agent; on the other hand, it will be a good reducing agent (it will have the ability to be oxidized or lose electrons).

Standard redox potentials are measured in volts or millivolts (mV), based on the reduction of hydrogen ions (protons) to form hydrogen gas, as shown in the following equation:

$$2 H^+ + 2e^- \rightarrow H_2$$

In this reaction, the hydrogen gas is at a pressure of 1 atmosphere (760 mmHg), and the $[H^+]$ is 1.0 M (pH = 0). Under these conditions, the standard reduction potential ($E°'$) is arbitrarily set at 0 V. When corrected to the biological pH of 7.0, the resulting $E°'$ becomes –0.42 V (–420 mV). Table 5.4 provides representative values for $E°'$ for a variety of common substances. The table illustrates the reduction of the substance, but it shows only half a reaction. As we have discussed, redox reactions involve something that is being oxidized and something that is being reduced, but this table focuses only on the relative ability of a single substance to be reduced.

From table 5.4, you can see that the easiest to reduce is oxygen and the hardest to reduce is

Table 5.3 Common Terms and Actions in Biological Redox Reactions

Process	What happens	Abbreviation	Term for what loses or gains electrons
Oxidation	Loss of electrons	LEO	Reducing agent Reductant Antioxidant
Reduction	Gain of electrons	GER	Oxidizing agent Oxidant Pro-oxidant

Table 5.4 Standard Redox Potentials for Common Biological Reactions

Reduction half-reaction	$E^{\circ\prime}$ (volts)
Acetyl CoA + CO_2 + H^+ + 2e^- → pyruvate + CoA	–0.48
2 H^+ + 2e^- → H_2	–0.42
α-Ketoglutarate + CO_2 + 2 H^+ + 2e^- → isocitrate	–0.38
NAD^+ + 2 H^+ + 2e^- → NADH + H^+	–0.32
Lipoic acid + 2 H^+ + 2e^- → dihydrolipoic acid	–0.29
GSSG + 2e^- + 2 H^+ → 2 GSH	–0.23
FAD + 2e^- + 2 H^+ → $FADH_2$	–0.22
Pyruvate + 2e^- + 2 H^+ → lactate	–0.18
Oxaloacetate + 2e^- + 2 H^+ → malate	–0.17
Fumarate + 2e^- + 2 H^+ → succinate	0.03
Ubiquinone + 2e^- + 2 H^+ → ubiquinol	0.04
Dehydroascorbic acid + 2e^- + 2 H^+ → ascorbic acid	0.08
Cytochrome c Fe^{3+} + e^- + 2 H^+ → cytochrome c Fe^{2+}	0.23
½ O_2 + 2e^- + 2 H^+ → H_2O	0.82

acetyl CoA (when it is converted to pyruvate). However, we know that the reverse reaction will take place readily—that is, when pyruvate is converted (oxidized) to acetyl CoA. With some of the other examples, you can see that they are written in the direction opposite to what occurs in the cell; remember that the only reason for this is that we are writing them as reductions. The reduction reactions for some common biological antioxidants are also shown. The reaction involving lipoic acid, a commonly purchased supplemental antioxidant, is written in the opposite way; it would work in the cell in an antioxidant function because an antioxidant donates electrons to something else. As discussed in chapter 1, GSH, representing reduced glutathione, is the common name for a tripeptide that plays an enormous role as an antioxidant in the cell. When it functions as an antioxidant, it donates electrons and becomes the oxidized form, shown as GSSG. Similarly, ascorbic acid (vitamin C), a common water-soluble biological antioxidant, donates electrons to become the oxidized form, or dehydroascorbic acid.

As already mentioned, table 5.3 shows only half reactions, but in the cell, one half reaction would be coupled to another half reaction to generate the real reaction that takes place in vivo. When we combine two half reactions, such that one substance is reduced and another is oxidized, we describe the overall reaction as a redox reaction. The overall change in redox potential for the combined reaction is given by the following:

$$\Delta E^{\circ\prime} = E^{\circ\prime}_a - E^{\circ\prime}_b$$

Let us see how this works. In the citric acid cycle, malate can be oxidized to oxaloacetate by transferring electrons to NAD^+, making it NADH. In this example, NAD^+ would behave just as shown in its half reaction from table 5.4. However, we would be doing the opposite to the reduction of oxaloacetate to make malate, so we would need to write this half reaction the opposite way. We would determine the overall standard redox ($\Delta E^{\circ\prime}$) for this combined reaction by algebraically combining the two half reactions to make the overall net reaction. Notice that when the first reaction is written the opposite way, we reverse the value of the $E^{\circ\prime}$ by changing the sign from minus to plus.

$$\text{malate} \rightarrow \text{oxaloacetate} + 2 H^+ + 2 e^-$$
$$E^{\circ\prime} = 0.17 \text{ V}$$
$$NAD^+ + 2 H^+ + 2 e^- \rightarrow NADH + H^+$$
$$E^{\circ\prime} = -0.32 \text{ V}$$

The net reaction is as follows:

$$\text{malate} + NAD^+ \rightarrow \text{oxaloacetate} + NADH + H^+$$
$$\Delta E^{\circ\prime} = -0.15 \text{ V}$$

If we think of the overall process in oxidative phosphorylation, electrons from substrates such as malate, isocitrate, α-ketoglutarate, and lactate are transferred to NAD^+, making it the reduced form, NADH. These electrons on NADH are then passed through the electron transport chain to oxygen, reducing it to water. We can show the overall $\Delta E°'$ for this by writing the two half reactions. We take the half reaction for NAD^+ reduction and reverse it to make it NADH oxidation. Then we take the $E°'$ value and change its sign. Finally, we take the half reaction for oxygen reduction. Combining these two reactions, we get the overall reaction in which electrons on NADH reduce oxygen.

$$NADH + H^+ \rightarrow NAD^+ + 2 H^+ + 2e^-$$
$$E°' = 0.32 \text{ V}$$
$$1/2\ O_2 + 2 H^+ + 2e^- \rightarrow H_2O$$
$$E°' = 0.82 \text{ V}$$
$$NADH + H^+ + 1/2\ O_2 \rightarrow NAD^+ + H_2O$$
$$\Delta E°' = 1.14 \text{ V}$$

This shows how we can describe the propensity to accept electrons measured in volts. The larger the value is in volts, the greater the propensity to accept electrons. Thus, the reduction of oxygen is the most powerful reduction process on the list. We could also say that oxygen is the strongest oxidizing agent (oxidant) in the table. Moreover, when electrons are transferred from NADH to oxygen, the overall reaction is even more spontaneous, measured in volts. The key point is that electrons move spontaneously to the compounds with the more positive redox potential.

We can put the $\Delta E°'$ values into more familiar terms by converting this redox drive, measured in volts, into energy units, measured in kilojoules per mole. We do this using the following equation:

$$\Delta G°' = -nF\ \Delta E°'$$

Here, n is the number of electrons transferred, F is the **Faraday constant**, with a value of 96.5 kJ per volt per mole, $\Delta E°'$ is the algebraic sum of the two $E°'$ for the two half reactions, and $\Delta G°'$ is the standard free energy change.

Working this equation out for the transfer of electrons from NADH to oxygen, we get this:

$$\Delta G°' = -2 \times 96.5 \text{ kJ/V} \times 1.14 \text{ V}$$
$$\text{or } -220 \text{ kJ/mol}$$

Considering that the standard free energy for hydrolysis of ATP is -30.5 kJ/mol, it is easy to understand how much free energy is released during electron transfer from substrates to oxygen. We can also learn that, when two half reactions are combined to create a redox reaction, a positive value for $\Delta E°'$ will signify that the reaction occurs with a negative value for standard free energy change ($\Delta G°'$). Such reactions are energetically favorable (i.e., downhill reactions). The larger the value of $\Delta E°'$ is, the less likely that reaction is to be an equilibrium (reversible) reaction.

The discussion so far has dealt with standard values, those expressed using the designation $\Delta E°'$, where everything is at 1 M concentration. In reality, concentrations of reactants and products in redox reactions in the cell are far less than 1 M, so we use the term ΔE, not $\Delta E°'$. This argument is similar to that made in the previous chapter regarding actual free energy changes for reactions in vivo (thus, the use of ΔG rather than $\Delta G°'$). Using the general redox reaction that follows as a model, we could determine the actual ΔE for the general reaction by using the equation that follows the reaction:

$$A_{red} + B_{ox} \rightarrow A_{ox} + B_{red}$$

$$\Delta E = \Delta E°' + \frac{RT}{nF} \ln \frac{[A_{ox}][B_{red}]}{[A_{red}][B_{ox}]} \quad (5.1)$$

This equation is known as the **Nernst equation**. R is the gas constant we saw in the previous chapter, with a value of 8.314 J/mol × K; F is the Faraday constant, with a value of 96,500 J per volt per mole; n is the number of electrons transferred in the redox reaction; and T is the absolute temperature (K). Since R is expressed in joules, not kilojoules, the Faraday constant is also expressed in joules. If we plug in the values for the two constants R (8.314×10^{-3} kJ per mole per degree K) and F (which we defined before as 96.5 kJ per volt per mole), use T = 310 K (i.e., 37 °C), and call the ratio of the oxidized and reduced products over reactants Q, we can simplify the equation to the following:

$$\Delta E = \Delta E°' - 0.0267/n \ln Q$$

OXIDANTS AND ANTIOXIDANTS

In our discussion on electron transfers, we noted that during electron transfers from NADH to coenzyme Q (ubiquinone) using complex I, a ubisemiquinone intermediate is formed by the acceptance of a single electron (see figure 5.11). This intermediate can accept another electron and two H^+ to become coenzyme QH_2 (ubiquinol), or it can transfer the single electron to oxygen, forming the superoxide anion (O_2^-). Normally, at least 95% of the O_2 consumed by the body accepts four electrons to make two molecules of water, catalyzed by cytochrome c oxidase; the remainder is reduced by one electron, forming superoxide (figure 5.19).

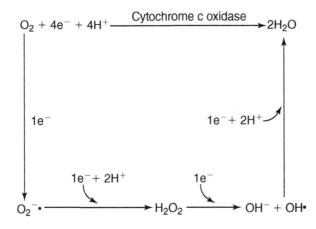

Figure 5.19 The stages of reduction of oxygen in cells. Most oxygen molecules are reduced to water as they accept four electrons and four protons in a reaction catalyzed by cytochrome c oxidase. One-electron reduction of oxygen forms superoxide anion, which can undergo a dismutation reaction, accepting another electron and two protons to form hydrogen peroxide. Reduction of hydrogen peroxide through acceptance of another electron forms the hydroxyl radical and the hydroxide anion. Addition of another electron creates water. The bold dot identifies the substance as a free radical with an unpaired electron.

Superoxide is known as a **free radical** because it contains an unpaired electron. We typically show its structure as $O_2^-·$, where the dot represents the unpaired electron. As a free radical, superoxide is quite reactive; not all the reactions it can undergo are helpful to the cell. For example, superoxide can be a reducing agent by donating its unpaired electron to ions such as Fe^{3+} (ferric ion), forming an oxygen molecule and the reduced Fe^{2+} (ferrous ion). It can also act as an oxidizing agent, accepting another electron and two H^+, forming hydrogen peroxide. We refer to highly reactive oxygen compounds like superoxide as **reactive oxygen species**, or ROS. They may or may not be free radicals.

The mitochondria are the major site of superoxide production in muscles. Superoxide can be formed at complex III, when electrons on coenzyme QH_2 or ubiquinol are transferred to cytochrome c. Again, the electrons are transferred one at a time from ubiquinol to cytochrome c, such that an intermediate ubisemiquinone is present. This intermediate can also undergo a one-electron transfer to oxygen to form superoxide. Current research tells us that the proportion of oxygen that is converted to superoxide at either complex I or III is highest when the rate of electron transport from substrates to oxygen is lowest. Under these state 4 conditions of electron transport, when ADP availability limits oxidative phosphorylation, proton motive force (the electrochemical gradient across the inner mitochondrial membrane) is high and the lifetime of the ubisemiquinone intermediate is longest. This provides a greater opportunity for one-electron reduction of oxygen, instead of the normal four-electron reduction that makes water. One of the helpful roles of uncoupling proteins is to allow electron transport without coupled phosphorylation, thus reducing the proton motive force, shortening the lifetime of ubisemiquinone, and reducing superoxide formation (see figure 5.13). During elevated or maximal rates of electron transport (state 3 conditions), less than 1% of cellular oxygen is reduced to form superoxide because the lifetime of the ubisemiquinone is shortest and there is less time for oxygen to accept just one electron.

☑ KEY POINT

Although superoxide is continually produced in the mitochondria in relatively low quantities, oxidative phosphorylation has evolved to limit superoxide and other ROS production by tightly coupling the stages of oxygen reduction in the cell via cytochrome c oxidase, such that H_2O is formed directly without the intermediary formation of hydrogen peroxide or the hydroxyl radical that would occur without the presence of cytochrome c oxidase. This is a critical evolutionary development that limited the production of ROS as intermediary compounds in the reduction of oxygen and accompanied the adaptation of organisms to a growing oxygen presence in the environment many millions of years ago.

Other Reactive Oxygen and Nitrogen Species

Superoxide is formed constantly in the body. Its presence allows other ROS to be created. Two superoxide radicals can undergo a spontaneous **dismutation reaction**, in which one of the superoxides donates its unpaired electron to the other. The former becomes an oxygen molecule, while the latter adds two protons and becomes *hydrogen peroxide*, H_2O_2. The spontaneous dismutation of superoxide is too slow to be of much value to the cell. However, both mitochondria and cytosol of cells are endowed with a **superoxide dismutase** (SOD) enzyme, which quickly catalyzes the removal of superoxide (figure 5.20). The cytosolic form of SOD contains a copper and a zinc ion at the active site and is known as Cu-Zn SOD. The mitochondrial SOD contains a manganese ion at the active site and is known as Mn SOD. Two isomers of SOD are needed because complex I forms superoxide on the matrix side of the intermembrane, whereas complex III forms and releases superoxide in the intermembrane space where it can be removed by Cu-Zn SOD. Superoxide is also produced during the enzymatic conversion of xanthine to hypoxanthine and of hypoxanthine to uric acid catalyzed by **xanthine oxidase** (figure 5.20). Superoxide is also formed by cells of the immune system and in endothelial cells of the blood vessels in a reaction catalyzed by **NADPH oxidase** (figure 5.20). Formation of superoxide by such white blood cells as **neutrophils** and tissue-scavenging **macrophages** is essential to their roles in protecting us. Not only is superoxide an important weapon for these immune cells, but superoxide can also be converted to hydrogen peroxide, which can react with a chloride ion in a reaction catalyzed by **myeloperoxidase** to form *hypochlorite* (figure 5.20). The latter compound, a component of bleach, is a particularly effective weapon by activated immune cells. The accelerated formation of superoxide by white blood cells in response to their activation is known as the **respiratory burst** and is important to their ability to kill invading bacteria and microbes.

Hydrogen peroxide is uncharged, so unlike the superoxide anion, it can cross cell membranes. Hydrogen peroxide is a strong oxidizing agent (easily reduced) and can cause damage to cell constituents. Perhaps its most damaging quality is that it can be split into the very dangerous and highly reactive **hydroxyl radical** and hydroxide ion when it accepts an electron from ferrous or cuprous (Cu^+) ions in a reaction known as the **Fenton reaction** (figure 5.20). Hydroxyl radicals are so reactive that their lifetime can be measured in billionths of second. Hydrogen peroxide is removed in reactions catalyzed by two enzymes (figure 5.20). **Catalase**, found in peroxisomes and in mitochondria in heart and skeletal muscle cells, converts hydrogen peroxide to water. **Glutathione peroxidase** (GPX) is a selenium-containing enzyme that plays very important roles in the cell. Using glutathione, it can rid mitochondria of hydrogen peroxide. It is also involved in removal of organic peroxides (figure 5.20). The sulfhydryl (–SH) group of GSH is very important as an electron donor (reducing agent), clearing cells of peroxides. In the process, glutathione is oxidized to the disulfide form, GSSG. The active, GSH form of glutathione is recovered by reduction of GSSG using electrons on NADPH, a molecule we will encounter later.

Many cells also produce a highly reactive nitrogen compound, **nitric oxide** (NO), that is an important signaling molecule. Nitric oxide is formed by an enzyme known as **nitric oxide synthase** (NOS), using the amino acid arginine as the source of nitrogen (figure 5.20). At least four forms of NOS exist. In endothelial cells, formation of NO by endothelial NOS (eNOS) leads to relaxation of smooth muscle in artery walls and increased blood flow. In cells of the immune system, an inducible form of NOS (iNOS) produces NO that can combine with other substances, such as superoxide, to generate more potent responses to invading microorganisms. A mitochondrial form of NOS (mtNOS) is similar to neuronal (nNOS) and eNOS in that it is activated by calcium ions. The matrix NO can combine with superoxide in a diffusion controlled reaction to make the more dangerous substance peroxynitrite (ONO_2^-). Another NOS enzyme (mtNOS) was recently discovered in mitochondria. A rationale for production of NO in mitochondria is hard to reconcile, given that NO is a competitive inhibitor to oxygen for cytochrome c oxidase. In effect, NO in mitochondria should diminish ATP production by oxidative phosphorylation. It now appears that the concentration of NO produced in a calcium-activated mitochondrion is insufficient to do more than partially inhibit cytochrome c oxidase. Its role as a competitive inhibitor is to raise the K_m for oxygen at the active site of cytochrome c oxidase, an effect that can be overcome when oxygen concentration is elevated. Further, it is hypothesized that the mild inhibitory effect of NO in mitochondria facilitates oxygen diffusion to mitochondria distanced farther from the cell membrane, through which oxygen first passes. In this way, NO acts to distribute ATP formation more homogenously in all parts of the cell. Because of the significance of reactive nitrogen-containing species such as NO and peroxynitrite, we often include the word *nitrogen* with ROS to provide the more inclusive term *reactive oxygen and nitrogen species* (RONS).

Cellular Damage From RONS

Reactive oxygen species (ROS) and reactive nitrogen species (RNS) are sometimes collectively referred to as reactive oxygen and nitrogen species (RONS). RONS, and particularly the highly reactive hydroxyl

Superoxide dismutase reaction
$$2O_2^{-}\bullet + 2H^+ \longrightarrow H_2O_2 + O_2$$

Catalase reaction
$$2H_2O_2 \longrightarrow 2H_2O + O_2$$

Reduction of ferric to ferrous ion by superoxide
$$O_2^{-}\bullet + Fe^{3+} \longrightarrow O_2 + Fe^{2+}$$

Fenton reaction
$$Fe^{2+} + H_2O_2 \longrightarrow OH\bullet + OH^- + Fe^{3+}$$

Glutathione peroxidase breaks down hydrogen peroxide using glutathione
$$2GSH + H_2O_2 \longrightarrow GSSG + 2H_2O$$

Glutathione removes organic peroxides (ROOH)
$$2GSH + ROOH \longrightarrow GSSG + ROH + H_2O$$

Myeloperoxidase in leukocytes produces cytotoxic hypochlorite
$$H_2O_2 + Cl^- \longrightarrow ClO^- + H_2O$$

Xanthine oxidase oxidizes hypoxanthine and xanthine, producing superoxide
$$\text{Hypoxanthine} + O_2 + H_2O \longrightarrow \text{Xanthine} + O_2^{-}\bullet$$
$$\text{Xanthine} + O_2 + H_2O \longrightarrow \text{Uric acid} + O_2^{-}\bullet$$

NADPH oxidase generates superoxide
$$NADPH + 2O_2 \longrightarrow 2O_2^{-}\bullet + NADP^+ + H^+$$

Nitric oxide synthase (NOS) produces signaling molecule nitric oxide (NO)
$$\text{Arginine} + O_2 + NADPH \longrightarrow NO + \text{Citrulline} + NADP^+$$

Nitric oxide and superoxide easily react together to form peroxynitrite ion
$$NO + O_2^{-}\bullet \longrightarrow ONO_2^-$$

Figure 5.20 Important enzyme-catalyzed reactions that play a major role in producing or removing reactive species in cells. Two GSH, or reduced glutathione, molecules donate a pair of electrons and two protons to become glutathione disulphide, or GSSG. Recipients of electrons and protons from two GSH molecules are organic peroxides (shown as ROOH), which after reduction become alcohols (ROH).

radical, can react with many molecules in cells, including DNA, proteins, and lipids. The slow buildup of threats to mitochondrial and nuclear DNA over years of oxidative damage is thought to be a factor in aging. The relatively rapid turnover (breakdown and resynthesis) of proteins is also partly due to their susceptibility to oxidative damage and their need to ensure structural integrity to maintain optimal function. The free radical theory of aging was first proposed in the 1950s by the American scientist Denham Harman. It has continued to receive scientific support over the years, and it is now well established as an integral component of the aging process. Polyunsaturated fatty acids, which are important components of the phospholipids found in cell membranes, are particularly susceptible to damage by RONS. It has been known for more than 200 years that exposure of fats to air or oxygen will turn them rancid. This rancidity is caused by **peroxidation** of the fatty acids by RONS. During peroxidation, RONS such as the hydroxyl radical can react with the phospholipids by abstracting a hydrogen from one of the double bonded hydrogen molecules in the unsaturated fat, and in so doing, forming water. In the following equation, LH represents the lipid and OH* represents the hydroxyl radical, with the * representing the electron that is donated:

$$LH + OH^* \rightarrow L^* + H_2O$$

The lipid radical (L*) can then spontaneously propagate a chain reaction, which can result in peroxidation and breakdown of a number of phospholipids in a cell membrane and damage and disruption of cellular function. Similar damage also results from RONS-induced oxidation of proteins and DNA. As will be discussed in the following section, endogenous antioxidants (e.g., glutathione) and antioxidant vitamins (e.g., vitamin E) can help stop these reactions and prevent peroxidative damage to membranes and other cellular molecules, including proteins and DNA. Limiting oxidative stress may be important in limiting various conditions associated with aging (Pourova et al. 2010)

Protection From Reactive Oxygen and Nitrogen Species

Our bodies produce a number of endogenous oxidants that we have categorized as RONS. Added to this is a host of exogenous substances, such as pollution, ozone, automobile exhaust, solvents, pesticides, and cigarette smoke. Foreign molecules that may pose a hazard to us are generally categorized as **xenobiotics**. Some xenobiotics are free radicals or can produce RONS. Three approaches to maintaining control over RONS are principal. First, reducing the formation of or exposure to RONS and xenobiotics is an essential step. Second, our endogenous scavenging systems, including antioxidant enzymes and other antioxidant molecules, can scavenge those RONS already present. Finally, mechanisms in cells can upregulate antioxidant defenses in the face of persistent problems (Halliwell and Gutteridge 1999). A failure to maintain a balance between the formation of oxidants (e.g., RONS) and their removal by our various antioxidant systems produces a state known as **oxidative stress**. Powerful oxidants such as those just discussed can damage proteins, DNA, and the unsaturated fatty acid molecules in membranes. It is no wonder, then, that aging and a variety of diseases, such as cancers, heart disease, neurodegenerative diseases, and type 2 diabetes, have a relationship to oxidative stress.

Figure 5.20 summarizes a number of important reactions that produce and remove ROS, such as SOD, catalase, and GPX. In addition to specific antioxidant enzymes, the body contains a number of nonprotein antioxidant molecules. For example, the essential nutrients vitamin C, vitamin E, and carotenoids such as β-carotene can by themselves scavenge free radicals. Minerals such as selenium, zinc, manganese, and copper are constituents of antioxidant enzymes. Finally, a host of non-nutrient antioxidants from a wide variety of foods of plant origin are important. The general name **phytochemical** is applied to those molecules that can function in our bodies as antioxidants. Fruits, vegetables, and grains, as well as common products derived from these such as tea and wine, are the major sources. Flavonoids, polyphenols, and lycopene are common examples of phytochemicals.

✓ KEY POINT

Humans are generally well endowed with antioxidant enzymes and other antioxidants, such as glutathione, that limit the degree of oxidative stress we experience. Nevertheless, a diet rich in antioxidants, such as those found in fruits, vegetables, and other sources, has been shown to reduce indices of oxidative stress and to provide many health benefits associated with reducing oxidative stress from inflammatory-related diseases and conditions.

Exercise and Oxidative Stress

It is extremely difficult to directly measure free radical formation in a human being. Rather, we use indirect measures to point to increased or decreased ROS formation; from this, we make inferences. From a theoretical perspective, during exercise, we should expect a greater formation of superoxide, and therefore hydrogen peroxide and hydroxyl radicals, simply because of an increased flux through the electron transport chain with the increased need for ATP. Since we can greatly increase oxygen consumption and oxidative phosphorylation during exercise, it has been postulated that exercise can increase the generation of oxygen radicals and the possibility of oxidative damage or stress in muscles and other tissues.

Even though the proportion of oxygen that undergoes one- instead of four-electron reduction should decrease with increased activity of the electron transport chain, the fact that the overall flux may increase 10- to 20-fold above resting levels when we exercise should lead us to suspect an increase in superoxide formation. Bailey and colleagues (2003) were the first to actually measure increased formation of free radicals in venous blood of humans during a single-leg exercise task, using an expensive technique known as electron paramagnetic resonance spectroscopy. In this study, the researchers noted that the outflow of free radical species in the venous blood increased as exercise intensity increased.

The active muscle is not the only source of increased RONS formation during exercise. It has been observed that RONS production by white blood cells increases with exercise. However, the effect of RONS production by white blood cells is increasingly blunted the more trained the person is for the particular exercise activity (Mooren, Lechtermann, and Völker 2004). This is likely a consequence of increased antioxidant enzyme activity in white blood cells with training (Elosua et al. 2003). Some types of exercise, such as intense isometric or eccentric contractions, can create inflammation in muscle. Part of the inflammatory response is caused by neutrophils (white blood cell subset) accumulating in the muscle. As mentioned earlier, these cells produce superoxide that can cause damage within the muscle and can also produce other reactive species (Nguyen and Tidball 2003). Obesity, a major public health problem, is characterized by elevated markers of oxidative stress. When people who are obese perform either aerobic or resistance exercise, they produce higher levels of lipid peroxidation products than normal-weight individuals (Vincent, Morgan, and Vincent 2004). Acute feeding of high-fat diets also increases RONS production in rodents. Such a response suggests either an enhanced production or a reduced ability to scavenge free radicals, or both, in the obese state. Aging is also characterized by increased RONS production and oxidative stress, and increased RONS are associated with various diseases of senescence (Pourova et al. 2010).

Rested muscle produces RONS. When muscle becomes more active, formation of RONS increases. Having some RONS in muscle confers a benefit, since they are often important signaling agents that help regulate acute responses to exercise as well as positive adaptations to training in the muscles. Animal studies have demonstrated that inhibiting RONS production during exercise training actually blunts the signaling pathways that regulate adaptations to training, such as increasing mitochondria and aerobic capacity in skeletal muscle (Gomez-Cabrera et al. 2005). Important signaling pathways such as NFκB can help regulate antioxidant adaptations and mitochondrial synthesis consequent to endurance training. The inhibition of RONS production during exercise can blunt the response of these pathways and, consequently, positive training-induced adaptations (Powers et al. 2010). However, a study involving endurance training in humans failed to demonstrate a short-term negative effect of antioxidant supplements on endurance training adaptations (Yfanti et al. 2010).

Dietary antioxidants such as vitamin E and vitamin C have been touted as important for minimizing muscular damage from exercise. Athletes have often been encouraged to supplement their diets with a variety of antioxidants. While eating a diet high in antioxidants would likely have health benefits and could influence longer term effects of oxidative stress such as aging, studies have demonstrated that it is unlikely that large intakes of antioxidants will have significant physiological effects in diminishing exercise-induced muscle damage; instead, they may inhibit positive adaptations to training (McGinley, Shafat, and Donnelly 2009; Gomez-Cabrera et al. 2008). In addition, studies have demonstrated that antioxidant supplements taken during training may also suppress health-promoting benefits such as improved insulin resistance (Ristow et al. 2009).

Unaccustomed exercise and overtraining lead to muscle soreness, inflammation, and damage. This can be reduced by repeated exposure to the activity or through training. Attempts to limit the amount of stress by taking anti-inflammatory medication such as NSAIDs can potentially reduce oxidative stress in muscles that results from infiltration of white blood cells, such as neutrophils and macrophages. However, since the presence of white blood cells, particularly macrophages in muscle following exercise, is obligatory for the activation of muscle satellite cells and their role in stimulating muscle hypertrophy, reducing inflammation may also limit the amount of muscle repair, adaptation, and hypertrophy that would be induced by the training.

RONS and Cell Signaling in Exercise and Training

As mentioned earlier, it is well known that resting skeletal muscle produces RONS and that muscular activity is associated with increased production of RONS. However, thanks to the pioneering work of Michael Reid and his colleagues, we now know that RONS actually play a true physiological role in skeletal muscle function (reviewed in Lecarpentier 2007). In a series of different experiments, Reid and his colleagues demonstrated that RONS, and specifically hydrogen peroxide, induce a biphasic effect on force development in unfatigued muscle; a moderate increase in hydrogen peroxide increases force development, whereas high concentrations cause impaired contractile function. In fact, they also showed that the low hydrogen peroxide levels present in skeletal muscle under resting conditions are necessary to preserve normal muscle performance, since addition of catalase, the antioxidant enzyme that dehydrates hydrogen peroxide to molecular oxygen and water, reduces force development. Low levels of hydrogen peroxide increase the calcium sensitivity of contractile proteins, which can explain the positive effects on muscle force production under those conditions. The negative effects of hydrogen peroxide at high concentrations could be prevented by antioxidant pretreatment or reversed by the addition of antioxidant agents. However, during strenuous exercise, RONS are generated faster than the buffering capacity provided by the body's antioxidants, so that muscle performance is impaired. In fatigued muscle, pretreatment by antioxidants can also blunt the negative effects of RONS on skeletal muscle contractile function. For example, studies that used free radical scavengers have found increased fatigue resistance in isolated muscle models (Allen, Lamb, and Westerblad 2008). Other studies have shown that RONS also play a physiological role in the regulation of SERCA protein content and activity (Tupling et al. 2007), as well as in that of muscle glucose uptake during exercise (Sandström et al. 2006). Collectively, it can be concluded from these studies that exposure to low levels of RONS represents an optimum state in skeletal muscle and any deviation from this optimum induces a loss in muscle performance.

The reports of antioxidant supplements diminishing training adaptations in skeletal muscle and limiting health-related benefits of exercise also highlight the importance of RONS as important signaling molecules (Powers et al. 2010). As previously noted, early research on RONS and exercise tended to focus on their potential to induce tissue damage. More recent research has highlighted their importance as signals for initiating positive adaptations to training. A number of important signaling pathways and transcription factors that promote positive adaptations to training are sensitive to and mediated by redox status of muscle and other tissues. These pathways and the mechanisms of the effects of redox status on their activation and inhibition are discussed in more detail in chapter 3 (see figure 3.26, p. 68). The important point to be remembered is that earlier discussions of limiting oxidative stress during training with high intakes of antioxidants were misplaced for the most part, since such interventions may actually inhibit beneficial adaptations to muscle performance as well as some of the health benefits associated with regular exercise.

KEY POINT

During exercise, RONS production increases markedly; however, severe muscular damage due to RONS-induced peroxidation is rarely seen, indicating that normal muscle is well endowed with antioxidant protection. Training also increases antioxidant protection by possibly increasing levels of antioxidant enzymes and other nonenzymatic antioxidant levels, such as glutathione, as well as reducing generation of superoxide. Athletes who attempt to mitigate oxidative damage in their muscles by ingesting large amounts of antioxidants or limiting postexercise muscle inflammation may actually be acting in a counterproductive manner, since reducing RONS signaling and inflammation in muscle may inhibit positive adaptations to training as well as positive health benefits.

NEXT STAGE

Effects of Aging and Exercise Training on Muscle Mitochondria

A great deal of recent research highlights mitochondrial function—specifically, its importance in aging and its adaptive responses to training. In this section, we highlight a few examples of this emerging research, such as mechanisms that may help maintain mitochondrial function during aging and the novel effects of short, high-intensity training on mitochondrial biogenesis in muscle.

Aging and Mitochondrial Function

A review by Dr. Russ Hepple (2009) of the University of Calgary summarized the effects of aging on muscle mitochondria. Mitochondrial dysfunction is a characteristic of aging skeletal muscle and may be a significant factor in the overall decline in function of all cells consequent to aging. A decline in mitochondrial function with aging is not simply a reduction in the amount of mitochondria per unit of skeletal muscle, which would result from a combination of aging and inactivity. Mitochondrial dysfunction in aging also involves a decline in the energy provision, or oxidative capacity, per unit of mitochondria. This decline in mitochondrial function with aging is due to a combination of decreases in mitochondrial biogenesis and in mitochondrial degradation, resulting in an overall decrease in mitochondrial turnover.

Mitochondrial proteins are continually broken down and resynthesized; the average half-life of mitochondrial enzymes is about seven days. It is postulated that this turnover is necessary to mitigate accumulation of oxidative damage in mitochondria due to RONS production. Aging results in reduced mitochondrial protein turnover due to less synthesis and degradation, which leads to an accumulation of mitochondrial oxidative damage and a consequent increase in dysfunctional mitochondrial proteins and enzymes associated with the citric acid cycle and oxidative phosphorylation. It appears that mitochondrial complex IV proteins, including cytochrome c oxidase, are the most susceptible to this aging-related dysfunction. In addition, the greatest increases in superoxide production with aging appear to occur within the mitochondrial matrix (Xu et al. 2010). The cumulative result is a decrease in relative mitochondrial ATP production and oxidative capacity, which is further exacerbated by additional increases in RONS production.

Calorie Restriction

One intervention that has proven to be successful in mitigating the aging-related effects on mitochondrial dysfunction, at least in animal models, is calorie restriction (Hepple 2009). This form of experimental calorie restriction (CR) in rodent models utilizes a diet that restricts food intake to 60% of the calories consumed by freely eating normal rodents, while still maintaining adequate nutrient intakes. This type of diet would be unrealistic to implement for most humans, but it does illustrate the mechanisms associated with the age-related decline in mitochondrial function. Rodents following a calorie-restricted diet live up to 40% longer than normal rodents and exhibit little decline in mitochondrial function with aging. The CR animals also have lower mitochondrial RONS production and maintain mitochondrial protein synthesis and degradation activities similar to younger animals. It is hypothesized that CR is able to mitigate mitochondrial dysfunction by upregulation of the silent information regulator-1 (SIRT1) signaling pathway in response to the chronic decrease in mitochondrial energy state (ATP/ADP ratio) induced by CR. The SIRT1 pathway can stimulate both the maintenance of mitochondrial degradation mechanisms and the expression of peroxisome proliferator–activated receptor γ coactivator-1α (PGC-1α), an important transcription factor that, as discussed in chapter 3, coordinates the expression of genes involved in augmenting mitochondrial biogenesis (Hepple 2009).

Another benefit of CR in this model is to enhance insulin sensitivity, thereby decreasing circulating insulin in the aging animals. Since CR of 40% is not a practical means of promoting mitochondrial health in humans, other means have been explored to determine if similar effects can be seen on factors influencing mitochondrial function with aging. One such example is resveratrol, a compound found in grape skin and consequently in wine, which has been shown to activate SIRT1 and enhance mitochondrial biogenesis (Hepple 2009). Whether this dietary supplement would actually benefit human mitochondrial function in aging or improve longevity has yet to be determined.

Physical Activity

Another mechanism to possibly preserve mitochondrial function with aging is exercise. A study by Safdar and colleagues (2010) demonstrated that a physically active lifestyle could at least partially conserve mitochondrial oxidative and antioxidant capacity and maintain mitochondrial repair capacity despite accumulation of age-related ROS-induced mutations in mitochondrial DNA. However, other studies have demonstrated that even with training, older people could not maintain mitochondrial capacity as well as younger ones, possibly due to the inability to fully preserve PGC-1α expression even when physically active (Lanza and Nair 2010). Hence, while physical activity appears to help preserve some aspects of mitochondrial function and oxidative capacity in the face of aging, it does not appear to be able to overcome all of the age-related declines in mitochondrial function. Research into factors that may delay or offset age-related mitochondrial dysfunction continues to be a hot topic, since mitochondrial decline is closely related to many of the cellular functioning declines seen with aging.

Interval Training

We have known for almost 50 years that classical endurance training stimulates mitochondrial biogenesis in skeletal muscle. More recently, research has demonstrated that high-intensity interval training of low volume also stimulates mitochondrial biogenesis and increases muscle oxidative capacity to a similar degree as endurance training of higher volume and lower intensity (Little et al. 2010). As noted earlier in this chapter, as few as six training sessions consisting of 8 to 12 60-s cycling intervals at $\dot{V}O_2$max, separated by 75 s of rest, over 2 weeks may elicit significant increases in muscle mitochondria, aerobic enzymes, and endurance performance that are comparable to the results of much longer, traditional aerobic training (Little et al. 2010)

As previously noted, the signaling for enhanced mitochondrial biogenesis is coordinated via increases in SIRT1, which modifies the activation of PGC-1α and its activation via acetylation (addition of an acetyl group). These signals then activate the appropriate protein transcription and translation necessary for mitochondrial biogenesis, including all of the mitochondrial enzymes associated with oxidative phosphorylation. Little and colleagues (2010) have demonstrated that low-volume, high-intensity interval training will result in elevations of more than 50% in muscle SIRT1 content, as well as increased content of PGC-1α activator in the muscle nuclei. In addition, a second signaling factor, mitochondrial transcription factor A (Tfam), was also elevated in muscle nuclei consequent to the training. These signaling changes also stimulate mitochondrial biogenesis and increase oxidative enzyme capacity in skeletal muscle following traditional endurance-training protocols. Upregulation of PGC-1α may also signal increases in muscle glycogen content and indirectly enhance muscle glucose uptake. We will discuss these effects further in chapter 6. It now appears that a number of training protocols may result in enhanced muscle mitochondrial content and oxidative capacity. Although more intense than traditional endurance training, the shorter time required to participate in low-volume, high-intensity interval training may appeal to some people, particularly if future studies demonstrate that this type of training may confer health benefits and mitochondrial maintenance during aging similar to those from more traditional endurance training.

SUMMARY

Oxidative phosphorylation is the synthesis of ATP from ADP and Pi in association with the transfer of electrons from fuel molecules to coenzymes to oxygen. Oxidative phosphorylation, responsible for generating the preponderance of ATP in our bodies, takes place in mitochondria, utilizing the citric acid cycle and the electron transport (respiratory) chain. The citric acid cycle is the primary source of reduced coenzymes. The electron transport system, represented by four protein–lipid complexes in the inner mitochondrial membrane, oxidizes the reduced coenzymes and transports the electrons to oxygen. During the process of electron transport through these complexes, free energy is released, which is utilized to transport protons across the mitochondrial inner membrane against a concentration and electrical gradient. Protons are allowed to flow down their electrical and chemical gradient through a specialized inner mitochondrial protein complex known as ATP synthase. Free energy released during proton flow is harnessed to ADP phosphorylation, making ATP. The exact stoichiometry of ATP produced per atom of oxygen consumed (P/O ratio) during electron transport in humans is less than 2.5. Although electron transport is normally tightly coupled to ADP phosphorylation, leakage of protons across the inner membrane (i.e., uncoupled oxidation) takes place using inner membrane uncoupling proteins.

The citric acid cycle (Krebs or TCA cycle) is the pathway that removes the last carbon atoms in the form of CO_2 from all the body's fuels, while electrons associated with the hydrogen atoms of these fuels are used to reduce the coenzymes NAD^+ and FAD. The citric acid cycle is a circular pathway catalyzed by eight enzymes, all of which are located in the mitochondrial matrix except SDH. Acetyl groups containing two carbon atoms enter the pathway attached to CoA and join with oxaloacetate to form citrate. Because one turn of the cycle yields two CO_2, the citric acid cycle neither produces nor consumes oxaloacetate or any other intermediate of the cycle.

Since most of the ATP is consumed in the cytosol of cells, whereas ATP is produced from ADP and Pi in mitochondria, special transporters move ATP from the matrix of the mitochondria while simultaneously bringing ADP in. The rate of ATP hydrolysis in the cytoplasm is closely matched to the citric acid cycle, the rate of electron transport, and ADP phosphorylation in mitochondria. Calcium ions regulate the activity of pyruvate dehydrogenase that converts pyruvate produced in the cytosol to acetyl CoA. Calcium also regulates two enzymes in the citric acid cycle. Oxidative phosphorylation in skeletal muscle, generating ATP, is tightly regulated to rates of ATP hydrolysis in the cytosol. Calcium control of both processes is essential. Rather than focusing on changes in the concentrations of the individual substrates for oxidative phosphorylation, exercise biochemists have adopted two ratios that better account for the coupling of ATP demand to ATP provision. Changes in muscle-tissue oxidation rates better match alterations in the energy demand based on the phosphorylation potential ([ATP]/[ADP] × [Pi]) and mitochondrial redox potential ([NADH]/[NAD^+]) generated by the regulated dehydrogenases in mitochondria. The oxygen tension at the level of the mitochondria ($P_{mito}O_2$) can limit oxidative phosphorylation. This can occur during exercise at altitude or when blood flow to exercising muscle is partially or completely occluded. The rate of NADH formation in the citric acid cycle may limit the overall rate of oxidative phosphorylation in skeletal muscle if carbohydrate available to exercising muscle is depleted, forcing the muscle to use fat as the exclusive fuel. For an athlete, this means reducing the pace of the activity.

The basis for oxidative phosphorylation is the coupling of oxidation of fuel molecules to reduction of oxygen. The ability of substances to be reduced can be measured against the reduction of hydrogen ions and given a specific value described as the reduction potential. Reduction potentials can be converted into free energy values. Partial reduction of oxygen in the electron transport chain can give rise to superoxide, a free radical with one unpaired electron. Superoxide is converted to hydrogen peroxide by the enzyme superoxide dismutase. Hydrogen peroxide can be converted to the dangerous hydroxyl radical. Nitric oxide, an important signaling molecule, can combine with superoxide to form highly reactive peroxynitrite. Reactive oxygen and nitrogen species are collectively known as RONS. Aging may be partially a consequence of continued low-level exposure to RONS and the cumulative buildup of damage resulting from this exposure in mitochondria and other organelles. Although exercise can result in increased production of RONS, this does not typically result in significant negative effects in normal healthy muscle, since endogenous antioxidant protection is high in muscle. In fact, small increases in certain RONS can have positive effects on muscle force production, SERCA activity, and glucose uptake. Exercise generation of RONS can also be important in signaling positive muscular and health-related adaptations to training.

REVIEW QUESTIONS

1. Determine the number of ATP molecules generated from the complete oxidation of a pyruvate, based on the P/O ratios used in this book. In earlier texts, P/O ratios were given as 3 for NADH and 2 for $FADH_2$. If you used these older numbers, how many ATP would be formed from the complete oxidation of pyruvate?

2. How many electrons are needed to reduce four molecules of oxygen? How many protons would be translocated (pumped) across the inner membrane during the reduction of four oxygen molecules?

3. A subject exercising at 110 W, breathing room air (21% O_2), has a $\dot{V}O_2$ of 2.0 L per minute. What would the $\dot{V}O_2$ be if the subject were performing at the exact same workload but breathing a gas mixture with only 14% O_2?

4. You have isolated mitochondria and are testing various substrates by measuring the oxygen consumption with a special electrode. You add the inhibitor rotenone. What must be your substrate? Why?

5. In figure 5.18, the $\dot{V}O_2$ is approximately 0.4 L per minute when the subject is merely sitting

on the cycle ergometer doing no exercise. Why?

6. Approximately how many kilocalories (or kilojoules) of energy are expended when a subject exercises for 2 h at an average $\dot{V}O_2$ of 3.0 L per minute?

7. You determine the $\dot{V}O_2$max of a group of sedentary students and cross country runners while they are running at constant speed on a treadmill. The treadmill grade is increased every 2 min until the subjects are no longer able to continue. For the untrained subjects, the $\dot{V}O_2$ increases progressively up to the point at which the subjects can no longer continue. For the cross country runners, the $\dot{V}O_2$ levels off about 2 min before they are forced to stop running. Explain why these differences exist.

8. Calculate the standard free energy when electrons on $FADH_2$ are transported to oxygen.

9. Among cytochrome c reduced form, cytochrome c oxidized form, pyruvate, and oxygen, which can dehydroascorbic acid reduce?

10. Determine the $\Delta E°'$ for the enzymatic oxidation of lactate to pyruvate.

11. Using the Nernst equation, determine the ΔE for the oxidation of lactate to pyruvate at 25 °C at the time when the concentrations of the substances in the reaction are (in millimoles) as follows: lactate, 10; pyruvate, 0.2; NAD^+, 2; NADH, 0.05. Ignore the $[H^+]$ in your calculation.

12. What are the potential benefits and drawbacks of adding large amounts of antioxidant supplements to an athlete's diet?

chapter 6

Carbohydrate and Related Metabolism

Most active people are aware of the importance of carbohydrate for physical performance. Many others are likely aware that a diet with a greater proportion of complex and high fiber sources containing carbohydrate instead of fat, particularly saturated fat, is healthier than one with an equal or lower proportion. Few people, however, know much about the chemical reactions in the body involving carbohydrate or how the body, especially skeletal muscle, adjusts its fuel utilization to spare the use of carbohydrate unless stores are high or the need for ATP production is acute. Carbohydrate is brain food, and our chemistry is set to favor scarce carbohydrate stores for brain use. This macronutrient has received a lot of bad publicity, as exemplified by the focus on diets low in carbohydrate. Obesity, particularly in combination with lack of physical activity, is one of the fastest growing health problems in the world. Obesity and inactivity are associated with a diminished ability to handle carbohydrate and the development of metabolic disorders collectively termed *metabolic syndrome*. In the extreme, this can develop into type 2 diabetes, a condition that endangers the health of an increasing number of people. Type 2 diabetes threatens to become an epidemic throughout the developed and developing worlds.

We begin this chapter by looking at the various carbohydrates and the ways in which glucose entry into cells is regulated. This chapter includes a detailed examination of the glycolytic pathway. We also look at the metabolism of glycogen and how the synthesis and breakdown of liver and muscle glycogen are tied to the priorities of carbohydrate storage and utilization. Lactate, a product of intense exercise, is a major focus, followed by the mechanisms whereby electrons generated in the cytosol can feed into the electron transport chain. Important routes to making carbohydrate are used when dietary sources are inadequate. Gluconeogenesis, or the synthesis of glucose from noncarbohydrate sources, is vitally important. The **pentose phosphate pathway** is often overlooked because it is not active in muscle; we will consider why it is important. This chapter also looks at some of the signaling pathways that control cellular metabolism. Material is organized into the basic pathways, followed by the mechanisms that regulate these pathways. Each section ends with a discussion of the role of exercise. The Next Stage section discusses the controversy around the mechanisms by which muscle glycogen depletion may contribute to muscular fatigue.

CARBOHYDRATES

We classify carbohydrates as monosaccharides, disaccharides, and polysaccharides. Examples of **monosaccharides** are glucose, fructose, and galactose; these are simple sugars called hexoses containing six carbon atoms—*hex* for six carbon atoms and *ose* meaning sugar. We will encounter two other monosaccharides that are pentoses. As discussed in chapter 3, ribose and deoxyribose, each containing five carbon atoms, are found in ribonucleotides and deoxyribonucleotides, respectively. Figure 6.1 shows the structures for the predominant forms of α-D-glucose, α-D-galactose, and α-D-fructose. The D refers to the configuration about carbon atom 5 in each molecule. Only the D forms of the monosaccharides are acceptable to glucose-metabolizing

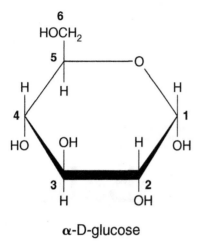

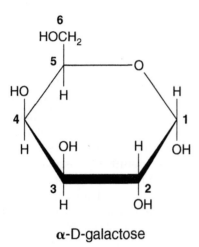

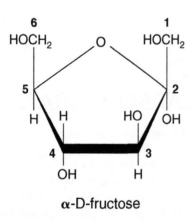

Figure 6.1 Structural formulas for the monosaccharides α-D-glucose, α-D-galactose, and α-D-fructose. The numbers identify the carbon atoms, and the D refers to the absolute configuration about carbon 5. The symbol α refers to the position of the OH groups on carbon atom 1 of D-glucose and D-galactose and on carbon 2 on D-fructose. These molecules are all isomers with the formula $C_6H_{12}O_6$. Note that D-glucose and D-galactose differ only in the configuration about carbon atom 4.

enzymes in animals and humans. In the remainder of this chapter, the D form of each monosaccharide is assumed; thus, the letter will be omitted. Recall that only L amino acids can be used by animals to make proteins. The α in front of the name refers to the configuration about carbon atom 1 in glucose and galactose and about carbon 2 in fructose. The other configuration about these carbons is β, in which the hydroxyl group (OH on carbon 1 for glucose and galactose and on carbon 2 for fructose) would be pointing up. Sugars such as glucose and galactose, in which carbon atom 1 is free (not attached to another molecule), are capable of reducing certain metal ions, producing a colored reaction product. They are said to be *reducing sugars*. Fructose is not a reducing sugar.

Disaccharides are formed when two monosaccharides join together. The common disaccharide sucrose is composed of the monosaccharides glucose and fructose, while lactose (milk sugar) contains glucose and galactose. Maltose, produced during the digestion of dietary starch, contains two glucose molecules joined together. Specific digestive enzymes located on the surface of intestinal cells hydrolyze disaccharides into their monosaccharide constituents. Thus, sucrase, lactase, and maltase digest sucrose, lactose, and maltose, respectively. In sucrose, carbon atom 1 of glucose is joined to the fructose, so sucrose is a nonreducing sugar. In maltose or lactose, there is a free carbon atom 1, so both are reducing sugars.

Glycogen and starch are polysaccharides, but only starch is significant as a dietary source of carbohydrate. From a nutritional perspective, starch is a complex carbohydrate. When it is completely digested, its products are glucose molecules. Glycogen is a highly branched polysaccharide, composed only of glucose molecules joined together.

☑ KEY POINT

Most of the world measures concentrations of glucose, cholesterol, lactate, and other substances in SI units—that is, millimoles per liter (mmol/L or mM). Conversion of a glucose concentration in millimolar units to mg/dL (the units commonly used for medical tests in the United States) can be worked out using a known relationship: A 1 mM glucose solution contains 18 mg of glucose per deciliter. For lactate, a 1 mM concentration equals 9 mg/dL. For cholesterol, a 1 mM concentration is equivalent to 38.5 mg/dL.

After a meal containing a variety of foods, the main products of carbohydrate digestion are glucose, some fructose, and galactose from milk sugar. These substances are absorbed into the blood and transported to the liver, where galactose is converted to glucose. Fructose can be utilized as a substrate for glycolysis in muscle, liver, and adipose tissue. It enters the pathway at several locations, but in the end, its products are no different from those of glucose.

CELLULAR UPTAKE OF GLUCOSE

The normal blood glucose concentration, referred to as **euglycemia**, is approximately 5 to 6 mM, equivalent to 90 to 108 mg of glucose per deciliter (100 mL) of blood. Following a meal, blood glucose is elevated above normal and can temporarily increase to 9 mM (160 mg/dL) or more. For normal people, postmeal blood glucose levels quickly return to euglycemic levels. However, in people with uncontrolled insulin-dependent **diabetes mellitus** (type 1 diabetes), blood glucose concentrations can reach 20 mM (360 mg/dL) or higher, and elevated levels can exist for more prolonged periods of time. People with established type 2 diabetes can have resting blood glucose concentrations in the range of 7 to 9 mM, or 126 to 162 mg/dL (Burtis and Ashwood 2001). We call elevated blood glucose concentrations **hyperglycemia**. Blood glucose concentration well below normal (about 2.5 mM or less, or less than 45 mg/dL) is called **hypoglycemia**. You can become hypoglycemic if you do not eat for a long period of time or if you exercise for hours without ingesting carbohydrate. People who do not pay attention to carbohydrate intake during prolonged competitions or exercise may be forced to stop because of hypoglycemia. In addition, people who control diabetes with insulin also need to be constantly aware of their blood glucose levels, since both hyper- and hypoglycemia can result from inadequate control of blood glucose. This can be a significant problem for some diabetics; the lack of close control of blood glucose levels can lead to many complications of diabetes, including impairment of vision and the loss of limbs as nerve damage resulting from a prolonged lack of blood glucose homeostasis progresses. A gradient exists for glucose entry into cells because the glucose concentration in the blood and extracellular fluid is much higher than inside cells. Glucose is a polar molecule with five hydroxy (OH) groups. It is therefore a poor substance for crossing the hydrophobic cell membrane. To get glucose inside cells, a transporter is needed—that is, a protein molecule that will allow glucose entry across the cell membrane. The process of transporting a substance down its concentration gradient across a membrane is known as facilitated diffusion. As mentioned in chapter 2, membrane transport is characterized by simple Michaelis–Menten kinetics, with a Vmax and a K_m. As such, membrane transport exhibits saturation kinetics similar to the effect of increasing substrate concentration on the activity of an enzyme. A family of membrane transporters exists for glucose. These **glucose transporters** are given the abbreviation GLUT and a number to identify the particular transporter isoform. These transporters exhibit tissue specific locations and unique kinetic parameters, as shown in table 6.1.

To be metabolized, glucose must diffuse into a cell through a glucose transporter. Entry into some cells is regulated; entry into others is unregulated. We would expect the unregulated entry of glucose, which depends only on the relative concentration gradient of glucose across the membrane, to occur in cells that rely primarily on glucose as an energy source. In fact, this is what happens for red blood cells, brain cells, and kidney cells. Liver cells,

Table 6.1 Mammalian Hexose Transporters

GLUT isoform	Approximate K_m (mM) for main substrate	Tissue specificity and characteristics
GLUT-1	5 for glucose	Erythrocytes, neurons, muscle, for resting glucose uptake in most cells
GLUT-2	10-15 for glucose	Liver, pancreatic beta cells, small intestine, and kidney
GLUT-3	1-2 for glucose	Brain mainly, but expressed to small extent in other cell types
GLUT-4	3-5 for glucose	Expressed in insulin-sensitive tissues (muscle, heart, adipose tissue)
GLUT-5	6 for fructose	Jejunum (part of small intestine)

which store excess glucose as glycogen, also have unregulated glucose uptake. However, as we will see, there is another level of control. Large tissues, such as skeletal muscle or fat, as well as those in the heart, have regulated glucose uptake. Glucose transport across cell membranes in regulated tissues occurs primarily by the GLUT-4 transporter, which, unlike the other transporter isoforms, is regulated by insulin. Skeletal muscle accounts for about 70% to 80% of the insulin-stimulated uptake of glucose.

Insulin, a polypeptide hormone secreted by the beta cells of the pancreas, is the main regulator of glucose transport. When blood glucose concentration is elevated (e.g., following a meal), blood insulin concentration increases to help glucose enter the regulated tissues. Insulin binds to an insulin receptor that spans the cell membrane. Through a complicated mechanism involving other intracellular subunits and protein kinases, GLUT-4 transporters are translocated from intracellular storage vesicles to the cell membrane to aid glucose entry. Thus, insulin increases the Vmax of glucose transport, but only in those cell types (i.e., muscle and fat) expressing the GLUT-4 transporter gene. The GLUT-4 transporters are inserted into the surface membrane but appear more prominently in the T-tubules, where the glucose that is taken up can be distributed more easily to the interior of the fiber. Muscle does contain GLUT-1 transporters that are consistently present in the sarcolemma. These do not respond to insulin. Because they are not present significantly in the sarcolemma, they have a limited effect on glucose uptake. A more detailed examination of the effects of insulin binding to cells is presented later in this chapter.

Exercising skeletal muscle also has an increased ability to take up glucose from the blood, independent of the effect of insulin. The exercise effect also involves a stimulation of GLUT-4 transporter translocation to the cell membrane from intracellular storage sites that are different from those affected by insulin. The mechanism for the exercise effect involves a different signaling pathway, which may be activated by the elevated calcium concentration, caused by activation of the muscle fiber through its motor neuron, or altered cellular energy status (i.e., increased [ADP]/[ATP]), caused by increased ATP consumption by ATPase enzymes or increased production of RONS. Interestingly, the effects of exercise and insulin are additive, supporting the fact that the signaling systems are different. This muscle contraction effect persists into the early postexercise period in order to rebuild depleted stores. During prolonged exercise tasks or games, we must ingest glucose to maintain blood levels because exercising muscle has an augmented capacity to take up glucose from

✓ KEY POINT

People who engage in endurance-exercise training programs increase the total content of GLUT-4 transporters in the trained muscle. This means that there is an increased maximal capacity for glucose transport in the trained muscle. Interestingly, when subjects are compared during the same exercise intensity after training versus before, they have more total muscle GLUT-4 transporters, but fewer are located in the sarcolemma and T-tubules, where they would aid in glucose transport. One of the adaptations that takes place after training is the use of less carbohydrate and more fat to fuel the same exercise level. This adaptation helps preserve glycogen stores and, consequently, prolongs the ability to maintain exercise for longer periods of time.

the blood. Failure to supply enough glucose to the body during prolonged physical activity can lead to problems associated with hypoglycemia.

Type 1 diabetes mellitus (T1DM), or insulin-dependent diabetes mellitus, is a condition in which blood insulin does not exist, due to an autoimmune destruction of the insulin-secreting beta cells of the pancreas. Type 1 diabetes mellitus is also known as juvenile diabetes, since it typically presents in childhood. People with T1DM need exogenous insulin to survive, but because of the effect of exercise on glucose uptake, they must carefully balance the type, intensity, and amount of exercise with blood glucose levels, carbohydrate intake, and insulin dose. Type 2 diabetes mellitus (T2DM), or non–insulin-dependent diabetes mellitus, is usually diagnosed in adults, is often associated with obesity, and is one of the most pressing health problems in wealthy countries. It is typically characterized by hyperglycemia, hyperinsulinemia (elevated blood insulin, at least in the early stages), and insulin resistance. Insulin resistance means that for a given blood insulin concentration, less glucose is taken up into muscle and fat cells. It also means that the insulin cannot suppress output of glucose from the liver, as it does in healthy people, when the blood glucose concentration is increased.

In early stages, T2DM can be characterized by excess secretion of insulin from the pancreas in response to an increase in blood glucose. As the condition progresses, insulin secretion from the beta cells of the pancreas is reduced, suggesting some failure in the beta cells. At least one-third of T2DM patients need to take insulin. Figure 6.2 illustrates blood glucose and insulin concentrations in response to an oral glucose challenge in the fasted state. For comparison purposes, representative data are shown from a healthy adult, a person in a pre-T2DM condition, and a person with a more advanced stage of T2DM. A major source of the problem with T2DM is resistance to insulin in muscle and adipose tissue, although abnormalities in lipid metabolism, such as increased levels of fatty acids in the blood as well as elevated levels of fat in skeletal muscle, also coexist. Research shows a common link between obesity and diabetes in that each is characterized by markers of cellular inflammatory stress that appears physiologically as insulin resistance and T2DM. Newer research links the development of obesity-related inflammatory responses in adipose tissues with skeletal-muscle insulin resistance. The mechanisms of these relationships are discussed in detail in chapter 7. Health problems with T2DM include increased risks for heart attack, stroke, some cancers, blindness, kidney failure, and amputation due to inadequate blood flow in the legs.

The term **metabolic syndrome** is prominently used by health professionals. Those diagnosed with metabolic syndrome have three or more of the following five disorders: fasting hyperglycemia, high blood pressure, elevated blood triglyceride levels, decreased levels of high-density lipoprotein (HDL) cholesterol, and obesity, particularly in the abdominal area (Armitage et al. 2004). This syndrome may progress to T2DM and cardiovascular disease. Although we typically associate T2DM with adults, evidence shows impaired glucose tolerance in obese children under 10 years of age (Rocchini 2002). Type 2 diabetes can be controlled to some extent with drugs. However, because of the strong association between T2DM and obesity, a reduction in body fat is an important nondrug approach to treatment. With an understanding of the mechanism for exercise-induced blood glucose transport, health authorities are proposing daily exercise for those predisposed to or experiencing T2DM (Tucker et al. 2004). Among the many benefits of exercise is an increase in muscle heat shock proteins (HSPs). We have previously noted that HSPs are important in aiding the folding and assembly of proteins; however, new research has ascribed many other functions to HSPs. Among these findings are the beneficial effects of HSPs on insulin sensitivity. Increases in muscle HSP levels as a result of exercise or heat exposure (as induced by hot tubs) in people with type 2 diabetes have been demonstrated to reduce inflammatory pathways known to inhibit insulin signaling and to enhance insulin sensitivity (Geiger and Gupte 2011). This indicates that HSPs have an important role to play in limiting the development of type 2 diabetes. It also highlights a mechanism by which exercise and other factors that raise HSPs in muscles may mitigate factors that promote the development of insulin insensitivity.

PHOSPHORYLATION OF GLUCOSE

Once glucose enters a cell, it is covalently modified by transfer of the terminal phosphate from ATP to carbon atom 6 of glucose to make glucose 6-phosphate (glucose 6-P), as shown in the following equation:

$$\text{glucose} + \text{ATP} \rightarrow \text{glucose 6-phosphate} + \text{ADP}$$

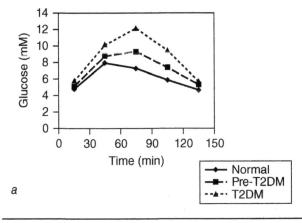
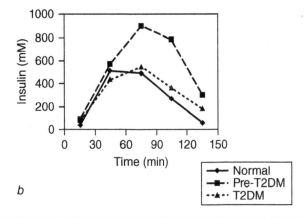

Figure 6.2 Representative responses in (*a*) blood glucose and (*b*) blood insulin concentrations in fasted subjects following an oral glucose challenge. Values are shown for a healthy person, a person with insulin resistance who is not yet classified as having type 2 diabetes mellitus (pre-T2DM), and a person with T2DM. Note that the blood glucose concentration increases more with T2DM, yet the insulin response is blunted.

A class of enzymes known as hexokinases catalyze this reaction. When glucose is phosphorylated, the product, glucose 6-P, is trapped inside the cell. This reaction is essentially irreversible because of the accompanying large free energy change, since ATP is hydrolyzed. The four hexokinase isoenzymes are identified as hexokinases I, II, III, and IV (HK I, HK II, HK III, and HK IV). Hexokinase IV, more commonly known as glucokinase, is found in the liver and beta cells of the pancreas.

The differences between glucokinase (HK IV, also referred to as GK) and the other hexokinase isozymes (HK I, HK II, and HK III) are as follows:

1. Glucokinase is found only in the liver and pancreas, whereas the other hexokinase isozymes are found in all cells.

2. The amount of hexokinases I, II, and III in cells remains fairly constant; they are thus constitutive enzymes. The amount of glucokinase in liver cells depends on carbohydrate content of the diet. Glucokinase is thus described as an inducible enzyme, because a diet high in carbohydrate induces liver cells to make more. This effect is attributed to insulin. Conversely, diabetes (with its attendant low insulin), starvation, or a low-carbohydrate diet will mean less glucokinase.

3. Hexokinase isozymes I, II, and III have a low K_m for glucose (0.02-0.13 mM), whereas glucokinase has a high K_m for glucose ($\sim$ 5-8 mM). Hexokinases I, II, and III are thus very sensitive to glucose, whereas glucokinase activity becomes important for phosphorylating glucose only when the intracellular concentration of glucose is elevated. (Figure 2.4, p. 22, illustrates the responses of the reaction velocities for glucokinase and hexokinase to changes in glucose concentration.)

4. Hexokinase isozymes I, II, and III, but not glucokinase, can be inhibited by the product of the reaction, glucose 6-P. This is an example of feedback inhibition. Thus, if the concentration of glucose 6-P increases inside a cell, it inhibits the activity of hexokinase; glucose will not get phosphorylated, and its concentration will increase in the cell. This increase reduces the gradient for transport, thus slowing down glucose entry into the cell.

5. In liver cells only, a glucokinase regulatory protein (GKRP) binds GK and inhibits its activity, effectively stopping glucose phosphorylation in the liver. This inhibition can be overcome by an increase in liver-cell glucose concentration, as would occur following a meal containing carbohydrate.

For glucose to be metabolized in a cell, two processes must occur: it must be transported into the cell using a glucose transporter and it must be phosphorylated to glucose 6-P. Which of these two processes limits subsequent metabolism of glucose can be hotly debated. However, under most normal physiological conditions, the concentration of free glucose inside skeletal muscle and most other cells except liver is very low. This leads us to believe that glucose transport is rate limiting to glucose metabolism, since glucose is phosphorylated about as fast as it enters the cell. However, in some situations during exercise, blood flow in muscle is high and GLUT-4 transporters appear in sufficient numbers to move blood glucose into the muscle cytosol. Under such conditions, glucose phosphorylation can limit glucose use (Fueger et al. 2003).

☑ KEY POINT

The fact that glucose transport is usually rate limiting to glucose entry into skeletal muscle helps to explain an important point regarding postexercise (postcompetition) feeding. Feeding carbohydrate to athletes after training or competition should take place as soon as possible. One reason is the effect of the prior muscle activity on the content of GLUT-4 transporters in the muscle sarcolemma. This exercise effect, plus the independent effect of an increase in blood insulin with the carbohydrate feeding, will promote glucose transport and hasten the synthesis of muscle glycogen for the next exercise session or competition.

Diets high in sucrose or high-fructose syrups, common in many sweetened products, can provide considerable fructose. Dietary fructose is absorbed in the jejunum of the small intestine and transported, like glucose, via the portal vein to the liver. The liver and kidney have a fructokinase enzyme that specifically phosphorylates fructose to fructose 1-phosphate, which can enter glycolysis, but does so at a step below the main regulatory enzyme. We will encounter this in the next section. The lack of regulatory control of fructose 1-phosphate is believed to promote elevated formation of a class of blood lipoproteins rich in triglycerides, the very low density lipoproteins (VLDLs). Unlike glucokinase activity, fructokinase activity is not regulated by insulin, so people with diabetes can metabolize fructose. Diets high in refined carbohydrate, particularly high-fructose syrups as found in soft drinks, have been associated with increases in the incidence of type 2 diabetes and obesity in North American populations (Gross et al. 2004). Fructose is metabolized primarily in the liver. Unlike glucose, which utilizes GLUT-4 for transport into cells, fructose relies on GLUT-5, which except for in the liver is found in very low quantities in most tissues. Metabolism of fructose in the liver favors lipogenesis. Studies have found that diets high in fructose contribute to increased body weight and higher circulating levels of fat and cholesterol (Bray 2007).

Glycemic index is a measure of the rate at which different carbohydrate foods are digested and absorbed, as well as how they contribute to elevating blood glucose levels. Foods containing quickly absorbed carbohydrate may be important for maintaining blood glucose levels during intense exercise. This is discussed later in this chapter. However, regular intake of high glycemic foods and drinks, particularly when combined with a diet low in fiber, are strongly associated with increased risk of developing type 2 diabetes.

GLYCOLYSIS

Glycolysis is the principal route for carbohydrate breakdown in all cells. As discussed in chapter 4, glycolysis is the breakdown of glucose or glycogen to pyruvate. Chapter 4 shows the two major starting materials for glycolysis, glucose and glycogen, and provides two reactions summarizing their conversion to pyruvate:

$$\text{glucose} + 2\text{ ADP} + 2\text{ Pi} + 2\text{ NAD}^+$$
$$\rightarrow 2\text{ pyruvate} + 2\text{ ATP} + 2\text{ NADH} + 2\text{ H}^+$$

$$\text{glycogen}_n + 3\text{ ADP} + 3\text{ Pi} + 2\text{ NAD}^+$$
$$\rightarrow \text{glycogen}_{n-1} + 2\text{ pyruvate} + 3\text{ ATP}$$
$$+ 2\text{ NADH} + \text{H}^+$$

We will further classify glycolysis as **anaerobic** if the pyruvate is reduced to lactate or as **aerobic** if the pyruvate is completely oxidized in mitochondria, first by being converted to acetyl coenzyme A (CoA), which enters the citric acid cycle (figure 6.3). Anaerobic glycolysis is the only source of ATP for erythrocytes (red blood cells) and is a major source of ATP for skeletal muscle under severe exercise conditions or when muscle blood flow is compromised. Although cardiac muscle can utilize carbohydrate as a fuel, it is not well endowed to produce ATP by anaerobic glycolysis; thus, it has relatively poor ability to survive under ischemic (low blood supply) conditions. Cancer cells use anaerobic glycolysis to a significant extent for their ATP supply.

Figure 6.3 summarizes carbohydrate metabolism in a skeletal muscle cell. Blood glucose is taken up through GLUT-4 transporters, whose content in the

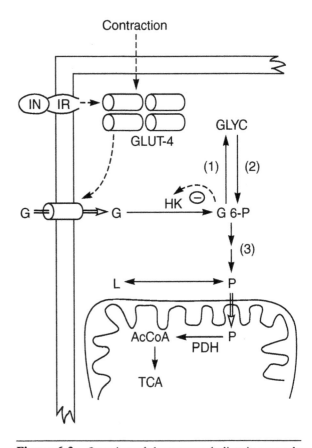

Figure 6.3 Overview of glucose metabolism in a muscle cell. Glucose uptake into muscle fibers through GLUT-4 transporters is greatly facilitated by insulin (IN) binding to the insulin receptor (IR) or by contractile activity. Through two mechanisms, GLUT-4 transporters are then translocated from internal storage sites to the muscle membrane. Once inside the cell, glucose is phosphorylated by hexokinase (HK) to glucose 6-phosphate (G 6-P), which can either be used to make glycogen through pathway 1 or can immediately be broken down to pyruvate (P) in the glycolytic pathway (3). Breakdown of glycogen (2) during exercise is accelerated to produce G 6-P for glycolysis. Pyruvate has two major fates: reduction to lactate (L) or entry into the mitochondrion and conversion into acetyl CoA (AcCoA) using pyruvate dehydrogenase (PDH). Acetyl CoA enters the citric acid cycle (CAC). The dotted arrow with the negative sign indicates that G 6-P, the product of the HK reaction, can inhibit HK if the concentration of G 6-P rises.

sarcolemma is increased by insulin binding to its receptor, or with active muscle contraction, or both. Once phosphorylated, the product glucose 6-P has two major fates in a muscle cell. Following a meal, and when the muscle fiber is inactive, glucose 6-P will be directed to the synthesis of glycogen. During exercise, stored glycogen is broken down to glucose 6-P and metabolized in the glycolytic pathway to pyruvate. Any glucose that enters an actively contracting muscle fiber is first converted to glucose 6-P, then broken down to pyruvate using the glycolytic pathway. Figure 6.3 emphasizes the important central position occupied by glucose 6-P. Because our focus is on skeletal muscle, and since there are two sources of glucose 6-P, we will consider the glycolytic pathway to begin with glucose 6-P. As mentioned previously, glucose 6-P can inhibit its own formation by hexokinase if it is not removed in an adequate way through glycolysis or glycogen formation. As we will see, this is important because it prevents unnecessary entry and phosphorylation of blood glucose if stored glycogen can provide glucose 6-P at an adequate rate. One of the interesting questions we will look at later is the importance of blood glucose as a source of glucose 6-P to fuel glycolysis during exercise.

Enzymes and Reactions of Glycolysis

The conversion of glucose 6-P to pyruvate is catalyzed by glycolytic enzymes found in the cytoplasm of cells. In the past, it was assumed that the enzymes of glycolysis are freely dissolved in the cytosol of the cell. This view is no longer held. Many, if not all, glycolytic enzymes may be bound to other structures in the cell, such as to structural filaments providing shape to the cell, to the endoplasmic reticulum (sarcoplasmic reticulum in muscle), to the outer membrane of the mitochondrion (as is the case for hexokinase), or to contractile proteins such as actin. Some of the enzymes catalyzing the reactions of glycolysis may also be physically linked to each other, such that the product of one enzyme is immediately passed to the next enzyme as its substrate. In this case, the actual concentrations of the intermediates in glycolysis, except glucose 6-P, fructose 6-phosphate (fructose 6-P), pyruvate, and lactate, would not increase much, even if glycolysis were proceeding rapidly.

> **KEY POINT**
>
> The term *flux* refers to how fast substrates enter a biochemical pathway and become products at the other end. In terms of glycolysis, its flux may be rapid, but there may not be a large increase in the concentrations of intermediates within the pathway. This is because they are acted on by the next enzyme in the pathway as fast as they are formed. We would expect the rate of formation of pyruvate and lactate to provide a useful index of the flux of glycolysis, since these are the major products.

Glycolysis has two major functions. One is to generate energy in the form of ATP. For red blood cells, glycolysis is the only energy-generating source; making ATP is its only role. The second function is to generate pyruvate for final oxidation in the mitochondrion, along with NADH. Figure 6.4 outlines the major reactions of the glycolytic pathway, showing all of the chemical structures. If we start from glucose 6-P, there is one early reaction that requires ATP, the one catalyzed by phosphofructokinase-1. If we start from glucose and use the hexokinase reaction to make glucose 6-P, ATP is used in an additional step. It may seem absurd to use ATP to make ATP, but for help with this, we can use the example of a roller coaster. The first part of the roller-coaster ride is a slow ascent to the highest point on the course, using a motor that catches the underside of the cars and moves them up. Thereafter, the potential energy of this high elevation is used to create the kinetic energy that makes a roller-coaster ride so thrilling and appealing. Overall, the $\Delta G°'$ for glycolysis for the sequence of reactions beginning with glucose 6-P and ending with pyruvate is –19 kJ/mol (–4.5 kcal/mol). Using the standard free energy change for each of these reactions, and putting in intermediate concentrations representative of normal physiological conditions, we find that the overall free energy (ΔG) change for this sequence of reactions is –39 kJ/mol, or –9.3 kcal/mol (Berg, Tymoczko, and Stryer 2002). From this sequence, three ATP are produced along with two NADH. The latter is a source of electrons for the electron transport chain, creating five more ATP.

> **KEY POINT**
>
> The prefix *bis* means that two phosphates are attached to separate locations on the same molecule. If three phosphates were attached to separate places on the same molecule, we would use the prefix *tris*. Note the difference between these prefixes and *di-* and *tri-*, which are used when two phosphate and three phosphate groups are attached to each other, as in adenosine diphosphate and triphosphate, respectively.

The glycolytic pathway, beginning from glucose 6-P and ending with pyruvate, is shown with full structures in figure 6.4. For those who find chemical structures daunting, the individual reactions are shown in table 6.2. Starting from glucose, glycolysis can be considered to involve three stages. In the first or priming stage, two phosphorylation reactions produce a hexose with two phosphate groups attached.

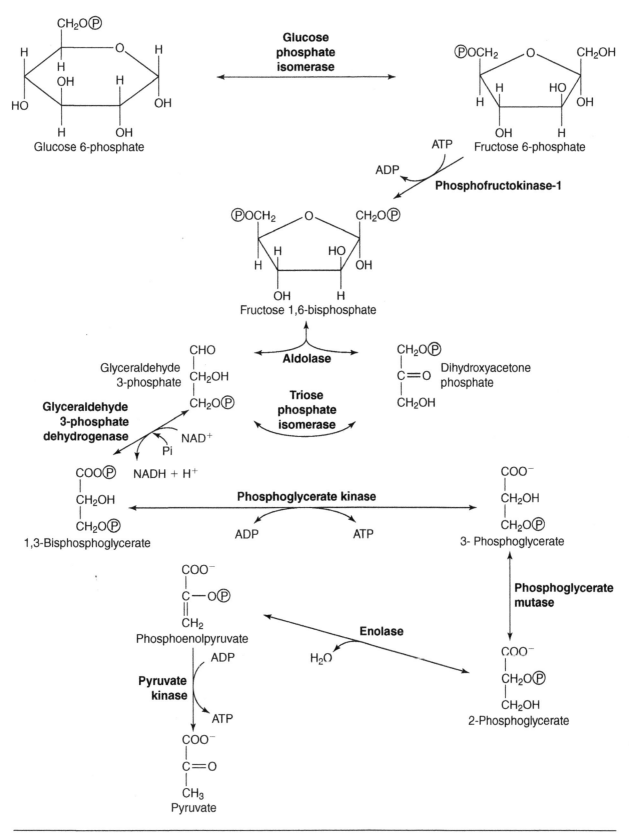

Figure 6.4 The pathway from glucose 6-phosphate to pyruvate (glycolysis), shown in complete structural detail.

Table 6.2 The Individual Reactions of Glycolysis—The Pathway From Glucose 6-Phosphate to Pyruvate

Number	Detailed reaction	Enzyme
FIRST STAGE: PRIMING		
1	Glucose 6-phosphate ↔ fructose 6-phosphate	Glucose-phosphate isomerase
2	Fructose 6-phosphate + ATP → fructose 1,6-bisphosphate + ADP + H⁺	Phosphofructokinase-1
SECOND STAGE: SPLITTING		
3	Fructose 1,6-bisphosphate ↔ glyceraldehyde 3-phosphate + dihydroxyacetone phosphate	Aldolase
4	Dihydroxyacetone phosphate ↔ glyceraldehyde 3-phosphate	Triose phosphate isomerase
THIRD STAGE: OXIDIZING AND REDUCING		
5	Glyceraldehyde 3-phosphate + Pi + NAD⁺ ↔ 1,3-bisphosphoglycerate + NADH + H⁺	Glyceraldehyde 3-phosphate dehydrogenase
6	1,3-bisphosphoglycerate + ADP ↔ 3-phosphoglycerate + ATP	Phosphoglycerate kinase
7	3-phosphoglycerate ↔ 2-phosphoglycerate	Phosphoglycerate mutase
8	2-phosphoglycerate ↔ phosphoenolpyruvate + H₂O	Enolase
9	Phosphoenolpyruvate + ADP + H⁺ ↔ pyruvate + ATP	Pyruvate kinase

This is called a hexose bisphosphate. Stage 2 is the splitting stage, conversion of the hexose bisphosphate into two triose phosphates—three carbon sugar molecules with an attached phosphate group. Stage 3 consists of oxidation–reduction reactions and the formation of ATP.

At the beginning of the first stage, the *glucose-phosphate isomerase* reaction converts glucose 6-P into another hexose phosphate, fructose 6-P. This is called an isomerase because glucose 6-P and fructose 6-P are isomers. The next step is the phosphorylation of fructose 6-P, using a phosphate group from ATP. It is catalyzed by the enzyme *phosphofructokinase-1* (abbreviated PFK-1). We add the number 1 to the enzyme because another enzyme, phosphofructokinase-2, will transfer a phosphate from ATP to the 2 position of fructose 6-P. As already mentioned, this is a priming reaction, increasing the chemical potential energy of the product fructose 1,6-bisphosphate. The reaction, which is strongly exergonic, is shown with the arrow going in one direction only. Phosphofructokinase-1 catalyzes the committed step of glycolysis, committing the cell to glucose degradation. Phosphofructokinase is under tight regulation, and its activity controls the flux of glycolysis. Phosphofructokinase is considered a rate-limiting enzyme in glycolysis. See chapter 2 for more information on rate-limiting enzymes.

Stage 2 begins with the splitting of a hexose bisphosphate. *Aldolase* splits fructose 1,6-bisphosphate into two triose phosphates, glyceraldehyde 3-phosphate and dihydroxyacetone phosphate. This reaction is freely reversible in the test tube, but it is driven toward hexose bisphosphate splitting in glycolysis. As we will see, some tissues can make glucose from three carbon precursors. When they do, some of the reactions of glycolysis go the opposite way, including that catalyzed by aldolase. Only glyceraldehyde 3-phosphate has a further role in glycolysis. Therefore, the enzyme *triose phosphate isomerase* catalyzes the reversible interconversion of the isomers dihydroxyacetone phosphate and glyceraldehyde 3-phosphate. In the pathway toward pyruvate, this enzyme ensures that all of the carbon atoms in fructose 1,6-bisphosphate are funneled through the glycolytic pathway via glyceraldehyde 3-phosphate. At this point in the pathway, all of the original carbon atoms in each glucose molecule are in the form of two molecules of glyceraldehyde 3-phosphate.

The third stage begins with a complicated reaction, producing 1,3-bisphosphoglycerate from glyceraldehyde 3-phosphate. *Glyceraldehyde 3-phosphate dehydrogenase* carries out the following: (a) It oxidizes the aldehyde group of glyceraldehydes (carbon 1) to a carboxylic acid group, and in the process reduces NAD⁺ to NADH + H⁺, and (b) it reacts the acid group with a phosphate (Pi) to make a mixed anhydride bond between a carboxylic acid and phosphoric acid. This bond is energy rich. The NADH generated in the glyceraldehyde 3-phosphate dehydrogenase reaction can be used to reduce pyruvate to lactate. The electrons can also be shuttled into the mitochondrion, where they will be used to reduce oxygen and generate ATP.

The next reaction involves capturing the energy-rich bond of the mixed anhydride bond in 1,3-bisphosphoglycerate. *Phosphoglycerate kinase* catalyzes a substrate-level phosphorylation reaction in which an ATP is generated from ADP via removal of the phosphate group from the mixed anhydride, producing 3-phosphoglycerate. This represents the first energy-generating reaction of glycolysis; based on a starting point of glucose, the balance for ATP used minus that created is zero. Remember, two 1,3-bisphosphoglycerates are obtained from each glucose.

Phosphoglycerate mutase catalyzes the movement of a phosphate group from carbon 3 to carbon 2 of the glycerate molecule. Mutases catalyze intramolecular phosphate transfer reactions. An intermediate appears in this phosphate transfer reaction, known as 2,3-bisphosphate glycerate (i.e., 2,3-BPG). In red blood cells, 2,3-BPG is important because it can reduce the affinity of hemoglobin for oxygen, thus allowing more oxygen to leave its binding to hemoglobin at the tissue level. When a person moves from lower to higher altitude, one of the adaptations that occurs over time is an increase in the content of 2,3-BPG in red blood cells. This allows the tissues to obtain more oxygen than they would normally, since the hemoglobin in red blood cells holds on to relatively less oxygen at the lower oxygen pressure present in the capillaries of muscle and other active tissues, thus partially compensating for the lower oxygen saturation of hemoglobin at altitude.

Enolase catalyzes the dehydration of 2-phosphoglycerate to form the energy-rich molecule phosphoenolpyruvate (PEP). Phosphoenolpyruvate is an energy-rich molecule because of the enol phosphate. As such, it is capable of phosphorylating ADP. Next, *pyruvate kinase* carries out the second substrate-level phosphorylation reaction, generating ATP from ADP and leaving the product pyruvate. Even though ATP is formed in this reaction, a large free energy change occurs, to the extent that the reaction is considered to be essentially irreversible.

As shown in figure 6.3, pyruvate has two major fates: reduction to lactate or entry into the mitochondrion for complete oxidation. If the former occurs, the NADH + H$^+$ generated in the glyceraldehyde-phosphate-dehydrogenase reaction is oxidized to NAD$^+$, and the pyruvate is reduced to lactate using the enzyme *lactate dehydrogenase*. If the pyruvate enters the mitochondrion, then the NADH + H$^+$ generated in the glyceraldehyde-phosphate-dehydrogenase reaction must be converted back to NAD$^+$ by one of two shuttle mechanisms. This must occur, or glycolysis will come to a halt due to a lack of NAD$^+$. We discuss the shuttles later in this chapter.

In the glycolytic sequence of reactions in which glucose 6-P is converted to pyruvate, a net of one proton (H$^+$) is generated for each glucose 6-P that we start out with (see table 6.2). One proton is produced when fructose 1,6-bisphosphate is formed in the PFK-1 reaction. A single proton is generated when glyceraldehyde 3-phosphate is converted to 1,3-bisphosphoglycerate. Finally a proton is used when PEP is converted to pyruvate and ATP. Since there are two of each of the three-carbon intermediates, beginning with glyceraldehydes 3-phosphate, for each glucose 6-P we start out with, the net is one proton formed. Interestingly, if the pyruvate is reduced to lactate, a proton is consumed for each lactate formed. Thus, the pathway from glucose 6-P to lactate actually removes a proton from the medium, in essence creating a buffer effect. We will talk about this later.

If the final product of glycolysis is lactate (as opposed to pyruvate, which enters the mitochondrion), the lactate is not a waste product. It can have a number of functions. For example, lactate may leave the cell in which it is formed and enter the blood, where it may be taken up by another type of tissue and oxidized to pyruvate, and then converted into acetyl CoA for terminal oxidation in the citric acid cycle. Similarly, lactate from one muscle fiber may be transported into an adjacent fiber and oxidized. Blood lactate may be taken up by liver cells, becoming a precursor for making glucose (i.e., gluconeogenesis), or it may remain in the cell where

☑ KEY POINT

In addition to the glucose and glycogen used in glycolysis, fructose may be a significant part of the carbohydrate in our diets. It is phosphorylated on carbon 1 to fructose 1-phosphate by a fructokinase in the liver. A specific aldolase, aldolase B, splits fructose 1-phosphate into dihydroxyacetone phosphate and glyceraldehyde. The former can now continue on in glycolysis. The glyceraldehyde is phosphorylated to glyceraldehyde 3-phosphate by a triokinase, and then it proceeds along the glycolytic pathway. Although fructose may be phosphorylated by hexokinase in other tissues to make fructose 6-P, the fructokinase pathway in liver is the major way we metabolize dietary fructose.

it is formed and either used as a source of energy or changed into glycogen by reversal of glycolysis.

Regulation of Glycolysis

Glycolysis is a pathway that can generate ATP. It also produces cytosolic-reducing equivalents in the form of NADH and plays a major role in providing substrate for the citric acid cycle as pyruvate. If pyruvate is the main fate of (aerobic) glycolysis, it produces a pair of electrons in the form of NADH for the electron transport chain. Glycolysis is a major energy-yielding pathway during three general states of muscle. (1) It is important during rapid transitions in muscle activity from rest to exercise or from one exercise intensity to a higher level of intensity. (2) It is important during exercise when oxygen is limited as a substrate for oxidative phosphorylation. This can occur during isometric exercise when intramuscular pressure occludes blood flow or during supramaximal exercise requiring energy expenditure above $\dot{V}O_2$max. (3) Aerobic glycolysis, producing pyruvate, is important during steady-state exercise when the $\dot{V}O_2$ is 60% or more of $\dot{V}O_2$max.

Glycolysis does not occur haphazardly; its rate is governed by the energy needs of the cell. In part, regulation depends on the concentration of substrates for the various reactions in the pathway: glucose, glycogen, glucose 6-P, and ADP. The need for glycolysis depends on the rate at which ATP is hydrolyzed to drive endergonic reactions. With ATP hydrolysis, ADP increases in the cytosol and becomes available for the two substrate-level phosphorylation reactions where it will be converted to ATP, the phosphoglycerate kinase and pyruvate kinase reactions. However, the major form of control is in regulating the rate of key enzymes.

Normally, regulated enzymes are those that catalyze key irreversible reactions near the beginning of a pathway. Three reactions in the pathway from glucose to pyruvate are irreversible: the hexokinase, the PFK-1, and pyruvate kinase reactions. Of these, only the hexokinase and phosphofructokinase reactions are major points of control, at least for skeletal muscle. In liver, the pyruvate kinase reaction is subject to control, as will be discussed later.

Hexokinase can be regulated by the product of its reaction, glucose 6-P. If the concentration of glucose 6-P increases too much, either because the pathway is being slowed down at another step or because the breakdown of glycogen is producing the glucose 6-P at a sufficiently rapid rate, then glucose 6-P binds to a site on the hexokinase enzyme and slows down the reaction rate. We call this process feedback inhibition or product inhibition. As mentioned previously, inhibition of hexokinase results in a sharp decrease in glucose uptake into a cell.

Phosphofructokinase (PFK-1) is the major regulatory enzyme for the pathway flux from glucose 6-P to pyruvate and is under complex control. We will look at two tissues in which control of PFK-1 is critical. In liver, PFK-1 is a tetramer, identified as L4. In skeletal muscle, it is identified as M4. In liver, the role of PFK-1 is primarily to control overall carbohydrate metabolism. We discuss the control of PFK-1 in liver later in this chapter. In skeletal muscle, PFK-1 must respond very quickly from its low level of activity at rest to rapid activity at the onset of vigorous contractions in order to provide ATP. As an enzyme catalyzing a nonequilibrium reaction, PFK-1 has only two substrates, ATP and fructose 6-P. Since ATP is present in cells in millimolar quantities, with little variation, it poses no limitation to PFK-1 activity as a substrate. On the other hand, fructose 6-P concentration can play a significant role as a substrate. It can be generated rapidly by the breakdown of glycogen, as we will see.

Phosphofructokinase-1 control is primarily based on changes in the concentrations of a number of allosteric effectors. Table 6.3 lists the allosteric effectors for PFK-1 and summarizes their sources.

Adenosine triphosphate is a substrate, but it can also bind to a negative allosteric site on the enzyme and inhibit its activity. Protons (H^+) enhance the inhibition by ATP. Citrate, the product of the citrate synthase enzyme in the mitochondrion, is also present in the cytosol. The level of citrate provides an index of fuel and ATP sufficiency. For example, in resting muscle, the principal fuel is fatty acids. When fatty acids are actively providing most of the needed ATP in a skeletal muscle under rest or during low-intensity activity, citrate concentration is elevated; this inhibits PFK-1 at a negative allosteric site and minimizes glucose utilization.

Adenosine monophosphate, ADP, and Pi are important positive allosteric effectors, and the changes in their concentrations reflect the intensity of muscle activity. We say that these metabolites provide a direct level of feedback control signaling the extent to which the glycolytic pathway must be adjusted. We have seen that the ATP concentration in most situations does not change markedly as a percentage of its rest value. However, even small changes in ATP can have much larger effects on ADP and AMP concentrations, as seen in table 4.3 (p. 98). Further, the magnitude of their changes reflects the need for glycolytic ATP production. We also know that, at the onset of vigorous contractile activity, phosphocreatine (PCr) is critically impor-

Table 6.3 Allosteric Modulators of Muscle Phosphofructokinase-1

Allosteric effector	Source	Effect on PFK
ATP	PCr, glycolysis, oxidative phosphorylation	Necessary substrate, but allosteric inhibitor
H^+	ATP hydrolysis	Enhances ATP inhibition
Citrate	Citrate synthase (TCA cycle)	Negative allosteric effector
ADP	ATP (ATPases)	Allosteric activator
AMP	ADP (adenylate kinase)	Allosteric activator
Pi	PCr (creatine kinase)	Allosteric activator
Fructose 2,6-P_2	PFK-2	Allosteric activator

tant to regenerate ATP, and the decline in PCr is inversely related to the rise in free Pi. Further, when the phosphate group on PCr is transferred to ADP, a proton is consumed (see chapter 4). As a result, an alkalization effect occurs early in intense contractions, signaling a decrease in [H^+] (i.e., increase in pH). This has the effect of relieving the inhibition by ATP on PFK-1, which is exactly what is necessary. The formation of ammonia by the enzyme adenylate deaminase also has an acid-buffering effect and, therefore, a positive effect on PFK-1 activity.

Phosphofructokinase-1 is also positively affected by an increase in *fructose 2,6-bisphosphate* (F2,6-P_2). This is formed from fructose 6-P by an enzyme known as *phosphofructokinase-2* (PFK-2), which transfers a phosphate group to carbon 2 of fructose 6-P. Epinephrine and muscle contraction increase the activity of PFK-2, which increases the F2,6-P_2 concentration. F2,6-P_2 binds to an allosteric site to increase the activity of PFK-1 for both the muscle PFK-1 isoform and the liver PFK-1 isoform (Rider et al. 2004). Other levels of control of PFK-1 may exist that involve its binding to certain proteins in the cell, which can alter its sensitivity to inhibitors. It is difficult to determine how effective such binding is by doing a laboratory assay. The actual contraction process itself seems to be very important in both initiating glycolysis at the start of exercise and rapidly turning it off at the end (Crowther et al. 2002a; Crowther et al. 2002b).

GLYCOGEN METABOLISM

Glycogen is a polysaccharide, composed of hundreds of glucose molecules (monosaccharides) joined end to end, with prevalent branches (figure 6.5). Glucose units are joined end to end through the alpha (α) configuration of carbon atom 1 of one glucose unit to carbon atom 4 of the next, producing an α 1-4 *glycosidic bond*. Branches occur when the carbon atom 6 of one glucose is joined to the carbon atom 1 of another glucose unit in the α configuration, producing the α 1-6 glycosidic bond. As shown in figure 6.5, a glycogen molecule consists of proteins and glycogen. It is known as a glycogen granule or a **glycosome** (Shearer and Graham 2002). The glucose units are referred to as glucosyl units. The glycogen part is built in tiers around a central protein molecule known as **glycogenin**. It is estimated that the maximum number of glucosyl units in the largest, 12-tiered glycogen particle is 5,500. The glycogen particles exhibit a range of sizes in rested skeletal muscle (Shearer and Graham 2002). Most glycogen molecules in muscle are in the 7- to 8-tier size range. When glycogen is synthesized in muscle, the preference appears to be to add more medium-sized glycogen particles rather than increasing the size of existing particles to the largest possible size (Graham et al. 2010). This suggests that medium-sized glycogen particles may be optimal for metabolic requirements in situations of varied energy demand encountered by muscle. In addition to glycogenin, the glycosome also contains enzymes that are involved in the formation and degradation of the glycogen particle, including enzymes that regulate these processes.

☑ KEY POINT

Skeletal muscle represents about 40% to 50% of the total mass of a lean person. As we have discussed, skeletal muscle can increase its rate of metabolism multifold during exercise. Therefore, controlling glycolysis is an important strategy for an organism, since the fuel for glycolysis is carbohydrate—brain food—and the body stores relatively small amounts of carbohydrate relative to fat stores. This becomes particularly important during exercise, when body carbohydrate stores are low.

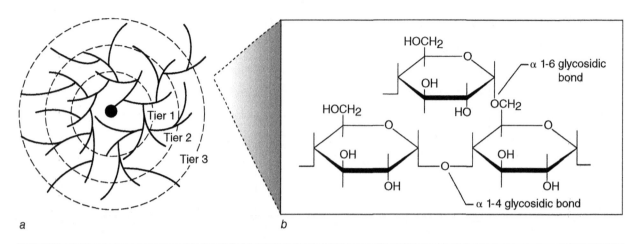

Figure 6.5 (a) A schematic of a glycogen particle showing the central glycogenin protein and three tiers of glycogen surrounding it. The glucose (glucosyl) units in the glycogen molecule are joined together by α 1-4 glycosidic bonds. Branches are formed by α 1-6 glycosidic bonds. (b) A section of the glycogen molecule with three individual glucosyl units. It illustrates the two kinds of glycosidic bonds that give rise to chains and branches.

Reprinted, by permission, from J. Shearer and T. Graham, 2002, "New perspectives on muscle glycogen storage and utilization during exercise," *Canadian Journal of Applied Physiology* 27(2): 179-203.

Glycogen is a convenient way to store glucose inside cells without affecting cell **osmotic pressure**. Osmotic pressure depends on the number, not the size, of dissolved substances. For example, one molecule of glycogen may contain 5,000 glucose units yet produce a miniscule influence on osmotic pressure compared to 5,000 individual glucose molecules. In this section, we look at the synthesis and breakdown of glycogen, integrating the latter with the glycolytic pathway in muscle. Glycogen contains a number of OH groups and therefore interacts with water in the cell. It is estimated that each gram of glycogen has between 2.5 and 3.0 g of associated water. This means that, on a weight basis, glycogen is a heavy fuel.

Mechanism of Glycogen Breakdown

Glycogenolysis, or the breakdown of glycogen, is the phosphorolytic cleavage of glucose units, one at a time, from glycogen molecules through the introduction of Pi. This way of breaking a bond is unusual, since the introduction of a water molecule or hydrolysis is what we typically see. The phosphate group breaks the α 1-4 bond between two glucosyl units, one of which is at the end of the chain. This releases the terminal glucosyl unit as glucose 1-phosphate (figure 6.6). *Glycogen phosphorylase*, often simply called phosphorylase, acts on chains of glucose units to produce glucose 1-P. This is converted into glucose 6-P by *phosphoglucomutase*. A *debranching enzyme* is necessary to remove branches because phosphorylase can cleave off only glucosyl units that are joined in a linear fashion. The fact that numerous branches appear in glycogen enhances its exposure to phosphorylase, increasing the rate at which it can be broken down. Glucose 6-P produced by glycogenolysis in muscle feeds into the glycolytic pathway. In the liver, glycogenolysis mainly provides glucose for the blood. Therefore, glucose 6-P from the phosphoglucomutase reaction is dephosphorylated to free glucose using the enzyme *glucose 6-phosphatase*. The glucose can be released from the liver to help maintain blood glucose levels.

Blood glucose levels are influenced by a number of hormones. With increasing intensity of exercise, there are increases in circulating levels of epinephrine, **norepinephrine**, growth hormone, cortisol, and glucagon, all of which act to stimulate liver glycogenolysis or gluconeogenesis. Circulating insulin levels also drop as exercise intensity rises. This change reduces the stimulus for storage of liver glycogen and complements the glycogenolytic stimuli of the other hormones. As chapter 7 notes, these hormone actions also stimulate adipose tissue's release of fatty acids. Increased fatty-acid oxidation during exercise can help spare glycogen oxidation and preserve limited carbohydrate stores.

Glycogen Synthesis and Storage

The major stores of glycogen are in liver and skeletal muscle. Table 6.4 provides approximate values for the amounts of glycogen in liver and skeletal muscle

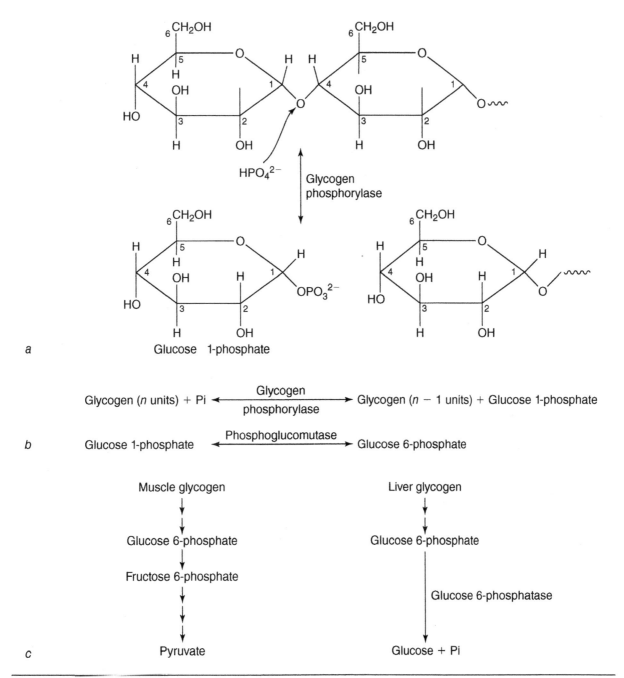

Figure 6.6 (a) The phosphorolytic cleavage of the α 1-4 bond between the last glucosyl residue and the second-to-last glucosyl residue. This generates glucose 1-phosphate and a glycogen molecule shortened by one glucose unit. (b) Summary of the two reactions of glycogenolysis, beginning with glycogen and ending with glucose 6-phosphate. (c) Summary of the two major fates of glycogen in muscle and liver. Glucose 6-phosphate in muscle is directed to glycolysis, whereas the phosphate group on liver glucose 6-phosphate is removed, and the free glucose can be released to the blood.

for men and women under three different dietary conditions. As shown, the amount of carbohydrate in the diet influences the amount of stored glycogen. The normal, mixed diet commonly eaten by North American people contains about 45% carbohydrate, but many people exist on a much higher proportion of carbohydrate in their diet. Glycogen is stored in both liver and muscle following a meal. However, after exercise, when muscle glycogen levels are reduced, glycogen is stored preferentially in the exercised muscles. Ingestion of large amounts of carbohydrate subsequent to exercise-induced depletion of muscle glycogen can stimulate an increase in muscle glycogen content by 50% to 80% above

Table 6.4 Approximate Glycogen Stores in Liver and Muscle for Adults Following Normal, High-Carbohydrate, and Low-Carbohydrate Diets

Storage site	Tissue weight (kg)	TOTAL GLYCOGEN CONTENT (G)		
		Normal diet	High-carbohydrate diet	Low-carbohydrate diet
Man, 70 kg				
Liver	1.2	40-50	70-90	0-20
Muscle	32.0	350	600	250
Woman, 55 kg				
Liver	1.0	35-45	60-70	0-15
Muscle	22.0	242	410	170

normal resting levels (Graham et al. 2010). This is partly attributable to an increase in insulin sensitivity in skeletal muscles with low glycogen content because of the exercise.

The concentration of glycogen in muscle is typically determined from muscle biopsies. The usual procedure is to hydrolyze the glycogen molecule with acid or an enzyme, amyloglucosidase, to produce glucose units. Then the glucose concentration is determined and the glycogen is expressed as millimoles of glucose or glucosyl units. As chapter 4 discusses, the concentration of glucose or glucosyl units can be expressed in three ways. We could make our measurement on the basis of the wet weight of tissue as extracted from the biopsy needle. We could also take the biopsy sample and freeze-dry it to remove the water, or we could express the glucose concentration in millimolar (mM) units using the relationship that the fraction of intracellular water in muscle is 0.7. An earlier technique precipitated the glycogen and then determined its mass by a reaction that generated a colored product. In this case, the glycogen content was expressed in grams of glycogen per 100 g of muscle tissue. Muscle glycogen concentration can also be measured noninvasively using nuclear magnetic resonance spectroscopy, based on the signal from the ^{13}C isotope naturally present in carbon compounds (Price et al. 2000).

In glycogen storage, or **glycogenesis**, glucose units are added one at a time to existing glycogen molecules, creating long unbranched chains with α 1-4 glycosidic bonds. A *branching enzyme* then creates α 1-6 bonds that lead to the highly branched final structure (see figure 6.5). Glucose enters the cell and is phosphorylated to glucose 6-P by hexokinase (muscle) or glucokinase (liver). Figure 6.7 illustrates the three-step process of glycogen formation, starting with an existing glycogen particle and glucose 6-P, the basic precursor for increasing the size of a glycogen molecule.

First, glucose 1-phosphate is made as the phosphate group on glucose 6-P is moved to the 1 position in a reaction catalyzed by *phosphoglucomutase*. To make glycogen, the glucose unit must next be activated by a reaction between glucose 1-phosphate and uridine triphosphate (UTP), making UDP-glucose. This reaction is catalyzed by *UDP-glucose pyrophosphorylase*, and the product is the precursor form of

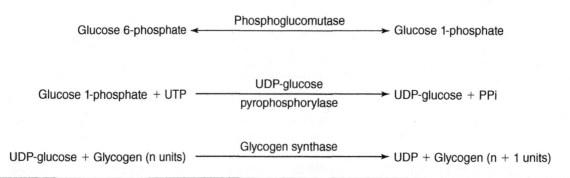

Figure 6.7 Glycogen synthesis. Glucosyl units in the form of uridine diphosphate (UDP) glucose are added to a preexisting glycogen primer molecule with n glucosyl units, creating a glycogen molecule with n + 1 glucosyl units. The precursor to make glycogen is typically glucose that is phosphorylated to glucose 6-phosphate.

glucose to be added to the glycogen primer. Finally, *glycogen synthase* adds glucosyl (glucose) units from UDP-glucose to the nonreducing end of the primer, releasing UDP and producing a glycogen molecule enlarged by an added glucose. Overall, glycogen synthesis is irreversible, due in part to subsequent hydrolysis of inorganic pyrophosphate (PPi) by *inorganic pyrophosphatase*. The glycogen synthase reaction is also irreversible. Glycogen synthase produces long chains of glucose molecules that the branching enzyme transforms into the treelike structure of glycogen found in vivo. The UTP needed to make glycogen is generated by transfer of the terminal phosphate from ATP on to UDP in a reaction catalyzed by *nucleoside diphosphate kinase*. Glycogen synthase has been reported to be able to translocate between cytosolic subcellular pools during different metabolic conditions to best respond to glycogen resynthesis requirements (Graham et al. 2010). The actual primer needed to make a glycogen particle is glycogenin. This is a self-glycosylating enzyme that transfers glucosyl units to itself from UDP-glucose until there are 7 to 11 glucosyl units. Glycogen synthase now adds glucosyl units from UDP-glucose to the glycogenin primer, and a branching enzyme creates the branched structure. New information suggests that other regulatory proteins also play a role in normal glycogen formation and metabolism. Although their exact roles in glycogen formation are still being studied, it is known that mutations in two of these regulatory proteins, laforin and malin, may lead to impairment of normal branching in glycogen synthesis, as well as to other effects that can disrupt regulation of glycogen metabolism (Graham et al. 2010).

Physiologically distinct forms of glycogen have been identified. Initially, these were thought to represent two distinct sizes of glycogen molecules, which were called *proglycogen* (smaller) and *macroglycogen* (larger). Newer information suggests that these differences are not due to size but to the location of the subcellular pools of glycogen. Fractions of muscle glycogen appear to exist that are more readily metabolized during exercise than others (Graham et al. 2010). Glycogen molecules are clustered in three distinct locations in skeletal muscle: the subsarcolemma region, the intermyofibrillar concentration at the I-band, and the intramyofibrillar pool near the Z-band that extends to the A-band. The inter- and intramyofibrillar locations constitute the majority of muscle glycogen granules. They are clustered proximal to mitochondria and sarcoplasmic reticular regions. These localizations support the energy requirements for myosin ATPase and SERCA ATPase activities, respectively (see previous chapters for more details regarding these enzymes). The intramyofibrillar glycogen pool appears to be preferentially utilized during endurance exercise, but it is also most rapidly resynthesized during recovery (Graham et al. 2010). Depletion of the different locations of muscle glycogen appear to be associated with different durations and intensities of exercise. They correlate with fatigue, indicating that glycogen localization within muscle may be as important as total glycogen stores in fatigue resistance during exercise.

Storage of glycogen following training or competition is a key consideration for athletes because glycogen is such an important fuel for exercising muscle. Moreover, stores of glycogen in muscle are limited. Research has provided us with some valuable information. For example, the synthesis of glycogen following exercise is most rapid in the first hour. Thereafter, the rate is elevated until glycogen stores at least reach preexercise levels. Enhanced insulin-stimulated glucose uptake and glycogen synthesis seem to be sensitive to glycogen content. Glycogen storage beyond preexercise levels, or supercompensation, can take place for several days after glycogen stores are severely depleted. The slower storage of glycogen beyond normal levels seems to be insulin independent. A controversial issue is whether carbohydrate alone or carbohydrate supplemented with protein is better in terms of postexercise glycogen replenishment. Ivy and colleagues (2002) reported that the inclusion of protein with carbohydrate significantly increased the synthesis of glycogen, compared to carbohydrate alone, by 4 h postexercise. However, this conclusion has been challenged by subsequent research, and current recommendations do not suggest the necessity of including protein with carbohydrate as a requirement to optimize postexercise synthesis of muscle glycogen. Nevertheless, as chapter 8 shows, timely ingestion of protein does enhance the synthesis of skeletal muscle protein following a resistance-training workout.

☑ KEY POINT

Glycogen supercompensation via dietary intake of large amounts of carbohydrate is a proven way of prolonging endurance exercise. Since muscle and liver glycogen stores are limited and carbohydrate utilization is critical for exercise performance, increasing stores of glycogen prior to beginning exercise will allow for performance of endurance exercise for longer periods before glycogen depletion begins to induce fatigue.

Regulation of Glycogen Metabolism

As mentioned earlier, glycogen is found as a particle or glycosome, including the large glycogen molecule along with the enzymes glycogen phosphorylase, synthase, branching and debranching enzymes, enzymes that regulate glycogen metabolism, glycogenin, and other proteins (Graham et al. 2010). In muscle, as noted earlier, most glycogen particles are found in close association with the sarcoplasmic reticulum, between myofibrils near mitochondria, and under the sarcolemma.

Since the enzymes to synthesize and degrade glycogen are found in the same particle, it would seem that glycogenesis and glycogenolysis could occur simultaneously. If this were the case, a futile cycle would be set up that accomplished nothing except the hydrolysis of energy-rich phosphates such as UTP. To avoid the futile cycling of glycogen, glycogen phosphorylase should be active when glycogen synthase is inactive, or vice versa. In the liver, phosphorylase should be inactive following a meal, but glycogen synthase should be active to store the glucose obtained from food. Between meals, liver phosphorylase should be active to provide glucose for the blood, whereas glycogen synthase should be inactive. In rested muscle, synthase should be active and phosphorylase inactive following a meal, but if the muscle starts to work, the phosphorylase should be active and the synthase inactive. In fact, we might expect the activity of phosphorylase to be graded during exercise; activity should be greatest during very hard exercise when carbohydrate is the most needed fuel and much less when exercise intensity is low enough for oxidation of fat to maintain ATP levels in the exercising muscle. Control of glycogen phosphorylase can thus be tied to the whole strategy of using carbohydrate for oxidative phosphorylation when plenty is available, or when the muscle demand for ATP can be adequately satisfied only by carbohydrate degradation.

One enzyme can be active and the other simultaneously inhibited if the two respond in opposite directions to the same stimulus. The main regulation of this type of enzyme pair is covalent attachment of a phosphate group, or phosphorylation, which requires a protein kinase. Removing the phosphate groups requires a phosphoprotein phosphatase. The regulation of glycogen metabolism in liver and muscle represents another form of signal transduction, introduced in chapters 2 and 3. For the control of glycogen metabolism, external signaling molecules (e.g., the hormones epinephrine, insulin, and glucagon) bind to specific cell membrane receptors and activate protein kinases that phosphorylate specific proteins—glycogen phosphorylase (GP) and glycogen synthase (GS) in this case. Phosphorylation activates phosphorylase (GPa form) but inhibits glycogen synthase (GSb form). Removal of the phosphate groups (dephosphorylation) by a phosphoprotein phosphatase inactivates glycogen phosphorylase (GPb) but activates glycogen synthase (GSa). An additional level of complexity in the control of glycogen metabolism lies in the fact that liver and muscle glycogen phosphorylase and synthase, as well as the enzymes that control them, are sensitive to the action of a number of molecules or ions that signal the nutritional and metabolic state of the cell.

Figure 6.8 summarizes the control of GP and GS by the phosphorylation and dephosphorylation mechanisms. Inactive glycogen phosphorylase (GPb) is phosphorylated on a single serine residue

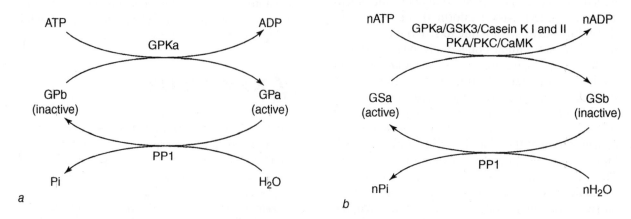

Figure 6.8 Regulation of glycogen phosphorylase and glycogen synthase by reversible phosphorylation. (*a*) The inactive form of glycogen phosphorylase (GPb) is phosphorylated by the active form of glycogen phosphorylase kinase *a* (GPKa) to make active GP (GPa). (*b*) Glycogen synthase is active when it is in the dephosphorylated state (GSa).

to make the active GPa form. Removal of the single phosphate group is accomplished by phosphoprotein phosphatase 1 (PP1). Control of glycogen synthase by phosphorylation and dephosphorylation is more complex. First, there are at least 10 sites on GS where a phosphate group can be attached. Second, there are at least seven protein kinases that can phosphorylate GS in total, including glycogen phosphorylase kinase (GPKa), glycogen synthase kinase-3 (GSK3), protein kinase A (PKA), protein kinase C (PKC), a calmodulin kinase (CaMK), and two casein kinases, I and II (casein K I and II). However, only a single phosphoprotein phosphatase (PP1) removes these phosphates. The ability to phosphorylate GS on multiple sites suggests that glycogen synthase activity can be graded more sensitively to cell needs.

Regulation of Glycogenolysis in Muscle

In our distant past, hundreds of thousands of years ago, we survived in a hostile world by being able to flee or fight. For both responses, rapid generation of ATP for muscles was essential. Glycolysis is indispensable for this purpose, and glycogen is the primary source of glucose 6-P to fuel the glycolytic pathway. Glycogenolysis is regulated at the level of glycogen phosphorylase. It can be controlled through alteration of the proportion of enzyme in the active (GPa) versus inactive (GPb) form. Two mechanisms accomplish this—one activated by the hormone epinephrine and the other based on changes in intracellular calcium concentration. Glycogenolysis can be regulated by substrate concentrations (glycogen and Pi), and it can be regulated by the concentrations of positive and negative allosteric effectors for both the GPa and GPb forms. Table 6.5 summarizes the regulation of glycogenolysis in muscle.

Role of Epinephrine

When the hormone epinephrine binds to its β-**adrenergic receptor** on a skeletal muscle fiber, it unleashes a cascade of activation events that result in the conversion of phosphorylase b to phosphorylase a. The overall scheme for regulation of glycogen phosphorylase by epinephrine in skeletal muscle is summarized in figure 6.9 and described in detail here as follows.

- When one is aroused by a sudden or startling event or anxiousness before a competition, epinephrine is released to the blood from the adrenal medulla.
- Epinephrine binds to β-adrenergic receptors (βAR) on skeletal muscle. The higher the receptor concentration and the higher the epinephrine concentration in the blood, the greater the binding.
- Epinephrine binding to the βAR results in a change in a specific membrane-attached G protein (G_s). The G refers to GTP (guanosine triphosphate), since the activated G protein binds GTP.
- The activated G protein interacts with a membrane protein, adenylyl cyclase (AC) (some call it adenylate cyclase), which increases its ability to change an ATP molecule into a cyclic AMP (cAMP) and inorganic pyrophosphate (PPi). Cyclic AMP has a single phosphate group that is joined to both the 5' and the 3' carbon of ribose. It is sometimes designated 3',5'-AMP.
- Cyclic AMP activates *protein kinase A*, which then transfers a phosphate group from ATP to glycogen phosphorylase kinase b, making it glycogen phosphorylase kinase a (GPKa).

Table 6.5 Mechanisms Regulating Glycogenolysis in Muscle

Mechanism of regulation	Inducing factors	Source	Effect
Phosphorylation of GPb	Calcium	Sarcoplasmic reticulum	Increases glycogenolysis
	Epinephrine	Adrenal medulla	Increases glycogenolysis
Substrate concentration	Glycogen	Glycogenesis	Increases glycogenolysis
	Pi	Creatine phosphate	Increases glycogenolysis
Allosteric effectors	AMP	Adenylate kinase	Increases GPb activity, smaller effect on GPa activity
	ADP	ATPases	Increases GPa activity a little
	IMP	Adenylate deaminase	Increases GPb activity
	ATP	Energy systems	Inhibits GPb

GPa, glycogen phosphorylase *a*; GPb, glycogen phosphorylase *b*.

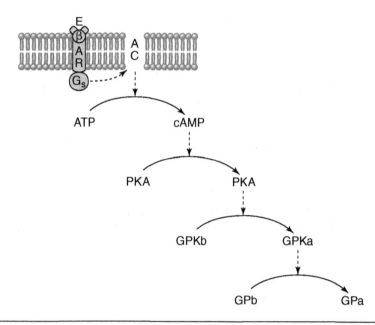

Figure 6.9 Conversion of inactive glycogen phosphorylase *b* (GPb) into the active (GPa) form by epinephrine (E). The beta-adrenergic receptor is βAR; G_s is an activating G protein intermediate; and AC, the enzyme adenylyl cyclase, converts ATP into cyclic AMP (cAMP). PKA or protein kinase A converts inactive glycogen phosphorylase kinase *b* (GPKb) into the active GPKa form.

- Glycogen phosphorylase kinase *a* transfers a phosphate group to a serine residue in each of the two subunits of glycogen phosphorylase *b*, making it the active, glycogen phosphorylase *a*.
- Activated glycogen phosphorylase *a* now breaks down glycogen to make glucose 6-P for glycolysis.

The role of epinephrine to increase the activity of glycogen phosphorylase even before any activity has taken place is an example of a *feed-forward* mechanism. In contrast to a feedback mechanism, which responds to a change in physiological condition, a feed-forward mechanism prepares the person for what will follow.

Role of Calcium

When a muscle is activated by a nerve, the activation induces calcium ions to be released from sarcoplasmic reticulum of skeletal muscle fibers. The Ca^{2+} ions can separately activate glycogen phosphorylase kinase, as shown in figure 6.10. Glycogen phosphorylase kinase has four different kinds of subunits: α, β, γ, and δ. The α and β subunits can each be phosphorylated by protein kinase A, so that the GPK is changed to the active GPKa form, as discussed previously. The δ subunit is **calmodulin**, a small protein that can bind up to four Ca^{2+} ions. When the δ subunit binds four Ca^{2+}, it becomes active and can phosphorylate phosphorylase *b* to make it phosphorylase *a*. The highest level of activity for glycogen phosphorylase kinase occurs when it has both α and β subunits phosphorylated and the δ subunit with four calcium ions, as shown in figure 6.10.

Reversing the Activation Steps

When the competition or fearful event ends and the need to maintain a high rate of glycogenolysis passes, the activation process is reversed quickly. We can summarize the various reversal steps as follows:

- Epinephrine concentration drops quickly after the event, so the activation of adenylyl cyclase ceases and cAMP formation stops.
- At the same time, an enzyme, cAMP phosphodiesterase, changes cAMP to 5'-AMP. This means that protein kinase A is no longer activated.
- Phosphoprotein phosphatase 1, which is inhibited when cAMP concentration is elevated, becomes active and immediately removes the phosphate groups from both glycogen phosphorylase *a* and glycogen phosphorylase kinase *a*.
- The $[Ca^{2+}]$ in the muscle fiber quickly drops as Ca^{2+} ions are pumped back into the sarcoplasmic reticulum.
- Overall, glycogenolysis is rapidly reduced.

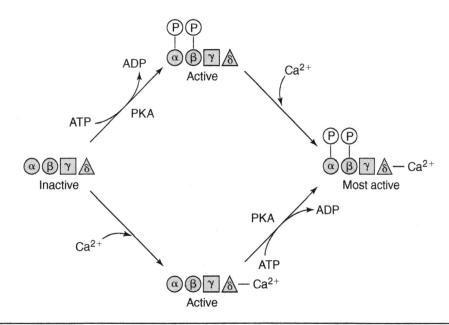

Figure 6.10 Glycogen phosphorylase kinase (GPK), which can be activated by the epinephrine cascade as well as by binding Ca^{2+}. GPK has four different kinds of subunits. The α and β subunits can be phosphorylated by protein kinase A (circled P). The δ subunit is calmodulin and can bind four Ca^{2+} ions. Activation of GPK is strongest when it is phosphorylated and when it contains bound calcium ions.

Regulation of Glycogenolysis by Substrate and Allosteric Effectors

Glycogen phosphorylase has two substrates, glycogen and Pi. We know that the K_m of glycogen phosphorylase for glycogen is approximately 1 mM, or about 1% of the actual concentration of glycogen in rested muscle (Rush and Spriet 2001). This tells us that phosphorylase is saturated with glycogen even when the concentration of glycogen is low as it is at the end of prolonged exercise. However, the K_m of phosphorylase for its other substrate, Pi, is approximately two times the resting level of Pi in muscle (e.g., about 2 mM). This suggests that phosphorylase activity is very sensitive to changes in [Pi], responding progressively to an increase in Pi that parallels the intensity of exercise (see table 4.3 on p. 98).

Some interesting experimental observations point to even more complexity in the control of glycogen metabolism in humans. For example, epinephrine-deficient humans can exercise and break down glycogen at rates similar to those seen in control subjects. If epinephrine-deficient patients are infused with epinephrine before exercise, no additional glycogen breakdown occurs for the same exercise task (Kjaer et al. 2000). Glycogen breakdown can occur in exercise even when the percentage of glycogen phosphorylase in the active form (GPa) is not higher than it is in rest (Kjaer et al. 2000). Such data suggest two possibilities. Phosphorylase b must play a role in glycogenolysis during exercise, or GPa must be capable of increasing its activity in response to changes in the concentrations of key metabolites in muscle (by a combination of lowered K_m for Pi and increase in reaction velocity), or both.

As evidenced by many exercise studies, the activation of glycogen phosphorylase to the a form (e.g., % of GPa) occurs early in exercise, but then exercise can proceed with the proportion of phosphorylase in the active form at levels only slightly above those at rest. This has been a puzzling observation, but it is now easy to rationalize. Glycogen phosphorylase b, while basically inactive at rest, can be allosterically activated by AMP levels, as would be seen in an active muscle (see table 4.3, p. 98). This activation can be such that rapid glycogenolysis can take place even when the proportion of glycogen phosphorylase in the a form is low. In addition, phosphorylase b activity is increased allosterically by inosine monophosphate (IMP) and decreased by glucose 6-P and ATP. During moderate-intensity exercise, neither ATP nor IMP concentrations are likely to change sufficiently from rest values to allosterically alter phosphorylase b activity. However, for high-intensity exercise, an increase in IMP and decrease in ATP could act together to enhance rates of glycogenolysis solely by an increase in phosphorylase b activity. Glycogen phosphorylase a activity can also be increased up to threefold by concentrations of AMP and ADP found in active muscle (Rush and Spriet 2001).

The actual concentration of glycogen in muscle can influence the rate of glycogenolysis when all other circumstances are the same. For example, differences of more than twofold can exist in the use of glycogen for the identical exercise task in subjects if they start out with low initial glycogen levels (compared to high levels) (Weltan et al. 1998; Arkinstall et al. 2004). As muscle glycogen content decreases during exercise, its use is proportionally reduced. This suggests that the rate of glycogenolysis is sensitive to the size of the glycogen molecule, perhaps because glycogen is more accessible to both the debranching enzyme and glycogen phosphorylase. Exercise training can reduce the reliance on glycogen as a fuel when the same exercise task is performed after, compared to before, training. After a seven-week training program, LeBlanc and colleagues (2004) found that glycogenolysis was decreased compared to pretraining levels. This was not due to the extent of activation of phosphorylase or to a training-induced change in phosphorylase activity. They found lower concentrations of ADP, AMP, and Pi in muscle samples after training, pointing to changes in substrate and allosteric effectors that could account for the glycogen-sparing action with training. In summary, the regulation of glycogenolysis can be rapid and proportional based on activation mechanisms that overlap and reinforce each other to ensure that the provision of glucose 6-P meets the energy needs of the fiber. Exercise training results in reduced levels of the substrate Pi and allosteric effectors ADP and AMP so that the ATP needs are met less by carbohydrate and more by fat oxidation.

Since muscle glycogen is the principal source of substrate for glycolysis, regulation of glycogen phosphorylase is critical to appropriately meet the demands for ATP provision during exercise. This is especially important during exercise at altitude, where the oxygen content of the inspired air is reduced and glycolytic flux is therefore increased. Exercise training has a glycogen sparing effect, with the result that less carbohydrate and more fat will be oxidized after training.

The loss of muscle glycogen during exercise is correlated with the onset of fatigue. The possible mechanisms of this association are discussed in the Next Stage section of this chapter. Athletes have attempted to slow down the rate of muscle glycogenolysis during prolonged exercise by ingesting carbohydrate during the exercise period. Research findings have been conflicted regarding the success of carbohydrate ingestion to attenuate the rate of muscle glycogen loss during prolonged exercise, with various studies reporting either no influence or a significant sparing effect when both animals and humans were involved (Karelis et al. 2010). The reasons for these conflicting findings are not clear. What is clearer is the temporal association of muscle glycogen depletion with exercise fatigue. In addition, despite the controversy, most endurance athletes still continue to supplement with carbohydrate during competition, since benefits are associated with the delay of fatigue onset beyond those necessarily associated with muscle glycogen depletion. Blood glucose is better maintained during long-term exercise with timely ingestion of carbohydrate. The concentration and composition of carbohydrate in sport drinks and gels have been tailored based on research to optimize the rate of carbohydrate absorption. The factors involved in optimizing absorption of glucose, electrolytes, and water are beyond the scope of this text; however, the use of higher glycemic index carbohydrate as simple sugars (which are more easily absorbed and digested and rapidly enter the blood) are favored for obvious reasons.

> ### ✓ KEY POINT
> The body contains limited supplies of carbohydrate but plenty of fat. The use of glycogen as a fuel is therefore graded to the intensity of activity, responding to activating calcium levels and to metabolites whose concentrations increase during exercise (ADP, Pi, and AMP). On the other hand, if muscle glycogen content is high in muscle, mechanisms to regulate its breakdown appear to be less sensitive. The result is that more glycogen is used.

Regulation of Glycogenolysis in Liver

The liver stores glucose for the blood by converting it into liver glycogen. Between meals, when there is no source of absorbed glucose from the gut to maintain blood glucose concentrations, the liver breaks down glycogen at a rate needed to maintain euglycemia. Regulation of glycogenolysis in liver is similar to that in muscle in that there are active (phosphorylase *a*) and inactive (phosphorylase *b*) forms of liver phosphorylase. However, notable differences exist. Liver phosphorylase is controlled more by phosphorylation and dephosphorylation and much less by allosteric mechanisms. Glucagon is the primary hormone that stimulates the rise in liver cAMP concentration and conversion of liver phosphorylase *b* to liver phosphorylase *a*. Glucagon is produced in and released from the alpha cells of the pancreas in response to a decrease in blood glucose. Therefore, a rise in glucagon in the blood signals a need to break down liver glycogen to glucose and release this to the blood. Glucagon binds to specific receptors on liver cells, and this binding results in a cascade of events, including a rise in cAMP concentration. Following a meal, when blood glucose and insulin concentrations are elevated, insulin activates protein phosphatase 1, increasing the conversion of liver phosphorylase a to liver phosphorylase b, while glucose binds to and inactivates any existing liver phosphorylase a.

Regulation of Glycogenesis

To store glycogen in liver and muscle, there must be an increase in glucose uptake into muscle and liver cells and glucose phosphorylation to glucose 6-P. In muscle, glucose transport plays a very important role. In liver, transport is less important than the activity of glucokinase. Like glycogen phosphorylase, glycogen synthase exists in two forms. Glycogen synthase a (GSa) is the unphosphorylated form that is normally active. Phosphorylation converts GSa to GSb, which is inactive. Glycogen synthase a has also been referred to as GSI, where the I stands for independent; that is, GSI is independent of the glucose 6-P concentration. Glycogen synthase b has also been referred to as glycogen synthase D (GSD); it is normally inactive but can become activated by an allosteric mechanism if the glucose 6-P concentration increases.

We saw in figure 6.8 that phosphorylation and, thus, inactivation of GSa can be performed by a number of protein kinases. Dephosphorylation, which activates glycogen synthase, is catalyzed by a single enzyme, PP1. Figure 6.11 summarizes the regulation of glycogen synthase, focusing on just the key regulators and enzymes that are responsible for controlling the synthesis of glycogen. Key signaling substances (cAMP and Ca^{2+}) that activate the breakdown of glycogen are responsible for enhancing phosphorylation of GSa to make it inactive in glycogen synthesis. Cyclic AMP concentration is increased by epinephrine in muscle and by glucagon in liver. Removal of the phosphate groups using PP1 is stimulated by insulin and opposed by cAMP. This is just as we should expect, because we don't want the synthesis and breakdown to be simultaneously active. Following a meal, when blood glucose and insulin are increased, we would expect glycogen synthesis to be favored, and this is how it is. The role of glucose 6-P to allosterically activate the GSb (GSD) form, overriding the inhibition by phosphorylation, is entirely appropriate. Let's see how. Suppose you are doing intermittent or interval training. You want the phosphorylase to be active during the activity and the glycogen synthase to be inactive. However, when you stop your exercise bout and are recovering, glucose 6-P concentration should still be elevated. During this rest period, you don't need glycogenolysis and glycolysis; however, it is a good time to allow some of the glucose 6-P to be converted to glycogen, and this is what happens.

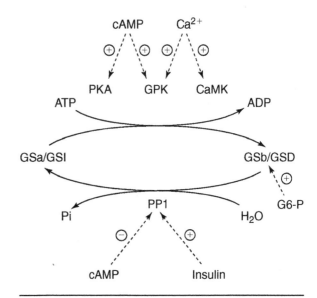

Figure 6.11 The major regulators of glycogen synthase. Conversion of the active, glucose 6-phosphate (G6-P)–independent form (GSa or GSI) to the inactive, glucose 6-phosphate–dependent (GSb or GSD) form takes place by phosphorylation of serine residues. Three key enzymes that phosphorylate GSa/GSI are protein kinase A (PKA), glycogen phosphorylase kinase (GPK), and calmodulin kinase (CaMK). These enzymes are activated by cAMP, whose formation is stimulated by epinephrine (muscle) or glucagon (liver). Dephosphorylation of GSb/GSI is accomplished by phosphoprotein phosphatase 1 (PP1), whose activity is enhanced by insulin and inhibited by cAMP. An increase in G6-P can activate GSb/GSD. Dotted lines and a positive or negative sign indicate activation or inhibition, respectively.

The regulation of glycogenesis is very complex. For simplicity's sake, the following summarizes the key control points:

- Provision of precursor in the form of glucose is critical to rapidly inducing glycogen resynthesis. An increase in blood glucose induces an increase in blood insulin concentration. Insulin enhances glucose transport and is known to increase PP1 activity, thereby enhancing the conversion of GSb to GSa.

- Muscle glycogen synthesis is most rapid immediately following exercise (Price et al. 2000). The proportion of glycogen synthase in the active form (GSa) is higher at low glycogen levels.
- Researchers have identified an early insulin-independent phase of glycogen synthesis, followed by an insulin-dependent phase. The former occurs in the immediate postexercise period, whereas the latter phase is associated with postactivity feeding.
- The more glycogen is depleted following exercise, the more rapid the resynthesis rate is, and the more this rate is sensitive to insulin and glucose. It is suggested that the smaller glycogen particle in the exercised muscle enhances GS and PP1 activity.
- Glycogen synthesis is acutely sensitive to glycogen concentration, but in a reciprocal way. This suggests that glycogen exerts a feedback inhibition on glycogen synthase such that there is an upper limit for glycogen storage in muscle.
- During glycogen resynthesis, regulation of glycogen molecule size takes place such that most glycogen is stored as medium-sized glycogen units. Muscle glycogen content is preferentially increased by adding further muscle glycogen units rather than increasing existing or remaining glycogen molecules to a maximum size.
- After glycogen-depleting exercise, trained people can resynthesize glycogen faster than the untrained (Hickner et al. 1997).

LACTATE METABOLISM

Lactate and lactic acid are probably more misunderstood than any other area in exercise metabolism. For example, statements such as the following are routinely made by those who really do not understand the metabolism: "Lactic acid is produced by anaerobic glycolysis; the lactic acid, being an acid, dissociates and generates a proton, thus acidifying muscle and blood and causing fatigue." It is also believed by many that lactate or lactic acid as a product of anaerobic glycolysis must be indicative of an oxygen insufficiency at the cellular level. Lactic acid has also been mistakenly linked to postexercise muscle soreness in the minds of some of the exercising public. Using a treadmill or cycle ergometer, equipment to measure oxygen and carbon dioxide levels and to provide $\dot{V}O_2$ and $\dot{V}O_2$ data, and a blood lactate analyzer, it is easy to generate data, as shown in figure 6.12. The data in the figure show $\dot{V}O_2$ and blood lactate concentration measures for a single subject during cycle ergometer exercise at progressively increasing intensity. Although $\dot{V}O_2$ increases linearly with exercise intensity, blood lactate concentration breaks from an early linear increase to become an exponential increase. The break point is

> **KEY POINT**
>
> Following glycogen-depleting exercise, when resynthesis of glycogen is critical for subsequent activity, early and frequent carbohydrate feeding is essential. The form of this carbohydrate is less critical than the total amount of carbohydrate made available for glycogen resynthesis. While this can ensure that glycogen replenishment starts early, when the rate of resynthesis is highest, there is a limit to the amount of glycogen that can be stored in muscle, despite the continued presence of high blood glucose and insulin. When some threshold amount of glycogen is reached, glycogen synthase activity is inhibited. Although elevated glucose transport into muscle still occurs, the glucose 6-P is channeled into the glycolytic pathway such that lactate is produced at an accelerated rate, even in a rested muscle. Similarly, pyruvate oxidation is increased at the expense of fat oxidation.

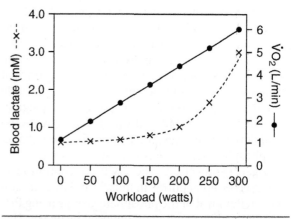

Figure 6.12 Blood lactate concentration and $\dot{V}O_2$. Values are for a single subject cycling against progressively greater resistance starting from loadless cycling (0 W). The resistance was increased every 2 min in 50 W increments, and measurements were made every 50 W. Physiologists define a lactate threshold or break point where lactate concentration changes from a linear increase with workload to an exponential increase. In this figure, blood lactate data are insufficient to define a specific break point.

often described as the "lactate threshold" (Robergs and Roberts 1997). It is important to make frequent measures of lactate and to keep the step increases in intensity small to clearly identify the break point.

One of the misleading assertions is that the lactate threshold represents the point where the muscle ATP yield from oxidative phosphorylation is compromised by insufficient oxygen, and that maintaining work output requires the addition of anaerobic glycolysis as a source of ATP. Thus, the lactate threshold has been defined as an "anaerobic threshold" (Wasserman et al. 1973). However, this reasoning does not agree with what we now know.

Lactate Formation, Accumulation, Acidification, and Disappearance

The summary that follows outlines what research has told us about lactate metabolism. Lactic acid is a modestly strong acid with a pK_a of 3.8. Thus, at physiological pH of ~7.0, it exists almost exclusively as the anion (La^-) and not as the undissociated acid (HLa). Moreover, as chapter 5 notes, the formation of lactate from either stored glycogen or glucose taken up from the blood is not the primary cause of net proton formation and exercise-induced acidosis, or a drop in pH (Robergs, Ghiasvand, and Parker 2004). An exception to this may occur under ischemic conditions in the muscle when lactic acidosis predominates as the source of hydrogen ion production (Marcinek, Kushmerick, and Conley 2010). Ischemia refers to situations where muscle blood flow is restricted, as in when a tourniquet may be applied to stop bleeding or under certain disease conditions, such as intermittent claudication.

Proton or hydrogen ion accumulation results in acidification due to hydrogen ion buildup. During intense exercise, this acidification has traditionally been attributed to lactic acid buildup and its dissociation into lactate and a hydrogen ion. When pyruvate and NADH are converted to lactate and NAD, a hydrogen ion is actually consumed from solution (see figure 6.13). We now know that this association of lactate accumulation with muscle acidification during intense exercise is more coincidental rather than causal. Instead, most of the acidification of skeletal muscle during intense exercise can be attributed to ATP breakdown to ADP, which generates a proton or hydrogen ion (see figure 6.14).

When aerobic metabolism predominates, generation of ATP in the mitochondria consumes most of the excess protons produced from ATP breakdown for use in the electron transport chain or in the formation of water. The regeneration of ATP from creatine phosphate also consumes a proton. However, when ATP is regenerated via the glycolytic pathway rather than via oxidative phosphorylation in the mitochondria or from creatine phosphate, the protons produced from ATP breakdown to ADP that occurs at the start of glycolysis are not as readily reused in mitochondrial respiration (Robergs, Ghiasvand, and Parker 2004). In addition, the glycolytic pathway reactions that produce ATP do not consume hydrogen ions in their reactions. When

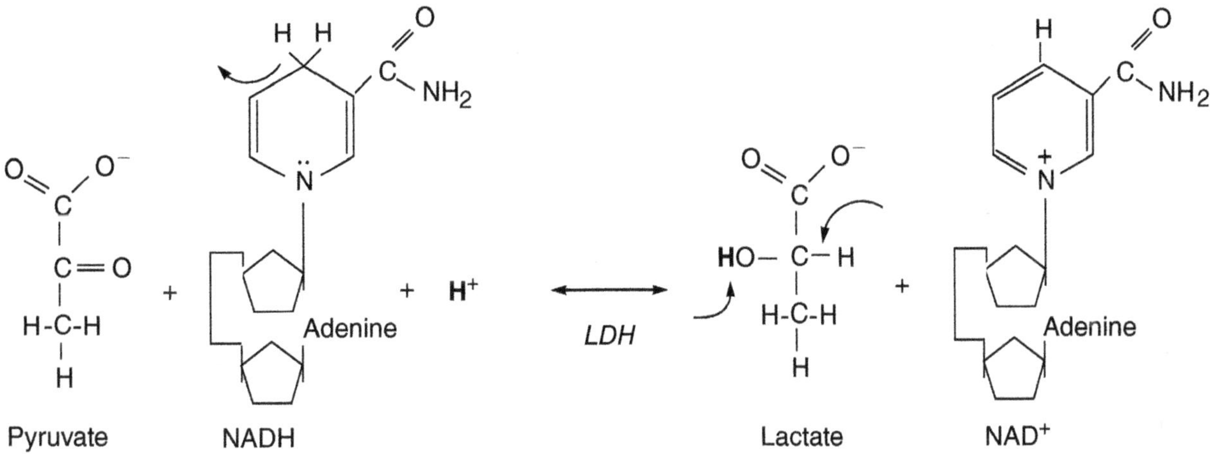

Figure 6.13 Substrates and products of the lactate dehydrogenase (LDH) reaction. Two electrons and a proton are removed from NADH and a proton is consumed from solution to reduce pyruvate to lactate. Arrows pointing away from a bond represent removal of a bond or a group. Arrows pointing to a bond represent addition of an atom or group.

Reprinted from R.A. Robergs, F. Ghiasvand, and D. Parker, 2004, "Biochemistry of exercise-induced metabolic acidosis," *American Journal of Physiology* 287: R502-R516. Used with permission.

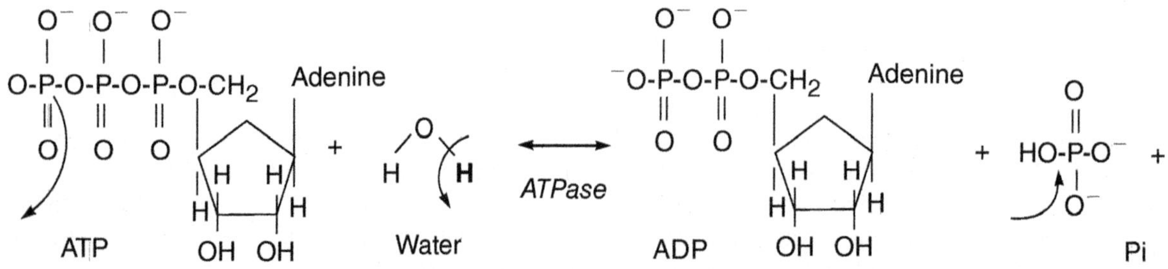

Figure 6.14 Substrates and products of the ATPase reaction. This reaction is referred to as a hydrolysis reaction (ATP hydrolysis) due to the involvement of a water molecule. An oxygen atom, two electrons, and a proton from the water molecule are required to complete the free inorganic phosphate product of the reaction. The remaining proton from the water molecule is released into a solution. Arrows pointing away from a bond represent removal of a bond or group removal. Arrows pointing to a bond represent addition of an atom or group.

Reprinted from R.A. Robergs, F. Ghiasvand, and D. Parker, 2004, "Biochemistry of exercise induced metabolic acidosis," *American Journal of Physiology* 287: R502-R516. Used with permission.

glycolysis accelerates to quickly regenerate ATP in times of intense exercise, a significant amount of ATP is quickly produced via the glycolytic pathway; it is then utilized by the various ATPase enzymes to power events that cause muscular contraction and relaxation. The amount of hydrogen ions produced via ATP breakdown by the ATPase enzymes during intense exercise actually exceeds the amount of hydrogen ions from dissociation of lactic acid into lactate. This concept is outlined in figure 6.15. Thus, the accumulation of protons and the resultant drop in pH associated with very intense exercise are not as attributable to lactate buildup as such; in fact, experimental evidence indicates that when lactate production is artificially inhibited during exercise, the drop in muscle pH is even more rapid (Robergs, Ghiasvand, and Parker 2004). Nevertheless, increased lactate production does coincide with increased glycolytic activity and its resultant acidification of muscle and blood; so, the common measurement of lactate levels in muscle and blood during and following exercise is still a reasonable indirect measure of anaerobic metabolic activity and acidosis. Figure 6.15 summarizes the differences in net hydrogen ion accumulation during aerobic versus anaerobic predominant metabolism and highlights the primary source of the hydrogen ion accumulation from accelerated ATP breakdown during intense activity, since not all hydrogen ions from ATP hydrolysis can be uptaken and utilized in the mitochondria.

It has also been assumed for years that lactate formation causes muscle fatigue. Part of this assumption is that lactate is elevated in fatigued muscles and that acidification of the muscle during intense exercise inhibits metabolic enzymes, reduces calcium sensitivity of the contractile proteins, and slows muscle relaxation, thus contributing to fatigue. Research has questioned the importance of exercise-induced acidosis as a major factor in muscle fatigue and has suggested that other metabolically related causes may be more critical (reviewed in Allen, Lamb, and Westerblad 2008). It has been reported that lactate and acidification may actually defend muscle from the fatiguing effects of a loss of intracellular potassium ion from very active muscle fibers by enhancing chloride channel activity (Allen and Westerblad 2004). A review by Westerblad, Allen, and Lännergren (2002) concludes that it is the increase in Pi, caused by a precipitous decline in PCr, that is a "major cause" of intracellular fatigue. Muscle fatigue is a very complex phenomenon, attributable to a variety of sites and factors, often depending on the nature of the muscle activity that caused it. Additional factors associated with muscle fatigue are highlighted in chapters 4 and 5. An accumulation of intracellular lactate, while correlated with fatigue, is usually not directly associated with its cause. The accelerated glycolysis that results in lactate accumulation actually allows us to continue exercise at high intensities. Lactate is transported

✓ KEY POINT

Although accumulation of muscle and blood lactate during intense exercise occurs at the same time as increases in hydrogen ion concentration and decreases in pH, it is not primarily the dissociation of lactic acid into lactate and a hydrogen ion that is responsible for this drop in pH. Nevertheless, measures of blood and muscle levels of lactate accumulation during intense exercise can provide a reasonable measure of glycolytic contribution to exercise performance.

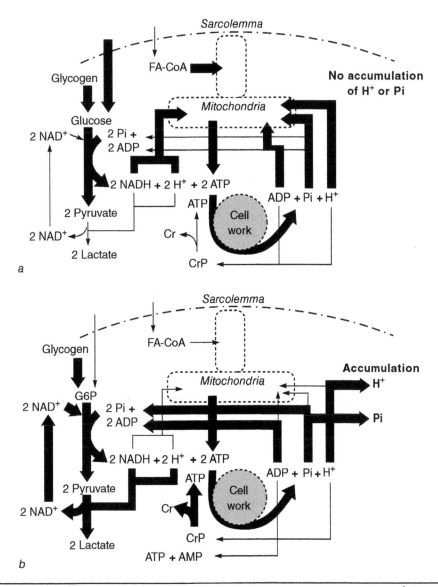

Figure 6.15 Energy metabolism in skeletal muscle during (a) steady-state exercise at ~60% $\dot{V}O_2$max. Macronutrients include blood glucose, muscle glycogen, blood-free fatty acids, and intramuscular lipid. Blood-free fatty acids and intramuscular lipolysis yield FA-CoA. Pyruvate, NADH, and protons produced from glycolysis are consumed for mitochondrial respiration. The same is true for the products of ATP hydrolysis (ADP, Pi, and H^+). Such a metabolic scenario is pH neutral. (b) Short-term, intense exercise is shown at ~110% $\dot{V}O_2$ max. Size of the arrows approximate relative involvement of that reaction and predominant fate of the products. Pi is also a substrate of glycogenolysis. Cellular ATP hydrolysis is occurring at a rate that cannot be 100% supported by mitochondrial respiration, increasing reliance on ATP regeneration from glycolysis and creatine phosphate. For every ADP that is used in glycolysis and in the creatine kinase reaction, a Pi and a proton are released into the cytosol. The magnitude of proton release is greater than that for Pi due to the need to recycle Pi as a substrate in glycolysis and glycogenolysis. The accumulation of protons is a balance between reactions that consume and release protons, cell buffering, and proton transport out of the cell. The biochemical cause of proton accumulation is not lactate production, but ATP hydrolysis.

Reprinted from R.A. Robergs, F. Ghiasvand, and D. Parker, 2004, "Biochemistry of exercise-induced metabolic acidosis," *American Journal of Physiology* 287: R502-R516. Used with permission.

from the fiber accompanied by a proton (see Lactate Transport and Lactate Shuttles on p. 182). Lactate does not cause the proton formation but does provide the mechanism through its cotransport with H^+ to decrease extracellular pH.

The common belief that lactic acid causes delayed muscle soreness also needs to be debunked. We know that muscle soreness following intense or unaccustomed exercise is due in part to the inflammatory response to muscle damage induced by overstretching muscle fibers and the breaking and stretching of muscle sarcomeres. The inflammatory response is a necessary and important step in activation of muscle healing and of muscle satellite cells that act to repair the muscle. Acidification of muscle does not cause muscle soreness. This has been demonstrated many times in experiments where downhill running or lengthening muscle contractions (which are energetically less expensive, producing much less muscle acidification and lactate buildup, but also inducing more muscle damage and inflammation) result in much more muscle soreness than uphill running or shortening muscle contractions (which are energetically more expensive, producing more lactate and muscle acidification). Metabolic acidosis and lactate buildup are generally reversed within 30 min following intense exercise. Hence, the assertion that muscle soreness that occurs 24 h after exercise can be treated by massaging lactic acid out of the muscle is also incorrect. In addition, it has been demonstrated that massage does not influence muscle soreness, alter muscle blood flow, or enhance lactate clearance from muscle; actually, it may hinder it (Wiltshire et al. 2010; Tiidus and Shoemaker 1995).

Lactate is generated by the reduction of pyruvate in the cytosol of cells. Addition of electrons on NADH to pyruvate produces lactate. We describe this as anaerobic glycolysis. Pyruvate can be produced in glycolysis at a rate far in excess of the ability of pyruvate to be transported down its concentration gradient into mitochondria, where it is oxidized by pyruvate dehydrogenase. The activity of the enzyme lactate dehydrogenase is generally high in skeletal muscle. In addition, on purely energetic grounds, reduction of pyruvate by NADH to make lactate is the favored direction. Therefore, whenever glycolysis produces pyruvate, lactate should also be produced. Pyruvate can also accept an amino group from the amino acid glutamic acid in a transamination reaction. The pyruvate is then changed to the amino acid alanine. Exercising muscle releases alanine to the blood, a topic considered further in chapter 8.

Glycogenolysis produces glucose 6-P for the glycolytic pathway. Agents representing the principal mechanisms activating glycogenolysis (epinephrine, calcium, Pi, AMP) are graded to the intensity of contractile activity. Moreover, AMP, ADP, Pi, and fructose 2,6-bisphosphate, the key regulators of phosphofructokinase, are also increased in muscle in proportion to the relative intensity of contractile activity. Therefore, we would expect an exercise-induced increase in glucose 6-P formation from glycogenolysis and a matching increase in the conversion of this glucose 6-P to pyruvate in the glycolytic pathway. Indeed, the activation of both glycogenolysis and glycolysis increases exponentially with a linear increase in exercise intensity. We have also discussed the fact that glycogenolysis and, therefore, glycolysis are more active if glycogen stores are elevated as opposed to lowered. Finally, as exercise becomes more intense, more muscle fibers become involved with higher glycogen phosphorylase activity and glycolytic potential and lower oxidative capacity. Therefore, we should expect glycolysis to become increasingly important early in exercise and as exercise increases in intensity. It is not necessary for the relative intensity of exercise to be even moderate for lactate to be produced. Exercise at only 40% of maximum aerobic capacity can increase the rate of glycolysis by more than 20-fold compared to rest (Katz and Sahlin 1990).

Lactate generated by anaerobic glycolysis can accumulate within the fiber, but it can also leave the fiber, passing through a **monocarboxylate transporter** (MCT) down a concentration gradient to the extracellular fluid. From here, it could enter an adjacent fiber with a lower intracellular lactate concentration or a nearby capillary, where it could be transported and dissolved in the blood. Lactate can also be produced and consumed within the same muscle fibers during exercise. Locations close to ATP-consuming myofibrils may have net lactate production, while locations proximal to mitochondria, which can use pyruvate derived from lactate for oxidation, have net lactate consumption (van Hall 2010).

Lactate is also produced by red blood cells, so blood lactate reflects erythrocyte metabolism and lactate generated through glycolysis, principally by muscle. As we have discussed, lactate is a fuel with the capability of generating six pairs of electrons for the electron transport chain and one GTP if the lactate is oxidized to pyruvate, the pyruvate is oxidized to acetyl CoA, and the acetyl group is metabolized by the citric acid cycle. Complete oxidation of one

> ### ✓ KEY POINT
> Lactate production begins to accelerate during exercise at intensities below $\dot{V}O_2$max due to accelerated glycolysis, which produces pyruvate and NADH faster than they can be taken up and metabolized by the mitochondria. It is this mismatch of glycolytic and aerobic metabolic capacity and activation that results in increased lactate generation during exercise, not a lack of oxygen available to muscle.

lactate can generate approximately 15 ATP. As the lactate circulates in the blood throughout the body, lactate can diffuse down a concentration gradient, enter cells, and be used by the heart, liver, kidney, brain, neurons, and nonworking skeletal muscle (van Hall 2010). As we will discuss in the later section on lactate shuttles, blood lactate levels during exercise reflect the balance between lactate-producing cells (predominantly active skeletal muscle fibers) and those tissues and cells that consume lactate as a fuel (heart and nonworking muscle) or convert it to glucose (principally liver and kidney). Any imbalance in the production or removal of lactate from the blood is reflected by an increase or decrease in its concentration.

Within a single muscle fiber, the primary fate of pyruvate produced by glycogenolysis and glycolysis is entry into the mitochondrion for oxidation or reduction to lactate. Which of these two processes predominates depends for the most part on the following:

1. *The rate of pyruvate formation or the glycolytic flux rate.* This may depend both on the intensity of the activity and on the availability of glycogen, as we have discussed.

2. *The cytosolic redox state.* This is the concentration ratio of [NADH] to [NAD$^+$]. NADH and NAD$^+$ participate in oxidation–reduction reactions as substrates; thus, major relative changes in their concentrations influence the direction of reversible redox reactions. An increase in the concentration ratio of NADH to NAD$^+$ would favor lactate formation. In the cytosol, the concentration ratio of NADH to NAD$^+$ is less than 0.01, as opposed to the situation in mitochondria, where this ratio is closer to 0.5 (Sahlin et al. 2002). Thus, the cytosol is highly oxidized, which helps to drive the reaction catalyzed by glyceraldehyde phosphate dehydrogenase and so to generate pyruvate. The more highly reduced mitochondrion favors electron transfer from NADH to oxygen.

3. *The number and size of mitochondria.* These reflect the potential for oxidative phosphorylation. More mitochondria would favor pyruvate entry and oxidation.

4. *The availability of oxygen.* Oxygen is the final acceptor of electrons in oxidative phosphorylation. Even though the cell can adjust to relatively low oxygen concentrations by increasing the phosphorylation and redox potential, if oxygen concentration decreases below a threshold level, it can limit oxidative phosphorylation, forcing the cell to rely on anaerobic glycolysis. At present, it is unclear what the actual concentration of oxygen at the cytochrome c oxidase complex must be to prevent oxygen-limited ATP formation.

5. *The total activity of lactate dehydrogenase (LDH) and the specific isozyme found in the cell.* (See figure 2.13 on p. 29.) This is because these two will be competing for pyruvate formed in glycolysis.

In chapter 4, we discuss the three common fiber types in human muscle based on their myosin composition (table 4.1 on p. 81). If we compare slow-twitch (ST or Type I) and two types of fast-twitch (FTA or Type IIA, and FTX or Type IIX) skeletal muscle fibers with heart muscle cells (cardiomyocytes), we learn some important metabolic lessons. First, their typical activities differ greatly. Heart muscle is continuously active because the heart is always working. Slow-twitch muscle fibers are involved in low-intensity activity; they are also active during strong contractions. For the same cross-sectional area, these fibers generate nearly the same tension as fast-twitch muscle fibers, but they cannot shorten as rapidly. Fast-twitch fibers are usually active during intense activity, but they are not normally active during low-intensity activity. They can shorten faster than ST muscle fibers, primarily because they have fast myosin-heavy chains.

Table 6.6 summarizes important differences between the heart muscle, ST, FTA, and FTX skeletal muscle fibers. Both types of FT fibers can hydrolyze ATP at a maximum rate that exceeds that of ST fibers and heart muscle. Thus, the regeneration of ATP must also be faster in FT fibers. However, ATP regeneration in FT fibers involves more glycolysis and less fuel oxidation than in the ST fibers and heart muscle because the FT fibers have a greater capacity for glycolysis and a poorer blood supply—thus, they

Table 6.6 Relative Comparisons for Heart Muscle (HM), Slow-Twitch (ST), and Two Fast-Twitch (FTA, FTX) Skeletal Muscle Fibers

Metabolic factor	Fiber-type comparison
Maximum rate of ATP hydrolysis	FTX > FTA > ST > HM
Maximum glycolytic flux rate	FTX > FTA > ST > HM
Blood supply or availability of oxygen	HM > ST > FTA > FTX
Fiber size	FTX > FTA > ST > HM
Maximum oxidative capacity	HM > ST > FTA > FTX

HM, heart muscle; ST, slow-twitch; FT, fast-twitch.

have less oxygen and a lower capacity for oxidative phosphorylation. As table 6.6 summarizes, there are also subtle differences in the metabolic capacities of the FT fibers, with FTA fibers having a higher oxidative capacity but a lower glycolytic capacity. In summary, either type of FT fiber would produce more lactate and consume less oxygen when regenerating ATP compared to an ST fiber or heart muscle.

Lactate Transport and Lactate Shuttles

During exercise with the blood flow occluded, as in modestly strong isometric contractions, glycolysis is the major source of ATP. As we have seen, glycolysis is also significant during exercise at intensities beyond $\dot{V}O_2max$. For these two exercise conditions, lactate concentration will increase in the muscle fiber and lactate will appear in the blood, well beyond the normal 1 mM level. Lactate can diffuse across cell membranes as the undissociated acid, lactic acid; however, this is of very minor importance, since there is little undissociated lactic acid at a pH of 7. Therefore, it is the lactate ion that must cross the cell membrane down its concentration gradient. Because it is a charged ion, lactate needs a lactate transporter. However, this lactate transporter has also been found to transport other negatively charged ions, such as pyruvate; therefore, it has been named monocarboxylate transporter (MCT). Recently, isoforms of the MCT have been observed and identified with numbers, much as in the numbering system for the glucose transporters. MCT-1 and MCT-4 are in skeletal muscle, with the former more prominent in slow-oxidative fibers and the latter more common in fast-twitch muscle fibers (Coles et al. 2004). The MCT acts as a symport, transferring lactate down its gradient, accompanied by a proton (H^+).

Lactate transport across the muscle sarcolemma can occur in both directions; the favored direction depends only on the lactate and proton gradient. Thus, we might expect that during exercise at an intensity near $\dot{V}O_2max$, an active FT fiber may be producing significant lactate and H^+, which are transported together through MCT-4 and MCT-1 into the extracellular fluid surrounding the fiber. As shown in figure 6.16, a neighboring ST fiber with a low lactate concentration may take up the lactate and a proton using an MCT isoform, because the lactate concentration outside this fiber is higher than inside.

New research, however, has suggested that interfiber lactate exchange is minor relative to lactate production and utilization within the fibers themselves. Hence, a significant portion of the lactate produced within both FT and ST fibers may be oxidized within the mitochondria of the fibers themselves (van Hall 2010). In the muscle fibers, the lactate generated proximal to the sites of rapid ATP utilization such as the myofibrils can be oxidized to pyruvate, which can enter mitochondria where pyruvate is converted to acetyl CoA for entry into the citric acid cycle. Thus, as well as intramuscular shuttling of lactate from a

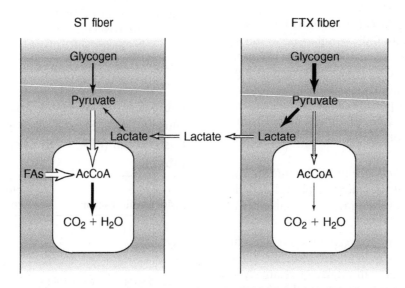

Figure 6.16 A lactate shuttle between two neighboring fibers in the same active muscle. The relative thickness of each arrow reflects the importance of the major pathways or membrane transport process to ATP formation. The low-oxidative FTX fiber has active glycogenolysis and glycolysis, producing pyruvate at a high rate. Most of the pyruvate is reduced to lactate, which can leave the fiber and get taken up by an adjacent high-oxidative ST fiber. In the ST fiber, lactate is oxidized to pyruvate and used as a source of acetyl CoA (AcCoA). Oxidation of fatty acids (FAs) occurs mainly in the ST fiber.

net-producing to a net-consuming fiber, lactate can also be moved within the muscle fiber itself from areas of production to places of oxidation, such as the mitochondria. Prior to entry to mitochondria, lactate will be converted to pyruvate via reversal of the LDH reaction in the cytoplasm, since this reaction does not occur within the mitochondria itself.

Lactate can also be shuttled between an active muscle that produces it and other tissues, such as heart, liver, or inactive muscle that can consume the lactate as a fuel. As we will see, lactate is an important source for glucose formation in the liver, a process known as gluconeogenesis.

Interaction of Exercise Intensity, Diet, and Training on Lactate Metabolism

In rested muscle, the activity of the glycolytic pathway is quite low, as evidenced by the observation that the RQ (respiratory quotient) of rested muscle is close to 0.75, indicating the preponderance of fat as the fuel for oxidative phosphorylation. During moderate-intensity dynamic exercise, the rate of glycolysis may increase by 30- to 40-fold or more above the resting level in some fibers, but less in others. Differences will reflect the metabolic profile of the fibers and the extent to which the fiber is active during the exercise. During moderate exercise intensity, lactate is generated in the active muscle; however, the concentration of lactate within the muscle and blood leaving it may not increase significantly. This is because much of the lactate is reconverted to pyruvate and oxidized in mitochondria of the muscle fiber it was formed in, or the lactate formed is consumed within other fibers of the same muscle, or both of these processes occur. As exercise intensity increases linearly beyond 50% to 60% of $\dot{V}O_2$max, the rate of glycolysis, expressed as the rate of pyruvate formation, increases at an even faster rate. Although exercise may still be described as submaximal (that is, at an intensity less than $\dot{V}O_2$max), pyruvate can be formed at such a rate that a significant fraction will be reduced to lactate. As we have discussed, this does not mean that there is a lack of oxygen to act as the electron acceptor, because measures of cytosolic PO_2 reveal that sufficient O_2 is present to meet the needs of the electron transport chain at exercise intensities approaching $\dot{V}O_2$max. During supramaximal exercise, glycogen utilization and lactate production can increase exponentially. In well-motivated subjects, muscle lactate and blood lactate concentrations can exceed 30 and 20 mM, respectively.

We noted in a previous section that the rate of glycogenolysis during submaximal exercise depends on the concentration of glycogen. Since the glucose 6-P produced during glycogen breakdown feeds into the glycolytic pathway, we might expect the amount of lactate formed during submaximal exercise also to be directly related to muscle glycogen concentration. Indeed, a number of experiments have demonstrated that the same person doing exactly the same submaximal exercise task can have a muscle lactate concentration significantly higher if starting muscle glycogen stores are high, compared to low. Figure 6.17 illustrates the relationship among starting glycogen concentration, the rate of glycogen utilization, and muscle lactate concentration during submaximal (70% of $\dot{V}O_2$ peak) exercise on a cycle ergometer. The relationships shown in figure 6.17 are configured from studies that have demonstrated that muscle glycogen utilization and lactate production can vary significantly if subjects carry out the same exercise task with high and low starting concentrations of muscle glycogen (Spencer, Yan, and Katz 1992; Baldwin et al. 2003; Arkinstall et al. 2004). Compared to values from the same exercise performed with elevated glycogen, overall carbohydrate oxidation is reduced and fat utilization is increased when preexercise glycogen stores are low (Spencer, Yan, and Katz 1992). Chapter 7 discusses the interaction between fat and carbohydrate as fuels in more detail.

Endurance training can have a profound effect on fuel utilization when the same absolute exercise task is performed after, compared to before, training. Considering that the same absolute exercise intensity means that the rate of energy expended would be the same, there are still profound changes in the fuels

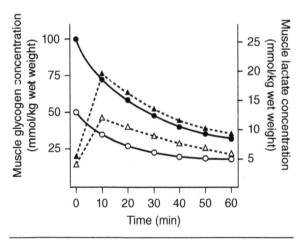

Figure 6.17 The effect of muscle glycogen concentration (high = solid, low = open, circles and triangles) in vastus lateralis muscle on glycogen utilization (circles) and lactate production (triangles) during ergometric exercise at 70% of peak.

utilized. At the same absolute intensity, there would be smaller reductions in PCr and glycogen, along with smaller increases in ADP, AMP, ammonia, Pi, Cr, and lactate. Given that ADP, AMP, Pi, and ammonia all promote glycogenolysis and glycolysis, the use of carbohydrate would be reduced. Training increases the content of citric acid cycle and electron transport chain proteins, as well as the enzymes involved in fat metabolism. This means that relative changes in ratios of NADH to NAD^+, ATP to ADP, and acetyl CoA to CoA would be less after training and that the stimulus for pyruvate dehydrogenase activation would be reduced (LeBlanc et al. 2004). In all, the adaptations with training would act to reduce carbohydrate utilization while enhancing fat oxidation. As few as five days of training can attenuate the decreases in PCr and glycogen and the increase in lactate during submaximal exercise (Phillips et al. 1996). These data demonstrate that the intracellular milieu can be altered in such a way as to spare carbohydrate utilization even when there is no evidence of a change in muscle oxidative capacity or whole-body $\dot{V}O_2max$.

OXIDATION OF CYTOPLASMIC NADH

During the formation of pyruvate in glycolysis, NAD^+ is reduced by the enzyme glyceraldehyde phosphate dehydrogenase, forming NADH. If the fate of pyruvate is anaerobic glycolysis, NADH is oxidized by lactate dehydrogenase, regenerating NAD^+. On the other hand, when pyruvate enters mitochondria and is oxidized there, or if pyruvate is converted to the amino acid alanine (a minor but still significant fate in muscle), no immediate regeneration of NAD^+ occurs. We have already mentioned that the cytosol is much more highly oxidized (i.e., $[NAD^+]$ is much greater than [NADH]) compared to mitochondria, and that this helps to drive glycolysis. This gives rise to the question: How is cytoplasmic NAD^+ regenerated if not by lactate dehydrogenase?

The simplest solution to this metabolic problem would be for cytoplasmic NADH to enter the mitochondrial matrix and to be oxidized in the electron transport chain. However, the inner mitochondrial membrane is impermeable to NADH; this is critical to maintaining a highly oxidized cytosol and more reduced mitochondrial matrix in order to facilitate the specific metabolism for each compartment. To get around this roadblock, two shuttle systems transfer electrons (reducing equivalents) on cytoplasmic NADH into the mitochondrion without actual crossing of the inner membrane by NADH.

Glycerol-Phosphate Shuttle

The glycerol-phosphate shuttle (see figure 6.18) transfers electrons on cytosolic NADH to FAD, then to ubiquinone (coenzyme Q) in the mitochondrial inner membrane. The cytosolic NADH forms during the glyceraldehyde phosphate dehydrogenase reaction of glycolysis (reaction 1 in figure 6.18). This and subsequent reactions in glycolysis are not affected by this shuttle system. In reaction 2, the cytosolic NADH, not reoxidized back to NAD^+ by the lactate

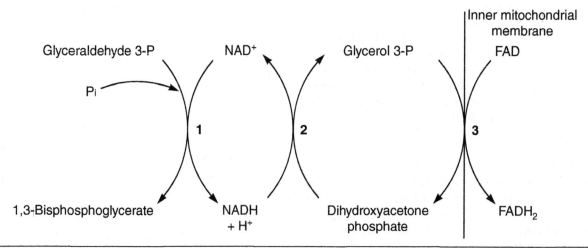

Figure 6.18 The glycerol-phosphate shuttle. This shuttle transfers electrons from cytosolic NADH to FAD in the inner mitochondrial membrane, generating $FADH_2$. The NADH generated in reaction 1, catalyzed by the glycolytic enzyme glyceraldehyde 3-phosphate dehydrogenase, is used to reduce dihydroxyacetone phosphate in reaction 2, catalyzed by a cytoplasmic glycerol phosphate dehydrogenase. The resulting glycerol 3-phosphate is oxidized by a FAD-dependent mitochondrial glycerol phosphate dehydrogenase, located in the inner mitochondrial membrane (reaction 3). Subsequently, electrons on $FADH_2$ are transferred to coenzyme Q (ubiquinone). This shuttle is irreversible because of reaction 3.

dehydrogenase reaction, transfers a hydride ion and proton to dihydroxyacetone phosphate, changing it to glycerol 3-phosphate (glycerol 3-P) and oxidizing NADH to NAD$^+$. The enzyme catalyzing this reaction (reaction 2 in figure 6.18) is *cytosolic glycerol phosphate dehydrogenase*. Glycerol 3-P diffuses to the outer side of the inner mitochondrial membrane, where it is oxidized by *mitochondrial glycerol phosphate dehydrogenase* (reaction 3). This enzyme is located in the inner membrane but faces the intermembrane space, so glycerol 3-P need not penetrate the inner membrane. Instead of using the NAD$^+$ coenzyme, the mitochondrial form of glycerol phosphate dehydrogenase uses FAD. The FAD binds tightly to the mitochondrial form of the glycerol phosphate dehydrogenase. It is another flavoprotein dehydrogenase similar to succinate dehydrogenase (complex II) in the electron transport chain. In reaction 3, the products are FADH$_2$ and dihydroxyacetone phosphate. The electrons on FADH$_2$ transfer to ubiquinone (coenzyme Q) in the electron transport chain and then to oxygen. The dihydroxyacetone phosphate is then able to accept electrons from cytoplasmic NADH.

The glycerol-phosphate shuttle exhibits no net consumption or production of dihydroxyacetone phosphate or glycerol 3-P. It simply transfers electrons from cytosolic NADH to the electron transport chain. This shuttle generates approximately 1.5 ATP per pair of electrons transferred from cytosolic NADH to oxygen. It is irreversible because reaction 3 goes in only one direction. Although not the more prevalent of the two major shuttle systems, this system is easier to explain. The glycerol-phosphate shuttle operates to a minor extent in a variety of tissues, such as brain, but it is very important in fast-twitch (Type II) skeletal muscle fibers.

Malate–Aspartate Shuttle

The malate–aspartate shuttle is more complicated and is reversible (see figure 6.19). It is the dominant shuttle for the liver, the heart, and slow-twitch (Type I) muscle fibers. This shuttle system transfers electrons (reducing equivalents) on NADH in the cytosol to NAD$^+$ in the mitochondria (as opposed to electron transfer to FAD in the glycerol-phosphate shuttle). As a result, for each two cytosolic electrons on NADH transferred to oxygen in the mitochondria, we get approximately 2.5 ATP.

Cytosolic NADH is converted to NAD$^+$ through reduction of oxaloacetate to malate. This allows the NAD$^+$ to be used in the glyceraldehyde phosphate dehydrogenase reaction in glycolysis. The malate is transported into the matrix using an antiport mechanism with α-ketoglutarate. In the matrix, the malate

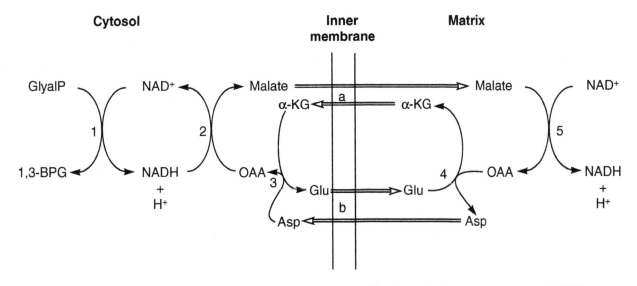

Figure 6.19 The malate–aspartate shuttle. This shuttle transfers electrons from cytoplasmic NADH to mitochondrial NADH. Reduction of NAD$^+$ in the cytosol by electrons on glyceraldehyde phosphate (GlyalP) is catalyzed by glyceraldehyde phosphate dehydrogenase (1). The resulting NADH is used to reduce oxaloacetate (OAA), producing malate and NAD$^+$ catalyzed by cytosolic malate dehydrogenase (2). Malate is transported into the matrix using an antiport (a) that simultaneously transports α-ketoglutarate (α-KG) from the matrix to the cytosol. The α-KG is converted into glutamate (Glu) by accepting an amino group from aspartate (Asp); the latter is converted into oxaloacetate using cytosolic aspartate aminotransferase (3). Glutamate is transferred from the cytosol to the matrix using the antiport (b) that simultaneously transfers Asp to the cytosol. In the matrix, NADH is produced by the oxidation of malate to OAA using mitochondrial malate dehydrogenase (5). The oxaloacetate is converted to aspartate using mitochondrial aspartate aminotransferase.

now transfers its electrons to NAD$^+$, generating NADH and oxaloacetate. In this way, cytosolic-reducing equivalents become matrix electrons on NADH. The complication of the malate–aspartate shuttle is that there is no mechanism to directly transport oxaloacetate from the matrix to the cytosol across the inner membrane. Instead, oxaloacetate is converted into the amino acid aspartic acid (aspartate) as the amino group on glutamic acid (glutamate) is moved to oxaloacetate. Amino group transfers are catalyzed by a class of enzymes known as **aminotransferases** (also known as transaminases). The resulting aspartate is transported across the inner membrane by an *aspartate–glutamate transporter* (antiport) that simultaneously transports glutamate from the cytosol to the matrix. As in the glycerol-phosphate shuttle, there is no net production or consumption of any of the intermediates in this shuttle. Direction is provided to this potentially reversible shuttle by the formation of NADH in the cytosol and its oxidation in the electron transport chain of the mitochondria.

GLUCONEOGENESIS

As mentioned previously, glucose is the principal fuel for the brain, erythrocytes (red blood cells), and some other tissues. In the brain, the glucose is almost exclusively oxidized. In red blood cells, anaerobic glycolysis is the fate of the glucose, since these cells lack mitochondria. The amount of glucose in the blood is relatively small, and it is used constantly by a variety of tissues. To maintain blood glucose levels, liver glycogen is broken down through glycogenolysis, and the glucose is released to the blood. At rest, in the **postabsorptive state**, glucose release from the liver is about 10 to 14 μmol per kilogram per minute (about 125-175 mg per min for a 70 kg [154 lb] person). For a person with 70 g of liver glycogen, the liver would be exhausted of glycogen in about 8 h if no food was eaten or if there were no other way to produce glucose for the blood. This R_a (rate of appearance) of glucose can increase up to 10-fold during hard exercise (Angus et al. 2000). However, there are limited supplies of glycogen in liver to maintain blood glucose. For example, an overnight fast can markedly reduce liver glycogen, while a 24 h fast can deplete it completely. Consider what can happen when exercise is performed without food intake. Without exogenous glucose to help maintain blood glucose, and in the face of an increased uptake by exercising muscle, blood concentration could fall to hypoglycemic levels during prolonged exercise, seriously impairing performance. Obviously, another mechanism must exist to make glucose for the body when glycogenolysis of liver glycogen becomes limited.

As the name suggests, gluconeogenesis is the formation of new glucose from noncarbohydrate precursors, such as lactate, pyruvate, glycerol, propionic acid, and particularly the carbon skeletons of amino acids. The carbon skeletons of amino acids are what remain of the amino acids when the amino group is removed. Humans cannot oxidize amino groups on amino acids. These are mainly converted to the molecule **urea** using the **urea cycle**, and the urea is excreted in the urine. The urea cycle is discussed in chapter 8. It is important to remember that while it is possible to convert carbohydrates into fatty acids, the reverse is not possible in humans or other mammals. Hence, while the body can store significant amounts of energy as triglycerides in adipose tissue, only the glycerol component of triglycerides (not the fatty acids, which contain high amounts of energy) is available for gluconeogenesis. Therefore, amino acids from the breakdown of body protein can, out of necessity, become an important additional source of precursors for gluconeogenesis in prolonged exercise, particularly during starvation conditions. Some implications of this for diet and exercise are discussed further in the next two chapters.

Gluconeogenesis occurs mainly in the liver and to a much smaller extent in the kidney. It becomes important when liver glycogenolysis cannot produce glucose for the blood at a necessary rate. The central nervous system (principally the brain) needs about 125 g of glucose each day, and other tissues that rely exclusively on glucose need an additional 30 to 40 g of glucose a day. This means we need about 160 g or so of glucose just for the glucose-dependent tissues. If we exercise or do work, we use blood glucose in the contracting muscles. Fat cells use glucose to act as a source of glycerol (as glycerol 3-phosphate) to make triglyceride molecules.

Whenever dietary carbohydrate (glucose, fructose, and galactose) is inadequate to supply the body with the glucose it needs, gluconeogenesis makes up the difference. This becomes particularly significant during fasting or starvation, when eating a low-carbohydrate diet, or during prolonged exercise, when the working muscles use a lot of blood glucose and glucose cannot be ingested or absorbed as quickly as it is utilized. Gluconeogenesis also occurs whenever the blood lactate concentration rises, as in moderate to severe exercise. Although lactate is used as a fuel by other skeletal muscles and the heart, a considerable amount is extracted from the blood by the liver and used to make glucose. The **Cori cycle** describes the process by which carbon cycles between liver and muscle. As shown in figure 6.20, lactate released from muscle circulates to the liver, where it is converted to glucose. The glucose is released to the blood and taken up by active muscle. Through glycolysis, glucose is converted to lactate, released from the muscle, and recycled again by the liver. Heart muscle also oxidizes lactate (figure 6.20).

☑ KEY POINT

During prolonged exercise, gluconeogenesis gradually increases in activity to compensate for the diminishing ability of liver glycogenolysis to produce glucose as the liver glycogen particles decrease in size. Amino acids from proteins can become an increasing source of gluconeogenic precursors in prolonged exercise, particularly in starvation conditions. The lean body mass loss that one sees in prolonged starvation is largely a result of muscle protein being used for gluconeogenesis.

Reactions of Gluconeogenesis

For the most part, new glucose is made from simple precursors by the reversal of the glycolytic pathway, as figure 6.21 reveals. In this figure, you can tell where the glycolytic pathway fits in, although only the reactions that contribute to glucose formation are shown. Most of the reactions of glycolysis are reversible; so, from this perspective, they can go in either direction, driven by mass action. However, there are three irreversible (nonequilibrium) glycolytic reactions: the pyruvate kinase reaction, the phosphofructokinase reaction, and the hexokinase or glucokinase reaction. These irreversible reactions act as one-way valves, making the glycolytic pathway flow only to pyruvate and lactate. In principle, glycolysis can go backward if there are alternate ways of getting around the three irreversible reactions.

Lactate is an important gluconeogenic precursor. It is taken up by liver cells and oxidized to pyruvate, as we have seen. However, the pyruvate must be converted to PEP. This cannot be done through

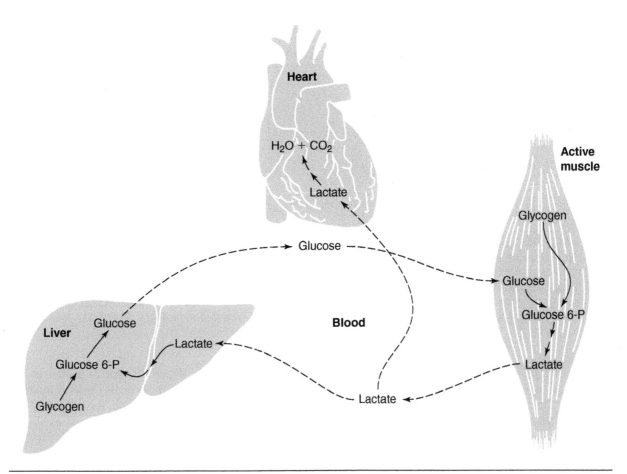

Figure 6.20 Lactate shuttling between tissues. Here, an active muscle releases lactate to the blood where it can be used as a fuel by the heart. The lactate can also be taken up by liver and converted to glucose in a process known as gluconeogenesis. The dotted arrows show lactate and glucose release to and uptake from the blood.

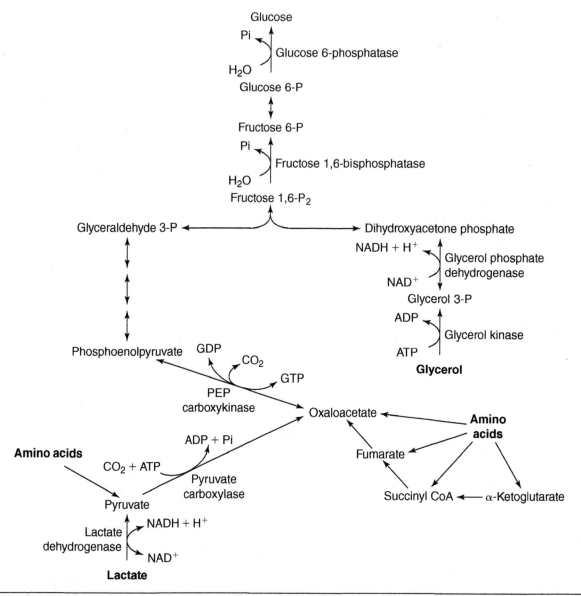

Figure 6.21 Formation of glucose from noncarbohydrate precursors (lactate, glycerol, and amino acids). A number of equilibrium (reversible) reactions of glycolysis are used; to bypass the remainder, new reactions are employed. Pyruvate carboxylase and PEP carboxykinase allow pyruvate to be converted into PEP. Fructose 1,6-bisphosphatase and glucose 6-phosphatase allow fructose 1,6-bisphosphate to be converted to glucose. Glycerol can be made into dihydroxyacetone phosphate using glycerol kinase and glycerol phosphate dehydrogenase. Many amino acids form citric acid cycle intermediates or pyruvate once they lose their amino groups. These can be converted to oxaloacetate using reversible citric acid cycle reactions.

driving the pyruvate kinase reaction backward because pyruvate kinase catalyzes a nonequilibrium reaction. Therefore, to convert pyruvate to PEP, an alternate route is needed. This requires two new reactions with two new enzymes, *pyruvate carboxylase* and *phosphoenolpyruvate carboxykinase* (most people simply use the acronym PEPCK). Pyruvate is first converted to oxaloacetate using pyruvate carboxylase; the oxaloacetate is then converted to PEP, catalyzed by PEPCK. The irreversible pyruvate kinase reaction is bypassed, but at a cost of the equivalent of two ATP. The process involves multiple steps and membrane crossings in the conversion of lactate or pyruvate (or both) to glucose, as shown in figure 6.22. This figure emphasizes the role of the mitochondrion in the conversion of lactate and pyruvate to glucose.

Pyruvate carboxylase catalyzes a carboxylation reaction, using the tightly bound coenzyme **biotin**. This is a B vitamin that acts to bind and transfer

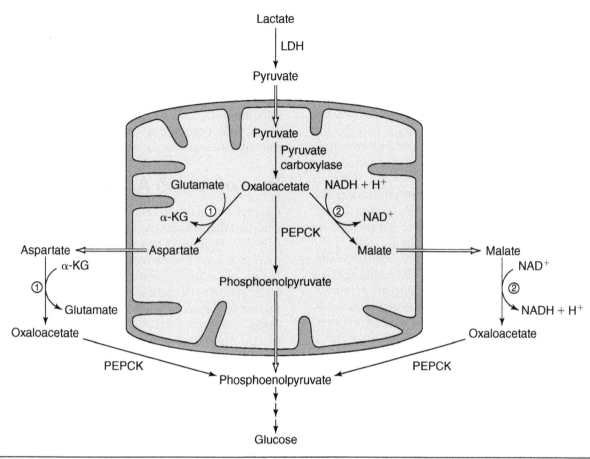

Figure 6.22 Specific steps and their cellular location as lactate and pyruvate are converted to glucose. The conversion of pyruvate to oxaloacetate takes place in the mitochondrial matrix. The next step is the formation of phosphoenolpyruvate, catalyzed by a mitochondrial PEPCK (phosphoenolpyruvate carboxykinase), which can be transported out of the mitochondrion. The oxaloacetate can also be converted to phosphoenolpyruvate by a cytosolic PEPCK. However, oxaloacetate cannot leave the mitochondria. It can undergo transamination using glutamate to form aspartate (reaction 1) or it can be reduced to malate (reaction 2). Both malate and aspartate can be transported out of mitochondria. In the cytosol, aspartate can undergo reverse transamination, forming oxaloacetate (reverse of reaction 1). Malate can be oxidized to oxaloacetate (reverse of reaction 2). Cytoplasmic oxaloacetate can be converted to PEP using cytosolic PEPCK.

CO_2 in the form of bicarbonate ion (HCO_3^-), which is needed in this reaction to change a molecule with three carbon atoms into one with four carbon atoms.

$$ATP + pyruvate + HCO_3^- \rightarrow oxaloacetate + ADP + Pi$$

This reaction takes place in the mitochondrial matrix.

The oxaloacetate is converted into PEP by phosphoenolpyruvate carboxykinase (PEPCK).

$$oxaloacetate + GTP \leftrightarrow phosphoenolpyruvate + GDP + CO_2$$

PEPCK is located in both the cytosol and mitochondrial matrix. If the oxaloacetate is converted to PEP in the matrix, the latter is then transported across the mitochondrial inner membrane, where it will continue toward glucose. If oxaloacetate is not converted to PEP by the matrix PEP carboxykinase, it must be a substrate for the cytosolic PEP carboxykinase enzyme. However, as noted earlier, oxaloacetate cannot be transported across the inner mitochondrial membrane. Therefore, oxaloacetate must be converted into something that can cross the inner membrane via its own transporter. It can be transaminated to aspartate, which can use the aspartate–glutamate transporter (antiport). Alternatively, it can be reduced to malate, and the malate can be transported by an antiport with α-ketoglutarate.

Once the pyruvate kinase reaction is circumvented, glycolysis can run backward up to the fructose 1,6-bisphosphate step. At this point, an enzyme known as *fructose 1,6-bisphosphatase* removes one phosphate group, creating fructose 6-P. After this latter compound is converted to glucose 6-P, *glucose*

6-phosphatase removes the phosphate group. The product, free glucose, is ready to be released by the liver (and, to a very small extent, the kidney) to the bloodstream.

Glycerol is also a key precursor for gluconeogenesis, as shown in figure 6.21. Glycerol is released from fat cells when triglyceride molecules are broken down in a process known as lipolysis. Two glycerol molecules are capable of making one glucose molecule. In liver, the glycerol is first phosphorylated by *glycerol kinase* to make glycerol 3-P, which is oxidized to form dihydroxyacetone phosphate. The enzyme that catalyzes this latter reaction is the cytosolic form of *glycerol 3-P dehydrogenase* that we saw in the glycerol-phosphate shuttle. During the oxidation of glycerol 3-P, NAD^+ is reduced to NADH. During starvation, glycerol produced during lipolysis of stored fat can act as a gluconeogenic precursor, providing 15% to 25% of the glucose produced by the liver. However, the majority of glucose produced by the liver during starvation is from lactate or amino acid precursors.

As noted above, humans do not have the ability to form glucose from acetyl CoA. Therefore, the carbon atoms in fatty acids that are converted to acetyl CoA cannot be used as a source of glucose. Most fatty acids have an even number of carbon atoms. As the next chapter shows, all of the carbon atoms in these fatty acids appear as 2-carbon acetyl groups, each attached to a CoA. Thus, an 18-carbon fatty acid (stearic acid) gives rise to nine acetyl groups. There are some odd-chain fatty acids with an odd number of carbon atoms. During the breakdown of these to acetyl units, there will be one leftover 3-carbon propionyl group, or *propionate*. This can be converted to glucose, so it is considered a gluconeogenic precursor. Though we will skip a detailed discussion of the metabolism of propionate (propionyl group), it is first converted to propionyl CoA at the cost of two ATP, then it is converted to succinyl CoA in three additional steps. As a citric acid cycle intermediate, succinyl CoA can be converted to oxaloacetate using the last four steps in the citric acid cycle.

Role of Amino Acids

Chapter 8 contains a detailed description of amino acid metabolism. For now, we will take a simple overview. Figure 6.21 reveals that amino acids can be used as a source of carbon atoms to make glucose. In fact, of the 20 common amino acids, 18 may have all or part of their carbon atoms directed to the formation of glucose; we call these **glucogenic amino acids**. The exceptions, in mammalian tissues, are leucine and lysine; these are said to be **ketogenic amino acids**.

Most adults are in a state that we call **protein balance** or, as it is also known, **nitrogen balance**. The former term means that over a long period of time, adults maintain a constant level of protein in their bodies. The latter term means that if the entire nitrogen intake into the body is measured (mainly in the form of amino acid amino groups) and all of the nitrogen lost by the body is measured (in urine, feces, sweat, and so on), these measures will balance. Whether we use the term protein balance or nitrogen balance, we are essentially talking about the same thing. The average adult takes in about 1 g or more of amino acids in dietary protein each day per kilogram of body weight. People in protein balance (or nitrogen equilibrium) get rid of the same amount of amino acids. We do not excrete these amino acids. Rather, the nitrogen is removed (in the form of the amino groups) and then excreted (mainly in the form of urea). The remaining parts of the amino acids, which can be called the carbon skeletons, are used for the following:

1. Immediate oxidation to generate energy
2. Conversion to glucose and later oxidation
3. Conversion into fat, which is stored and then oxidized later

For all of these options, the final disposal of the carbon skeletons is oxidation, with the carbon atoms appearing as carbon dioxide and the hydrogen and electrons as water.

When the amino groups on most of the amino acids are removed, we get *citric acid cycle intermedi-*

> ### ☑ KEY POINT
>
> If you start from pyruvate and make a molecule of glucose, you can see that this process requires energy. Adenosine triphosphate is consumed to make oxaloacetate, GTP (equivalent to ATP) is consumed in the PEPCK reaction, and an ATP is consumed when the phosphoglycerate kinase reaction goes backward. Add to this the loss of NADH when the glyceraldehyde 3-phosphate dehydrogenase reaction is driven backward. Now, it takes two pyruvate to make one glucose molecule, so the overall process is energetically expensive. Hence, while gluconeogenesis can contribute to glucose supply during times of glucose and glycogen depletion, such as starvation or very long endurance exercise, it is energetically expensive and it only serves as a stop gap measure until glucose stores can be more effectively restored by dietary intake.

ates such as oxaloacetate, α-ketoglutarate, fumarate, and succinyl CoA, as well as pyruvate. Of course, these four tricarboxylic acid cycle intermediates and pyruvate can be used to make glucose, as figure 6.21 reveals. If you review the citric acid cycle, you can see that α-ketoglutarate, fumarate, and succinyl CoA can all produce oxaloacetate, which is the precursor to forming PEP. The net formation of citric acid cycle intermediates through removal of the amino groups from amino acids is known as **anaplerosis**. Conversion of the amino acid glutamate to aspartate does not produce a net citric acid cycle intermediate because while it does generate α-ketoglutarate, it also consumes an oxaloacetate. Although our focus here is on the formation of glucose in liver, it is appropriate to mention that these citric acid cycle intermediates can be formed in exercising muscle, where they may increase the flux capacity for the cycle.

The carbon skeletons of amino acids (including the hydrogen and oxygen) are not wasted but are treated as fuel. The amino groups can be removed from most of the amino acids through transfer to other molecules. An example of this is shown in figure 6.23 for the **branched-chain amino acids** (leucine, isoleucine, and valine). The amino groups on these amino acids are removed, primarily in muscle, through transfer to α-ketoglutarate to make branched-chain keto acids (BCKAs) from the branched-chain amino acids plus the amino acid glutamic acid (glutamate). Then, the amino group on glutamic acid is transferred to pyruvate, regenerating the α-ketoglutarate and forming alanine, which is released from muscle to the blood. Liver extracts the alanine and removes the amino group by transferring it to oxaloacetate, making aspartate (aspartic acid). The carbon skeleton remaining, which is pyruvate, can then be used to make glucose.

Figure 6.24 shows the cycling of amino groups from muscle to liver by way of alanine. The amino group ends up as urea, and the carbon skeleton (pyruvate) is converted into glucose in the liver. This has been described as the *glucose-alanine cycle*.

During exercise, the rate of this cycle increases. The glucose-alanine cycle shows the importance of amino acids as a source of glucose. It also reinforces the fact that muscle protein can be broken down more rapidly under certain conditions, generating amino acids that can be a source of glucose. Muscle, the primary protein-containing tissue, is wasted to make glucose if the diet does not contain enough food energy or carbohydrate.

☑ KEY POINT

Diets high in protein are favored by a number of athletes. Because there is a limit to how much protein can be synthesized in skeletal muscle and other tissues, and because there is no way to store excess amino acids, they are oxidized or used to make glucose. When the need for gluconeogenesis is minimal, the excess amino acids are oxidized. Thus, like carbohydrate, which promotes its own oxidation when present in excess, diets excessively high in protein promote protein oxidation.

Regulation of Gluconeogenesis

Glucose is brain food, so we should not be surprised that gluconeogenesis is very carefully and comprehensively controlled. The conditions that favor the onset of gluconeogenesis are those in which blood glucose levels are threatened because liver glycogenolysis is declining as a result of low liver glycogen stores. To help explain the regulation of gluconeogenesis and show how the manner of its control is the opposite of that for the glycolytic pathway, the two are integrated in figure 6.25. Gluconeogenesis and glycolysis share the same enzymes in the middle portion of the glycolytic pathway, with the flux of gluconeogenesis going from pyruvate toward glucose and glycolysis going the opposite way (figure 6.25). As on a railway with east–west and west–east routes that share the same section of

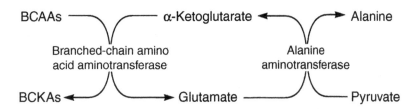

Figure 6.23 Removal of amino groups from amino acids. In muscle, amino groups on the branched-chain amino acids (BCAAs), such as leucine, isoleucine, and valine, can be transferred to α-ketoglutarate, making glutamate and branched-chain keto acids (BCKAs). The amino group on glutamate can be transferred to pyruvate, making alanine and regenerating α-ketoglutarate. Alanine can exit the muscle cell and travel through the blood to the liver.

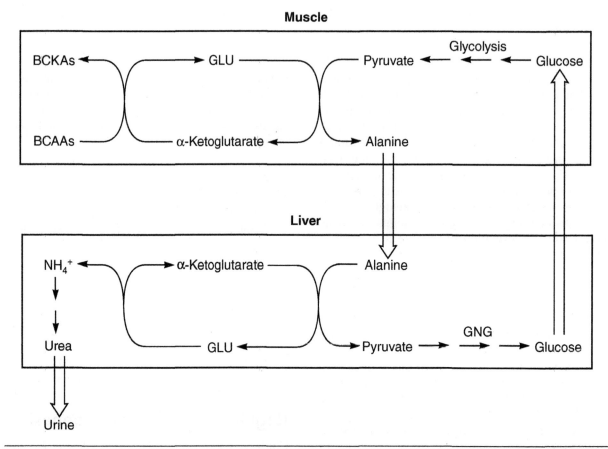

Figure 6.24 The glucose-alanine cycle. This cycle transfers amino groups (NH_2) and carbon in the form of alanine from muscle to liver. In the liver, the amino group is used to make urea, whereas the carbon skeleton is used to make glucose by gluconeogenesis (GNG). Urea is removed from the body in the urine. The glucose leaves the liver and is a carbon source for pyruvate in muscle glycolysis. BCAAs are the branched-chain amino acids; BCKAs are the corresponding branched-chain keto acids that result from the loss of amino groups. GLU represents glutamic acid.

track, traffic cannot move in both directions simultaneously. Thus, the control of gluconeogenesis and glycolysis is integrated, ensuring that one direction is favored while the other is inhibited. Regulation is complex, involving expression of genes for key enzymes in liver, control by protein phosphorylation, and allosteric regulation of enzymes. The net effect is to simultaneously depress glycolysis and stimulate gluconeogenesis under circumstances in which blood glucose levels are threatened and to stimulate glycolysis and inhibit gluconeogenesis under carbohydrate-feeding conditions.

At rest, in the postabsorptive period, the brain utilizes 60% or more of the glucose output from the liver. Of the remainder, most can be accounted for by skeletal and heart muscle. During prolonged exercise, utilization of glucose by skeletal muscle can increase almost 20-fold. This helps to explain the need for careful control of liver glucose output between meals, particularly under exercise conditions. Fortunately, exercise results in an increase in two key precursors for gluconeogenesis, lactate and glycerol. The rate of appearance of lactate in the blood increases exponentially with exercise intensity. As the next chapter discusses, glycerol released from fat depots during exercise increases as stored triglyceride is broken down to provide fatty acids as a fuel. Thus, the accelerated provision of key substrates for gluconeogenesis helps to maintain blood glucose concentration.

Role of Hormones

Under conditions in which gluconeogenesis is needed, specific hormones play a direct or participatory role in turning on and increasing the rate of gluconeogenesis. Glucagon is a polypeptide hormone secreted by the alpha cells of the pancreas. Its concentration rises as the blood glucose concentration decreases below normal (< 5 mM). It acts to stimulate the breakdown of liver glycogen in the postabsorptive period. When liver glycogen concentration is low, glucagon stimulates gluconeogenesis.

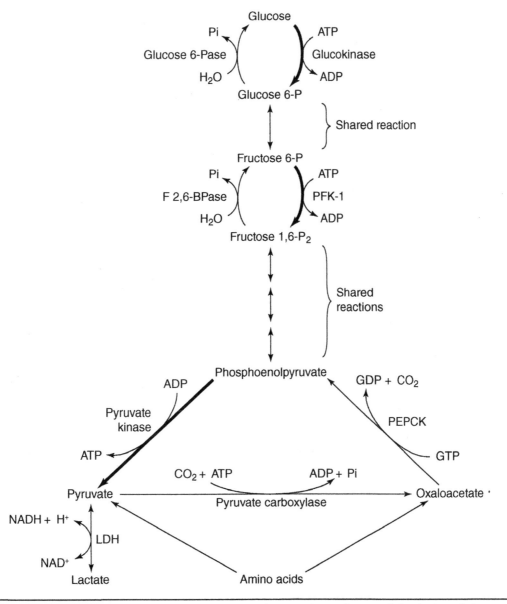

Figure 6.25 Glycolysis and gluconeogenesis in the liver. In the opposing pathways of glycolysis and gluconeogenesis in liver, the potential exists for the two processes to be active simultaneously. As the means of regulation, the three substrate cycles between glucose and glucose 6-phosphate (glucose 6-P), between fructose 6-P and fructose 1,6-P_2 (P_2 means bisphosphate), and between phosphoenolpyruvate and pyruvate operate in one direction only at any time. The glycolytic enzymes (glucokinase, phosphofructokinase-1 [PFK-1], and pyruvate kinase) and the gluconeogenic enzymes (glucose 6-Pase [glucose 6-phosphatase]), fructose 1,6-BPase [bisphosphatase], PEPCK [phosphoenolpyruvate carboxykinase], and pyruvate carboxylase) can be regulated by allosteric effectors or protein phosphorylation, or through regulation of the expression of their genes to promote either glycolysis or gluconeogenesis, but not both at the same time.

Glucagon has a variety of actions in liver to promote gluconeogenesis. It stimulates the formation of gluconeogenic enzymes. It promotes gluconeogenesis while inhibiting glycolysis by protein phosphorylation and allosteric mechanisms. Glucagon binds to its specific cell membrane receptor. Through a G protein mechanism, it activates the enzyme adenylyl cyclase, which produces cAMP. As we have seen already, cAMP activates protein kinase A, which increases glycogenolysis in liver. Cyclic AMP, acting through protein kinase A, also stimulates gluconeogenesis.

As mentioned earlier in this chapter, insulin is released by the beta cells of the pancreas. Its concentration rises when blood glucose concentration increases beyond normal levels (> 5 mM). Insulin promotes glucose uptake into insulin-sensitive tissues, thereby helping lower blood glucose concentration. Insulin has a negative effect on gluconeogenesis, opposing not only the secretion of glucagon but also its biochemical effects on liver. The mechanism of action of insulin is very complex. We will mention some of this at the end of the chapter. Briefly, insulin opposes the effects of glucagon by activating cAMP phosphodiesterase, which breaks down cAMP. It activates specific phosphoprotein phosphatases that remove phosphate groups added by protein kinase A.

Researchers have concluded that it is not so much the absolute levels of insulin and glucagon that play a role in determining glucose production by the liver, but the insulin to glucagon ratio (I/G). During exercise, we know that there is a decline in blood insulin concentration and an increase in blood glucagon, thus magnifying the decline in the I/G ratio (Trimmer et al. 2002).

Under more extreme conditions of physical stress, the glucocorticoid cortisol may play an important role in gluconeogenesis. Cortisol is a steroid hormone secreted by the adrenal cortex under catabolic conditions. We might suspect that its concentration would increase under severe or prolonged exercise conditions, particularly in the fasted state. Cortisol has a catabolic effect in that it promotes net protein breakdown in skeletal muscle. In this way, it increases the availability of amino acids to act as gluconeogenic precursors. In the liver, cortisol stimulates the expression of genes coding for gluconeogenic enzymes. If cortisol levels do not increase, the presence of even a low level of cortisol in the blood acts to assist the process of gluconeogenesis; persons with cortisol deficiency can develop hypoglycemia more easily than others.

✓ KEY POINT

Trimmer and colleagues (2002) reported that the maximal capacity of gluconeogenesis to support glucose release from the liver peaks at about 1.5 mg per kg per min. This level of glucose production is insufficient to maintain blood glucose during exercise if there is not a substantial contribution from liver glycogenolysis. Accordingly, blood glucose concentration declines, leading to a decrement in performance. This emphasizes the need for people to eat some hours before prolonged exercise and to support blood glucose through regular carbohydrate intake during the activity.

Control by Altering Gene Expression

Glycolysis and gluconeogenesis share the same equilibrium (reversible) enzymes in the central portion of the glycolytic pathway. Control of gluconeogenesis and glycolysis takes place outside of this central portion through three substrate cycles that regulate the overall rates of glycolysis and gluconeogenesis in the liver (see figure 6.25). These substrate cycles involve the intermediates glucose and glucose 6-P, fructose 6-P, and fructose 1,6-bisphosphate, as well as PEP and pyruvate. If you focus on these substrate cycles, you should be able to see that if the enzymes involved are simultaneously active, the net effect is no flux in either direction, but there would be a loss of energy-rich phosphates. The enzymes involved in these substrate cycles catalyze reactions that are irreversible (nonequilibrium) with large energy changes.

One strategy to ensure that one direction is favored is to alter the expression of the genes for the key reactions in the substrate cycles, shown in figure 6.25. Glycolysis is favored by increased activity of glucokinase, phosphofructokinase-1, and pyruvate kinase via stimulation of the transcription of their genes and the simultaneous depression of transcription of genes for the gluconeogenic enzymes. On the other hand, gluconeogenesis can be favored through increased transcription of glucose 6-phosphatase, fructose 1,6-bisphosphatase, pyruvate carboxylase, and PEPCK, as well as the simultaneous depression of transcription of the three nonequilibrium glycolytic enzymes. The hormones insulin and glucagon play a major role in controlling gene expression, with glucagon promoting gluconeogenic enzymes and insulin promoting glycolytic enzyme formation (glucokinase, PFK-1, and pyruvate kinase) and suppressing expression of genes for PEPCK and glucose 6-phosphatase. Factors that can influence gene expression are also discussed in chapter 3.

The initial stages of type 2 diabetes are accompanied by insulin resistance. This insulin resistance is not just with muscle; insulin's suppressive effect on gluconeogenic enzymes is also reduced, so that activities of glucose 6-phosphatase, fructose 1,6-bisphosphatase, PEPCK, and pyruvate carboxylase are increased. The net effect is increased glucose formation by gluconeogenesis, release of glucose from the liver, and hyperglycemia.

As discussed in detail in chapter 3, a number of signaling pathways are involved in the specific regulatory roles of glucagon and insulin. We discussed the activation of protein kinase A by glucagon earlier. Protein kinase A can phosphorylate a nuclear protein known as *cyclic AMP response element binding protein* (CREB) (figure 6.26). Phosphorylated CREB now binds to a control region of a gene whose protein product is a coactivator in the formation of key liver enzymes. We know this coactivator to be *peroxisome proliferator–activated receptor γ coactivator 1* or PGC-1γ. PGC-1γ now works with a special liver gene transcription regulator known as *hepatocyte nuclear factor-4* (HNF-4). Together, PGC-1γ and HNF-4 upregulate genes that are coding for gluconeogenic enzymes. Although this suggests enormous complexity in the regulation of gene expression, it helps to fine-tune the levels of enzymes that need to respond to the metabolic changes taking place in our bodies over a day.

PGC-1 regulation of expression of liver gluconeogenic enzymes can also be influenced by caloric restriction. You may recall from the Next Stage section in chapter 5 that caloric restriction in animal models can lead to a significant increase in lifespan, due in part to enhanced preservation of mitochondrial function. Among other effects of caloric restriction are the enhancement of expression of gluconeogenic enzymes and the downregulation of expression of glycolytic enzymes. These coincide with the caloric restriction effects on mitochondria (discussed in chapter 5) and are part of a pattern of metabolic changes that also include increased fatty-acid oxidation. New research has demonstrated that caloric restriction in mice will result in

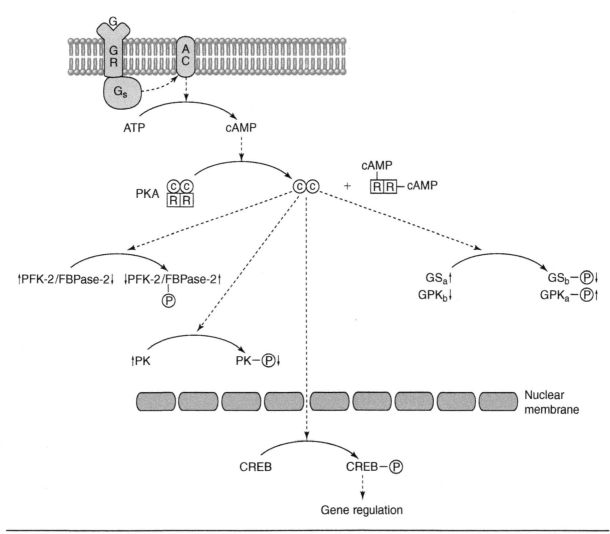

Figure 6.26 Roles of glucagon in the liver. Glucagon (G) binds to its receptor (GR), activating G protein (G_s), which activates adenylyl cyclase (AC). In turn, AC converts ATP into cyclic AMP (cAMP). Protein kinase A is inactive as a tetramer with two catalytic (C) and two regulatory (R) subunits. The binding of cAMP to the R subunits allows the catalytic subunits to phosphorylate protein substrates, changing their activity, shown as active (up arrow) or inactive (down arrow). A key substrate is the enzyme phosphofructo-2-kinase/fructose 2,6-bisphosphatase (PFK-2/FBPase-2), where phosphorylation decreases the kinase activity while increasing the phosphatase activity, thus decreasing the overall concentration of the allosteric regulator fructose 2,6-bisphosphate. Other substrates are pyruvate kinase (PK), glycogen synthase *a* (GSa), phosphorylase kinase *b*, and cyclic AMP response element binding protein (CREB). A circled P indicates the attached phosphate.

increased levels of an important regulator of lifespan in mammals, the histone deacetylase silent information regulator 1, or SIRT1, which can influence the activation of a number of important signaling pathways. SIRT1 will be induced in liver in response to caloric restriction. It acts by deacetylation (removal of an acetyl group) from the PGC-1 molecules. This will result in the PGC-1 signaling of the increased expression of liver gluconeogenic enzymes (Rodgers et al. 2005). These mechanisms are also likely to be important in regulating human adaptations to changes in diet and caloric intake. Evolutionarily, they probably help maintain glucose availability by limiting glycolysis, enhancing gluconeogenesis, and upregulating fatty-acid metabolism in times of limited availability of food.

Control of Enzyme Activity

Control of gluconeogenesis and glycolysis through regulation of the transcription of genes for key enzymes plays a powerful role. However, there is a considerable lag in time before changing the expression of specific genes can lead to significant changes in the amounts of functional enzymes. Therefore, rapid-acting mechanisms must be available to control glycolysis and gluconeogenesis.

Pyruvate kinase, the nonequilibrium glycolytic enzyme, can be inhibited by phosphorylation. The increase in glucagon under conditions favoring hepatic glucose output increases the activity of protein kinase A through an increase in cAMP. Among its many substrates (e.g., CREB), protein kinase A phosphorylates *pyruvate kinase*, leading to its inhibition (figure 6.26). This effectively blocks glycolysis at the end of the pathway yet has no influence on gluconeogenesis. A glucagon-induced rise in hepatic-cell cAMP concentration results in the phosphorylation of another remarkable bifunctional enzyme known as *6-phosphofructo-2-kinase/ fructose 2,6-bisphosphatase* (PFK-2/FBPase-2). On this single polypeptide protein, there are two distinct enzyme activities. The PFK-2 part is responsible for converting fructose 6-P to fructose 2,6-bisphosphate ($F2,6-P_2$). We discussed $F2,6-P_2$ earlier in the context of the regulation of glycolysis. $F2,6-P_2$ is a potent activator of PFK-1, which is the glycolytic enzyme. As well as activating glycolysis, $F2,6-P_2$ allosterically inhibits fructose 1,6-bisphosphatase. The other half of this bifunctional enzyme, FBPase-2, breaks down fructose 2,6-bisphosphate to fructose 6-P. Here is where glucagon comes in. Phosphorylation of PFK-2/FBPase-2 at a single serine residue results in the instantaneous inhibition of the kinase part of the bifunctional enzyme and activation of the phosphatase part with the result that the concentration of fructose 2,6-bisphosphate falls. The net effect of the phosphorylation of PFK-2/FBPase-2 is to inhibit glycolysis and allow gluconeogenesis. In summary, phosphorylation of two key enzymes by a cAMP-dependent protein kinase A acts in switchlike fashion to inhibit glycolysis and to activate gluconeogenesis. This is summarized in figure 6.27.

Under conditions in which gluconeogenesis needs to be stimulated, the acetyl CoA level in the liver increases. This arises from an increase in beta-oxidation of fatty acids, a topic covered in the next chapter. Acetyl CoA from fatty acids is a substrate for the citric acid cycle. Acetyl CoA can also act as an allosteric effector in two ways. First, an increase in cytosolic acetyl CoA inhibits pyruvate kinase at a negative allosteric site. Secondly, acetyl CoA is an obligatory activator for pyruvate carboxylase, which carboxylates pyruvate to make oxaloacetate. The rise in acetyl CoA signals that there is plenty of fuel for the citric acid cycle, so carbon should be shunted to make glucose.

✓ KEY POINT

From a biochemical perspective, the brain could be characterized as the body's most important tissue. It is hard to argue with this statement, given the complex yet complementary mechanisms that regulate the level of glucose in the blood to act primarily as a fuel for the brain.

PENTOSE PHOSPHATE PATHWAY

As the next chapter discusses, most cells require a constant source of reducing equivalents for biosynthesis. These reducing equivalents are found in **NADPH**, a molecule differing from NADH in the fact that the former contains an additional phosphate group. As we have seen, electrons on NADH are used in the electron transport chain to reduce oxygen and generate ATP. Electrons on NADPH are used to synthesize new molecules, such as fatty acids, amino acids, and steroids. The most common name for the process of generating NADPH is the pentose phosphate pathway, but it is also known as the hexose monophosphate shunt or the 6-phosphogluconate pathway. Besides NADPH, the pentose phosphate pathway generates another important substance used in biosynthetic reactions. Ribose 5-phosphate is

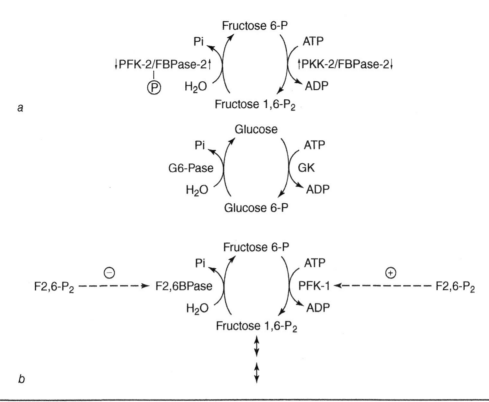

Figure 6.27 The regulation of glycolysis and gluconeogenesis in liver. (a) The concentration of the allosteric effector, fructose 2,6-bisphosphate (F2,6-P$_2$), is determined by the relative activity of the kinase and phosphatase activities of the bifunctional enzyme phosphofructo-2-kinase/fructose 2,6-bisphosphatase (PFK-2/FBPase-2). Phosphorylation of PFK-2/FBPase-2 by protein kinase A inhibits the kinase part while increasing the phosphatase part of the activity, leading to a decrease in F2,6-P$_2$ concentration. (b) The role of F2,6-P$_2$ in increasing PFK-1 activity while decreasing the activity of F2,6BPase. The net effect of an increase in F2,6-P$_2$ in liver is to increase glycolysis while simultaneously decreasing gluconeogenesis. PFK-1 is the glycolytic enzyme phosphofructokinase-1. F2,6BPase is the gluconeogenic enzyme fructose 2,6-bisphosphatase.

used in nucleotide biosynthesis and, thus, is needed to make ATP, DNA, RNA, FAD, coenzyme A, NAD$^+$, and NADP$^+$.

Figure 6.28 summarizes in full structural detail the major reactions of the pentose phosphate pathway as they take place in tissues such as the liver, adipose tissue, adrenals, testes, and ovaries, but not in skeletal muscle. The reaction begins with glucose 6-P, which is also a beginning point for glycolysis and glycogen synthesis. However, there are huge differences. In the pentose phosphate pathway, the glucose 6-P first undergoes an oxidation, catalyzed by *glucose 6-phosphate dehydrogenase*, generating NADPH plus H$^+$. This first reaction is the key reaction in the pentose phosphate pathway, and it is under tight regulation by one of its products, NADPH. NADPH is also formed in the reaction catalyzed by *6-phosphogluconate dehydrogenase*. The pentose phosphate pathway is regulated entirely at the first step. NADPH is an allosteric inhibitor for glucose 6-phosphate dehydrogenase. When the ratio of the reduced to oxidized form of this coenzyme is elevated (NADPH/NADP$^+$), glucose 6-phosphate dehydrogenase is inhibited.

In many tissues, the pentose phosphate pathway stops at ribose 5-phosphate, having generated sufficient NADPH for biosynthetic reactions and sufficient ribose 5-phosphate for nucleoside and coenzyme synthesis. This is known as the *oxidative stage* of the pentose phosphate pathway. In some cells, the need for NADPH exceeds that for ribose 5-phosphate. In these instances, the *nonoxidative* part of the pathway can continue in a different direction from ribulose 5-phosphate. In this portion of the pathway, three ribulose 5-phosphates can be converted into two fructose 6-phosphates and one glyceraldehyde 3-phosphate. In this way, the carbons in ribulose 5-phosphate are conserved as glucose precursors, or the fructose 6-P and glyceraldehyde 3-phosphate can enter the glycolytic pathway for pyruvate formation. If we include both the oxidative and nonoxidative portions of the pathway and

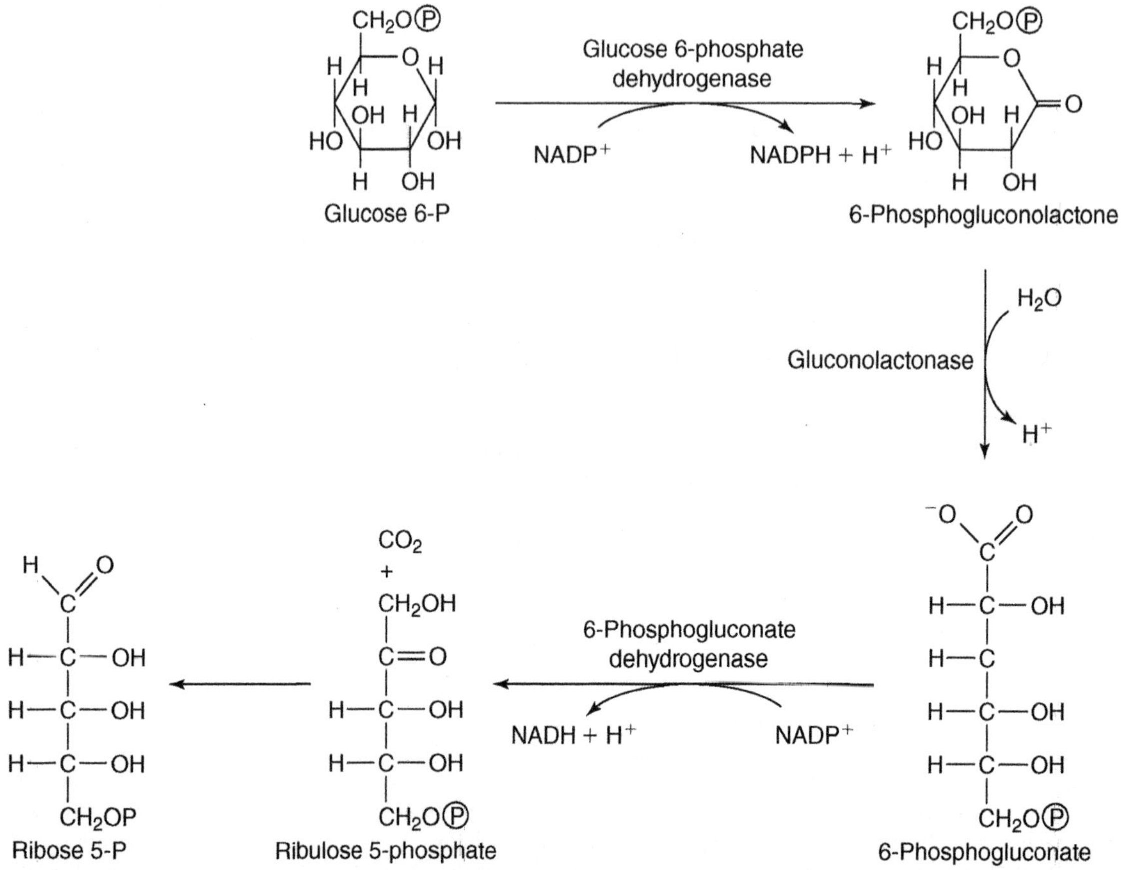

Figure 6.28 The pentose phosphate pathway (hexose monophosphate shunt), which can be divided into oxidative and nonoxidative stages. Detailed reactions reflecting the oxidative portion of the pathway are shown. Beginning with glucose 6-phosphate, this portion of the pathway generates two NADPH for biosynthesis and ribose 5-phosphate for nucleoside and coenzyme synthesis.

remember that we lose one CO_2 for each glucose 6-P we start out with, we could summarize the pentose phosphate pathway as follows, starting with six molecules of glucose 6-P:

6 glucose 6-P + 12 $NADP^+$ + 7 H_2O
→ 5 glucose 6-P + 12 NADPH + 12 H^+ + Pi

SIGNALING PATHWAYS

Chapters 2 and 3 introduce the concept of signal transduction, the mechanism by which external signals are communicated to and within cells in order to alter their metabolism. The preponderance of signaling pathways involve phosphorylation of proteins by protein kinases and removal of the attached phosphates by phosphoprotein phosphatases. To date, we know of at least 500 genes in humans that code for protein kinases (Hardie 2004). In our discussion of carbohydrate so far, we have noted specific signaling mechanisms related to control of glycogenolysis and gluconeogenesis. In this section, we investigate in more detail how two signal transduction pathways affect carbohydrate metabolism.

AMP-Activated Protein Kinase

One protein kinase, activated by 5'-AMP, plays a major role in metabolic regulation in a host of cell types. Its activity can be affected by nutrient supply and stress. It can influence the metabolism of carbohydrate, fat, and protein, as well as ion channels and cell survival. It occupies a special niche in metabolism. This protein kinase is introduced in chapter 3, but we will look at it in more detail here and will also refer to it in chapter 7.

Adenosine monophosphate–activated protein kinase, or AMPK, consists of three different kinds of protein subunits. The α subunit is the catalytic subunit; it acts to transfer a phosphate group from ATP to a protein substrate. The β and γ subunits play a regulatory role. In all tissues surveyed to date, there are two α subunits (α_1 and α_2), two β subunits (β_1 and β_2), and three γ subunits (γ_1, γ_2, and γ_3). As the name implies, regardless of the subunit composition (e.g., $\alpha_2\beta_2\gamma_2$, common in skeletal muscle), the protein kinase activity is increased when AMP binds to the γ subunit. The enzyme can also be activated when it is phosphorylated on its α subunit by the enzyme *AMP kinase kinase* or AMPKK. AMPKK is also activated by AMP. Adenosine monophosphate binding to AMPK makes it a better substrate for phosphorylation by AMPKK, thus reinforcing the activity of AMPK. While AMPK activity can be stimulated by about 10-fold with an increase in AMP concentration, phosphorylation by AMPKK can increase AMPK activity nearly 100-fold.

In skeletal muscle, the increase in [AMP] is brought about by *adenylate kinase*, which converts two ADP into an AMP and an ATP. AMPK is inhibited by PCr and ATP. Adenosine triphosphate is a substrate for AMPK. It also appears to competitively inhibit AMP from binding to its allosteric site unless the latter increases significantly. Thus, the activation state of AMPK reflects the energy status of the cell. At rest, with [AMP] low and [PCr] and [ATP] elevated, we would expect the activity of AMPK to be low. With the onset of muscle activity, and in a manner directly reflecting the intensity of contractile activity, the concentration of AMP rises; ATP and especially PCr fall, making AMPK activity proportional to the metabolic stress induced by the contractile activity. AMPK activation is less when muscle glycogen stores are elevated, suggesting that a site exists, possibly on subunit β, where glycogen can bind to inhibit the AMPK. Researchers have discovered that AMPK and AMPKK can be activated artificially by a molecule referred to as 5-amino-imidazole-4-carboxamide-riboside (AICAR). When given to animals, AICAR is taken up by cells and phosphorylated to make an analog of AMP, known as ZMP. ZMP is a powerful activator of AMPK and AMPKK. Studies using it have led to a great understanding of the array of effects that AMPK has in cells and whole organisms. Regulation of AMPK is summarized in figure 6.29.

AMPK has other important regulatory roles in skeletal muscle. When active, it enhances glucose transport by increasing the translocation of GLUT-4 transporters to the cell membrane. The mechanism for enhancing glucose transport is not the same as that caused by insulin, as we will discuss next. Active AMPK also increases the oxidation of fatty acids in muscle, as covered in the next chapter. AMPK increases the content of fructose 2,6-bisphos-

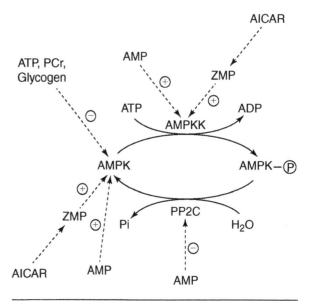

Figure 6.29 Activation of adenosine monophosphate–activated protein kinase (AMPK). AMPK can be activated allosterically by AMP or artificially by AICAR, which produces ZMP, an AMP analog. ATP, PCr, and glycogen all have a negative effect on AMPK activity. Phosphorylation of AMPK by AMPK kinase (AMPKK) results in the most active form of the enzyme. AMPKK is also activated by AMP. Removal of the phosphate group on AMPK is catalyzed by phosphoprotein phosphatases, such as PP2C.

sphate, thus increasing glycolysis. AMPK activity is also reported to be sensitive to endurance training, with increases in total AMPK activity and alterations in the subunit composition (Frøsig et al. 2004). In summary, AMPK activation parallels the cells' need to make ATP, thereby reinforcing some of the mechanisms we have already discussed. The drug metformin, one of the most widely prescribed drugs for type 2 diabetes mellitus, increases AMPK activity.

Metabolic stress induced by endurance exercise training is also an important regulator of AMPK. Along with several other signaling pathways, AMPK is an important regulator of the gene expression of slow-twitch muscle fibers, muscle fiber-type transformation, mitochondrial biogenesis, and the change to a more oxidative and carbohydrate-conserving metabolic profile, which is induced by endurance training (Yan et al. 2011). AMPK controls the expression of genes involved in oxidative and carbohydrate metabolism in conjunction with SIRT1. It can also partially regulate the activation of SIRT1 (Canto et al. 2009). These and other effects of exercise on signaling pathways are discussed in chapter 3.

Two Mechanisms for Regulating Glucose Uptake

In the beginning of this chapter, we discussed the necessity to regulate glucose uptake in skeletal muscle, adipose tissue, and cardiac muscle. This regulation is based on the availability of GLUT-4 transporters in the cell membrane to transport glucose into the cell down its concentration gradient. In skeletal muscle, the number of GLUT-4 transporters active in transporting glucose into fibers is regulated both by insulin and by the contractile process itself. These operate by two different mechanisms, as briefly reviewed next.

Role of Insulin

Insulin has potent metabolic effects in a host of tissues, including skeletal muscle. It has a predominantly anabolic role, stimulating glucose uptake and glycogen storage, fat synthesis and storage, and protein synthesis. Here, we briefly review the role of insulin in stimulating glucose uptake in skeletal muscle.

The insulin receptor in the cell membrane consists of four subunits. Two α subunits protrude into the extracellular space and are responsible for insulin binding. When insulin binds to the α subunits, the two intracellular β subunits become active as a **protein tyrosine kinase**, such that they phosphorylate tyrosine residues on their own polypeptides (autophosphorylation). Once phosphorylated, the β subunits become active in phosphorylating tyrosine residues on other intracellular proteins, known as *insulin receptor substrates* (IRS). At least four of these exist, but IRS-1 is principally involved in insulin signaling in skeletal muscle. Phosphorylated IRS-1 becomes a docking site for an enzyme known as phosphatidylinositol 3-kinase, or PI3K, which becomes active, creating a molecule known as phosphatidylinositol 3,4,5-trisphosphate (PI3,4,5-P_3). The PI3,4,5-P_3 activates two other protein kinases, protein kinase B, more commonly known as Akt, and an atypical member of the *protein kinase C* family, PKCξ. Both of these protein kinases can phosphorylate the protein AS160, which leads to translocation of GLUT-4 transporters from intracellular storage sites to the sarcolemma. The overall process of insulin-regulated glucose uptake is summarized in figure 6.30.

Role of Exercise

Exercising skeletal muscles can take up glucose in an insulin-independent way. As we have mentioned, exercise leads to a decrease in blood insulin con-

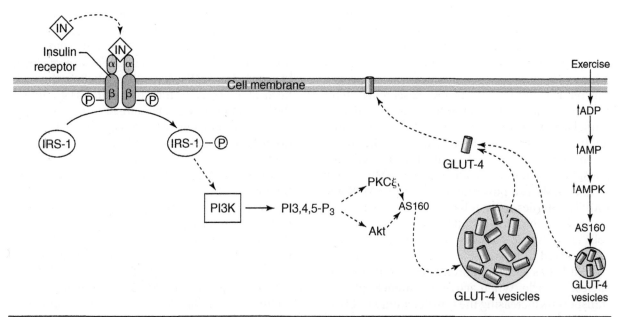

Figure 6.30 The role of insulin signaling in glucose transport in a muscle cell. Insulin (IN) binds to its receptor, consisting of two extracellular subunits, inducing tyrosine autophosphorylation in the two intracellular β subunits. This activates a protein kinase activity in the subunits that phosphorylate insulin receptor substrate-1 (IRS-1), which now provides a docking site for phosphatidylinositol 3-kinase (PI3K). PI3K produces phosphatidylinositol 3,4,5-trisphosphate (PI3,4,5-P_3), which leads to activation of two additional protein kinases, Akt and protein kinase Cξ, and release of GLUT-4 transporters from intracellular vesicles. The GLUT-4 transporters embed in the cell membrane, allowing glucose to enter the cell. Circled P represents phosphate groups, and dotted arrows show the flow of the signaling steps.

centration. Furthermore, the exercise effect does not operate through IRS-1 or PI3K, suggesting that the mechanism for the exercise effect must operate in a different way from the insulin-dependent pathway. Indeed, the effect of exercise is not just during the actual contractile activity but extends to at least several hours after exercise. Although much is unknown, and the detailed mechanism awaits additional research, we do know some factors. First, exercise leads to an increase in GLUT-4 transporters in the sarcolemma, but these are mobilized from a different storage site than those of the insulin-dependent pathway. Second, AMPK is involved. This is activated by exercise in an intensity-dependent fashion because the level of AMP reflects the increase in ADP, which depends on the rate of ATP hydrolysis. We now know that AS160 can also be phosphorylated by AMPK. Therefore, although insulin activates the PI3K pathway and muscle contraction activates the AMPK pathway, AS160 is a downstream target of both pathways and is involved in activating glucose uptake in response to both signals (see figure 6.30).

As previously noted, regular endurance training results in downregulation of glycolytic capacity, partially as a consequence of exercise induction of AMPK signaling pathways. Some evidence exists that short-term endurance training with untrained subjects in conditions of low levels of muscle glycogen may enhance the rate of mitochondrial biogenesis more than when training with normal muscle glycogen levels (Hawley and Burke 2010). This effect may in part be mediated by a greater induction of AMPK signaling in this metabolic environment. However, the longer term limitations to training capacity by chronic low levels of muscle glycogen and associated fatigue would preclude its general use in training of athletes. Strength training with low muscle glycogen levels is generally reported to have negative effects on adaptation to that form of training.

NEXT STAGE

Carbohydrate, Exercise Performance, and Fatigue

At various points in this chapter, we have noted the correlation of muscle glycogen depletion with the onset of fatigue, as well as the ability of glucose ingestion to delay the onset of fatigue during exercise. These relationships have been well documented by exercise scientists for more than 40 years. As well as improvements in duration of endurance exercise and total work, carbohydrate ingestion may also enhance repeated high-power work output as well as performance in multiple bouts of resistance exercise (Karelis et al. 2010). The most research, however, has been done demonstrating the effects of carbohydrate supplementation on improving maximal endurance performance of at least 45 min.

Despite these established associations, the exact mechanisms by which carbohydrate ingestion delays muscle fatigue are still not fully established. As we have previously noted, fatigue is a complex and multifaceted occurrence that is influenced by a myriad of events during different intensities and durations of exercise. The maintenance of blood glucose levels and muscle carbohydrate oxidation, or glycogen stores by carbohydrate administration, may influence a number of these factors. A review by Karelis and colleagues (2010) suggests that carbohydrate administration during exercise may influence at least three important factors that result in delayed fatigue and enhanced work capacity: (1) enhanced neural drive and attenuation of central fatigue, (2) protection from disruption of muscle cell homeostasis and integrity, and (3) maintenance of sodium–potassium pump ATPase activity.

Although experimental evidence of the effects of hypoglycemia on exercise performance is conflicting, maintenance of carbohydrate availability to the brain and central nervous system, which, as we have noted earlier, are highly reliant on carbohydrate metabolism, could enhance and maintain the central nervous system's function and ability to continue to fully activate muscle contraction. In support of this, several studies have found that when assessing the total amount of work accomplished by elite athletes over a 1 h period, performance when fasted is improved by both carbohydrate ingestion and the sipping and spitting out of a carbohydrate beverage. This suggests that sensors in the mouth can detect carbohydrate and signal its imminent arrival to the brain. The brain, which would be starting to invoke fatigue mechanisms to prevent the imminent depletion of carbohydrate and blood glucose levels, delays this intervention in the expectation of more carbohydrate; therefore, it allows exercise to continue for a period of time at a high intensity, delaying "central fatigue" (Karelis et al. 2010). These findings suggest that in some cases, actual depletion of carbohydrate and glycogen stores are not the primary cause of fatigue, but that central fatigue can occur to prevent their imminent depletion.

Studies also suggest that carbohydrate ingestion may be effective in attenuation of exercise-induced immune suppression, synthesis of heat shock proteins, oxidative stress, and positive regulators of inflammation, such as inflammatory cytokines and cortisol (Karelis et al. 2010). These effects may delay fatigue by attenuating some of the acute negative effects of inflammation-associated muscle disruption. Other studies have suggested that the sodium–potassium ATPase, which hydrolyzes ATP to maintain muscle membrane polarization during muscle contraction and relaxation, may be inhibited when the ability to maintain ATP homeostasis during exercise is compromised. Carbohydrate administration may be able to delay this by providing a source for muscle metabolism when muscle and liver glycogen levels are being depleted (Karelis et al. 2010). Other possible mechanisms that may delay fatigue onset during endurance exercise in association with carbohydrate administration include the maintenance of calcium homeostasis and SERCA ATPase activity, maintenance of the rate of ATP production via glycolysis and carbohydrate oxidation that may not be matched by aerobic metabolism of fatty acids, and the maintenance of appropriate levels of other high-energy-related metabolic intermediates, including creatine phosphate, IMP, and inorganic phosphate levels. However, Karelis and colleagues (2010) concluded in their review that these possible mechanisms have less empirical support.

It is well recognized that one of the main purposes of muscle fatigue during exercise is to downregulate muscular contraction and metabolic demand for ATP in order to preserve muscle ATP levels and energy status before they can drop to catastrophic levels. Enhanced carbohydrate availability during endurance exercise may be able, in a number of ways, to prolong ATP synthesis rates for a period of time, thus delaying the need for muscle contraction to be attenuated by fatigue mechanisms. The review by Karelis and colleagues (2010) highlighted the myriad of possible factors involved in maintaining and limiting muscular activity and physical performance during endurance exercise, as well as the extent to which current research findings support or refute some of these mechanisms as being critically influenced by carbohydrate administration. The fact that significant amounts of conflicting findings and uncertainty still exist with regard to how carbohydrate ingestion and maintenance of muscle glycogen stores attenuate muscle fatigue highlights the future challenges of research into the biochemical and physiological mechanisms associated with carbohydrate metabolism and the range of physiological functions they influence. Karelis and colleagues (2010) conclude their review by suggesting that once mechanisms are more clearly identified and the causes of muscle and central fatigue that can be influenced by carbohydrate metabolism and ingestion have been unequivocally demonstrated, this knowledge could have important implications for enhancing exercise performance, particularly in elite athletes.

SUMMARY

Most dietary carbohydrate is broken down in the intestinal tract to glucose, fructose, and galactose, but glucose is by far the most important carbohydrate. In the form of blood glucose and liver and muscle glycogen, body carbohydrate is a major fuel, particularly for the brain. Blood glucose is carefully regulated so that its concentration is approximately 5 to 6 mM (90-110 mg per 100 ml of blood). Glucose from the blood enters cells down a concentration gradient through a specific glucose transporter. The GLUT-4 transporter is found in skeletal and cardiac muscle and fat cell membranes, and its content in these membranes is greatly increased by insulin. Exercise also increases the uptake of glucose into muscle. The principal metabolic pathway for breaking down carbohydrate is glycolysis. Pyruvate, a product of glycolysis, has two major fates. It can be reduced to lactate by the enzyme lactate dehydrogenase, or it can enter the mitochondrion to be oxidized to acetyl CoA to enter the citric acid cycle.

Glycolysis is a carefully regulated process, with most of the control exerted by the enzyme phosphofructokinase (PFK). This enzyme is sensitive to the binding of a number of positive and negative modulators to allosteric sites. Rest or low-level muscle activity is associated with a low activity of glycolysis, whereas vigorous exercise greatly increases its rate in order to produce ATP. When pyruvate is oxidized in the mitochondrion and not reduced to lactate during glycolysis, NADH can potentially build up to the extent that there is too little NAD^+ to permit glycolysis to continue. Oxidation of cytoplasmic NADH, other than by lactate dehydrogenase, occurs through the action of two shuttle systems. The glycerol-phosphate shuttle and the malate–aspartate shuttle transfer electrons from the cytosol to the mitochondria for use by the electron transport chain. The production of lactate by the LDH reaction during glycolysis allows for the regeneration of NAD^+, leading to the continuation of glycolysis and consequent rapid ATP regeneration required to support intense exercise. Thus, the production of lactate facilitates rather than hinders the continuation of intense exercise. In addition, the production of lactate is coincident with a drop in pH during high-intensity exercise; it is not its cause.

Glycogen, the storage form of glucose in liver and muscle tissue, is synthesized from glucose taken up from the blood in a process that is greatly accelerated following a carbohydrate-rich meal. Liver glycogen is a reservoir of glucose for the blood. In muscle, the glycogen is ready to be fed into the glycolytic pathway when the muscle fiber becomes active. Since glycogen synthesis and glycogenolysis are opposing processes taking place in the cytosol of cells, one process must be inactive when the other is active in order to avoid a futile cycle. Two regulatory enzymes, glycogen phosphorylase and glycogen synthase, are thus affected in opposite ways by the same signal, which leads to their simultaneous phosphorylation. In addition to covalent modification, the activities of both synthase and phosphorylase are influenced by the binding of regulatory molecules. Following a meal, when blood insulin concentration is increased, the synthase is active and the phosphorylase inactive in both liver and muscle. Between meals, liver phosphorylase is activated by a rise in glucagon and a fall in insulin so that the glycogen is slowly broken down to glucose. During exercise, when epinephrine concentration is increased, muscle phosphorylase is active and synthase is inhibited. The tight regulation of glycogen metabolism emphasizes the importance of controlling body carbohydrate stores.

When body carbohydrate content is decreased and glucose output from the liver is threatened, glucose is made from a variety of noncarbohydrate sources, such as lactate, pyruvate, and the carbon skeletons of amino acids. This process, gluconeogenesis, takes place in liver and kidney and involves the reversible reactions of glycolysis. New reactions that bypass irreversible glycolytic reactions are catalyzed by pyruvate carboxylase, phosphoenolpyruvate carboxykinase, fructose 1,6-bisphosphatase, and glucose 6-phosphatase. These reactions allow the liver to produce glucose during times of need while glycolysis is virtually shut down. The pancreatic hormone glucagon promotes gluconeogenesis. Glycerol, produced when triglycerides are broken down, is an important gluconeogenic precursor in liver. In addition, when the amino group of any of the 20 common amino acids is removed, the carbon skeletons of 18 of these can be converted to glucose. The fact that skeletal muscle can be severely wasted during diets providing inadequate food energy demonstrates the importance of the brain as a tissue; the brain's need for a continuous supply of glucose requires the body to derive glucose from amino acids obtained from muscle proteins.

The pentose phosphate pathway is not important in muscle. However, in breaking down glucose 6-P in a variety of tissues, this pathway produces reducing equivalents in the form of NADPH and ribose 5-phosphate, a precursor for making nucleotides. Carbohydrate and lipid metabolism are regulated by the action of signaling systems that operate through protein kinases and phosphatases to alter enzyme activity within the cell. The AMP-activated protein kinase system is stimulated under conditions in which ATP provision is threatened. Thus, exercise is a potent stimulus for AMP kinase. Insulin operates by a signal transduction pathway, involving insulin receptor substrates and the phosphatidylinositol 3-kinase activation. The net effect of insulin on muscle is to increase GLUT-4 transporters in muscle cell membranes. Muscle activity per se also increases GLUT-4 membrane transporters, likely through an AMP kinase pathway as well as other signaling pathways.

REVIEW QUESTIONS

1. Stored fat is considered an ideal fuel from an efficiency perspective because, being stored in an anhydrous form, it provides about 9 kcal (38 kJ) per gram, as opposed to 4 kcal (17 kJ) per gram for glycogen. From the same perspective, we could say that glycogen is

not an efficient fuel because it is stored with water and has a lower fuel value. Calculate the approximate fuel value of glycogen in kilocalories and kilojoules per gram weight as stored, assuming that each gram of stored glycogen contains 3 g of associated water.

2. We have identified a variety of exercise and competition activities in which elevations in muscle glycogen concentration can enhance performance. Can you think of activities in which glycogen is an essential fuel but where supercompensated levels may be detrimental to performance?

3. The concentrations of glucose and lactate in the blood may be described in units such as millimoles or in units of milligrams per deciliter of blood. If the molecular weight of glucose is 180 and that of lactate is 90, convert the following concentrations of glucose and lactate to millimolar (mM) units: lactate concentration, 990 mg/dL; blood glucose concentration, 108 mg/dL.

4. From RER and other measures, it is determined that a distance runner is oxidizing carbohydrate at the rate of 4 g per minute and fat at the rate of 0.5 g per minute. If he is running at 4 min per kilometer, what percentage of his fuel is derived from fat if his total energy expenditure is 3.43 megajoules (MJ)? How far does he run?

5. A muscle glycogen concentration of 500 mmol of glucosyl units per kilogram of dry muscle weight is considered high. Convert this into millimoles of glucosyl units per kilogram wet muscle weight and into millimolar units using the relationship that a muscle sample as removed from a biopsy needle contains 77% water. Also, the proportion of intracellular water is 0.7 of the total muscle weight.

6. MCT-1 is found more in slow, high oxidative fibers and MCT-4 is found more in Type II fibers. If you knew that the K_m values of these transporters for lactate differed by a factor of 5, which fiber type would have the lower K_m transporter? Give a rationale for your answer.

7. Lactate has had a bad reputation in exercise. List the many ways in which lactate production can be beneficial for exercise performance and as a fuel source.

chapter 7

Lipid Metabolism

Most people are well aware that fats (or lipids) are very important fuels in the body. Most have probably heard that too much fat in the diet is harmful and that storage of excess lipid is a major health concern. Many may also know that lipids are not soluble in water. Few are aware, however, that lipids are also important structural elements and are involved in cell signaling. Fat is stored as **triacylglycerol** (triglyceride) in fat cells. Scientists describe the tissue containing fat cells as **adipose tissue**, usually white adipose tissue (WAT), to distinguish it from the brown adipose tissue (BAT) used as a heat source in small animals and infant humans, as well as human adults. Stored fat is considered a long-term energy store, in contrast with glycogen, which is considered a short-term energy store.

To understand the importance of lipids, one must learn about the various kinds of lipids and their metabolism. We will examine the storage of fat and how it and ketone bodies are oxidized to make ATP. We will review how humans can synthesize new fat. Fat stored in adipose tissue and muscle plays a key role in exercise metabolism. Moreover, competition exists between the oxidation of fat and of carbohydrate, and we will address how this is regulated. We will also look at the ability of fat tissue to secrete key regulatory peptides that act like hormones. The links between obesity (fat overload) and the development of insulin resistance in peripheral tissues such as liver and skeletal muscle will also be discussed. The chapter also includes a discussion of cholesterol and concludes with a Next Stage section dealing with the relationships among exercise, fat oxidation, and weight loss.

TYPES OF LIPIDS

Lipids are categorized as fatty acids, triacylglycerols, phospholipids, and sterols, whose roles in the body are largely determined by their chemical structure. We will address the role of cholesterol, the dominant sterol, at the end of this chapter. Lipids are hydrophobic, containing mostly nonpolar components. Some are amphipathic, meaning that they have some polar groups, although they are largely nonpolar. We have already considered fatty acids in chapter 5, where we noted that the beta-oxidation of fatty acids provides acetyl CoA for the citric acid cycle. Fatty acids are therefore important fuel molecules.

Fatty Acids

Fatty acids are also sometimes known as **free fatty acids** (FFAs) or non-esterified fatty acids (NEFAs). They are carboxylic acids, containing a long alkyl chain with a carboxylic-acid group at one end. Of the fatty acids in our body and food, the majority contain at least 16 carbon atoms in total. This means they have a long hydrocarbon tail and a carboxyl group for the head. Short-chain fatty acids are generally 4 to 6 carbon atoms in length, medium-chain fatty acids are 8 to 12 carbon atoms in length, and long-chain fatty acids have 14 or more carbon atoms. Fatty acids are usually known by their common or trivial names. Normally, they have an even number of carbon atoms. They can be saturated, with no carbon-to-carbon double bonds, or unsaturated, with one or more carbon-to-carbon double bonds. Figure 7.1 gives examples of a saturated and an unsaturated fatty acid. Because the fatty acids have pK_a values in the range 4.5 to 5.0, making them very weak acids, they exist predominantly as **anions**, or negatively charged ions, at neutral pH. Thus, the fatty acids in figure 7.1 are named palmitate (from palmitic acid) and oleate (from oleic acid).

A shorthand notation for describing fatty acid structure has been devised. The first number represents the number of carbon atoms in the fatty

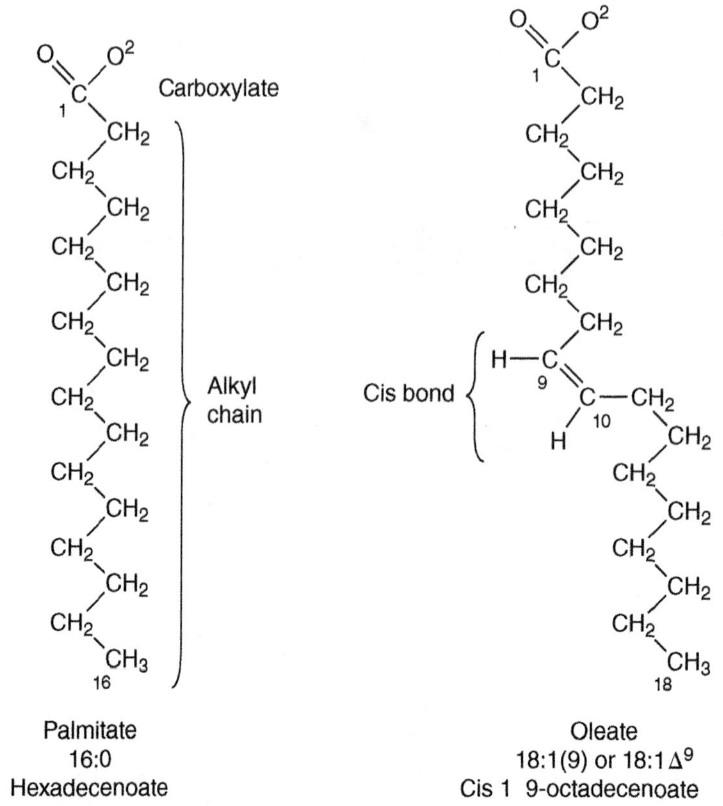

Figure 7.1 Detailed structures for the saturated fatty acid palmitic acid and the unsaturated fatty acid oleic acid. Both are named as anions because this is how they would exist at pH 7. The carboxylate is carbon atom 1, and other key carbon atoms are also shown with small numbers. The location of the double bond in oleate begins at carbon 9. Beneath each structure is the shorthand way it can be described.

acid. This is followed by a colon and a number that identifies the number of double bonds. After this, the actual location of the double bonds is shown either in parentheses or using the Greek letter Δ, with one or more superscript numbers to show the location of the double bonds. For example, linoleic acid is 18:2 (9, 12), or 18:2 $\Delta^{9,12}$ if we are using the system with the Greek letters. The configuration of the double bonds in naturally occurring fatty acids is almost always cis, as shown in figure 7.1. If a trans double bond occurs, it must be identified. Trans double bonds can be introduced in the manufacturing process, in which unsaturated fatty acids are hydrogenated, to make them more solid and less liquidlike. We know that these trans fatty acids are not good for us.

Most of the fatty acids in the body come from our diet, since we have a limited capacity to synthesize fatty acids. From a human nutrition perspective, the three most common fatty acids are palmitic acid (palmitate at pH 7.0), oleic acid (oleate), and stearic acid (stearate). Fatty acids also are classified into families depending on the location of the last double bond and the number of carbon atoms. As shown in table 7.1, α-linolenic acid is [18:3 (9, 12, 15)] and is called an omega-3 (ω-3) or n-3 fatty acid; that is, the last double bond begins three carbons from the end carbon. Note that γ-linolenic acid is an ω-6, or n-6, fatty acid. The n-3 (or omega-three) fatty acids are reputed to offer many health benefits, helping to lower bad blood lipid concentrations. They may also help our nervous systems. Linoleic acid and α-linolenic acid are **essential fatty acids** that we must get in the diet. They are also known as **polyunsaturated fatty acids**, or PUFAs, with two (or more) double bonds. Note that linoleic and α-linolenic acid represent the ω-6 and ω-3 families. The brain is particularly rich in PUFAs. Oleic acid, a monounsaturated fatty acid, is not considered an essential fatty acid, but a diet rich in this fatty acid also confers important health benefits. Flaxseed and flaxseed oil, as well as the oils from fish found in cold waters, are rich sources of PUFAs. The omega-3 PUFAs found in fish oils, particularly eicosapentaenoic acid (EPA) and docosahexaenoic acid (DHA), when included in quantity in the diet through cold-water fish or dietary supplements,

Table 7.1 Common Fatty Acids of Biological Significance

Common name	Systematic name	Shorthand notation	Family
Stearic acid	Octadecenoic acid	18:0	NA
Linoleic acid	9, 12-Octadecadienoic acid	18:2 (9, 12)	ω-9 or n-9
α-Linolenic acid	9, 12, 15-Octadecatrienoic acid	18:3 (9, 12, 15)	ω-3 or n-3
γ-Linolenic acid	6, 9, 12-Octadecatrienoic acid	18:3 (6, 9, 12)	ω-6 or n-6
Arachidonic acid	5, 8, 11, 14-Eicosatetraenoic acid	20:4 (5, 8, 11, 14)	ω-6 or n-6
Timnodonic acid	5, 8, 11, 14, 17-Eicosapentaenoic acid	20:5 (5, 8, 11, 14, 17)	ω-3 or n-3
Cervonic acid	4, 7, 10, 13, 16, 19-Docosahexaenoic acid	22:6 (4, 7, 10, 13, 16, 19)	ω-3 or n-3

have been strongly associated with reduced risks of cardiovascular disease (Wang et al. 2006). The other PUFAs, including those found in flaxseeds, are not thought to convey these specific health benefits to as great an extent.

The essential fatty acids and PUFAs are important components of phospholipids, found in cell membranes. They are also precursors for the synthesis of a family of hormonelike compounds known as the eicosanoids. Eicosanoids are synthesized in cells from the fatty acid arachidonic acid [20:4 (5, 8, 11, 14)]. The eicosanoids are known as local hormones because they are released from cells where they may have an effect on the secreting cell itself (autocrine effect) or on the cells in the immediate environment of the secreting cell (paracrine effect). Examples of eicosanoids are prostaglandins, thromboxanes, and leukotrienes. EPA and DHA are precursors to specific anti-inflammatory eicosanoids; when their concentration is elevated in cell membranes due to dietary supplementation, an environment that is more anti-inflammatory is created. This anti-inflammation promotion by EPA and DHA is thought to contribute to their attenuating influence on cardiovascular disease risks.

Triacylglycerols

Although the term *triglyceride* is more commonly used, we will use the more accurate *triacylglycerol* or its abbreviation TAG. An older term, *neutral fat*, has also been used. Figure 7.2 shows these molecules as triesters, made from the combination of a three-carbon molecule with three alcohol groups, known as glycerol, and three fatty acids. Remember that the combination of an acid with an alcohol is known as an ester. In the body, stored triacylglycerols rarely have the same fatty acid attached at all three positions on the glycerol. Stored fats have both saturated and unsaturated fatty acids. **Diacylglycerols** (also known as diglycerides) have only two fatty acids attached to the glycerol. Monoacylglycerols (also known as monoglycerides) have only one. It is important to note that the major fatty acids we have just discussed are found in foods as triacylglycerols.

The physical state of a TAG depends on the length of the carbon chain in the fatty acids and the number of double bonds. Shorter-length fatty acids, as well as more double bonds, lower the melting point. Thus, vegetable fats with polyunsaturated fatty acids are liquids, whereas the fat on the side of a steak, which contains more saturated fatty acids, is a solid. Palm oil is a liquid, despite having mainly saturated fatty acids, because the fatty acids have 10 to 12 carbon atoms, as opposed to 16 to 18 in most other triacylglycerols. Triacylglycerols and the fatty acids found in triacylglycerols are insoluble in water due to the large degree of hydrophobic hydrocarbon components. The hydrophobic nature of triacylglycerols makes them ideal for storing energy because, by weight,

> ### ✓ KEY POINT
> Omega-3-polyunsaturated fatty acids, particularly the forms found in fish oils, have been associated with reduced incidents for heart and circulatory diseases when elevated in the diet. The health benefits of these forms of unsaturated fatty acids are thought to be due to their effects on cell membranes and their ability to suppress inflammation through acting as precursors for the synthesis of anti-inflammatory eicosanoids. Increasing dietary intake of omega-3-PUFAs and reducing intake of omega-6-PUFAs may accentuate these benefits. It is not yet certain whether dietary intakes of these fatty acids may act to significantly suppress inflammatory responses in muscle following damage from unaccustomed exercise.

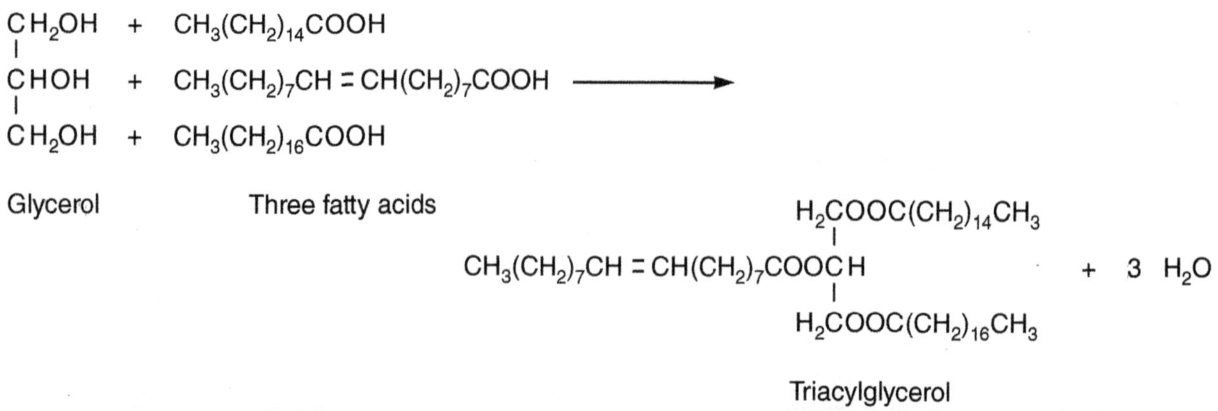

$$\begin{array}{l}\text{CH}_2\text{OH} \\ | \\ \text{CHOH} \\ | \\ \text{CH}_2\text{OH}\end{array} \quad \begin{array}{l}+ \quad \text{CH}_3(\text{CH}_2)_{14}\text{COOH} \\ \\ + \quad \text{CH}_3(\text{CH}_2)_7\text{CH}=\text{CH}(\text{CH}_2)_7\text{COOH} \\ \\ + \quad \text{CH}_3(\text{CH}_2)_{16}\text{COOH}\end{array} \quad \longrightarrow$$

Glycerol   Three fatty acids

$$\text{CH}_3(\text{CH}_2)_7\text{CH}=\text{CH}(\text{CH}_2)_7\text{COOC}\begin{array}{l}\text{H}_2\text{COOC}(\text{CH}_2)_{14}\text{CH}_3 \\ | \\ \text{H} \\ | \\ \text{H}_2\text{COOC}(\text{CH}_2)_{16}\text{CH}_3\end{array} \quad + \quad 3\ \text{H}_2\text{O}$$

Triacylglycerol

Figure 7.2 A triacylglycerol molecule, consisting of three fatty acids each joined by ester bonds (–COOC–) to each of the three carbons in glycerol.

they have more chemical potential energy than other fuel molecules, such as carbohydrate or protein. In fact, because the fatty acids in triacylglycerols are so reduced, the oxidation of 1 g of triacylglycerol has a standard free-energy change of more than 9 kcal per gram (38 kJ/g).

Phospholipids

Many **phospholipids** are derivatives of **phosphatidic acid** (or phosphatidate, since it is in anion form), whose structure is illustrated in figure 7.3. You should be able to see the glycerol part, with two fatty acids joined to it by ester bonds. Also attached is a phosphate group. Different groups can be attached to the phosphate in phosphatidic acid. If choline is attached, the molecule is called phosphatidylcholine or, commonly, lecithin (see figure 7.3). Phospholipids have a hydrophilic (water liking) part and a hydrophobic part. The hydrophilic part is due to polar chemical bonds and charged groups on the phospholipid. The hydrophobic part is the long hydrocarbon tail of the fatty acids, which can contain more than 16 carbon atoms. Phospholipids are major components of cell membranes and are shown in figures using a circle at the top, representing the hydrophilic part, with two squiggly lines running downward to represent the long, hydrophobic alkyl parts of the fatty acids. Cell-membrane phospholipids are constructed primarily from fatty acids derived from the diet and not from de novo synthesized fatty acids. Hence, as noted previously, diet influences the fatty-acid composition of membrane phospholipids, and diets high in omega-3-PUFAs raise the content of these fatty acids in cell phospholipid membranes.

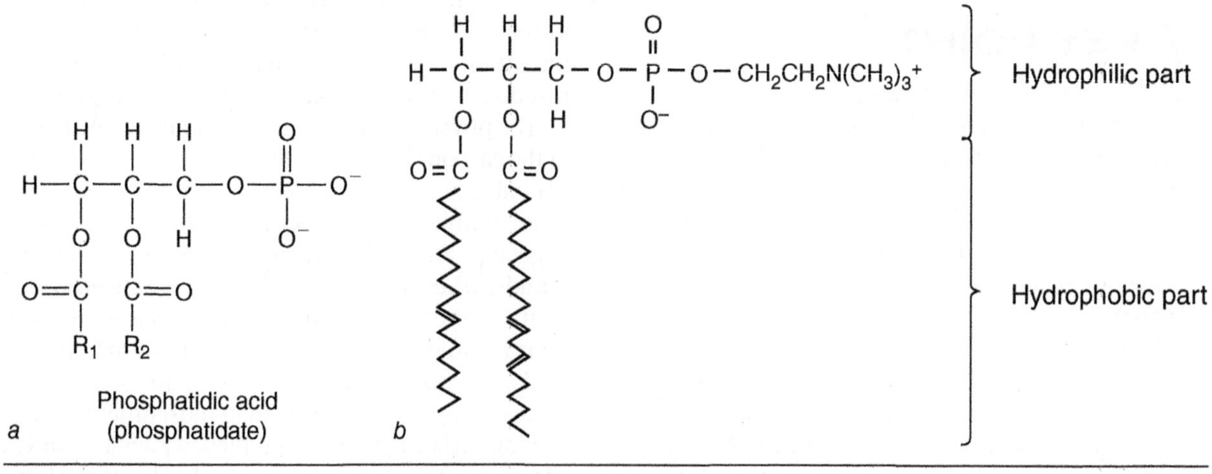

Figure 7.3 Phosphatidic acid and phosphatidylcholine. (*a*) Phosphatidic acid (or phosphatidate, since it is charged) consists of one glycerol with two fatty acids (R_1 and R_2 represent the long alkyl chains of each fatty acid) and a phosphate attached to the third carbon of glycerol. (*b*) Phosphatidylcholine is created when choline is combined with phosphatidic acid. Phospholipids such as phosphatidylcholine have a hydrophilic part, based on the positive and negative charges and the polar ester bonds.

Phosphatidylinositols are found in membranes, where they play an important role in cellular regulation as part of signaling systems. Figure 7.4 shows the structure of inositol. When phosphorylated at carbons 4 and 5 and attached to phosphatidic acid, it is known as phosphatidylinositol 4,5-bisphosphate, abbreviated PI 4,5-P_2. Figure 7.5 shows the hydrolysis of PI 4,5-P_2 by phospholipase C, producing inositol 1,4,5-trisphosphate (1,4,5-IP_3) and a diacylglycerol (DAG). The latter remains within the membrane, but 1,4,5-IP_3 is mobile in the cytosol to some extent. Hydrolysis of PI 4,5-P_2 by phospholipase C plays an important role in signal transduction. The 1,4,5-IP_3 can act on calcium channels located in the endoplasmic reticulum of the cell (called IP_3 channels) to release calcium ions, which can then activate other enzymes. Calcium ions and DAG can activate typical protein kinase C (PKC) enzymes that phosphorylate other proteins to alter their function. In the previous chapter, we noted the involvement of an atypical PKC, which can lead to translocation of GLUT-4 transporters to the cell membrane of skeletal muscle fibers when activated in an insulin-dependent mode.

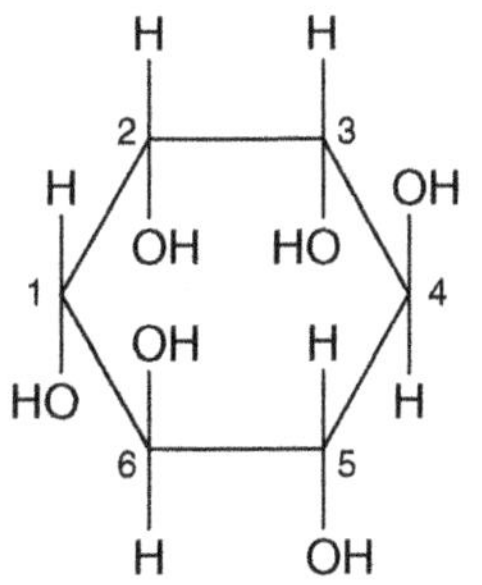

FAT STORAGE AND METABOLISM

Fat is stored as triacylglycerol in fat cells. It is considered a long-term energy store in contrast to glycogen, which is considered a short-term energy store. A normal adult can store about 8,000 to 12,000 kJ (2,000 to 3,000 kcal) of glycogen in liver and muscle, but even a lean man or woman has more than 300,000 kJ (75,000 kcal) stored as triacylglycerol in fat cells. As shown in figure 7.6, the major component of a fat cell (adipocyte) is a liquid droplet of triacylglycerol occupying much of the volume of the cell. The majority of fat stores are located in subcutaneous adipose tissue located below the skin. You can pinch this fat layer between your fingers. Other metabolically important fat stores can be found in skeletal muscle. Fat is also stored between organs in the abdominal region. This is known as visceral fat,

Figure 7.4 The chemical structure for inositol. The numbers identify its six carbon atoms.

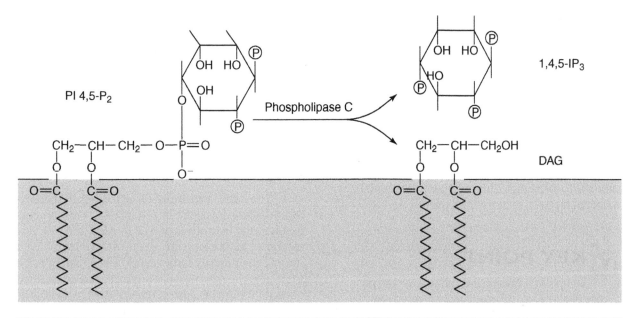

Figure 7.5 Hydrolysis of PI 4,5-P_2 by phospholipase C. Phosphatidylinositol 4,5-bisphosphate (PI 4,5-P_2) has its two fatty-acid chains buried in the hydrophobic core with the hydrophilic part, including the phosphorylated inositol, pointing into the interior of the cell. When activated, phospholipase C hydrolyzes PI 4,5-P_2, producing a diacylglycerol (DAG), which remains in the membrane, and inositol 1,4,5-trisphosphate (1,4,5-IP_3), which has finite mobility in the cytosol. The circled P represents phosphate groups.

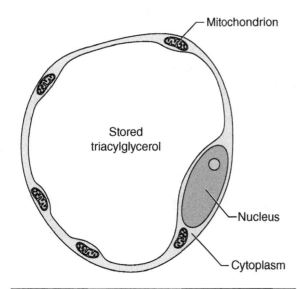

Figure 7.6 Cross section of a roughly spherical fat cell or adipocyte. Most of the volume of the cell is composed of a droplet of triacylglycerol. Cytoplasm is pushed to the periphery by the stored lipid. Fat cells have mitochondria and a single nucleus.

which is less metabolically active. As people age or store excess fat during weight gain, a portion of the stored fat goes to increase visceral fat. Increases in the amounts of visceral fat, as well as adipose tissue located around the waistline, are associated with increased risk of cardiovascular diseases.

Fatty acids are obtained mainly from food fat. After the fatty meal has been digested and the fatty acids have been absorbed, triacylglycerols are formed inside intestinal cells, combined with some protein and phospholipid, and released into the lymphatic system as lipoprotein particles known as **chylomicrons**. Another source of triacylglycerol is the liver. The liver makes and secretes triacylglycerol-rich **lipoproteins** known as **very-low-density lipoproteins** (VLDLs). The fatty acids in chylomicrons and VLDLs are released from their triacylglycerols through hydrolysis by an enzyme known as *lipoprotein lipase* (LPL). This enzyme is synthesized in adjacent fat cells, secreted from the cell, and attached to the endothelial lining of a nearby capillary. Fatty acids freed in the capillaries by LPL can flow down a concentration gradient into adjacent adipocytes with the aid of a specific carrier. Lipoprotein lipase is also present in capillaries in skeletal muscle, where it can generate fatty acids for muscle fibers.

Formation of Triacylglycerols

Fat synthesis is favored following a meal, when chylomicrons and blood glucose levels are elevated. An elevated insulin concentration in the blood facilitates entry of glucose into fat cells using the GLUT-4 transporter. In humans, excess dietary intake of carbohydrate resulting in excess caloric intake does lead to increased storage of fat. However, most of this fat storage is a result of increased storage of dietary fat, not the conversion of dietary carbohydrate into fatty acids in fat cells (Flatt 1995). Increasing dietary carbohydrate intake results in an increase in carbohydrate oxidation in muscle and other tissues and a reduction in fat oxidation. This consequently results in more dietary fatty acids being available for storage, since fewer fatty acids are oxidized for energy. It is these dietary fatty acids that are then preferentially used for increased triacylglycerol synthesis, while the excess carbohydrates are preferentially oxidized for energy. Since fatty-acid synthesis from nonfat sources is energetically expensive, limiting carbohydrate as a source of fatty-acid synthesis reduces the energy costs of fatty-acid formation and storage. As chapter 6 notes, an exception to this is dietary fructose, which is primarily converted to fatty acids in the liver. Thus, high intakes of dietary fructose can promote fat storage.

Triacylglycerol is also stored inside muscle cells, although the quantity is much less than in fat cells. Triacylglycerol stored in muscle is an important source of energy for muscle during rest, as well as in exercise conditions. Formation of TAG is a simple process, as shown in figure 7.7. We will consider this as three main steps.

1. The fatty acid must be activated by becoming attached to coenzyme A (CoA). This process involves the hydrolysis of ATP to AMP and PPi. It is catalyzed by the enzyme *acyl CoA synthetase*. Even though the product fatty acyl CoA is an energy-rich molecule, the reaction that produces it is considered nonequilibrium because of the irreversible hydrolysis of inorganic pyrophosphate by the enzyme *inorganic pyrophosphatase*.

2. The synthesis of triacylglycerols requires glycerol 3-phosphate. In adipose tissue, the major pathway is to form dihydroxyacetone phosphate starting

> ☑ **KEY POINT**
>
> Lipoprotein lipase activity shows a strong positive relationship with regularly performed physical activity. Low levels of LPL as a result of inactivity are associated with chronic diseases such as type 2 diabetes, obesity, and coronary heart disease (Hamilton, Hamilton, and Zderic 2004).

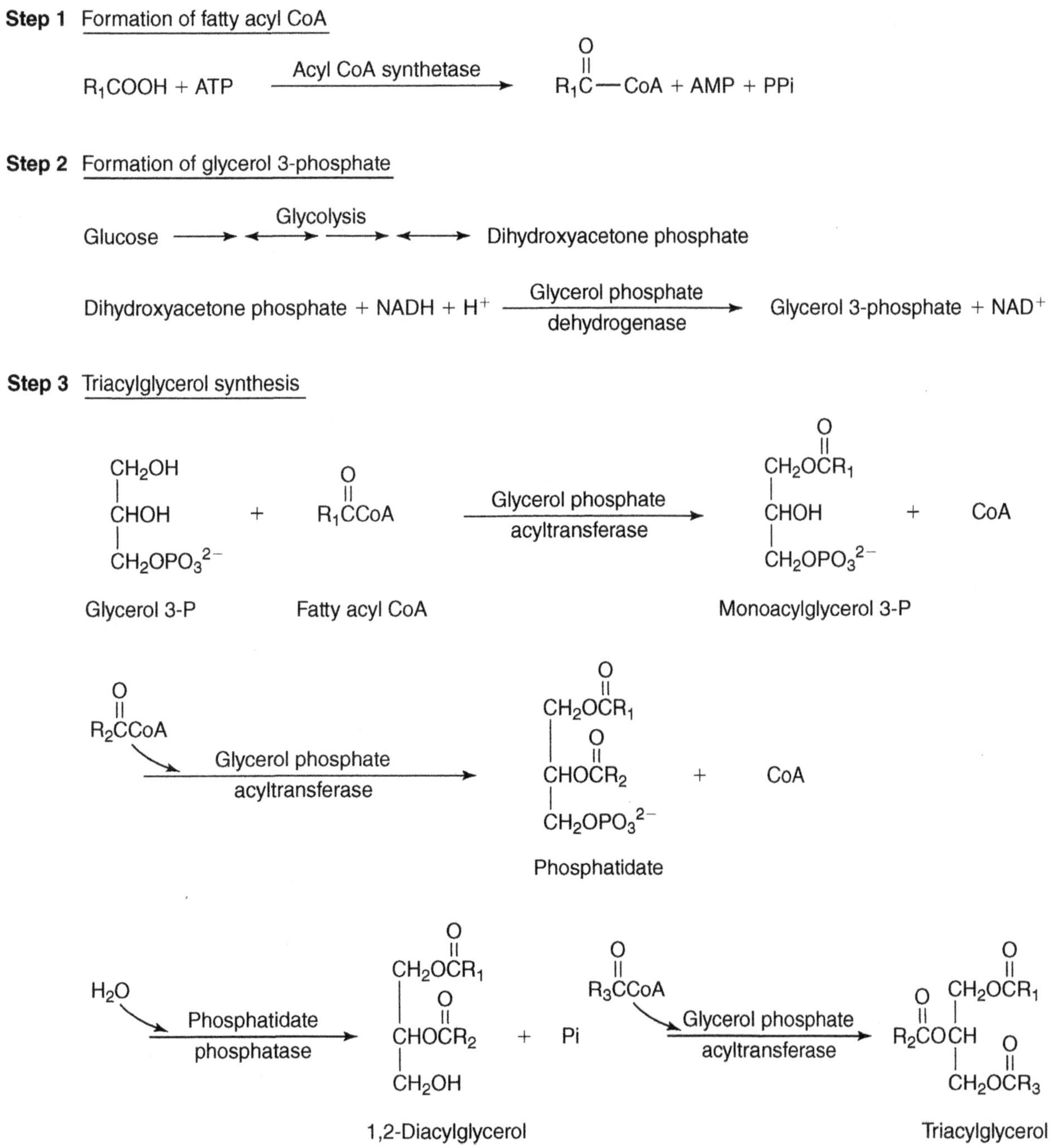

Step 1 Formation of fatty acyl CoA

$$R_1COOH + ATP \xrightarrow{\text{Acyl CoA synthetase}} R_1\overset{O}{\overset{\|}{C}}-CoA + AMP + PPi$$

Step 2 Formation of glycerol 3-phosphate

$$Glucose \xrightarrow{\text{Glycolysis}} Dihydroxyacetone\ phosphate$$

$$Dihydroxyacetone\ phosphate + NADH + H^+ \xrightarrow{\text{Glycerol phosphate dehydrogenase}} Glycerol\ 3\text{-phosphate} + NAD^+$$

Step 3 Triacylglycerol synthesis

$$\begin{array}{c} CH_2OH \\ | \\ CHOH \\ | \\ CH_2OPO_3^{2-} \end{array} + R_1\overset{O}{\overset{\|}{C}}CoA \xrightarrow{\text{Glycerol phosphate acyltransferase}} \begin{array}{c} CH_2O\overset{O}{\overset{\|}{C}}R_1 \\ | \\ CHOH \\ | \\ CH_2OPO_3^{2-} \end{array} + CoA$$

Glycerol 3-P   Fatty acyl CoA   Monoacylglycerol 3-P

$$R_2\overset{O}{\overset{\|}{C}}CoA \xrightarrow{\text{Glycerol phosphate acyltransferase}} \begin{array}{c} CH_2O\overset{O}{\overset{\|}{C}}R_1 \\ | \\ CHO\overset{O}{\overset{\|}{C}}R_2 \\ | \\ CH_2OPO_3^{2-} \end{array} + CoA$$

Phosphatidate

$$H_2O \xrightarrow{\text{Phosphatidate phosphatase}} \begin{array}{c} CH_2O\overset{O}{\overset{\|}{C}}R_1 \\ | \\ CHO\overset{O}{\overset{\|}{C}}R_2 \\ | \\ CH_2OH \end{array} + Pi \xrightarrow[\text{acyltransferase}]{R_3\overset{O}{\overset{\|}{C}}CoA\ \text{Glycerol phosphate}} \begin{array}{c} CH_2O\overset{O}{\overset{\|}{C}}R_1 \\ | \\ R_2\overset{O}{\overset{\|}{C}}OCH \\ | \\ CH_2O\overset{O}{\overset{\|}{C}}R_3 \end{array}$$

1,2-Diacylglycerol   Triacylglycerol

Figure 7.7 Storage of triacylglycerol as a three-step process. First, fatty acids are activated through attachment of a CoA. In the second step, glucose is partially broken down in glycolysis to dihydroxyacetone phosphate, which is subsequently reduced to make glycerol 3-phosphate. In the third step, three fatty acyl groups are added using the enzyme glycerol phosphate acyltransferase. Note that the phosphate group must be removed by a specific phosphatase.

from glucose and using glycolytic reactions, followed by reduction of dihydroxyacetone phosphate using the enzyme *glycerol phosphate dehydrogenase*. This is an enzyme we encountered in the previous chapter in the glycerol-phosphate shuttle. Other sources of glycerol 3-phosphate will be described in subsequent sections.

3. The formation of a triacylglycerol requires three fatty acyl CoA molecules and a glycerol 3-phosphate. The actual formation of triacylglycerol is a simple four-step process involving the sequential addition of two fatty acyl groups to glycerol 3-phosphate using a *glycerol phosphate acyltransferase* enzyme (or simply acyltransferase) to form phosphatidate

(anionic form of phosphatidic acid), hydrolysis of the phosphate group catalyzed by a *phosphatidate phosphatase*, and then addition of another fatty acyl group.

Lipolysis

The hydrolysis of triacylglycerols is best known as **lipolysis**. In this process, a triacylglycerol molecule yields glycerol and three fatty-acid molecules. Mobilization of fat (lipolysis) is favored under conditions of increased energy need, such as exercise, low-calorie dieting, or fasting (starvation), or when the body is cold. We usually view the process as one in which triacylglycerol in adipose tissue is broken down and the fatty acids and glycerol are released to the blood, where they are available to the energy-requiring cells. Skeletal muscle fibers also contain stored triacylglycerol, known as intramuscular triacylglycerol. When hydrolyzed, the fatty acids generated are oxidized within the contracting fiber. Enzymes inside adipose tissue fat cells and other fat-storing cells (e.g., muscle) are responsible for triacylglycerol hydrolysis. Lipolysis involves three hydrolysis reactions, each catalyzed by specific hormones. In adipose tissue, the removal of a fatty acid from triacylglycerol to form 1,2-diacylglycerol and a free fatty acid is catalyzed specifically by the enzyme desnutrin, or adipose triglyceride lipase (ATGL) (Ahmadian, Duncan, and Sul 2009). In skeletal muscle, this step is likely catalyzed by both desnutrin/ATGL and hormone-sensitive lipase, or HSL (Alsted et al. 2009). The second step, which removes a further fatty acid and leaves a 2-monoacylglycerol, is catalyzed by HSL. The last hydrolysis reaction is also catalyzed by HSL, assisted by a *monoacylglycerol lipase* (see figure 7.8). The numbers refer to the carbon atoms of glycerol. For example, a 2-monoglyceride (2-monoacylglycerol) has a single fatty acid attached to the glycerol molecule at the middle carbon atom.

Lipolysis takes place in locations other than in fat cells. For example, hydrolysis (digestion) of dietary triacylglycerol occurs in the small intestine, catalyzed by **pancreatic lipase**. As already mentioned, hydrolysis of triacylglycerol in blood lipoproteins is catalyzed by lipoprotein lipase. Finally, intramuscular triacylglycerol is hydrolyzed by a muscle-specific HSL, which generates fatty acids that are immediately available as a fuel for that fiber.

Regulation of Triacylglycerol Turnover

Formation of triacylglycerol molecules and their degradation take place in the cytosol of adipose tissue and other types of cells (skeletal muscle). Although different enzymes are used for synthesis and degradation of triacylglycerol, both processes could theoretically be fully active at the same time. This would result in ATP hydrolysis and glucose consumption, but little else. Although some advantages to continuous lipolysis and **esterification** exist, as we will see, one process is typically favored over the other, depending on circumstances in the body. For example, during exercise, when fatty acids are needed as a fuel, esterification is reduced, whereas lipolysis is accelerated. Following a meal, when there is a supply of fatty acid precursors to make triacylglycerol, lipolysis is depressed, whereas esterification is accelerated. As shown in figure 7.9, classical regulation takes place primarily at the level of HSL, controlled by phosphorylation and dephosphorylation, although other mechanisms can also operate in the regulation of lipolysis. For example, glucocorticoids, which are elevated during fasting, can upregulate transcription of desnutrin/ATGL, the primary enzyme that catalyzes the first step in the hydrolysis of triacylglycerol in adipose tissue. In contrast, refeeding and insulin will downregulate the transcription of desnutrin (Ahmadian, Duncan, and Sul 2009).

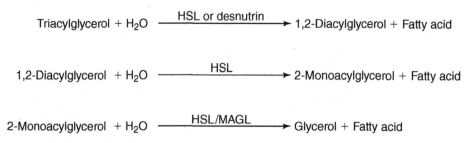

Figure 7.8 The breakdown of stored triacylglycerol in fat cells, involving three steps. Hormone-sensitive lipase (HSL) can catalyze all three hydrolyses. Desnutrin can also catalyze the first step, but a monoacylglycerol lipase (MAGL) plays a major role in the last hydrolysis.

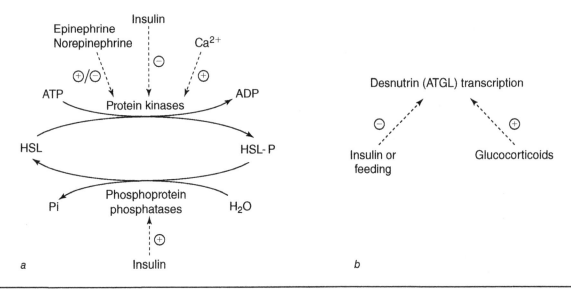

Figure 7.9 The major regulation of lipolysis through control of hormone-sensitive lipase (HSL). Phosphorylation of sites on HSL by a variety of protein kinases makes the enzyme active (HSL-P). Dephosphorylation by phosphoprotein phosphatases deactivates it. The influence of some important hormones is shown with dotted arrows and a plus or negative sign to indicate the effect. Epinephrine and norepinephrine can both activate or block HSL activity.

> ### ✓ KEY POINT
>
> The processes of triacylglycerol synthesis (esterification) and breakdown (lipolysis) take place in the cytosol of fat cells and other fat-containing cells, including skeletal muscle. Since these are opposing processes occurring in the same location, one would think they should not be simultaneously active—that if they were, the result would be wasteful ATP hydrolysis in the formation of fatty acyl CoA and waste of glucose to produce glycerol 3-phosphate. However, the two processes can be active simultaneously, which can be advantageous, although not to the same extent as individual activity. As when a drag racer guns the engine with a foot on the brake in order get away faster, if both processes take place simultaneously (although not maximally), a faster response can occur when fatty acids are needed, such as during exercise.

Regulation of Lipolysis in Adipose Tissue

Mechanisms behind the regulation of lipolysis in adipose tissue are shown in figure 7.10. At rest, HSL exists mostly in an inactive unphosphorylated form, and lipid droplets in a fat cell are surrounded by a protein family—the **perilipins** that render the droplet inaccessible to HSL. When both HSL and the perilipins are phosphorylated, the HSL becomes active and the perilipins cannot block lipolysis. Phosphoprotein phosphatases remove the phosphate groups from both HSL and the perilipins, and lipolysis is attenuated. The overall process of lipolysis is controlled primarily by the action of two hormones and a neurotransmitter. Epinephrine, released from the adrenal medulla, activates lipolysis when it binds to a β-adrenergic receptor on the fat cell. Norepinephrine, a neurotransmitter in the sympathetic nervous system, generates a positive effect on lipolysis by binding to β-adrenergic receptors ($β_1$ or $β_2$ subtypes). Both epinephrine and norepinephrine can bind to $α_2$-adrenergic receptors on fat cells, where the effect is to reduce lipolysis. Finally, insulin can bind to its own receptor on fat cells. Besides increasing glucose transport into fat cells by mobilizing GLUT-4 receptors, insulin can activate an enzyme that functions to decrease lipolysis. The details are as follows.

The catecholamines, epinephrine and norepinephrine, bind to a $β_1$- or $β_2$-adrenergic receptor and activate adenylyl cyclase through an activating G protein (G_s). As we saw in the previous chapter, *adenylyl cyclase* catalyzes the conversion of ATP into cyclic AMP (cAMP) and inorganic pyrophosphate. The cAMP binds to the regulatory subunits of protein kinase A, releasing the catalytic subunits. These catalytic subunits are now active as protein kinases, phosphorylating substrates such as HSL and perilipins, which coat lipid droplets in fat cells. Once phosphorylated, perilipins can no longer inhibit

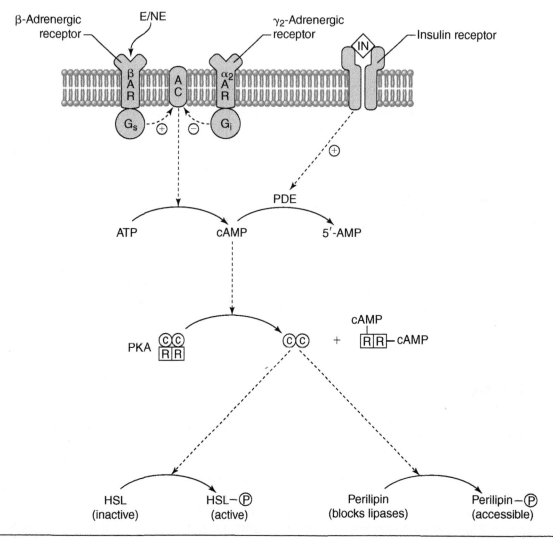

Figure 7.10 Mechanisms behind the regulation of lipolysis in fat by the actions of the catecholamines epinephrine (E) and norepinephrine (NE), as well as insulin (IN). Binding of E and NE to a β-adrenergic receptor (βAR) activates adenylyl cyclase (AC), mediated by a stimulatory G protein (G_s), leading to an increase in cyclic AMP (cAMP). When cAMP binds to the regulatory subunits (R) of protein kinase A (PKA), catalytic subunits (C) are released to phosphorylate hormone-sensitive lipase (HSL) and perilipin. E and NE can also bind to an α_2-adrenergic receptor (α_2AR), leading to a decrease in cAMP mediated by an inhibitory G protein (G_i). Insulin opposes lipolysis by activating cAMP phosphodiesterase (PDE), which breaks down cAMP.

the action of HSL on the stored triacylglycerols; perhaps phosphorylation causes the perilipin to dissociate from the lipid droplet. Hormone-sensitive lipase, phosphorylated on three serine residues, is translocated to the lipid droplet, where it catalyzes the hydrolysis of diacylglycerols. Phosphorylation of adipose tissue HSL alone increases its activity about twofold. However, the combination of HSL and perilipin phosphorylation increases lipolysis more than 90-fold, pointing to the concerted interaction between these two regulatory proteins. Interestingly, in older animals, HSL translocation from cytosolic sites to the lipid droplet under conditions favoring lipolysis is depressed. This may explain the reduced ability of older people to mobilize lipids (Holm 2003; Yeaman 2004).

Insulin inhibits this process by activating an enzyme, *cAMP phosphodiesterase*, which changes cAMP to 5'-AMP. As the previous chapter shows, occupancy of the insulin receptor by insulin leads to the activation of other protein kinases. In the case of adipose tissue, insulin activates protein kinase B (also known as Akt), which phosphorylates cAMP phosphodiesterase, making it active in breaking down cAMP. Thus, insulin acts to reduce the activation of protein kinase A. Insulin also activates phosphoprotein phosphatases, which remove phosphate from glycogen synthase a and glycogen phosphorylase a in both liver and skeletal muscle. It is quite likely that phosphoprotein phosphatases in fat cells are also activated by insulin, removing phosphates from HSL and perilipins and thus inhibiting lipolysis by another mechanism (Holm 2003).

Inhibition of lipolysis through catecholamine binding to the α_2-adrenergic receptors is mediated by an inhibitory G protein (G_i) that decreases the activity of adenylyl cyclase, thus reducing the level of cAMP. The balance between the prolipolysis β-adrenergic receptors and the antilipolysis α_2 receptors determines how easily fat can be mobilized from each adipocyte. For example, we know that adipose tissue just beneath the skin (subcutaneous adipose tissue, SCAT) is less responsive to exercise-induced adrenergic stimulation in obese people compared to men with normal weight (Stich et al. 2000). Such a condition is not permanent, since overweight men who exercise can markedly improve exercise-induced lipolysis by decreasing the antilipolytic effect of α_2-adrenergic receptors (de Glisezinski et al. 2003).

Protein kinase A is not the only enzyme that phosphorylates and activates HSL. The mitogen-activated protein kinase, or MAPK family, is involved in a host of regulatory processes in cells. These processes are discussed in chapter 3. However, here we should mention its role in HSL phosphorylation. One of the MAPK signaling pathways involves an *extracellular signal-related kinase*, or ERK. Specifically, ERK1/2 is activated when it is phosphorylated by cAMP-dependent kinase (protein kinase A). Once activated, ERK1/2 can phosphorylate sites on HSL. The exact role that phosphorylation by protein kinase A and ERK1/2 plays in terms of activation of HSL is difficult to establish in fat cells because phosphorylation leads to both an increase in HSL activity and HSL translocation to the lipid droplet. Moreover, phosphorylation of perilipin also plays a huge role in activating the overall process of lipolysis.

Since its discovery was fairly recent, less is known about the control of desnutrin/ATGL activity. As previously noted, desnutrin/ATGL is the primary enzyme that catalyzes the first step in the hydrolysis of triacylglycerol in adipose tissue. Until recently, this step was also thought to be regulated by HSL; however, studies with HSL-null mice have demonstrated that desnutrin/ATGL is the controlling enzyme for the first step of triacylglycerol hydrolysis in adipose tissue (Ahmadian, Duncan, and Sul 2009). It is thought that the regulation of desnutrin/ATGL differs from that of HSL. Although desnutrin/ATGL is likely a phosphoprotein, its phosphorylation is not mediated by protein kinases (Jaworski et al. 2007). It is also located on the lipid droplet (unlike HSL, which does not reside there) and therefore does not need to be activated by translocation to the droplet. Further research will be needed to complete our understanding of desnutrin/ATGL regulation (Jaworski et al. 2007).

Other hormones can also stimulate lipolysis. Growth hormone from the anterior pituitary and *cortisol* from the adrenal cortex are secreted in response to stressful conditions, such as fasting and exercise. Both increase the rate of whole-body lipolysis in an additive fashion, likely by increasing the cAMP concentrations beyond the levels induced by catecholamines. It is possible that these hormones decrease the activity of the inhibitory-G-protein pathway or increase insulin resistance so that the effect of insulin to downregulate lipolysis is reduced. One of the reasons athletes inject themselves with growth hormone is to reduce the size of adipose tissue stores while promoting growth of skeletal muscle. *Testosterone* and related anabolic steroids have strong effects on skeletal muscle growth, but they also have much milder effects to increase lipolysis. Castration is associated with a blunted adrenergic-dependent response and often an increase in adipose tissue mass. This can be reversed by testosterone administration. Androgens influence fat-cell metabolism by binding to intracellular receptors.

In addition to insulin, there are other negative effectors for lipolysis. *Adenosine*, produced locally from ATP, can bind to its own receptor on the adipocyte membrane. This receptor is coupled to the G_i, the inhibitory G protein that depresses the activity of adenylyl cyclase, resulting in inhibition of lipolysis. Estrogens also function by binding to intracellular receptors. The number of these receptors varies in fat deposits from different sites, pointing to the role of estrogen in promoting fat deposition at specific locations. For example, the gluteofemoral fat deposits are augmented with the rise in estrogen in pubescent females. Estrogens lead to a relatively small reduction in the lipolytic effect of catecholamines on fat cells.

Regulation of Lipolysis in Skeletal Muscle

Skeletal muscle contains variable amounts of intracellular lipid droplets that tend to be located between myofibrils, close to mitochondria (Watt, Heigenhauser, and Spriet 2002). Good evidence exists that this TAG provides a source of fatty acids when the muscle fiber is active. Intramuscular triacylglycerol

(IMTG, also known as intramyocellular lipid or IMCL) is hydrolyzed by an HSL, but HSL activity in skeletal muscle is regulated differently from adipose tissue HSL. Muscle HSL activity is increased by up to 100% at the start of exercise, but its activity tends to decline as moderate exercise is prolonged. However, adipose tissue HSL remains elevated throughout the exercise duration, albeit to a lesser extent than the maximal activation seen in muscle (Watt et al. 2006). We know that skeletal muscle HSL is activated when it is phosphorylated at sites similar to those in the HSL found in the adipose tissue. Like adipose-tissue HSL activity, skeletal-muscle HSL activity is increased by elevated epinephrine through a cAMP-dependent mechanism involving protein kinase A. However, there is no perilipin in skeletal muscle, so the extra level of regulation of lipolysis afforded by this protein is absent in muscle.

Compared to one at rest, an active muscle fiber has a calcium concentration increased by approximately 100-fold, as well as an increased AMP concentration. As we have seen, elevated Ca^{2+} levels can activate several kinases, including protein kinase C. The increase in AMP concentration in active muscle affects AMP-activated protein kinase (AMPK) activity, which is in part responsible for the contraction-induced increase in muscle-glucose uptake. In addition, exercise, as a stressor, can lead to phosphorylation and activation of the MAP kinase, ERK. Recent evidence from human studies indicates that β-adrenergic and other exercise-related changes are responsible for activation of AMPK at the start of exercise. In addition, AMPK activity declines as exercise progresses. These changes in AMPK activity are identical to changes in HSL activity in skeletal muscle during exercise. AMPK is the prime regulator of muscle HSL phosphorylation; thus, it is seen as the prime regulator of changes in muscle HSL activity during exercise (Watt et al. 2006). HSL content and activation via phosphorylation is higher in muscles of women than men (Roepstorff et al. 2006). As this chapter discusses in more detail in subsequent sections, this may partly explain sex differences in fat oxidation.

Recently, the presence of desnutrin/ATGL has also been identified in human skeletal muscle, although at lesser concentrations than found in adipose tissue. Unlike in adipose tissue, where it is likely the primary enzyme responsible for the first step in triacylglycerol breakdown, desnutrin/ATGL likely shares this responsibility with HSL in skeletal muscle. Desnutrin/ATGL is found only in Type I muscle fibers (Jocken et al. 2008). HSL activity in skeletal muscle is not influenced by training. However, research has found that eight weeks of endurance training can double the protein content and activation of desnutrin/ATGL in human muscles, indicating a training effect on this stage of fat oxidation in muscle (Alsted et al. 2009).

The hormone leptin, derived from adipose tissue, is important in weight regulation and has significant effects on muscle lipid metabolism. Specifically, leptin can stimulate de novo lipogenesis in muscle from glucose when glucose levels are high, and can simultaneously stimulate lipid oxidation. This simultaneous synthesis and breakdown of triacylglycerols in adipose tissue is noted earlier in this chapter. In muscle, this mechanism can serve as a glucose sink, shunting glucose toward fat synthesis when greater amounts of glucose are present than what is needed. Interestingly, it has been suggested that the simultaneous stimulation of fatty-acid synthesis and metabolism by leptin in skeletal muscle as a futile cycle could also be a source of thermogenesis (Dulloo et al. 2004). Similar mechanisms exist in brown adipose tissue. The importance of these cycles and of leptin in regulating their function relative to metabolic rate changes and obesity are still being investigated. In addition, these effects of leptin may also help prevent the development of insulin insensitivity in muscle. The mechanisms and implications of these effects are discussed later on in this chapter.

Fate of Fatty Acids and Glycerol

Let us review the role of fat cells in the human body. These represent the principal storage site for triacylglycerol in the body. Triacylglycerol (TAG) is a long-term energy source, and it turns over very slowly. If a person has 10 kg (22 lb) of TAG in the body, only about 50 to 70 g per day is turned over. This means that the lifetime of a TAG molecule stored in a fat cell is more than six months (Strawford et al. 2004). Following feeding, dietary fatty acids are taken up and esterified to make triacylglycerol. Esterification is reduced and lipolysis is accelerated when the body is under stress, as during exercise. This leads to increased net release of fatty acids to the blood. Lipolysis and esterification are active simultaneously, although at any time, one of the two is usually dominant through, as we have seen, the action of hormones. Lipolysis produces fatty acids and glycerol, but a portion of the fatty acids can be reesterified back to TAG after they are first activated by attaching a CoA at the expense of ATP (figure 7.11). This continuous circle of lipolysis and reesterification, or recycling within a cell, constitutes an intracellular triacylglycerol–fatty acid cycle (Reshef et al. 2003). Adipose tissue is the principal source of blood fatty acids that can be used for

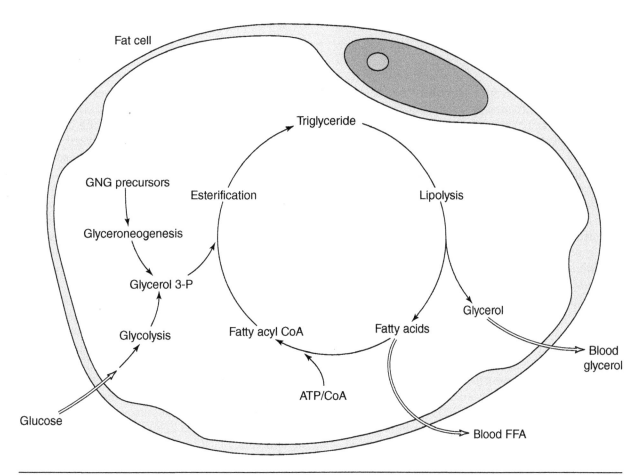

Figure 7.11 Recycling of triacylglycerol in a fat cell (adipocyte). This occurs when the opposing processes of lipolysis and esterification take place at the same time. Triacylglycerol molecules are broken down to fatty acids and glycerol during lipolysis. The glycerol is released from the fat cell to the blood, while the fatty acids have two fates. They may leave the fat cell and enter the blood, or they may be converted to fatty acyl CoA molecules within the fat cell, where they can be used to make new triacylglycerol. Glycerol 3-phosphate (3-P) is derived from glycolysis using glucose taken up from the blood. Open arrows show membrane transport. In the blood, fatty acids travel bound to the protein albumin and are known as FFA (free fatty acids).

energy by other tissues, including the heart and skeletal muscles. Fat cells also produce and secrete some other important substances, such as leptin, that can influence our overall metabolism and food intake behavior. Abnormalities in fat metabolism can underlie conditions such as obesity, type 2 diabetes, and cardiovascular disease.

Figure 7.11 summarizes the opposing processes of lipolysis and esterification in an adipocyte in the postabsorptive state. In this state, fat cells play a major role to provide fatty acids for oxidation by other tissues. Virtually all of the glycerol generated by TAG lipolysis is released to the blood because the gene for *glycerol kinase* is scarcely expressed in fat cells. Glycerol kinase is needed to attach a phosphate from ATP to glycerol to make glycerol 3-phosphate, essential for reesterification. Accordingly, glycerol release from adipose tissue can be used as a marker for the rate of lipolysis. If only lipolysis, and not reesterification, were active, we would expect to see three fatty molecules released for each glycerol because each TAG molecule has three fatty acids attached to a glycerol. However, a ratio of three fatty acids released to the blood for every glycerol is not present under any circumstances, since some fatty acids always stay in the fat cell, being reutilized to make TAG or oxidized. Indeed, in the fasted state, it is estimated that about 30% of the fatty acids released during lipolysis undergo reesterification.

Figure 7.11 shows that two major sources of glycerol 3-phosphate (glycerol 3-P) exist for reesterification. The classical view of TAG formation in the fasting state is that glucose is the source of glycerol 3-phosphate needed for reesterification. This is not likely. During fasting, both blood glucose and insulin concentrations are low, and insulin is needed to stimulate glucose transport into fat cells in just the same way it does in muscle cells. The question, then, is what is the source of glycerol 3-phosphate to reesterify fatty acids? As shown in figure 7.11, a process has been recently identified that converts gluconeogenic precursors (lactate, pyruvate, and some amino acids) into glycerol 3-P. This process, known as **glyceroneogenesis**, operates in much the same way gluconeogenesis does in the fasted state. The importance of this process is that it provides a source of glycerol 3-phosphate under conditions when blood glucose should be directed to those tissues that need it as a critical energy source—that is, the brain. Detailed biochemical studies have shown that if the glyceroneogenesis pathway operates during fasting conditions, then a limiting enzyme, *phosphoenolpyruvate carboxykinase* (PEPCK), must be expressed to a greater extent, allowing glycerol 3-phosphate to be made. Indeed, this is the case, since PEPCK gene expression can be turned on rapidly in the postabsorptive period and turned off when glucose is available. This response to feeding and fasting in adipose tissue is much like the response in liver for the genes coding for enzymes of gluconeogenesis (GNG). However, differences exist between GNG in liver and fat cells. Cortisol, a stress hormone from the adrenal cortex, upregulates the PEPCK gene in liver, stimulating GNG, but inhibits the PEPCK gene in adipose tissue. This differential response is exactly as we might predict, because the net effect of cortisol is to stimulate glucose production in liver during stress, while allowing more fatty acids to be released from the adipose cells by downregulating their reesterification.

The dominant fate for the free glycerol after a triacylglycerol is hydrolyzed is release to the blood. Since glycerol has three OH groups, it is quite soluble in the blood. It is metabolized by tissues that contain a *glycerol kinase*, an enzyme that phosphorylates the glycerol to form glycerol 3-phosphate. The major tissue in which this occurs is the liver. As we saw in the previous chapter, glycerol is a source of glucose during periods of fasting or starvation, providing 15% to 25% of the glucose produced. When fatty acids leave a fat cell and enter the blood, they become attached to the blood protein albumin, since they are not soluble in the aqueous plasma. We call fatty acids attached to albumin in the blood FFA (free fatty acids). The FFA circulate in the blood and may enter other cells where they can be used as an energy substrate, as we shall soon see. Transfer of FFA from adipose tissue to skeletal muscle during exercise requires sufficient perfusion of adipose tissue by blood. Therefore, adipose-tissue blood flow could pose a limitation to the delivery of fatty acids from fat tissue to exercising skeletal muscle.

Not all FFA circulating in the blood are taken up by cells and immediately oxidized. The liver has a high capacity to take up fatty acids from the blood. Some FFA are used to make other lipid compounds, as we will see, but others can be reesterified in the liver and released as part of the VLDL particles secreted by the liver. Circulating in the blood, the triacylglycerol-rich VLDLs are hydrolyzed by lipoprotein lipase in capillaries, and the fatty acids can enter an adjacent cell. If this is an adipocyte, the fatty acids are likely to be incorporated into TAG molecules and stored as part of the lipid droplet in the fat cell. Even a skeletal-muscle fiber stores fatty acids as TAG, although when the fiber is active, the more likely fate is oxidation to support the contractile activity. Thus, a triacylglycerol–fatty acid cycle also operates between cells in different organs.

Fatty acids are an ideal fuel for oxidation since they have a high energy density. Thus, the levels of FFA in the blood are elevated during stressful circumstances, such as exercise. Obese people with a larger adipose tissue mass have higher [FFA] under most circumstances. However, a consistently high [FFA] has drawbacks, not only because fatty acids are toxic in high concentrations, but also because a convincing relationship exists between elevated blood FFA and insulin resistance. As discussed previously, insulin resistance is a hallmark of type 2 diabetes. Elevated blood FFA levels lead to increased fatty-acid uptake and storage as TAG in skeletal muscle, which ultimately renders muscle less sensitive to insulin. It is now well known that insulin resistance is mediated in part through factors associated with obesity-related inflammatory events occurring in adipose tissue, as well as through changes to adipose-tissue-derived regulatory peptides, such as leptin, adiponectin, and resistin. The mechanisms, which induce insulin resistance in skeletal muscle and are related to obesity, are discussed in later sections of this chapter.

OXIDATION OF FATTY ACIDS

Skeletal muscle is a major site for oxidation of **long-chain fatty acids** (LCFAs), those fatty acids containing 14 or more carbon atoms, which are the dominant fatty acids stored in TAG. Capillaries in skeletal muscle also contain lipoprotein lipase, so that the fatty acids in the capillaries surrounding muscle fibers are those bound to albumin as FFA and those immediately obtained through hydrolysis of VLDL from liver by lipoprotein lipase. In order for the free energy stored in FFA molecules to be obtained, they must be transported from the blood, across the cell membrane of the utilizing cell, through the cytosol of the cell, across the inner mitochondrial membrane, and into the matrix. There, the fatty acids are broken down into two-carbon acetyl groups attached to CoA by beta-oxidation. We will look at these steps in detail. In skeletal muscle, fatty acids derived from intracellular TAG lipolysis must also be transported into mitochondria for oxidation.

Intracellular Transport of Fatty Acids

Because LCFAs are poorly soluble in an aqueous environment, it was assumed in the past that they could cross cell membranes by simple diffusion. This is a mechanism for transport, but we know now that fatty-acid transporters also exist that allow the fatty acids to be moved down their concentration gradient into the cytosol of fat-utilizing cells. With its ability to increase its metabolic rate to high levels, muscle is of primary importance for oxidizing fatty acids. Several transporters have been identified in skeletal muscle, including a plasma membrane fatty acid–binding protein ($FABP_{pm}$) and a fatty-acid translocase known as FAT/CD36. These transporters act at both the sarcolemma and mitochondrial sites to regulate fatty-acid transport across membranes. A diet high in fat induces an increase in both $FABP_{pm}$ and FAT/CD36. Endurance and high-intensity interval training increase FAT/CD36 content in mitochondria and sarcolemma by 50% and 10%, respectively, and sarcolemma $FABP_{pm}$ activity by 23%, with no change in mitochondrial $FABP_{pm}$ (Talanian et al. 2010). These changes highlight the importance of fatty-acid transporters as mediators of the increase in muscle fatty-acid oxidation capacity induced by exercise training. Studies have also established that intracellular stores of FAT/CD36 transporters can be mobilized to the muscle sarcolemma and to mitochondria with the onset of exercise (Holloway et al. 2009; Bonen et al. 2007). Fatty-acid uptake responds to acute exercise in much the same way as glucose uptake—that is, with an increase in transporters in the sarcolemma. However, additional fatty-acid transporters are also required and regulated at the mitochondrial membrane to facilitate fatty-acid transport across the mitochondrial membranes for metabolism.

Because of their insolubility, fatty acids in the cell cytoplasm of many tissues are attached to a cytosolic fatty acid–binding protein ($FABP_c$). In muscle, fatty acids attached to $FABP_c$ can be those originating in adipose tissue and taken up from the blood or those released when IMTG is hydrolyzed by skeletal muscle HSL and desnutrin/ATGL. To be oxidized in the mitochondria, fatty acids must first be activated, forming a fatty acyl CoA, as in previous sections of this chapter. This reaction, catalyzed by *acyl CoA synthetase*, is as follows:

$$\text{fatty acid} + \text{ATP} + \text{CoA} \rightarrow \text{fatty acyl CoA} + \text{AMP} + \text{PPi}$$

Acyl CoA synthetase is an enzyme of the outer mitochondrial membrane. The reaction it catalyzes is essentially irreversible because the PPi is hydrolyzed by inorganic pyrophosphatase to two Pi (not shown), which drives the reaction to the right. The fatty acyl CoA that is formed is an energy-rich molecule, much like acetyl CoA. Formation of a fatty acyl CoA costs two ATP because two phosphates are removed from ATP.

Transport as Acylcarnitine

Fatty acyl CoA (often simply called acyl CoA) must be transported across the inner mitochondrial membrane to the matrix, where it will be broken down into acetyl CoA units by the process of beta-oxidation. However, the mitochondrial inner membrane is impermeable to CoA and its derivatives, which permits separate regulation of CoA compounds in mitochondrial and cytosolic compartments. Transport of fatty acyl CoA into the mitochondrial matrix occurs using three different proteins and the small molecule **carnitine** (see figure 7.12). Only fatty acids attached to carnitine are able to cross the inner mitochondrial membrane. Therefore, the first step is to exchange a CoA for carnitine to create a fatty acylcarnitine. This is catalyzed by an enzyme in the outer membrane known as *carnitine palmitoyl transferase I* (CPT I). New research indicates that CPT I works together with FAT/CD36 in regulating and facilitating fatty acyl CoA transport across mitochondrial membranes. The exact mechanisms of these interactions and their control are not yet known; however, FAT/CD36 is found in the outer mitochondrial membrane, and greater FAT/CD36 increases mitochondrial fat oxidation independent of changes in CPT I (Holloway et al. 2009).

Carnitine, formed from the amino acids lysine and methionine, can cross the inner membrane, and a fatty acyl form of carnitine can also cross in the opposite direction. The *carnitine–acylcarnitine translocase* is also known as the carnitine–acylcarnitine antiport because this membrane protein transfers two different substances across the inner membrane in opposite directions.

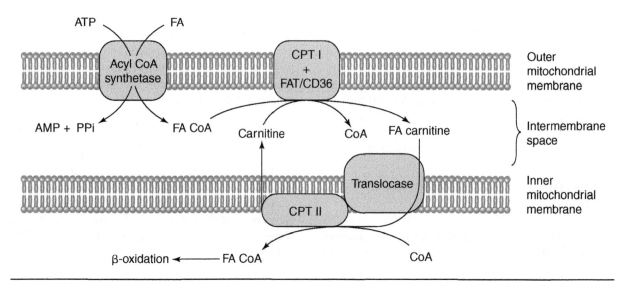

Figure 7.12 Transport of fatty acyl CoA into the mitochondrial matrix. Fatty acids are activated through attachment of a CoA using an outer mitochondrial membrane, acyl CoA synthetase. The CoA is then exchanged for a carnitine using an outer-membrane protein, carnitine palmitoyl transferase I (CPT I). The fatty acylcarnitine is conveyed across the intermembrane space in one direction, while a carnitine moves the opposite way using a carnitine–acylcarnitine translocase. The carnitine attached to fatty acylcarnitine is exchanged for CoA using an inner-membrane carnitine palmitoyl transferase II (CPT II).

On the matrix side of the inner mitochondrial membrane, a reverse exchange occurs in which the carnitine is exchanged for CoA using *carnitine palmitoyl transferase II* (CPT II), located on the matrix side of the inner membrane. Like acyl CoA, acylcarnitine is an energy-rich molecule, so the two reactions catalyzed by CPT I and CPT II are equilibrium reactions with equilibrium constants near 1. The net transfer of acyl groups attached to CoA in the matrix is given by the fact that the acyl CoA undergoes the process of beta-oxidation. Transport of fatty acyl units into the mitochondria for oxidation is regulated at the CPT I step. This enzyme is inhibited by malonyl CoA, which can place a limit on fat oxidation in skeletal muscle. This is discussed in more detail later in this chapter.

Carnitine deficiency, due to the body's inability to convert lysine into carnitine, is not an unusual metabolic disease. Patients with this deficiency have muscle weakness and poor exercise tolerance due to accumulation of triacylglycerol in muscle and the inability to oxidize fatty acids. Endurance athletes have commonly used carnitine supplementation in the belief that it will enhance their ability to oxidize fatty acids. However, little empirical evidence exists for such a metabolic or ergogenic effect in normal, healthy people.

Beta-Oxidation of Saturated Fatty Acids

The initial process in the oxidation of fatty acids is known as beta-oxidation, which occurs in the matrix of the mitochondrion. Beta-oxidation begins as soon as the fatty acyl CoA appears in the matrix, using repeated cycles of four steps. With each cycle, the fatty acyl CoA is broken down to form a new fatty acyl CoA, shortened by two carbon atoms, plus an acetyl CoA. Figure 7.13 summarizes the four steps in detail. Reactions 1 to 3 are designed to change carbon 3 (the β-carbon) from a methylene group (CH_2) to

> **KEY POINT**
>
> The carnitine transport system is necessary for the long-chain fatty acids (14 or more carbon atoms) that are typically stored as triacylglycerol in adipose tissue and skeletal muscle. **Medium-chain triacylglycerols** contain fatty acids that have a length of 6 to 12 carbon atoms. Compared to triacylglycerols with long-chain fatty acids, the medium-chain triacylglycerols are more rapidly digested and their fatty acids absorbed because these are not incorporated into chylomicrons. In addition, the medium-chain triacylglycerols can enter mitochondria for oxidation without the need for carnitine. The benefits of medium-chain triacylglycerols for exercise, if any, have not yet been defined.

a carbonyl group (C=O). Reaction 4 introduces a CoA group to carbon 3, cleaving off an acetyl CoA and leaving an acyl CoA shortened by two carbon atoms. This process is repeated until all the carbon atoms in the original fatty acyl CoA are in the form of acetyl CoA. For example, starting with palmitoyl CoA (a 16-carbon fatty acyl group attached to CoA), the four reactions of beta-oxidation would be repeated seven times to generate a total of eight acetyl CoA.

The first reaction is catalyzed by the enzyme acyl CoA dehydrogenase, located on the matrix side of the inner membrane of the mitochondrion. Using FAD as a coenzyme, acyl CoA dehydrogenase removes two electrons as hydrogen atoms, one from carbon 2 and one from carbon 3. In the process, FAD is reduced to $FADH_2$. The electrons are then passed through an electron transfer flavoprotein to coenzyme Q, resulting in the formation of coenzyme QH_2. The removal of the two hydrogen atoms in step 1 results in the formation of a carbon-to-carbon double bond between carbons 2 and 3 of the fatty acyl group. The double bond is trans because the

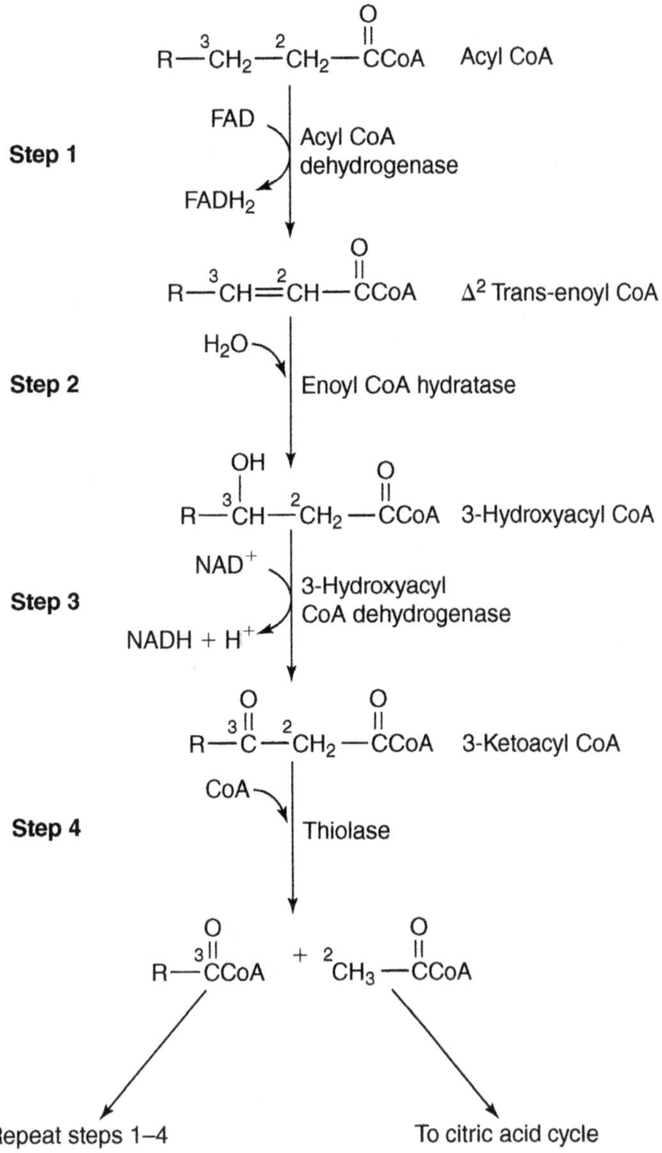

Figure 7.13 Beta-oxidation of fatty acids, a four-step process that removes acetyl CoA units one at a time from a fatty acyl CoA. The starting material is a fatty acyl group in which carbon 1 is attached to the CoA, carbon 2 (or α-carbon) is the next carbon, and the one beyond that is carbon atom 3 or β-carbon. Steps 1 to 3 are designed to change carbon 3 from a methylene (CH_2) to a carbonyl (C=O). The fourth step splits the original fatty acyl CoA between carbons 2 and 3 by introducing a CoA, resulting in an acetyl CoA and a new acyl CoA shortened by two carbons. This process is repeated until the original fatty acid is completely changed to acetyl CoA groups.

two hydrogens are on opposite sides, although this is not shown. The free energy released in the electron transfer from acyl CoA to FAD to coenzyme Q is insufficient to transport six protons from the matrix to the cytosolic side of the inner membrane. In this regard, the reaction is like that of succinate dehydrogenase or mitochondrial glycerol phosphate dehydrogenase.

In the second reaction, trans enoyl CoA is hydrated, accepting a water molecule. The enzyme catalyzing this reaction, *enoyl CoA hydratase*, is located in the matrix. The OH part of the water is added to carbon 3 (β-carbon), while the other hydrogen atom of water is added to carbon atom 2. The product is a 3-hydroxyacyl CoA (or β-hydroxyacyl CoA). In the third reaction, catalyzed by the matrix enzyme *3-hydroxyacyl CoA dehydrogenase*, the β-carbon (carbon 3) of the 3-hydroxyacyl CoA is oxidized, losing a hydride ion and a proton. Such oxidation reactions always use the nicotinamide coenzymes, NAD^+ in this case. This reaction changes the OH group to a carbonyl. In the last step, CoA attaches to carbon 3 (the β-carbon), which allows carbons 1 and 2 to come off as an acetyl CoA, leaving a new acyl CoA shortened by two carbon atoms. The enzyme responsible for the fourth reaction is *thiolase* (called thiolase because the CoA contains a terminal SH [thiol] group).

The reactions in figure 7.13 are shown as if they are irreversible. Actually, the overall changes of free energy are modest, and the reactions could rightfully be described as equilibrium (reversible). Direction is given to beta-oxidation by the fact that the acetyl groups feed into the citric acid cycle. Energy yield from beta-oxidation of fatty acids is considerable. For example, from the 16-carbon palmitic acid, the yield is 8 acetyl CoA, 7 $FADH_2$, and 7 NADH. If you do an ATP yield analysis from the oxidation of a fatty acid, remember to account for the fact that formation of fatty acyl CoA costs the equivalent of two ATP.

> ### ✓ KEY POINT
> The steps in the citric acid cycle from succinate to oxaloacetate are remarkably similar to the first three steps in beta-oxidation. For both, there is a dehydrogenation, generating a carbon-to-carbon double bond (fumarate in the citric acid cycle), a hydration (forming malate), and then a further dehydrogenation generating the keto group in oxaloacetate.

Oxidation of Unsaturated Fatty Acids

Many fatty acids stored in body fat are unsaturated. These fatty acids are also sources of energy in the form of acetyl CoA units and reduced coenzymes during beta-oxidation. Recall that the naturally occurring unsaturated fatty acids in foods and those stored in our bodies are in the cis configuration. However, step 2, catalyzed by enoyl CoA hydratase, needs a trans double bond for its substrate. Therefore, an additional step is required that is catalyzed by an enzyme, *enoyl CoA isomerase*, which converts the cis to a trans carbon-to-carbon double bond.

Other unsaturated fatty acids, such as linoleic acid, have their double bonds in the wrong position as well as in the wrong (cis) configuration for beta-oxidation. For these, *enoyl CoA isomerase* converts the cis to a trans double bond, but this is not enough. An additional enzyme known as a *reductase* converts the carbon-to-carbon double bond in the wrong position into a carbon-to-carbon single bond by the addition of hydrogen in the form of NADPH and H^+. As we have seen, NADPH is closely related to NADH, and it serves to reduce double bonds during synthetic reactions.

Oxidation of Odd-Carbon Fatty Acids

Most of the fatty acids found in land animals contain an even number of carbon atoms. However, plants and marine animals contain fatty acids with an odd number of carbon atoms. Oxidation of these odd-carbon fatty acids occurs through the normal route of beta-oxidation. In the last cycle of beta-oxidation, the starting substrate is a five-carbon acyl CoA. When this is split by the thiolase enzyme, the result is acetyl CoA and propionyl CoA, which contains the three-carbon propionyl group attached to CoA. Propionyl CoA is not treated as a fat. In fact, it is converted into the citric-acid intermediate, succinyl CoA, in a three-step process. This means that propionyl CoA generates a gluconeogenic precursor (see chapter 6).

OXIDATION OF KETONE BODIES

Ketone bodies are sometimes described as water-soluble, energy-providing lipids. Figure 7.14 shows the structure of the three ketone bodies. D-3-hydroxybutyrate and acetoacetate are formed primarily in the

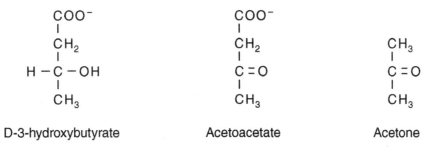

Figure 7.14 The three members of the ketone body family. D-3-hydroxybutyrate and acetoacetate are water-soluble lipids that can be oxidized in cells. Acetone, formed by spontaneous decarboxylation of acetoacetate, is present only in trace amounts.

matrix of liver mitochondria and, to a minor extent, in the kidneys. These are soluble in aqueous medium and are the major representatives of the ketone body family. Acetoacetate undergoes slow, spontaneous decarboxylation to yield acetone and carbon dioxide. The acetone is formed in only very small amounts, but because it is volatile, its presence, like that of alcohol, can be smelled in the breath.

Formation of Ketone Bodies

Ketone body formation accelerates in normal people when the body's carbohydrate content is extremely low—for example, during starvation or self-controlled fasting, when extremely low-carbohydrate diets are eaten, and during prolonged exercise with insufficient carbohydrate ingestion. In all of these conditions, carbohydrate content in the body is low, and, thus, carbohydrate utilization and blood insulin concentration are low. Accelerated ketone body formation also occurs during uncontrolled diabetes mellitus. Although blood glucose concentration is high because insulin is lacking, carbohydrate utilization by insulin-dependent tissues is low.

With starvation or fasting, low-carbohydrate diets, prolonged exercise, or uncontrolled diabetes mellitus, the adipose tissue releases large quantities of fatty acids due to an imbalance between triacylglycerol formation and lipolysis caused by low blood insulin concentration. Recall that insulin promotes triacylglycerol formation in fat cells and inhibits lipolysis. Thus, the net effect of low blood insulin is that lipolysis greatly exceeds triacylglycerol formation, resulting in a large increase in blood FFA concentration. For most healthy people, blood FFA concentrations range between 0.25 to 0.50 mM over the day. In uncontrolled diabetes mellitus, for example, blood FFA concentrations can reach toxic levels of 4 mM.

Under normal conditions, the liver is able to extract about 30% of the FFA that pass through it. With high blood FFA concentration, the liver extracts even more. The fate of the extracted fatty acids by liver is formation of fatty acyl CoA, and then subsequent formation of triacylglycerol or phospholipid, or entry into the mitochondrial matrix. During conditions favoring ketone body formation, entry of fatty acyl CoA into liver mitochondria is accelerated. Beta-oxidation of the fatty acyl CoA is greatly augmented, forming acetyl CoA at a rate that far exceeds the capacity of the liver mitochondria to oxidize it by the citric acid cycle. Moreover, conditions favoring ketone body formation are characterized by low matrix concentrations of oxaloacetate in the liver. Remember that a small amount of oxaloacetate synthesis has to occur continually to replace the small amounts of citric acid cycle intermediates that are used in other metabolic pathways. As noted in chapter 4, carbohydrates are required for oxaloacetate synthesis. Hence, in times of starvation or low carbohydrate diets, when carbohydrate availability is curtailed, lower oxaloacetate levels will prevail in the liver. When the combination of low oxaloacetate levels and high fat utilization limit complete entry of all fats into the citric acid cycle for oxidation, a portion of the excess acetyl CoA is directed to the formation of acetoacetate. Some acetoacetate is reduced to D-3-hydroxybutyrate. Figure 7.15 summarizes ketone body formation. The two major ketone bodies formed are acetoacetate and D-3-hydroxybutyrate. The ratio of 3-hydroxybutyrate to acetoacetate depends on the [NADH]/[NAD$^+$] ratio in the liver.

Figure 7.16 shows in more detail the steps between acetyl CoA and acetoacetate. In the first step, two acetyl CoA combine to form acetoacetyl CoA, catalyzed by *thiolase*, the same enzyme catalyzing the fourth reaction in beta-oxidation. This reaction is freely reversible. Acetoacetyl CoA combines with another acetyl CoA to form 3-hydroxy-3-methylglutaryl (HMG) CoA, catalyzed by *HMG CoA*

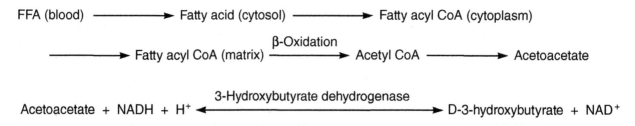

Figure 7.15 The steps in the formation of ketone bodies, starting with increased blood FFA (free fatty acid) concentrations and leading to the formation of acetoacetate in liver mitochondria. Acetoacetate can be reduced to 3-hydroxybutyrate; thus, liver produces both acetoacetate and 3-hydroxybutyrate.

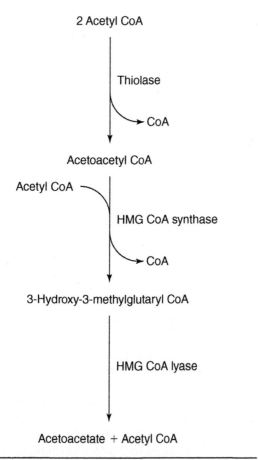

Figure 7.16 The formation of acetoacetate from acetyl CoA in three steps, each catalyzed by a separate enzyme. HMG represents 3-hydroxy-3-methylglutaryl.

synthase. The last step is splitting off an acetyl CoA, leaving acetoacetate. This reaction is catalyzed by *HMG CoA lyase*.

Fate of Ketone Bodies

Ketone bodies are used as a fuel for mitochondria in extrahepatic tissues (i.e., nonliver tissues), chiefly skeletal muscle, heart, and the brain. Normally, the brain uses glucose as its primary fuel because FFA cannot pass the blood–brain barrier. However, as blood glucose concentration falls and blood ketone-body concentration rises, the brain can extract and use 3-hydroxybutyrate and acetoacetate. Ketone-body oxidation in tissues (see figure 7.17) occurs as follows:

1. Ketone bodies are taken up by extrahepatic cells and transported into the mitochondrial matrix.
2. D-3-hydroxybutyrate is oxidized to acetoacetate using the enzyme *3-hydroxybutyrate dehydrogenase*. This reaction generates acetoacetate and NADH.
3. Acetoacetate takes a CoA from succinyl CoA and forms acetoacetyl CoA.
4. Acetoacetyl CoA is split into two acetyl CoA using the last enzyme of beta-oxidation of fatty acids, *thiolase*.
5. The acetyl CoA is oxidized in the citric acid cycle.

Ketosis

Prolonged depletion of body carbohydrate stores, or uncontrolled diabetes mellitus, leads to a condition known as ketosis. This results from accelerated ketone-body formation. Ketosis is characterized by the following:

1. **Ketonemia**—an increase in ketone-body concentration in the blood
2. **Ketonuria**—loss of ketone bodies in the urine
3. Acetone breath—loss of volatile acetone in expired air
4. Elevated blood FFA concentration, resulting in accelerated formation of ketone bodies and low blood insulin
5. Acidosis (or ketoacidosis)—a fall in blood pH due to the formation of H^+ when acetoacetate is formed and the loss of cations (e.g., Na^+) when ketone bodies are excreted (this can be fatal in uncontrolled diabetes mellitus)

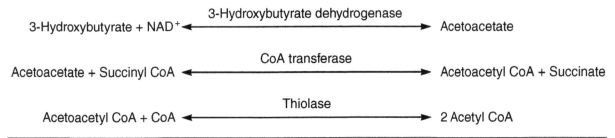

Figure 7.17 Acetoacetate and 3-hydroxybutyrate as fuels for mitochondria. The 3-hydroxybutyrate must first be oxidized to acetoacetate.

6. Hypoglycemia (in normal persons) or hyperglycemia (in uncontrolled diabetes mellitus)

The consumption of high-fat, low-carbohydrate, and severely caloric-restricted diets (ketogenic diets) has at times in the past been misguidedly promoted in the belief that by inducing ketosis and losing some of the ketone bodies and the energy they contain through ketonuria (loss of ketones in the urine), a significant amount of calories and body weight could be lost without energy expenditure. It is now clear that the amount of ketones lost via ketonuria under these dietary conditions is minimal, and that this process does not contribute significantly to weight loss. In addition, the many and significant undesirable health effects of high-fat diets and ketosis greatly outweigh any of the potential benefits of such diets as a means of weight loss.

> ### ✓ KEY POINT
> Ketones can be a useful fuel during submaximal exercise, sparing the use of muscle glycogen and blood glucose. Use of ketone bodies to fuel exercise performance is enhanced with a week or more of a ketogenic diet. This increases the activity of enzymes needed to form acetyl CoA in mitochondria from ketone bodies. The net effect is to reduce the need to provide the citric acid cycle with acetyl CoA from pyruvate. Despite these potential benefits, the long-term drawbacks of a ketogenic diet, which restricts carbohydrate intake, on exercise and training performance tend to make it impractical for most athletes.

SYNTHESIS OF FATTY ACIDS

Most fatty acids used by humans come from dietary fat. However, humans can synthesize fatty acids from acetyl CoA in liver, mammary gland, muscle, and adipose tissue, although fatty-acid synthesis is not a major process in humans. The acetyl CoA comes from amino acids, carbohydrate, and alcohol. Remember, as noted previously in this chapter, when humans are eating a mixed diet that is positive in calories and induces weight gain, the vast majority of the additional fatty acid and triacylglycerol deposited in adipose tissue are synthesized from dietary fats, since carbohydrate is preferentially either oxidized or stored in glycogen deposits. The synthesis of fatty acids is often described as **de novo lipogenesis** (DNL), which means synthesis of fatty acids starting from acetyl CoA. The 16-carbon palmitic acid is synthesized first. It can be extended in length or desaturated to make some unsaturated fatty acids (but not the essential fatty acids needed in the diet). Synthesis of palmitic acid occurs in the cytosol, whereas oxidation occurs in the mitochondria. Synthesis and degradation of fatty acids also use different enzymes, permitting separate regulation of these two opposing processes.

Pathway to Palmitic Acid

Synthesis starts with the carboxylation of acetyl CoA to make a three-carbon molecule known as malonyl CoA (see figure 7.18). The cytosolic enzyme *acetyl CoA carboxylase* catalyzes this reaction, which involves the carboxylation of acetyl CoA. The actual substrate is the bicarbonate ion (HCO_3^-), which is transferred to acetyl CoA by the B vitamin biotin. This reaction, which is the committed step in fatty-acid synthesis, is positively affected by insulin. We have also seen biotin involved in the pyruvate carboxylase reaction in gluconeogenesis. The malonyl CoA units are used to make fatty acids in the cytosol.

To make malonyl CoA in the cytosol, a continuous supply of acetyl CoA is necessary. Recall that acetyl CoA is formed in the mitochondrial matrix, primarily from pyruvate in the pyruvate-dehydrogenase reaction and from beta-oxidation of fatty acids. Thus, a mechanism must exist to get acetyl CoA from the mitochondrial matrix to the cytosol.

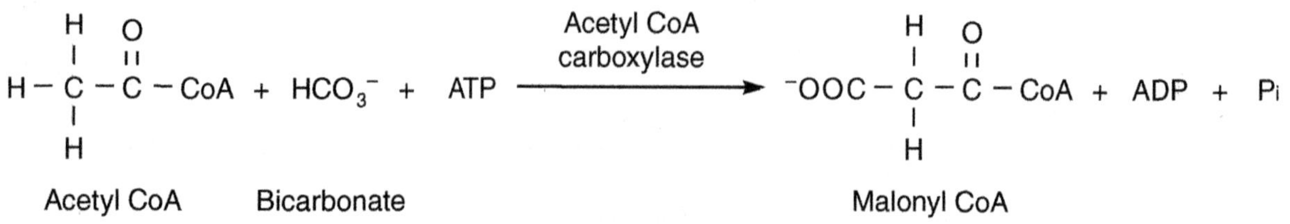

$$\text{H}-\underset{\underset{\text{H}}{|}}{\overset{\overset{\text{H}}{|}}{\text{C}}}-\overset{\overset{\text{O}}{\|}}{\text{C}}-\text{CoA} + \text{HCO}_3^- + \text{ATP} \xrightarrow{\text{Acetyl CoA carboxylase}} {}^-\text{OOC}-\underset{\underset{\text{H}}{|}}{\overset{\overset{\text{H}}{|}}{\text{C}}}-\overset{\overset{\text{O}}{\|}}{\text{C}}-\text{CoA} + \text{ADP} + \text{P}_i$$

Acetyl CoA   Bicarbonate   Malonyl CoA

Figure 7.18 Creation of malonyl CoA, the precursor for the synthesis of fatty acids, by the carboxylation of acetyl CoA.

Moreover, NADPH is needed in fatty-acid synthesis, so there must be a way of producing this in the cytosol. NADPH can arise from the pentose phosphate pathway (see chapter 6). It can also be produced by *malic enzyme*, as shown in the following reaction.

$$\text{malate} + \text{NADP}^+ \leftrightarrow \text{pyruvate} + \text{CO}_2 + \text{NADPH} + \text{H}^+$$

As noted previously, even though carbohydrate is not a major contributor to fatty-acid synthesis, it can be converted to fat in certain circumstances. Figure 7.19 outlines how an excess of glucose can be used to make malonyl CoA, the precursor for fatty-acid synthesis. Glycolysis converts glucose to pyruvate. It is translocated into the matrix and converted to acetyl CoA by pyruvate dehydrogenase. Acetyl CoA combines with oxaloacetate to produce citrate, using the first enzyme of the citric acid cycle, *citrate synthase*. The citrate is transported into the cytosol by an antiport, with malate simultaneously going from the cytosol to the matrix. In the cytosol, citrate is cleaved into oxaloacetate and acetyl CoA by *ATP-citrate lyase*. This enzyme uses the free energy of hydrolysis of ATP, not only to split citrate but also to attach a CoA. The acetyl CoA in the cytosol is now a substrate for acetyl CoA carboxylase. The oxaloacetate is reduced to malate and transported back into the matrix by the citrate–malate antiport.

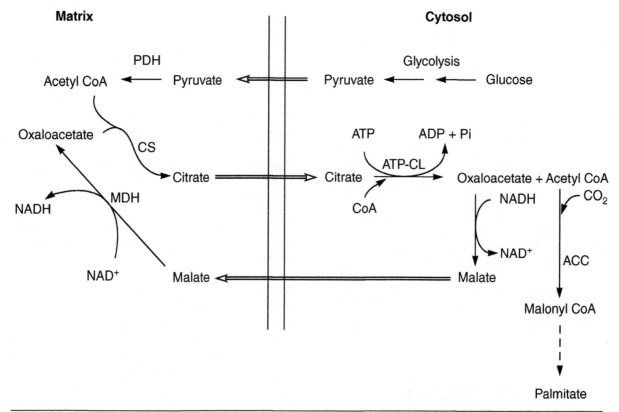

Figure 7.19 Use of excess glucose to make malonyl CoA. Pyruvate from glucose is converted to acetyl CoA by pyruvate dehydrogenase (PDH). Acetyl CoA combines with oxaloacetate to form citrate using citrate synthase (CS). Citrate crosses to the cytosol using an antiport, with malate going the other way. Citrate is split into acetyl CoA and oxaloacetate by ATP-citrate lyase (ATP-CL). The acetyl CoA is carboxylated to make malonyl CoA by acetyl CoA carboxylase (ACC). The oxaloacetate is reduced to malate by a cytosolic malate dehydrogenase, and malate returns to the matrix.

Palmitic-acid synthesis occurs using a large enzyme, *fatty-acid synthase* (FAS), which is composed of two very long, multifunctional polypeptide chains. The FAS has seven distinct enzyme activities. We will describe fatty-acid synthesis as it proceeds in stages. Refer to figure 7.20 for the details. In the loading stage, an acetyl CoA and a malonyl CoA are each transferred by separate *transacetylase* enzymes to the thiol end (–SH) of an *acyl carrier protein* (ACP) part of the FAS. These transfers result in an acetyl-ACP and malonyl-ACP. In the condensation stage, the acetyl group attaches to the end of the malonyl-ACP, and a CO_2 is eliminated. This creates a four-carbon acetoacetyl group attached to one of the ACP. This contains a carbonyl group (C = O) that must be converted to a methylene (CH_2), which is accomplished in three stages. In the first stage, the carbonyl is reduced to an alcohol group by using NADPH. The next stage involves a dehydration, which eliminates the alcohol OH group and another H, creating a carbon-to-carbon double bond. Finally, another reduction, again using NADPH, eliminates the double bond. The reduction, dehydration, and reduction are exactly opposite to the first three reactions in beta-oxidation, where instead of elimination of a carbonyl group (as occurs in fatty-acid synthesis), the carbonyl is created by dehydrogenation, hydration, and then dehydrogenation.

Figure 7.20 shows the formation of a four-carbon acyl group only. The FAS complex can form a 16-carbon palmitoyl group, which is removed as palmitate. To go from the four-carbon group attached to an ACP, as shown in figure 7.20, to a palmitoyl ACP, we need to add 12 more carbons. The first stage in elongating our butyryl ACP in figure 7.20 would involve, first, attachment of a malonyl CoA to the free ACP group. This malonyl group would condense with our butyryl ACP with the elimination of a CO_2, and we would create a new six-carbon ACP with a carbonyl group that we would need to eliminate by reduction, dehydration, and reduction. This process continues until a palmitoyl ACP is formed. At this point, the palmitoyl ACP is split by a *thioesterase* enzyme, releasing palmitate (palmitic acid).

To make one palmitate requires the following precursors: one acetyl CoA, seven malonyl CoA, and 14 NADPH + 14 H^+. It is obvious that a supply of reducing equivalents in the form of NADPH is necessary, emphasizing the importance of the pentose phosphate pathway and the malic enzyme as suppliers of NADPH. The need for malonyl CoA is obvious and, as we will see in the next section, the formation of malonyl CoA is a crucial point for regulation.

Regulation of Fatty-Acid Synthesis

As we have seen, control of a number of enzymes in liver depends on our diets. For example, the genes coding for enzymes of gluconeogenesis are stimulated by fasting or a low-carbohydrate diet but are repressed if the diet is high in carbohydrate. On the other hand, genes for the enzymes of glycolysis are repressed by fasting or a low-carbohydrate diet and are activated if the diet is rich in carbohydrate. Accordingly, we might suspect that genes coding for enzymes of de novo lipogenesis (DNL) should depend on the energy and macronutrient content of the diet, with maximal expression if there is an excess of food energy in the form of carbohydrate. This is the case, since a carbohydrate-rich diet leads to an increased expression of the genes for acetyl CoA carboxylase, which makes the key precursor, and FAS, which produces palmitate. Since the concentration of insulin in blood is also influenced by feeding, especially carbohydrate, it should not be a surprise to learn that insulin plays a key role in promoting DNL. On the other hand, demonstrating DNL has been difficult in humans because of the size of our fat depots compared to the amount of palmitate actually formed by DNL each day. For example, a person with 15% body fat and a body weight of 70 kg (154 lb) has approximately 10 kg (22 lb) of stored TAG. Assuming that the person mobilizes a maximum of 100 g of fat (a large amount) each day, this represents only 1% of the total stored fat. In addition, if the person ingests about 100 g of fat, one can easily understand how difficult it is to demonstrate that 20 g of palmitate is formed each day through DNL by the liver and adipose tissue. The reality is that DNL from glucose is of minor importance to the overall energy balance of the average person on a typical mixed diet.

With its role in providing malonyl CoA, the enzyme acetyl CoA carboxylase (ACC) is a prime site for regulation of DNL. Formation of malonyl CoA provides the precursor to synthesize palmitate. Moreover, malonyl CoA is a potent inhibitor of CPT I, which (together with FAT/CD36) coregulates the oxidation of LCFAs. Thus, the activation of ACC both promotes palmitoyl synthesis and inhibits fatty-acid oxidation. Acetyl CoA carboxylase is controlled by allosteric and phosphorylation mechanisms. Citrate is a positive allosteric effector for ACC, whereas fatty acyl CoA and malonyl CoA allosterically inhibit ACC. The potency of the inhibition by the fatty acyl CoA molecules is proportional to the carbon length of the fatty acyl units. An abundance of fatty acyl CoA and malonyl CoA signals a surfeit of products;

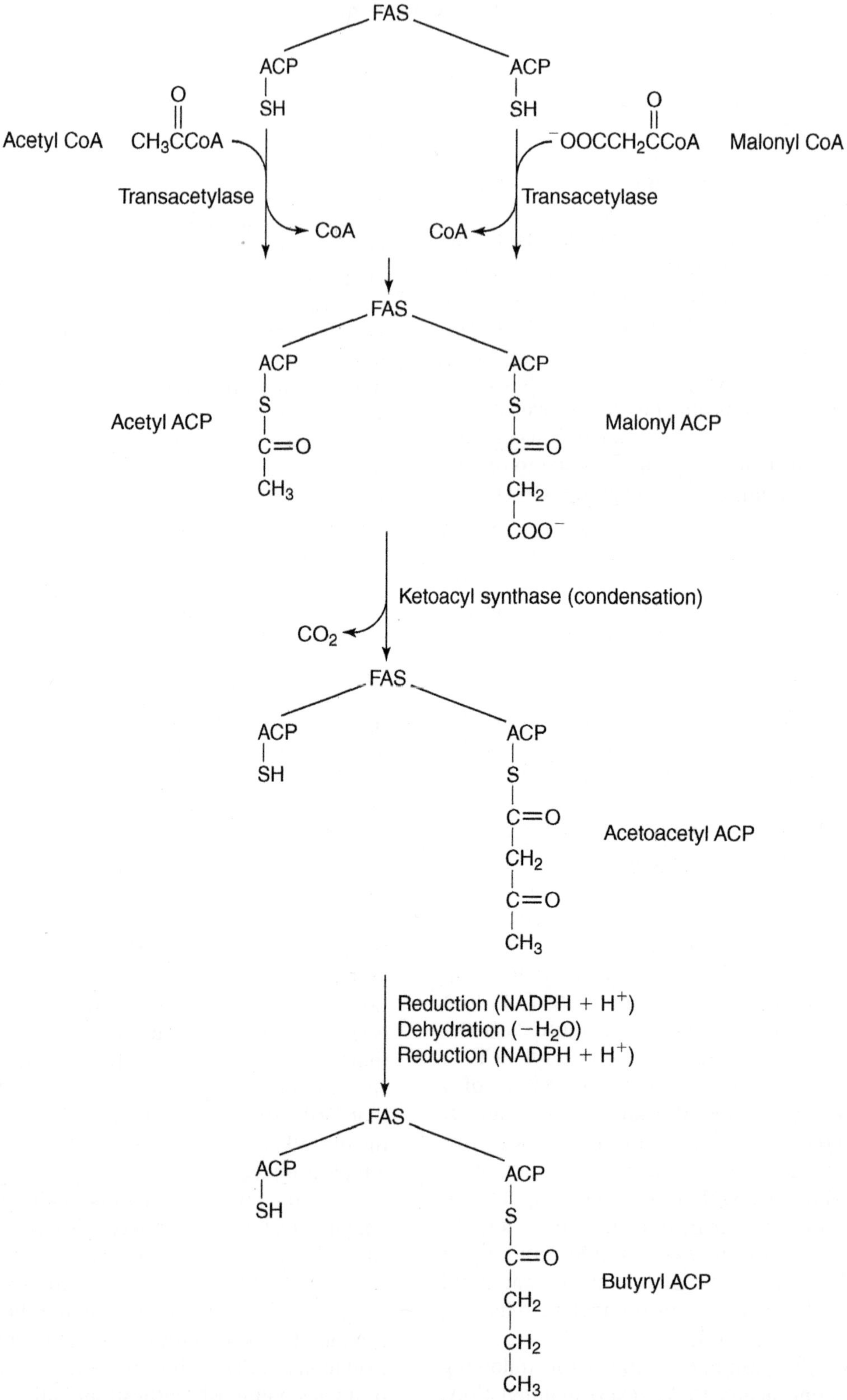

Figure 7.20 Synthesis of fatty acids on a large, multiunit enzyme, fatty-acid synthase (FAS), containing acyl carrier protein (ACP) subunits with a free thiol (SH) group. The process begins as an acetyl CoA and a malonyl CoA each attach to an ACP; CoA is released. The acetyl group leaves its ACP and attaches to the malonyl group with the elimination of a CO_2. This forms a four-carbon acetoacetyl ACP. The carbonyl group (C=O) is eliminated in three steps involving reduction using NADPH, a dehydration, and another reduction using NADPH. The resulting product is butyryl ACP. The process continues when another malonyl CoA is attached to the free ACP. Ultimately, a 16-carbon palmitoyl ACP is produced, and then a palmitic acid is released.

therefore, the need to make more fatty acids is no longer present. On the other hand, an increase in citrate concentration indicates an overabundance of acetyl CoA and a need to synthesize fatty acids. An abundance of carbohydrate means that a small amount of fatty-acid synthesis from carbohydrate could take place. An accumulation of fatty acyl CoA reflects a lack of carbohydrate substrate; thus, fatty-acid oxidation should predominate.

Acetyl CoA carboxylase activity is inhibited by covalent phosphorylation, directed by a cAMP-dependent protein kinase (protein kinase A, or PKA) and AMP-activated protein kinase (AMPK). The previous chapter shows that AMPK is active under stressful conditions in which ATP is increasingly needed. Dephosphorylation by protein phosphatases activates ACC. Phosphorylation and dephosphorylation of ACC depend on the relative levels of insulin and glucagon, with insulin promoting the unphosphorylated (active) form of ACC and glucagon promoting its phosphorylation. The latter results in inhibition of ACC. A relative rise in insulin concentration leads to a decrease in cAMP concentration by activation of cAMP phosphodiesterase; insulin also activates phosphoprotein phosphatases, which dephosphorylate ACC. AMPK will also increase translocation of the sarcolemma and mitochondrial fatty-acid cotransporters, FAT/CD36 and $FABP_{pm}$. In addition, AMPK indirectly inhibits malonyl-CoA production, thereby reducing the inhibitory effects of malonyl-CoA on CPT I and further facilitating fatty-acid entry into mitochondria. As also noted elsewhere in this chapter, signaling factors derived from adipose tissue, such as leptin and adiponectin, can also influence muscle-lipid turnover by activation of AMPK.

Malonyl CoA is also a potent inhibitor of the oxidation of fatty acids. In tissues where fatty acids can be oxidized to acetyl CoA in the mitochondria and where fatty acids can be simultaneously synthesized in the cytosol (primarily liver), malonyl CoA can inhibit the initial transfer of the fatty acyl group from CoA to carnitine at the cytosolic side of the inner membrane using the enzyme carnitine palmitoyl transferase I (see figure 7.12). This process prevents futile cycling in which fatty acids are actively broken down in the mitochondria while being simultaneously synthesized in the cytosol. Such regulation is active in the liver and adipose tissue, the main sites for fatty-acid synthesis in humans. On another level, a relative increase in the intake of carbohydrate in the diet induces the transcription of genes coding for lipogenic enzymes. A high-fat diet, on the other hand, results in a decrease in the transcription of genes for lipogenic enzymes.

KEY POINT

If de novo lipogenesis is not a prominent feature of human metabolism, one might ask how people can become obese on a high-carbohydrate and relatively low-fat diet. People become obese whenever they have a prolonged excess of food energy intake, whether it is carbohydrate or fat. If the diet provides excess calories as carbohydrate, the body uses carbohydrate more as an energy source so that fat stores and dietary fat are oxidized less and stored more readily. The excess carbohydrate, beyond what is needed to restore liver and muscle glycogen, can in part be converted to fatty acids or used as a source of glycerol 3-phosphate to help store even the small amount of dietary fat.

FAT AS FUEL FOR EXERCISE

Human fat stores are huge compared to carbohydrate stores. Moreover, considering the importance of glucose as fuel for the brain, it is important to use as much fat as possible, instead of carbohydrate, during exercise. The lipid used to fuel muscular work comes from fatty acids released from adipose tissue, traveling to muscle as FFA. Intramuscular triacylglycerol is also a significant source of fatty acids. A third possibility is fatty acids released from plasma triacylglycerols by lipoprotein lipase, although the evidence suggests that this can provide at most 10% of fat oxidized during exercise lasting an hour or more.

Free Fatty Acids and Exercise

Blood glucose levels are well maintained during exercise lasting up to 60 min. This is due to the careful match between uptake of glucose from the blood and release of glucose to the blood from the liver. The relative constancy of blood glucose concentration during exercise is not observed with the concentration of FFA. With exercise, there is an increase in lipolysis, brought about mainly by the increase in epinephrine in the blood, increased sympathetic nervous activation of adipocytes through norepinephrine, and the decrease in plasma-insulin concentration. In addition, reesterification of fatty acids in fat cells decreases, so more are released to the blood. Aiding this is a small increase in adipose-tissue blood flow that allows a greater proportion of fatty acids released from fat cells to enter the general circulation (Jeukendrup 2003). Figure 7.21 illustrates the effect of 90 min of exercise at a workload of 60% of $\dot{V}O_2$max on arterial FFA concentration for a subject in the postabsorptive state, and then for the same subject, on a separate occasion, 2 h after eating a meal containing 50% of the food energy as carbohydrate. Overall, the shapes of the two curves are similar, but the curve for the fasted state is displaced higher. The arterial FFA concentration reflects the relative balance between release of fatty acids from adipose tissue (R_a, rate of appearance) and FFA uptake by exercising muscle (R_d, rate of disappearance).

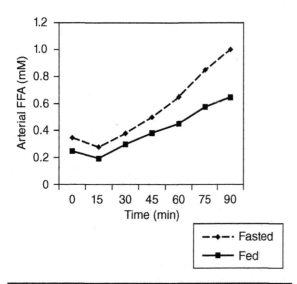

Figure 7.21 The arterial free fatty acid (FFA) concentration during 90 min of exercise at 60% of $\dot{V}O_2$max when a subject had fasted before the exercise and then 2 h following a meal containing 50% of the energy from carbohydrate.

At the start of submaximal exercise, there is an immediate increase in the rate of FFA uptake by the exercising muscle (R_d). This exceeds the more slowly responding increase in adipose tissue lipolysis (R_a), such that the plasma FFA initially falls. With the gradual increase in lipolysis, the rate of release of FFA from adipose tissue meets, then exceeds, the rate of uptake of plasma FFA ($R_a > R_d$). As a result, the concentration of plasma FFA tends to rise over the course of the exercise period. In the fed state, the adrenergic effect to stimulate lipolysis is blunted by the effect of the previous meal, especially the concentration of insulin. Thus, plasma FFA concentration begins at a lower level and remains lower throughout the 90 min exercise period. As we will show in a subsequent section, doing the same exercise protocol in the fed state, compared to the fasted state, is characterized by a higher respiratory exchange ratio (RER), reflecting greater oxidation of carbohydrate.

During more intense exercise, release of fatty acids from adipose tissue is reduced, even though there is a greater adrenergic stimulus for lipolysis. Two mechanisms have been proposed to explain this. At higher intensities of exercise, there is a higher concentration of lactate in blood. Researchers have proposed that lactate inhibits lipolysis, but this is a controversial hypothesis. It is possible that lactate increases reesterification of fatty acids into adipose tissue triacylglycerols through a process of glyceroneogenesis, discussed earlier. This means that rather than being released from the fat cell to enter the blood, the fatty acid is esterified to a new triacylglycerol molecule. Another more likely factor that may contribute to reducing fatty-acid release from adipose tissue is that adipose-tissue blood flow is reduced at higher intensities of exercise, since flow regulation aims to direct blood to the active muscles by decreasing flow to the liver, kidney, and adipose tissues. This effect is related to the relative intensity of exercise. Therefore, fatty acids released by lipolysis are less likely to enter the general circulation due to a lower adipose-tissue blood flow (Jeukendrup, Saris, and Wagenmakers 1998).

Focusing only on changes in plasma FFA concentration as an indication of fat oxidation during exercise can be very misleading, since this does not tell us what is happening with these fatty acids. We also could measure the RER to determine the relative use of fat, since we know that the lower the value, the more that fat is being oxidized. Again, this does not indicate which tissues are using the fat; although, with a large increase in metabolic rate because of the exercise, it is safe to assume that exercising skeletal muscle is the major tissue contributing to the RER value. We could further measure the concentration of FFA in arterial and venous blood across an exercising limb to determine the relative use of plasma FFA. However, accurate measures of FFA uptake into skeletal muscle require good values for blood flow, and this is not a simple technique. Moreover, subcutaneous adipose tissue can be an active source of FFA to the blood of the exercising limb, again leading to complications in the interpretation of the differences in arterial and venous FFA concentration. A technique that is proving quite helpful is to infuse a labeled fatty acid into a vein at a precise rate. Sampling blood before and during exercise for changes in the concentration of the labeled fatty acid allows one to make predictions about the rate of disappearance of fatty acids from the blood (Bülow 2003).

Use of Intramuscular Triacylglycerol

As we discussed previously, skeletal muscle contains intracellular lipid drops, called intramuscular triacylglycerol (IMTG) or intramyocellular lipid (IMCL). Skeletal muscle also has HSL. Although some variability exists in research findings, it is estimated that up to 10% of total energy expenditure in exercise lasting 1 h or longer can be derived from IMTG. While the use of IMTG can be an important source of energy during exercise, it still represents only a small fraction of the total amount of fat oxidized by muscle. The majority of fat oxidized by muscle during exercise is mobilized from adipose tissue stores. Nevertheless, trained people have higher levels of IMTG than the untrained, as well as higher activity levels of muscle HSL. Although some research findings are conflicting, trained athletes can probably oxidize greater amounts of IMTG during exercise than untrained people (Melanson, MacLean, and Hill 2009).

> ### ☑ KEY POINT
> While the body contains limited carbohydrate stores, there is an abundance of stored triacylglycerol. Carbohydrate is a better fuel for intensely exercising muscle for three reasons: (1) It can generate acetyl CoA for the citric acid cycle at a much higher rate than fatty acids from adipose tissue and intramuscular triacylglycerols can, (2) more ATP per unit of oxygen is produced with carbohydrate, and (3) carbohydrate can generate ATP in the absence of oxygen via anaerobic glycolysis. Fat oxidation is favored in less intense exercise when the rate of energy utilization does not require a significant amount of carbohydrate use. The use of fats as a fuel source whenever possible serves to preserve the limited carbohydrate stores in muscle and liver.

METABOLISM DURING EXERCISE: FAT VERSUS CARBOHYDRATE

At rest, during the postabsorptive state, the RER is about 0.75 to 0.82, indicating that lipid is the predominant fuel being oxidized by the body. The respiratory quotient (RQ) across rested muscle is even lower, demonstrating that inactive muscle uses fat as its primary fuel. If a person in the rested state eats a carbohydrate-rich meal, blood glucose concentration is elevated, blood insulin concentration is increased, blood FFA concentration is reduced, and whole-body RER is increased; the RQ across rested muscle increases, as well as the lactate released from it (Didier et al. 2000). In a prolonged fasted state, when body carbohydrate stores are severely reduced, blood FFA levels are elevated and overall body RER is reduced. Observations like these were reported by Randle and his colleagues in the 1960s; from these studies, the authors proposed a *glucose-fatty acid cycle*. The proposition regarding this cycle is that a reciprocal relationship exists between carbohydrate and fat in terms of oxidation. As already mentioned, an increase in carbohydrate boosts its oxidation and depresses that of fat. If fatty-acid concentration in the blood is elevated, fat oxidation is promoted and carbohydrate oxidation is depressed (Roden 2004). The use of carbohydrate and fat during exercise follows a similar trend, but limitations do exist, which we will discuss.

Effects of Exercise Intensity

As noted in the previous chapter, it has been convincingly demonstrated that in well-nourished people during exercise, the RER increases in proportion to the increase in exercise intensity. This means that as exercise intensity increases, the contribution of carbohydrate oxidation to ATP formation also increases, whereas that of lipid oxidation decreases. This is illustrated in figure 7.22, which shows the relative release of FFA and glucose into the blood and the utilization of glycogen during exercise at different intensities. Several points are immediately clear. The release of fatty acids into the blood from adipose tissue stores rises in parallel with exercise intensity to approximately 50% of $\dot{V}O_2max$, and then gradually declines. On the other hand, release of glucose into the blood from the liver increases with exercise intensity. As we have noted, glycogen utilization increases exponentially with increasing exercise intensity.

George Brooks has proposed a **crossover concept** to explain fuel utilization during exercise in terms of the balance between carbohydrate and fat. The crossover point is that relative exercise intensity at which ATP formation from the use of carbohydrate exceeds that of lipid. Exercise at power outputs beyond this point will rely more and more on carbohydrate oxidation and less and less on fat. This is illustrated as shown in figure 7.23, which represents changes in

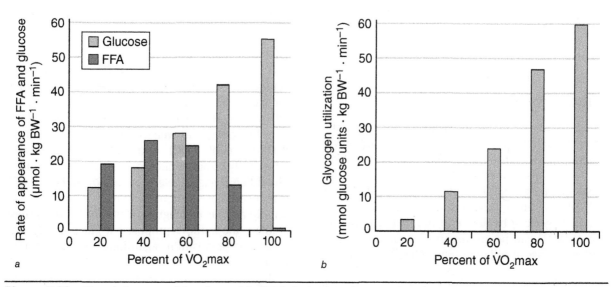

Figure 7.22 Use of glucose and free fatty acids (FFA) during exercise at various intensities, expressed as percentages of maximum oxygen uptake ($\dot{V}O_2$max). The use of FFA and glucose is based on their rates of appearance in the blood.
Data from Brooks 1997.

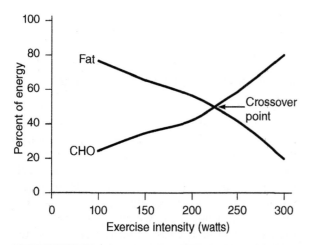

Figure 7.23 Relative contribution of lipid (fat) and carbohydrate (CHO) oxidation to energy needs at different intensities of work (watts). The crossover point is the exercise intensity beyond which energy production by carbohydrate oxidation exceeds that of lipid. The lines represent results from a moderately trained subject, and the crossover point could occur at higher or lower exercise intensities depending on the training state of the subject.
Data from Austin and Seebohar 2011.

fuel utilization with increasing work intensity in a moderately trained subject.

Effects of Diet

In the previous chapter, we showed (see figure 6.17 on p. 183) that preexercise diet, and hence muscle glycogen concentration, could play an important role in fuel utilization during submaximal exercise. With elevated muscle glycogen stores, more glycogen is utilized and more lactate is produced compared to values from the identical exercise task performed with low glycogen stores. Elevated muscle glycogen stores are even associated with lower blood FFA levels at rest (Didier et al. 2000). Although this point was not directly addressed in chapter 6, any reduction in carbohydrate use during exercise is made up by lipid oxidation because the only other possible fuel, protein, provides only about 5% of total energy needs at most. Thus, the reciprocal relationship between carbohydrate and lipid oxidation at rest extends to exercise up to 90% of $\dot{V}O_2$max. Beyond this intensity, exercise relies almost exclusively on carbohydrate. At rest, carbohydrate ingestion raises blood glucose and insulin levels. Insulin promotes glucose uptake by skeletal and cardiac muscle, as well as adipose tissue. It is also a potent inhibitor of lipolysis in adipose tissue, thereby depressing blood FFA concentrations.

A diet rich in fat may alter metabolism during exercise. It is important to distinguish acutely elevating plasma FFA concentration and its effects on exercise metabolism and performance from the effects that accrue as a result of seven days or more of a high-fat diet. Acutely increasing plasma FFA levels can be accomplished by infusion of triacylglycerol into a vein along with heparin, which is known to increase lipoprotein lipase activity. Such a protocol depresses blood glucose utilization at rest and during submaximal exercise and increases the use of fat during exercise, as evidenced by a lower RER. A number of studies have shown that a high-fat diet for seven or more days results in an increase in IMTG and an increased reliance on fat oxidation during submaximal exercise (Vogt et al. 2003; Zderic et al. 2004). During exercise, blood FFA and glycerol concentrations are higher, revealing a higher rate of adipose tissue lipolysis. A single high-fat meal consumed a few hours before exercise also increases fat utilization during exercise. This upregulation of fat utilization and consequent downregulation of carbohydrate utilization during exercise appears to be mediated by attenuation of PDH kinase activation, which reduces PDH activation and slows the PDH-mediated production of acetyl from carbohydrate sources in the mitochondria (Bradley et al. 2008).

Not all types of fats are oxidized to the same extent during exercise. When a diet of mixed types of fats are consumed before exercise, saturated fats are preferentially diverted for storage while polyunsaturated fats (PUFAs) are preferentially oxidized to supply the energy needed for the performance of the exercise. Omega-6 PUFAs appear to be the most preferred fat for oxidation during exercise (Bradley et al. 2008). This hierarchy of types of fats that are preferentially oxidized or stored has important implications for dietary recommendations regarding fat intake for exercise performance, as well as for body weight regulation.

> **☑ KEY POINT**
>
> Greater consumption of dietary fat increases fat utilization during exercise. Exercise metabolism also tends to favor the oxidation of dietary PUFAs, while dietary saturated fats have a greater tendency to be shunted for storage in adipose tissue. This has implications for dietary recommendations for exercise and weight loss.

Effects of Feeding During Exercise

We noted that if people perform a given amount and intensity of exercise with full muscle and liver glycogen stores, they will use more glycogen and produce a higher level of muscle and blood lactate compared to levels with the exact same exercise performed when glycogen stores are low. This tells us that if all things are equal, muscle will choose to use more muscle glycogen and blood glucose, if they are available. If a person doing prolonged submaximal exercise does not ingest glucose, the RER during exercise will gradually decline; after about 1 h, blood glucose levels will also decline. This decline in RER reflects the decreased availability of liver and muscle glycogen as exercise progresses and the limited carbohydrate stores are used up. In this situation, longer term, lower intensity exercise can be sustained for some time by increased fatty-acid oxidation, which can partially compensate for reduced carbohydrate availability and oxidation. However, for reasons noted in this and previous chapters, increased fat oxidation is less efficient, and it may be associated with the eventual onset of fatigue as carbohydrate and glycogen stores fall to critical levels. If the exercising subject takes in glucose during prolonged exercise, blood glucose levels are better maintained; also, the RER does not decline to the same extent. In addition, insulin concentrations are elevated, blood FFA levels are depressed, and the total oxidation of carbohydrate is increased and that of fat decreased (Watt, Krustrup, et al. 2004). This is another example of the effect of carbohydrate on promoting its own oxidation at the expense of fat.

Effects of Training

Regularly performed endurance exercise can play a significant role in fuel utilization during exercise. The trained muscle has a much larger capacity for oxidative metabolism than untrained muscle, as we have already addressed. Compared to before training, during exercise at the same absolute intensity (e.g., 200 W), fat oxidation is increased, carbohydrate oxidation is decreased, carbohydrate and glycogen depletion is delayed, and the time available to exercise before fatigue and exhaustion set in is vastly increased. However, if athletes increase their $\dot{V}O_2$max by 20% following an endurance training program, they are obviously capable of exercising at a much higher intensity. If we compare the oxidation of carbohydrate and fat during exercise at the same relative intensity posttraining versus pretraining, we see a different metabolic response. During exercise at, say, 60% of posttraining $\dot{V}O_2$max, the workload may be 180 W, while at 60% of pretraining $\dot{V}O_2$max, it may be 144 W. Therefore, we can say with absolute certainty that after training, the exercise $\dot{V}O_2$ will be higher at the same relative workload, and more total fuel will be oxidized during exercise. The relative use of carbohydrate and fat during exercise with the greater energy demand is more controversial. Some studies show that the relative use of that fuel is in about the same proportion as before training. Other studies show that the RER is lower, suggesting that relatively more fat is being oxidized to support the higher $\dot{V}O_2$. Even if there is no change in the relative amount of fat or carbohydrate used, by working at a higher intensity and consuming more fuel, the trained athlete would expend more calories and burn more fat in absolute terms than an untrained person who maintains exercise for the same duration.

Trained athletes have larger IMTG stores, and they use these stores to a greater extent than the untrained. Trained muscle has a higher lipoprotein-lipase activity, so the ability to hydrolyze and use fatty acids derived from VLDLs is enhanced (Morio et al. 2004). As previously noted, studies have also shown that key enzymes involved in fatty-acid transport (FAT/CD36 and CPT I) are increased with endurance training. Trained people also have up to twice the concentration of muscle mitochondria and mitochondrial enzymes relative to untrained ones. Having more muscle mitochondria favors greater oxidative metabolism and, specifically, fat metabolism during any exercise intensity. With more mitochondria, relatively more ATP can be produced aerobically from fat metabolism; this can happen more rapidly as well. This allows for an increase in fat utilization at higher intensities of exercise without compromising the rate of ATP regeneration. These types of training adaptations favoring lipid metabolism are also experienced by older subjects (Pruchnic et al. 2004). As we will emphasize in a later section on mechanisms regulating carbohydrate and lipid metabolism in muscle, AMPK plays a major role in skeletal-muscle fuel selection. Endurance training increases total AMPK activity in skeletal muscle and the basal activity of this enzyme (Frøsig et al. 2004).

Role of Body Composition

Despite the fact that overweight and obese people store a much greater proportion of triacylglycerol than their lean counterparts, the extra fat does not predispose overweight people to its use, compared to carbohydrate, during exercise (Mittendorfer, Fields, and Klein 2004). Indeed, overweight and obese subjects demonstrated lower levels of adipose tissue lipolysis and oxidation of FFA during exercise. This suggests a blunted response to catecholamines with increasing adiposity, which may contribute a greater initial resistance to weight and fat loss for obese people starting an exercise program.

Sex Differences

On average, women have greater total body fat than men, as well as different body-fat distribution. Venables and colleagues (2005) tested a cross section of 300 trained and untrained men and women in terms of fuel utilization, using indirect calorimetry during a graded exercise test to exhaustion. Compared to men, women had higher rates of fat oxidation and a later shift to carbohydrate oxidation as exercise intensity increased. One factor that may predispose women for greater fat metabolism is estrogen. Estrogen has been shown to be a potent factor in promoting fat metabolism in animals and humans. Research indicates that estrogen promotes fat oxidation via activation of AMPK (Oosthuyse and Bosch 2010). As previously noted, AMPK increases translocation of the sarcolemma and mitochondrial fatty-acid cotransporters, FAT/CD36 and FABP$_{pm}$. In addition, AMPK indirectly inhibits malonyl-CoA production, thereby reducing the inhibitory effects of malonyl-CoA on CPT I and further facilitating fatty-acid entry into mitochondria. Malonyl-CoA effects on muscle-fat utilization are discussed in more detail later in this chapter. Women are also reported to have higher IMTG stores, higher muscle content of FAT/CD36, FABP$_{pm}$, and CPT I, and higher HSL activity than men. Moderately active women also have higher muscle levels of β-oxidative enzymes than similarly active men (Maher et al. 2010). As discussed in chapter 3, estrogen also increases production of factors such as PPARα and PPARγ, which enhance gene expression of enzymes involved in fat metabolism (Maher et al. 2010).

Taken together, these differences, primarily mediated by estrogen, appear to predispose women for a greater metabolism capacity of fatty acids than men. The implications of these sex differences to exercise performance, particularly to ultra-long endurance performance where fat metabolism predominates, are not yet clear. Some researchers, however, have suggested that the menstrual cycle may have some positive effects on endurance exercise performance in females, particularly at the late-follicular stage, where estrogen is elevated and progesterone, which can counteract some of the metabolic effects of estrogen, is suppressed (Oosthuyse and Bosch 2010). The metabolic implications for older postmenopausal women, who have reduced estrogen levels, are not yet fully documented.

Mechanisms of the Glucose–Fatty Acid Cycle

We have discussed the effects of fuel availability on metabolism. In this section, we will look at mechanisms to account for these observations.

Increased Fat Availability and Carbohydrate Utilization

Randle and his colleagues (1963; 1964) offered a mechanism to account for the effects of increased fatty acids and ketone bodies on depressing carbohydrate oxidation (figure 7.24). They proposed an inhibition of glucose utilization in the muscle cell, blocked at two sites. The first step is at the pyruvate dehydrogenase reaction, due to an increased concentration of acetyl CoA and NADH in the mitochondrial matrix (see Regulation of Pyruvate Oxidation in chapter 5, p. 132). Second, the elevated matrix acetyl-CoA concentration would lead to an increase in citrate in the matrix and, therefore, an increase in cytosolic citrate. As discussed in chapter 6, phosphofructokinase, the prime regulatory enzyme of glycolysis, is inhibited by citrate, leading to a rise in the concentration of its substrate, fructose 6-phosphate, and thus glucose 6-phosphate. Since glucose 6-phosphate inhibits hexokinase, this would reduce the gradient for glucose transport into the muscle cell. Moreover, glycogen utilization would be depressed, since glycogenolysis is depressed if glucose 6-phosphate concentration is elevated. Overall, muscle carbohydrate oxidation would be depressed, and the difference would be made up by fatty-acid oxidation. Ketone bodies, if elevated by very low-carbohydrate diets or prolonged fasting, are able to inhibit glucose uptake and oxidation by the same mechanisms.

While the original observations of Randle and his colleagues have been supported by numerous studies since, the mechanisms they proposed have not been embraced (Roden 2004). For example, intracellular citrate concentrations have not been shown to be elevated to the extent needed to account for phosphofructokinase inhibition. While glucose uptake is depressed if blood FFA concentrations are increased sharply, this is not associated with marked increases in glucose 6-phosphate concentrations. Finally, the PDH activity is decreased in the postabsorptive state, but a dramatic increase in blood FFA concentration may not depress PDH activity further. In summary, the glucose–fatty acid cycle operates, without doubt. However, the mechanisms proposed by Randle and colleagues were incomplete, failing to account in total for their observations.

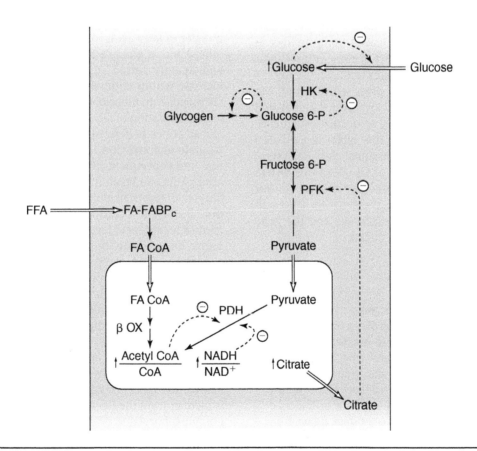

Figure 7.24 The mechanism for the glucose–fatty acid cycle. Increased blood FFA leads to increased FFA uptake, increased intracellular fatty acids (shown as FA-FABP$_c$), increased transport into mitochondria, and increased beta-oxidation (β OX), raising both the acetyl CoA/CoA and NADH/NAD$^+$ ratios, inhibiting pyruvate dehydrogenase (PDH) and pyruvate oxidation. Increased fatty-acid oxidation leads to increased citrate concentration (which inhibits phosphofructokinase, or PFK) and, subsequently, an increased intracellular glucose concentration. Thus, glucose uptake increases as well. An increase in glucose 6-phosphate inhibits glycogenolysis.

We know that a high-fat diet can raise blood FFA concentrations but, surprisingly, can increase blood glucose concentrations. This is characteristic of insulin resistance; as discussed in the previous chapter, it is a marker for the early type 2 diabetes condition. Researchers have also abruptly increased blood FFA levels by infusing triacylglycerol into a vein and then adding heparin, a known activator of the enzyme lipoprotein lipase. This results in a sharp rise in blood FFA, but it also produces hyperglycemia, even in healthy people. The blood glucose concentration in the postabsorptive state reflects the balance between glucose output from the liver, derived from liver glycogenolysis and gluconeogenesis, and glucose disposal by peripheral tissues; skeletal muscle is responsible for more than 70% of whole-body glucose disposal. Therefore, the FFA-induced hyperglycemia could arise from an increased glucose output from the liver, a suppression of glucose disposal by skeletal muscle, or both. We know that an elevation in FFA can counteract the normal suppressive effect of insulin on liver glucose output, with the result that too much glucose is released from the liver. It is also known that the FFA (or some intracellular metabolite from FFA) interferes with glucose uptake or its phosphorylation (or both), which could also lead to hyperglycemia. It is now clear that elevated blood FFA plays a major role in decreasing glucose uptake. In the short term, this effect is mediated through the insulin-receptor-signal cascade system that regulates the number of GLUT-4 transporters in the muscle-cell membrane (see chapter 6). For prolonged elevations in FFA, the genes for fatty acid transport are increasingly expressed and the gene for GLUT-4 transporter is reduced.

> ### ✓ KEY POINT
>
> The composition and quantity of food energy taken following exercise can play a significant role in the body's preparation to handle subsequent exercise tasks. Horowitz and colleagues (2005) reported that, on the day after a strenuous exercise task, subjects at rest had much higher blood FFA levels and lower carbohydrate oxidation if they were in an energy-deficient state. Being in an energy-deficient and carbohydrate-deficient state would inhibit subsequent exercise performance. Data such as these support the body's strategy to defend its carbohydrate reserves unless there is an excess.

Malonyl-CoA and Regulation of Fatty-Acid Oxidation in Muscle

In an earlier section of this chapter, we discussed the formation of malonyl CoA by the ATP-dependent carboxylation of acetyl CoA, catalyzed by the enzyme ACC (see figure 7.18). This reaction is important in liver, adipose, and mammary gland tissue, which use the malonyl CoA to synthesize palmitate. We also noted that malonyl CoA can inhibit the activity of the enzyme CPT I, which is responsible for the exchange of carnitine for long-chain fatty acyl CoA at the outer mitochondrial membrane (see figure 7.12). Inhibition of CPT I blocks entry of LCFAs into the matrix, thereby preventing their oxidation.

Skeletal and cardiac muscle are important consumers of fatty acids, and skeletal muscle also has some ability to use glucose for de novo lipogenesis. Malonyl CoA helps regulate fatty acid metabolism by muscle. We know that the oxidation of fatty acids increases during fasting and light exercise. Under both conditions, the concentration of malonyl CoA in muscle decreases, which would help facilitate the transfer of fatty acids into the mitochondria. On the other hand, if glucose and insulin levels are rapidly and acutely raised, muscle malonyl CoA levels increase within minutes. This would block entry of fatty acids into muscle mitochondria, sharply reducing the oxidation of fatty acids. The peptides, leptin, and adiponectin derived from adipose tissue can also regulate malonyl CoA levels through AMPK activation and consequent ACC inhibition. As previously noted, this can be an important factor in regulating fatty-acid levels in muscle and protecting muscle from the negative consequences of chronic elevation of muscle fatty-acid levels, such as the development of insulin insensitivity (Dulloo et al. 2004).

Malonyl CoA is synthesized in muscle by an isozyme of ACC, known as ACC_β or ACC_2. This isozyme of ACC is different from ACC_α, also known as ACC_1, which is found in the lipogenic tissues (liver, muscle, and adipose tissue). Unlike that of the liver isozyme, the ACC_β content in muscle does not depend on the composition of the diet (e.g., high or low in carbohydrate), nor is it sensitive to insulin and glucagon concentrations. In this way, it is distinct from the liver isoform. Moreover, ACC_β is not phosphorylated by the cAMP-dependent protein kinase. It is, however, phosphorylated by an AMPK that inactivates ACC_β. Citrate is a positive allosteric effector for ACC_β. Regulation of the activity of ACC_β is shown in figure 7.25. An increase in AMP concentration increases the activity of AMPK, which phosphorylates and inactivates ACC_β. This leads

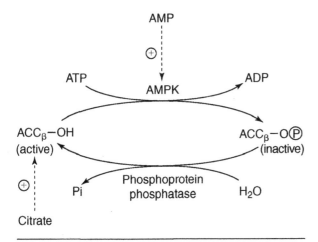

Figure 7.25 Regulation of acetyl CoA carboxylase beta (ACC_β) in skeletal and cardiac muscle through reversible phosphorylation and dephosphorylation. In the unphosphorylated state, ACC_β is active. The activity can be further increased allosterically by citrate. When phosphorylated (shown by the circled P), the enzyme is inactive. AMP-activated protein kinase (AMPK) phosphorylates ACC_β, while a phosphoprotein phosphatase removes the phosphate, leading to the active enzyme.

to decreased formation of malonyl CoA. Removal of the phosphate is catalyzed by a phosphoprotein phosphatase. We discussed this protein kinase in detail in the previous chapter. AMPK can be activated by an increase in the concentration of AMP, which comes about when adenylate kinase converts two ADP to AMP and ATP. The AMP both allosterically activates AMPK and activates an upstream kinase AMPK kinase (AMPKK) that activates AMPK. Recall that we previously noted that estrogen can also increase fat utilization by activation of AMPK and thereby inhibit malonyl-CoA synthesis.

If malonyl CoA is synthesized in skeletal and cardiac muscles, it must also be degraded, because its only use in skeletal and cardiac muscle is regulatory. An enzyme known as malonyl CoA decarboxylase (MCD) breaks down malonyl CoA to acetyl CoA (Sambandam et al. 2004). Figure 7.26 shows how malonyl CoA is created by ACC_β and degraded by MCD, as well as how some physiological conditions influence the concentration of malonyl CoA. The steady-state concentration of malonyl CoA in cytosol is therefore based on the relative activities of ACC_β and MCD. It is apparent that exercise simultaneously decreases ACC_β activity and increases the activity of MCD. The net effect of this is to decrease malonyl CoA concentrations, to relieve the inhibition of malonyl CoA at CPT I, and to increase the oxidation of LCFAs by mitochondria. On the other hand,

an abundance of the alternate fuel glucose leads to activation of ACC_β and an increase in the formation of malonyl CoA, thus decreasing the oxidation of LCFAs. The mechanism for the glucose effect is based on an increase in cytosolic citrate, which acts as a positive allosteric effector for ACC_β. It has been shown in rat and human skeletal muscle that an increase in glucose availability leads to an increase in cytosolic citrate.

Figure 7.27 summarizes how an increased availability of glucose can lead to an increase in malonyl CoA concentration and a block in the oxidation of LCFAs, whether they are derived from the breakdown of IMTG or are taken up as FFA from the blood. Evidence exists that an elevated skeletal-muscle glycogen concentration acts to reduce AMPK activity by binding to one of its subunits. Such an effect would lead to less phosphorylation of ACC_β and, thus, greater activity. This would lead to a higher concentration of malonyl CoA and subsequent greater inhibition of LCFA oxidation. On the other hand, low muscle glycogen would be associated with greater AMPK activity, more phosphorylation of ACC_β, decreased formation of malonyl CoA, and more LCFA oxidation.

Convincing evidence that changing concentrations of malonyl CoA can account for all differences in fuel oxidation in human skeletal muscle during

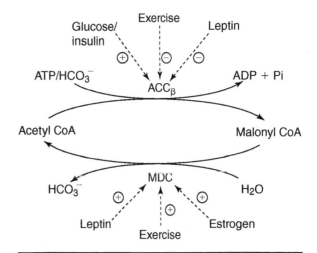

Figure 7.26 Creation and degradation of malonyl CoA. In skeletal and cardiac muscle, the concentration of the regulatory molecule malonyl CoA is based on the relative activities of acetyl CoA carboxylase beta (ACC_β), which makes malonyl CoA, and malonyl CoA decarboxylase (MCD), which breaks it down. Exercise and the attendant increase in the concentration of AMP lead to inhibition of ACC_β but activation of MCD. Citrate is a positive allosteric activator for ACC_β.

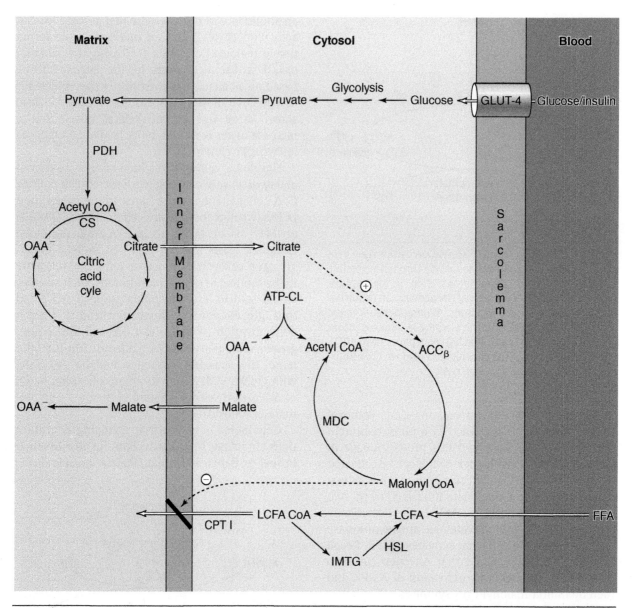

Figure 7.27 An increase in the availability of glucose to a skeletal or cardiac muscle cell, leading to a shift in fuel oxidation from fatty acids to carbohydrate. The increased breakdown of glucose by glycolysis and the increase in the conversion of pyruvate to acetyl CoA in mitochondria through the action of pyruvate dehydrogenase (PDH) result in overproduction of citrate. Some of the citrate is transported to the cytosol, where it is broken down to acetyl CoA and oxaloacetate (OAA$^-$). Acetyl CoA is converted to malonyl CoA in the cytosol by acetyl CoA carboxylase beta (ACC$_\beta$). Citrate is a positive allosteric effector for ACC$_\beta$. Malonyl CoA can block the entry of long-chain fatty acyl CoA into the mitochondria by inhibiting the first key enzyme involved in LCFA transport, carnitine palmitoyl transferase I (CPT I). Blockage of LCFA entry to mitochondria means that the entry of LCFAs as FFA (free fatty acids) in blood is decreased. It also leads to a reduction in lipolysis of intramuscular triacylglycerol (IMTG) by the muscle hormone-sensitive lipase (HSL). Open arrows refer to transport across a membrane.

exercise is scarce. Instead, data now suggest that a deficit in the availability of carnitine may act to limit fat oxidation during exercise with elevated levels of muscle glycogen or glucose (or both) and insulin (Roepstorff et al. 2005). Figure 7.28 summarizes how a deficit in carnitine could limit fat oxidation. During steady-state exercise, ATP needs are met by oxidation of carbohydrate and fatty acids. If one fuel is more dominant, the other is used less. An excess capacity to oxidize carbohydrate—brought about by high blood glucose concentrations, elevated muscle glycogen stores, or both—results in high rates of glycolysis and, therefore, pyruvate formation. The major fate of pyruvate is to enter the mitochondria and be converted to acetyl CoA. When there is a surfeit of carbohydrate, pyruvate dehydrogenase (PDH) activity is elevated and acetyl CoA is formed at a rapid rate. If the concentration of acetyl CoA exceeds the ability to form citrate in the citric acid cycle, an enzyme, carnitine acetyl transferase (CAT), can reversibly exchange a CoA for carnitine, producing acetyl carnitine. It has been shown that under carbohydrate-loaded exercise conditions, considerable carnitine can be tied up as acetyl carnitine (Roepstorff et al. 2005). This could lead to a limitation of transport of long-chain fatty acyl groups from the cytosol to the matrix, limiting the rate of beta-oxidation and the production of fat-derived acetyl CoA. Thus, with high carbohydrate stores, the increase in malonyl CoA concentration and its effect to limit long-chain fatty acyl group entry into the mitochondrion could be supplemented by an inadequate amount of carnitine.

> **☑ KEY POINT**
>
> Clearly, the body regulates its use of carbohydrate very carefully, especially during exercise. Blood glucose is critical to the function of the central nervous system and other tissues. However, these carbohydrate-sparing mechanisms are overridden and carbohydrate use is increased if there is a surfeit of blood glucose or muscle glycogen. As previously noted, this increased carbohydrate oxidation spares dietary fatty acids; instead, they are used for triacylglycerol synthesis in adipose tissue, such that an excess caloric intake of carbohydrate increases fat deposition in adipose tissue without greatly increasing carbohydrate use for synthesis of new fatty acids.

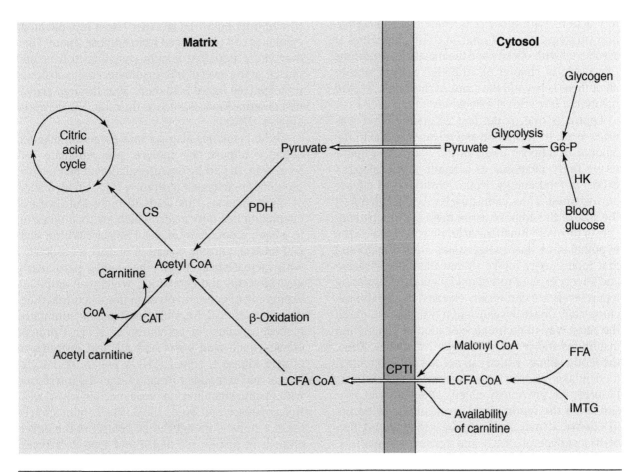

Figure 7.28 How a deficit in carnitine can limit fat oxidation. An increase in pyruvate formation via glycolysis because of elevated muscle glycogen or blood glucose (or both) results in an increase in conversion of pyruvate to acetyl CoA in mitochondria. When acetyl CoA is formed beyond the capacity of the citric acid cycle to use it, acetyl groups can be transferred to carnitine, forming acetyl carnitine, catalyzed by carnitine acetyl transferase (CAT). A decrease in carnitine, needed to allow long-chain fatty acyl groups (LCFA) to cross into the mitochondrion, can supplement the blockade caused by elevated levels of malonyl CoA at the level of CPT I. Abbreviations: G 6-P, glucose 6-phosphate; HK, hexokinase; PDH, pyruvate dehydrogenase; IMTG, intramuscular triacylglycerol. Open arrows refer to transport across a membrane.

ADIPOSE TISSUE AS AN ENDOCRINE TISSUE

Lyon and colleagues (2003) stated, "Fat is both a dynamic endocrine organ, as well as a highly active metabolic tissue." So far, our emphasis has been on the role of adipose tissue as a critical store of triacylglycerol. However, it is now well known that adipose tissue secretes a number of important peptide regulators, known as cytokines. These substances may be released to the blood where they may act on other cells in an endocrine fashion. Alternatively, they may be released locally, acting on cells in their immediate environment in a paracrine way. Expression of the genes for many of these adipose-secreted cytokines (also known as adipocytokines or adipokines) and their secretion from the fat cell are regulated by feeding and fasting. **Adipokines** interact with a variety of cell types, regulating metabolic, neural, or secretory actions with those cells. Inappropriate regulation (dysregulation) of some of these adipokines is associated with obesity and the metabolic syndrome (discussed in chapter 6). Although a discussion of all of these is beyond the scope of this book, we will mention a few critical adipokines.

Leptin is one of the first adipokines that was discovered. Its formation and release are tied to the amount of adipose tissue in the body. As adipose tissue mass increases as a result of a prolonged excess of food energy, leptin synthesis and release are increased. Thus, circulating levels of leptin reflect the size of the adipose tissue mass for that person. One of the main functions of leptin is to signal to the hypothalamus that energy stores are too large and that food energy intake should therefore decrease and energy expenditure should increase. Circulating leptin levels are chronically elevated in obesity; the chronically obese develop leptin resistance in much the same way that chronic elevation of insulin and insulin resistance is seen in type 2 diabetics. Thus, the leptin signal, which can act to reduce appetite, food intake, and increase muscle-fat oxidation, is blunted. As previously noted, when muscle sensitivity to the leptin signal is impaired, as occurs in obesity, it may lead to increased accumulation of triacylglycerol (TAG) and related metabolites, particularly diacylglyerol and **ceramides**, in the muscle, which contributes to the development of insulin resistance and type 2 diabetes (Dyck, Heigenhauser, and Bruce 2006). Regular exercise or a diet that includes the omega-3 PUFAs from fish oils (EPA and DHA) will help to partially reverse the leptin resistance in muscle (Dyck 2005).

Guilherme and colleagues (2008) reviewed the interactions between adipose tissue dysfunction and development of muscle insulin resistance. Adipose tissue plays an important role in regulating whole-body metabolism and sequestering fatty acids. Adipose tissue also produces leptin, adiponectin, and other adipokines that enhance insulin sensitivity of muscle and other tissues. In times of prolonged high-caloric intake leading to positive caloric balance and increased weight gain, TAG synthesis and levels of the enzymes involved in the process in adipose cells both increase. As adipose cells increase in size, they boost production of a number of peptides that affect TAG metabolism. One of these is monocyte chemoattractant protein-1 (MCP-1). A major role of MCP-1, as its name implies, is to enhance macrophage infiltration into the adipose cells. This can ultimately lead to a proinflammatory state in adipose tissue, which results in production of proinflammatory cytokines such as tumor necrosis factor-α (TNF-α) and interleukin-1β (IL-1β). These inflammatory cytokines can be released from adipose tissue. They may cause impaired insulin signaling in liver and muscle at the level of IRS (insulin receptor substrate) proteins (see figure 6.30 on p. 200) through activation of serine kinases, such as the c-Jun NH_2-terminal kinases (JNKs).

These cytokines also act in a paracrine manner, enhance adipose cell lipolysis, and attenuate TAG formation, in part by downregulation of peroxisome proliferator–activated receptor γ (PPARγ) mediation of these actions. This leads to increased levels of circulating free fatty acids, which are then taken up in greater amounts by skeletal muscle. Within skeletal muscle, chronic elevation of TAG is associated with increased levels of other lipids, particularly diacylglycerol and ceramides, which can attenuate expression of genes involved in mitochondrial function, such as PGC-1α, and inhibit insulin-stimulated glucose uptake via activation of several protein kinases, including novel and atypical isoforms of protein kinase C (PKC). TNF-α enhances diacylglycerol and ceramide formation, thereby interfering with insulin signaling via these mechanisms (Dyck, Heigenhauser, and Bruce 2006). The combination of these actions is thought to be central to the development of insulin resistance and type 2 diabetes, since it provides a link between obesity and its development (see figure 7.29). Interestingly, saturated fatty acids (e.g., palmitate) that accumulate in muscle cells are more prone to conversion into diacylglycerols and ceramides, whereas unsaturated fatty acids (e.g., linoleate) are preferentially stored as TAG (reviewed in Bilan et al. 2009). Figure 7.30

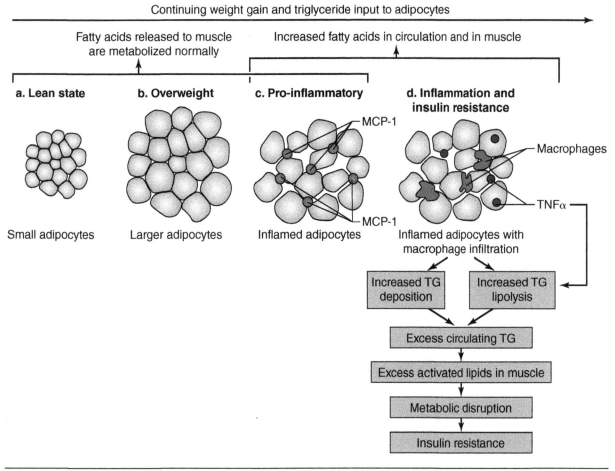

Figure 7.29 In the lean state (a), adipocytes store triglycerides (TG) normally, and there is normal mobilization of TG from adipocytes and metabolism of TG in skeletal muscle. (b) A positive caloric imbalance results in increased TG storage and adipocyte enlargement, but in non-diabetic individuals normal mobilization and metabolism of TG continues. (c) With a continued positive caloric imbalance, further increases in TG storage and adipocyte enlargement result in adipocyte production of monocyte chemoattractant protein-1 (MCP-1), which causes macrophage infiltration of adipocytes and inflammation. With ongoing inflammation in enlarged (obese) adipocytes (d), infiltrating macrophages secrete tumour necrosis factor-α (TNF-α), TG deposit and storage are impaired, and there is increased lipolysis and elevated circulating TG and fatty acids. This leads to increased fatty acid and lipid deposition in muscle, which disrupts mitochondrial metabolism and insulin mediated glucose uptake, causing insulin resistance.

illustrates how ceramides are synthesized, starting with the saturated fatty acyl CoA, palmitoyl CoA, and the amino acid, serine.

As already mentioned, many of the harmful adipokines are produced and released in response to a prolonged positive-calorie balance and oxidative stress, underlying many of the problems of obesity. *Resistin* is another example of an adipokine that is increased in the blood with obesity. It promotes peripheral insulin resistance in liver and muscle, likely due to its inhibitory actions on AMPK activity in those tissues. On the other hand, *adiponectin* is a polypeptide released from adipose cells that opposes the effects of the proinflammatory adipokines. From a health perspective, adiponectin provides beneficial effects, such as increasing insulin sensitivity in peripheral tissues and reducing atherosclerosis in arteries by reducing vascular inflammation. Serum adiponectin levels tend to be lower in obese subjects due to the effects of TNF-α on PPARγ expression and decreased transcription of the adiponectin gene in adipocytes. Weight loss is associated with increased serum adiponectin concentrations. Adiponectin also activates AMPK and has positive effects similar to those attributed to leptin on fat metabolism and protection against insulin resistance in skeletal muscle.

Besides secreting important cytokines, adipose tissue cells express receptors that recognize other cytokines in addition to the β ($β_1,β_2$)- and $α_2$-adrenergic and insulin receptors. **Ghrelin** is a polypeptide

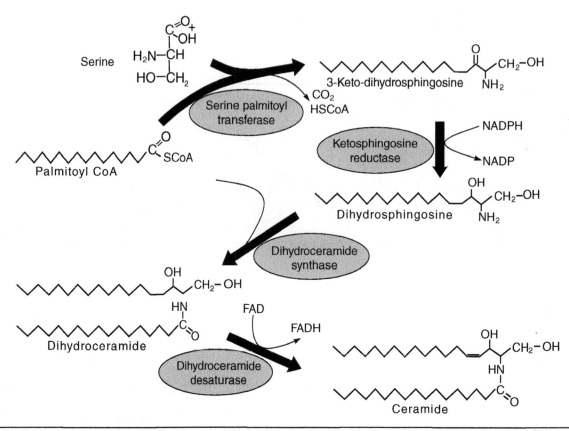

Figure 7.30 Ceramide biosynthesis pathway. De novo synthesis of ceramide begins with the formation of 3-keto-dihydrosphingosine via the linkage of serine with palmitic acid derived from palmitoyl CoA. This condensation reaction is catalyzed by serine palmitoyl transferase. Next, 3-ketodihydrosphingosine is reduced to form dihydrosphingosine, catalyzed by ketosphingosine reductase and coupled with the oxidation of NADPH. Dihydrosphingosine is then acylated by dihydroceramide synthase to form dihydroceramide. Dihydroceramide desaturase can subsequently desaturate dihydroceramide to form ceramide. The rate of de novo ceramide synthesis is regulated by the availability of the precursors, palmitoyl CoA and serine. Extracellular cytokines increase ceramide synthesis by upregulating the expression of serine palmitoyl transferase.

secreted by the gastric cells. In adipose tissue, ghrelin signals the formation of new fat cells, increased TAG formation, and decreased lipolysis. Its effect is to increase the storage of fat in the body, including the liver (Barazzoni et al. 2005).

CHOLESTEROL

Cholesterol is an important lipid, but we seem to hear only about its bad properties. Because of its hydrocarbon content (figure 7.31 illustrates the chemical structure), cholesterol is not soluble in water. In the blood, cholesterol either has a fatty acid attached to the cholesterol OH group, forming a cholesterol ester (about 70%), or appears as simple cholesterol (about 30%). Both cholesterol and cholesterol esters exist in lipid-protein complexes called lipoproteins. From a health perspective, the two main cholesterol-containing lipoproteins are the low-density lipoproteins (LDL) and the high-density lipoproteins (HDL). Chylomicrons represent another important lipoprotein fraction in the blood, transporting dietary fat. These are formed in intestinal cells, secreted into the lymphatic system, which empties into the bloodstream, and circulated throughout the body. Lipoprotein lipase hydrolyzes chylomicron and VLDL triacylglycerols, releasing the fatty acids.

Determinations of the concentrations of cholesterol and cholesterol-rich lipoproteins in the blood are important clinical tests. Total blood cholesterol concentration is reported in three different ways: milligrams cholesterol per deciliter, milligrams cholesterol per liter, and millimolar. In the United States, the first method predominates (mg/dL); many other parts of the world use the SI system (mM units). If the molecular weight of cholesterol is 387, you should be able to convert a blood value of 150 mg/dL to the value of 3.88 mM. More important in deter-

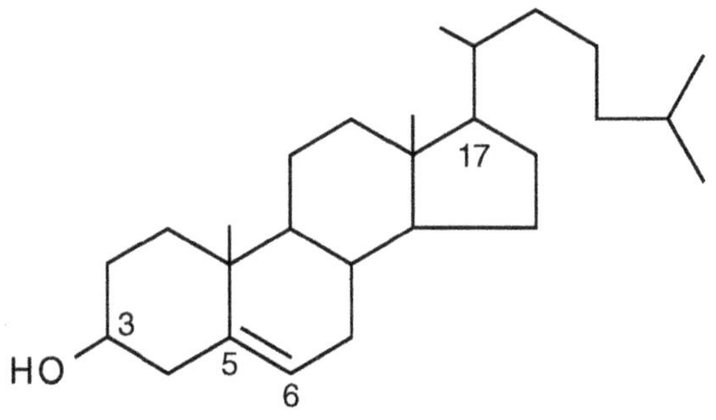

Figure 7.31 Shorthand structure for cholesterol. The molecular formula is $C_{27}H_{47}O$. Can you find all of the carbon atoms? The numbers illustrate the location of the hydroxyl group, the double bond, and the alkyl side chain.

mining a health profile is to ascertain the cholesterol concentration in the LDL and HDL fractions. From this perspective, it is best if the LDL cholesterol level is low while the HDL cholesterol is high.

Cholesterol is an important component of membranes, a precursor for the synthesis of steroid hormones (e.g., cortisol, testosterone, estrogen, and so on), bile salts, and vitamin D. It is also a major component of myelin in nerves. Cholesterol is synthesized from acetyl CoA units. Although present in all cell types, cholesterol synthesis is most important in the liver, intestines, and adrenal and reproductive glands. For most people, about 60% to 70% of the body's cholesterol is synthesized; the remainder comes from the diet. Although no clear evidence exists that regular physical exercise decreases blood cholesterol levels, it clearly alters the type of lipoprotein carrying cholesterol in the blood, raising the level of HDL. However, weight loss in obese people, whether accomplished by diet, exercise, or a combination of the two, has been shown to help decrease blood cholesterol levels. Certain drugs called statins have also been shown to markedly decrease total blood cholesterol levels. Decreased blood cholesterol levels, along with high HDL levels, significantly reduce the risks of heart and cardiovascular diseases. Studies have demonstrated that previous sedentary people who start physical activity of even very low intensity, which does not improve $\dot{V}O_2$max, will experience marked improvement of blood lipid profiles, increase their blood HDL levels, and reduce risk factors for development of type 2 diabetes (Helge 2010).

NEXT STAGE

Exercise, 24-Hour Fat Oxidation, and Weight Loss

By increasing fat oxidation, exercise is commonly assumed to lead to a negative fat balance and, consequently, loss of fat and weight. You may also encounter advocacy for the use of exercise at low intensities that maximize fat oxidation, or the so-called *fat burning zone*, as the best means by which optimal fat oxidation, fat loss, and weight loss can be achieved. In reality, the relationship among exercise, long-term fat oxidation, fat loss, and weight loss is much more complex than this. A review that summarizes a decade of work in this area (Melanson, MacLean, and Hill 2009), newer findings (Melanson et al. 2009), and support from studies in other laboratories (Smith 2009) have demonstrated some of these complexities.

As noted previously, fat oxidation during exercise is affected by exercise intensity: fat oxidation decreases as exercise intensity increases. In trained athletes, a maximal fat-oxidation rate of about 0.6 g per min occurs at intensities of about 60% to 65% $\dot{V}O_2$max; for the untrained, the maximal fat-oxidation rate of about 0.4 to 0.5 g per min is reached at about 45% to 50% $\dot{V}O_2$max. In the hours following exercise, the balance of fat versus carbohydrate oxidation is affected by (1) the caloric balance of energy intake versus expenditure and (2) the macronutrient content of the postexercise diet and the nutrient intake and glycogen stores that were present prior to the exercise bout. Contrary to the common belief that exercise will result in a net increase in fat oxidation over the exercise and 24 h recovery period, this is not typically the case. Melanson and colleagues (Melanson et al. 2009; Melanson, MacLean, and Hill 2009) performed a series of controlled studies using a room calorimeter chamber in which male and female, young and old, trained and untrained, as well as obese and nonobese subjects performed endurance, high-intensity, or weight-training exercises; 24 h calorimetry and RQ measurements were recorded. Resting 24 h control measures were also made on all subjects. It is important to note that for all subjects, 24 h energy intake as part of a mixed diet was exactly matched with energy expenditure. In all cases, exercise, whether it was of high or low intensity or involved cycling or weight training, did not significantly affect total 24 h fat oxidation relative to total fat oxidation by the same people during the 24 h nonexercised control conditions. The increase in total 24 h energy expenditure observed in the exercise conditions was accounted for by increased carbohydrate metabolism.

These results point out that exercise itself in conditions of caloric balance does not alter 24 h fat oxidation. While fat and carbohydrate oxidation were raised during the exercise, subsequent to the exercise and feeding of a mixed diet, carbohydrate oxidation increased relative to control and fat oxidation decreased. As discussed earlier in this chapter, the most likely reason for these results is that carbohydrate intake following exercise causes an insulin-induced suppression of fat oxidation (Melanson, MacLean, and Hill 2009). These results, as well as results from other similar studies, highlight the fact that diet is a more potent regulator of 24 h substrate oxidation during energy balance than exercise is. Fat oxidation is highest in the morning following an overnight fast, and carbohydrate oxidation is highest following a meal (Smith 2009).

How is it then that exercise can be a factor in overall weight loss and weight control? How do trained people maintain low body fat? It is likely that people who expend significant amounts of energy during exercise are not in caloric balance; they may not completely replace all calories expended during exercise. This practice has long-term consequences for fat balance and weight. In addition, the effects of consecutive days of exercise on fat metabolism have not been rigorously researched. For example, if carbohydrate stores are not completely replenished before a subsequent exercise bout, fat oxidation may be increased both acutely and over 24 h (Melanson, MacLean, and Hill 2009). This flexibility in substrate utilization is blunted by diabetes, obesity, and a lack of fitness (Smith 2009). Training increases muscle capacity for fat oxidation and the ability to enhance glucose oxidation after a meal. When dietary fat is increased in conditions of energy balance, untrained people are slower to turn on fat oxidation than trained ones are. Physical activity also immediately increases the rate of fat oxidation. These may be reasons why regular training can buffer any positive dietary fat balance that may occur from day-to-day variations in diet and can perhaps limit weight gain (Smith 2009).

The bottom line to these findings is that exercise will only result in fat loss when combined with a negative caloric balance. This may seem like an intuitive conclusion. However, exercise guidelines often overlook the need for dietary regulation, and exercise is often accompanied by an increase in dietary caloric intake (Smith 2009). Exercise recommendations cannot therefore be made without considering dietary and macronutrient intake (Melanson, MacLean, and Hill 2009). While the evidence that exercise is an important component of weight loss (particularly of its maintenance) is strong, evidence for the exact nature of these interactions is still being gathered. In particular, Melanson and colleagues (2009) suggest that future studies should "go beyond assessing the type and amount of energy intake in response to exercise and the impact of exercise on macronutrient oxidation rates. This will require studies where exercise and energy intake are combined in ways that reflect how exercise is in daily life."

In addition, Melanson, MacLean, and Hill (2009) note that the notion that light exercise or exercise in the fat burning zone, or at any other intensity, for that matter, will not affect fat balance or weight loss unless changes are also made to dietary energy and fat intake.

SUMMARY

In the body, lipids exist primarily as fatty acids, triacylglycerols (triglycerides), phospholipids, and cholesterol. Triacylglycerols consist of three long-chain fatty acids containing saturated or unsaturated carbon-to-carbon bonds, attached by ester bonds to the three-carbon alcohol glycerol. Triacylglycerols are mainly stored in specialized cells called fat cells or adipocytes. In these cells, triacylglycerols are made as fatty acids are joined to glycerol. The reverse reaction, lipolysis (or lipid mobilization to yield fatty acids and glycerol), is regulated by hormone-sensitive lipase and desnutrin/ATGL. Triacylglycerol formation is favored and lipolysis is inhibited by insulin, whereas epinephrine from the adrenal medulla and norepinephrine from the sympathetic nervous system play a major role in enhancing the rate of lipolysis.

Fatty acids released from the fat cell during lipolysis travel in the bloodstream attached to the protein albumin. The fatty acids can be used by other cells as fuel. For this to happen, the fatty acid is transported across the cell membrane using fatty-acid transport proteins, converted to a fatty acyl CoA, and then transported into the mitochondrial matrix through attachment to carnitine. In a four-step process known as beta-oxidation, long-chain fatty acids are broken down into two-carbon acetyl units attached to CoA. These acetyl CoA units can then feed into the citric acid cycle. When lipolysis is increased because body carbohydrate stores are low or during poorly controlled diabetes mellitus, fatty acids can be a source of carbon for the liver to make ketone bodies. Ketone bodies are also a source of energy, but unlike fatty acids, they can be used by the brain as a fuel. Many of the problems associated with diabetes mellitus relate to uncontrolled formation of ketone bodies. Acetyl CoA units produced by the partial breakdown of excess carbohydrate or amino acids are used to make fatty acids.

We obtain most of our fatty acids from our diet, but the liver and adipose tissue have a limited ability to make new fatty acids. Unlike lipolysis, the synthesis of new fatty acids (de novo lipogenesis) occurs in the cell cytosol and involves an intermediate known as malonyl CoA. Malonyl CoA is formed by acetyl CoA carboxylase, whose activity is regulated by allosteric and phosphorylation mechanisms. The oxidation of fat can play a major role in ATP formation by an active muscle. However, because of the limited rate at which fatty acids can be oxidized, their use is more important to a rested muscle or a muscle that is moderately active. The relative use of carbohydrate and fatty-acid oxidation by muscle depends on a variety of factors, including exercise intensity, exercise training, and diet. A reciprocal relationship exists between the availability of fatty acids and carbohydrate: Fatty acids are able to reduce glucose utilization, and elevated glucose or muscle glycogen is able to reduce the use of fatty acids. Thus, the same exercise task performed with elevated muscle glycogen stores is performed with greater use of carbohydrate than would be the case if muscle glycogen stores were lower. Malonyl CoA and potentially carnitine (if available) play roles in determining fuel oxidation by muscle.

Adipose tissue is more than a collection of cells that make and store triacylglycerol. New research shows that adipocytes produce and secrete a variety of biologically active peptides or cytokines. A number of these adipokines are produced in response to increased storage of fat due to a calorie imbalance. The increased fat content of adipocytes may lead to oxidative stress and the upregulation of expression of adipokine genes. A number of the adipokines (interleukin-6, tumor necrosis factor α, resistin, plasminogen activator inhibitor-1) are released from enlarged fat cells; they affect the adipose tissue and specifically muscle cells, contributing to such conditions as type 2 diabetes.

REVIEW QUESTIONS

1. Draw the chemical structures for arachidonic acid [20:4 (5, 8, 11, 14)] and elaidic acid, which is the trans form of oleic acid.

2. In the blood draining an adipose tissue site, the ratio of FFA to glycerol is 1.5 to 1. What proportion of triacylglycerol molecules being broken down is resynthesized?

3. During steady-state exercise, the average nonprotein RER is 0.87 over 60 min. According to tables, such an RER means that 58.5% of the energy comes from carbohydrate and 41.5% from fat, and that each liter of oxygen has the energy equivalent value of 5.043 kcal (21.09 kJ). If the average $\dot{V}O_2$ during 60 min is 2.4 L/min, what is the energy expended by the body during this time? Assuming an energy value of 4 kcal per gram (16.7 kJ/g) and 9 kcal per gram (37.6 kJ/g) for carbohydrate and fat, respectively, how many grams of carbohydrate and fat are oxidized over this period of time? Are we overlooking any other substance that may contribute to the energy expended?

4. Assuming that the average molecular mass of stored triacylglycerol is 860 g/mol, what percentage of this is glycerol?

5. During rhythmic leg kicking exercise, it is determined that 3 kg of quadriceps muscle is active over a 1 h period. Based on arterial and venous blood sampling across the active leg, it is determined that the average $\dot{V}O_2$ for this exercise is 1.0 L per minute. Before the exercise, there are 12 mmol per kilogram wet muscle weight of stored intramuscular

triacylglycerol; after 60 min of exercise, the value is 10 mmol per kilogram wet weight. Assume that the metabolic response during this exercise, and due entirely to the active muscle, was three times that of the rest of the body. Approximately how many kilocalories (kcal) of energy generated were due to the exercise itself? What is the approximate percentage of energy devoted to kicking that was derived from intramuscular triacylglycerol? What is the total mass of fat oxidized? What would be the other sources of energy? Assume that the RER was the same as in question 3 and that the average molecular weight of the stored triacylglycerol was the same as in question 4.

6. How might exercise, caloric intake, dietary macronutrient balance, and timing of eating and exercise influence fuel selection during exercise, postexercise fuel oxidation, energy, and fat balance?

chapter 8

Amino Acid and Protein Metabolism

We first looked at the chemistry of amino acids in chapter 1. In the present chapter, we view amino acids from a different perspective. Amino acids are obtained in dietary protein and are considered macronutrients. Protein provides approximately 17 kJ/g (4 kcal/g) of physiological energy, compared with the same value for carbohydrate and 38 kJ/g (9 kcal/g) for lipids. Humans typically ingest about 10% to 15% of their calories in the form of dietary protein. We eat protein to obtain the amino acids that are used to make body protein and other specialized substances. Amino acids are also used as a source of energy. We begin the chapter with an overview of amino acid metabolism. This is followed by sections on the breakdown of amino acids, the body's means of disposal of amino acid nitrogen groups through formation of urea, and the fate of the major portion of amino acids (their carbon skeletons) when they are catabolized. Amino acids can be fuels, and the ways in which they are used during exercise are important. We complete the chapter by reviewing functions of amino acids beyond their roles as fuels or components of proteins. In the Next Stage section, we also take a brief look at new research on dietary amino acids and their role in protein synthesis and degradation in muscle following resistance exercise.

OVERVIEW OF AMINO ACID METABOLISM

Figure 8.1 provides a summary of amino acid metabolism. Our principal source of amino acids is in the form of food protein. During digestion, protein is broken down to free amino acids that are absorbed into the blood. The amino acids in the blood and extracellular fluids represent an **amino acid pool**. Although the size of this pool is very small compared to the mass of body protein, the amino acids turn over very rapidly. Amino acids from dietary proteins enter this pool from the gut. They also enter the pool following release from cells. Amino acids are taken up from the pool when they are transported into cells to make cell protein. Each cell makes its own specific proteins, using amino acids released when proteins break down and taking up the rest from the pool.

The body of a normal 60 kg (132 lb) woman has approximately 10 kg of protein and about 170 g of free amino acids. As mentioned in chapter 1, there is a constant protein turnover, and protein synthesis balances protein breakdown for much of our lives. When put into numbers, the pace of protein synthesis in the human body is truly staggering. An average-sized adult needs to synthesize about 1 million billion hemoglobin molecules each second just to replace those degraded as red blood cells are broken down. In fact, there are other proteins, such as myosin, in greater abundance and with shorter half-lives whose rates of synthesis far outpace that of hemoglobin. A significant portion of our resting metabolic rate (RMR) is accounted for by the energy used in such protein and amino acid synthesis and turnover pathways.

The liver and, to a smaller extent, the kidneys are responsible for much of the amino acid metabolism in the body. More than half of the amino acids absorbed in the intestines following digestion of a protein meal are taken up by the liver (Gropper,

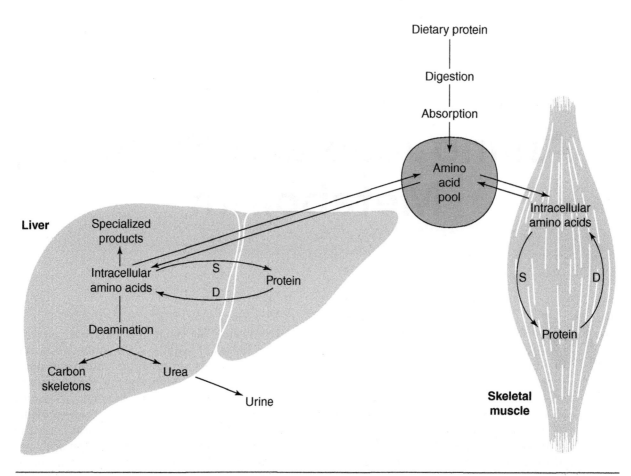

Figure 8.1 An overview of amino acid metabolism in the body showing two representative tissues—skeletal muscle and liver. Our principal source of amino acids is ingested protein. After digestion and absorption, amino acids enter an amino acid pool, which is in the blood and extracellular fluids. Cells take up amino acids from the pool to synthesize proteins (S). Degradation (D) releases amino acids that may be used again in protein synthesis or released to the amino acid pool. The liver makes a variety of specialized products from amino acids and breaks down most amino acids. The nitrogen on amino acids is converted to urea; the remaining carbon skeletons can be used as a source of glucose, oxidized, or converted into lipid.

Smith, and Groff 2005). It is believed that the liver regulates the composition of the amino acid pool, ensuring that the mix of amino acids is balanced. The liver is capable of synthesizing some of the 20 amino acids needed to make our body proteins. These are described as *nonessential amino acids* (dispensable amino acids), as opposed to the **essential amino acids** (indispensable amino acids) that cannot be synthesized and must be obtained in the diet from a variety of food sources. Remember that animal source proteins are typically *complete* proteins in that they contain all of the amino acids that are essential for human diets. Many plant based proteins are *incomplete* proteins in that they are very low in or missing one or more of the amino acids needed to make complete proteins in humans. Vegetarians can easily still meet all of their dietary protein needs by ensuring that they eat a wide variety of plant based foods with complementary amino acid contents. The amino acids can be grouped as follows.

- *Nutritionally nonessential:* alanine, asparagine, aspartate, cysteine, glutamate, glutamine, glycine, proline, serine, and tyrosine
- *Nutritionally essential:* arginine, histidine, isoleucine, leucine, lysine, methionine, phenylalanine, threonine, tryptophan, and valine

The categorization of amino acids as essential or indispensable and nonessential or dispensable really applies to healthy adults eating a sound diet. Some of the nonessential amino acids can become indispensable if they cannot be produced at a rate that is rapid enough. Special conditions such as infant growth, disease, organ failure, or the presence of drugs can make it more important to obtain the amino acids arginine, cysteine, glutamine, and proline through dietary sources. Amino acids that are synthesized in the liver are used within the liver or are released to the blood. The kidneys also play a role in the synthesis of some amino acids, as well as in the exchange of amino acids to and from different tissues and organs. Thus, the blood and extracellular fluids contain a pool of amino acids resulting from dietary intake, catabolism of cellular protein, and amino acid synthesis in the liver.

Skeletal muscle is the largest repository of free and protein-bound amino acids in the body, but in terms of the scope of metabolism of amino acids, muscle is rather limited. About 45% of the body weight of an adult is composed of skeletal muscle. This proportion is higher in people who are lean and well trained. This can represent an important nitrogen reserve during stressful conditions such as trauma, infection, and starvation. A variety of hormones and growth factors, such as insulin, insulin-like growth factor-1 (IGF-1), growth hormone, and testosterone, promote skeletal muscle growth. Other hormones, such as cortisol, help increase protein breakdown and metabolism in times of stress and starvation.

We have no ability to store amino acids as we can with carbohydrate (glycogen) and fatty acids (triacylglycerol). Moreover, the protein content of our adult bodies is remarkably constant, and so we might expect that we would oxidize amino acids at a rate commensurate with our intake—that is, about 10% to 15% of our daily energy expenditure. In other words, those amino acids not used in protein synthesis or converted to other substances (e.g., heme groups; hormones such as serotonin, adrenaline, and noradrenaline; nucleotides; glutathione; carnitine; and creatine) are simply used as fuels. As mentioned in chapter 6, amino acids first lose their amino groups or other N atoms. The resulting carbon skeleton can be oxidized directly, used to make glucose (gluconeogenesis), or converted into fat for storage.

Not shown in figure 8.1 is the transport of amino acids from the amino acid pool into cells. Because amino acids have charged groups, they need protein transporters to move them from the extracellular to the intracellular compartment or from the intracellular to the extracellular compartment. A number of amino acid transporters exist, and they fall into two broad categories. The transporters can be sodium-ion dependent or sodium independent. If the transporter is sodium dependent, the amino acid moves into a cell down a Na^+ concentration gradient. Therefore, this is a symport transport system, and the amino acids can be moved against their concentration gradient. Amino acid transporters may have broad specificity, recognizing a number of amino acids, or narrow specificity, recognizing only one or two closely related amino acids. (For a review of the various categories of amino acids, refer to figure 1.4 on p. 8.) Amino acid transporters are also subject to regulation by hormones and growth factors. Insulin, in particular, is important to stimulate amino acid uptake, particularly into skeletal muscle fibers.

☑ KEY POINT

Although the carbon skeletons from excess amino acids can be degraded to acetyl CoA, converted to malonyl CoA, and used to make palmitate in the liver (see chapter 7), this is not very significant. Immediate oxidation of amino acid carbon skeletons or conversion to glucose (gluconeogenesis) is the principal fate for excess amino acids. However, in times of starvation or glycogen depletion, such as toward the end of long endurance activity, the oxidation of amino acids increases.

DEGRADATION OF AMINO ACIDS

For a well-nourished person, the supply of amino acids arising from the degradation of intracellular proteins and those provided through the diet are normally sufficient to meet the immediate needs for protein synthesis. However, alterations in the balance of amino acids can be achieved through conversion reactions, in which one amino acid is changed into another by transfer of an amino group. This process is called **transamination**. In addition, excess amino acids are catabolized, but the amino groups must be removed beforehand. In chapter 6, we talked about the potential for 18 of the 20 amino acids to provide all or part of their carbon atoms to the formation of glucose in the process of gluconeogenesis. These are sometimes described as glucogenic amino acids, and the other two are described as ketogenic amino acids. As noted in our discussion of gluconeogenesis, amino acids consist of carbon skeletons and amino groups. Humans cannot use the nitrogen in the amino groups as an energy source, although it is needed to form some specialized products. Most amino nitrogen is excreted. The carbon skeletons are very valuable. As chapter 6 points out, after removal of the amino groups, many of the carbon skeletons become intermediates or are easily converted into intermediates in the citric acid cycle or pyruvate.

Amino acids undergo constant oxidative degradation under the following metabolic circumstances:

- During normal synthesis and degradation of proteins in the body, amino acids released in the constant breakdown process may not be immediately reused in synthesis. Since we cannot store amino acids, they will be degraded.
- When we ingest more amino acids than our bodies can use to make proteins or to convert to other substances, these amino acids are degraded. In North America, most people eat more protein than they need.
- During starvation, fasting, dieting, or uncontrolled diabetes mellitus, when carbohydrates are either not available or are not used due to an absence of insulin, amino acid catabolism accelerates.
- When we overtrain, an imbalance in protein turnover occurs that favors protein degradation, and amino acids are degraded.

Catabolism of individual amino acids has two major stages. First, the amino acids must lose their nitrogen atoms because we cannot obtain usable energy from nitrogen groups. Second, the resulting carbon skeletons are fed into specific energy-yielding pathways so that their chemical energy can be retrieved. The liver removes the amino groups from most amino acids, although skeletal muscle has a significant capacity for **deamination** of the branched-chain amino acids (BCAAs).

Transamination Reactions

Transfer of amino groups takes place with all of the amino acids except threonine and lysine. Even the other essential amino acids can undergo amino group transfer, but only in the direction of removing the amino groups because the body cannot synthesize their carbon skeletons. Transamination is carried out by *aminotransferase enzymes* (formerly known as transaminases), which contain vitamin B6. Most of the aminotransferase enzymes transfer the amino group of an amino acid to α-ketoglutarate (also known to many as 2-oxoglutarate), making glutamate and a new carbon skeleton in the form of a keto acid (see figure 8.2). The vitamin B6 is a precursor for the coenzyme pyridoxal phosphate (PLP; see table 2.1 on p. 25). Notice in figure 8.2 that the ketone group in α-ketoglutarate is next to the carboxylate; thus it is in the 2 or α position, making it an α-keto acid.

Additional aminotransferase enzymes are specific for amino acids other than glutamate. For example, *alanine aminotransferase* (see figure 6.23 on p. 191) involves the reversible transamination between the amino acid alanine and its α-keto acid, pyruvate, along with α-ketoglutarate and glutamate. The key word is reversible, since an excess of alanine is converted to glutamate or an excess of glutamate is converted to alanine. The actual concentrations are based on the equilibrium for each aminotransferase. *Aspartate aminotransferase* is involved in the reversible transamination between the amino acid aspartate and its α-keto acid, oxaloacetate, as well as α-ketoglutarate and glutamate. Alanine aminotransferase and aspartate aminotransferase are prominent enzymes that are known clinically as ALT (formerly known as glutamate pyruvate transaminase, GPT) and AST (formerly glutamate oxaloacetate transaminase, GOT), respectively. Serum ALT and AST activities are commonly used to determine if there has been acute trauma or disease of the liver.

Transamination reactions are freely reversible, with equilibrium constants of about 1 and standard free energy change values near zero. Their net direction depends on the relative concentrations of the four reactants. The overall effect is to transfer

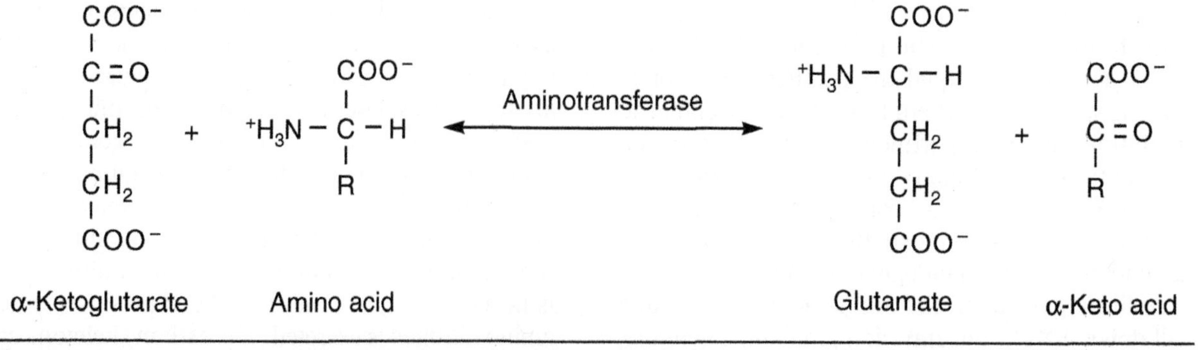

$$
\begin{array}{c}
COO^- \\
| \\
C=O \\
| \\
CH_2 \\
| \\
CH_2 \\
| \\
COO^-
\end{array}
+
\begin{array}{c}
COO^- \\
| \\
{}^+H_3N-C-H \\
| \\
R
\end{array}
\xrightarrow{\text{Aminotransferase}}
\begin{array}{c}
COO^- \\
| \\
{}^+H_3N-C-H \\
| \\
CH_2 \\
| \\
CH_2 \\
| \\
COO^-
\end{array}
+
\begin{array}{c}
COO^- \\
| \\
C=O \\
| \\
R
\end{array}
$$

α-Ketoglutarate   Amino acid   Glutamate   α-Keto acid

Figure 8.2 The aminotransferase reaction. Aminotransferases are a class of enzyme that transfers amino groups reversibly between the amino acid glutamate and its α-keto acid (α-ketoglutarate, also known as 2-oxoglutarate). The other participants are specific to the other individual amino acid and its carbon skeleton, an α-keto acid.

amino groups from a variety of amino acids to α-ketoglutarate (2-oxoglutarate), making glutamate. We will discuss the BCAAs as a special case, for their metabolism is primarily in skeletal muscle.

Deamination

Amino groups can be shuttled from one amino acid to α-ketoglutarate to make glutamate or from glutamate to another keto acid to create a new amino acid, as we have seen. However, most of us excrete an amount of amino nitrogen each day that equals our intake. Therefore, the body must rid itself of amino groups by forming urea. This means that there must be processes to remove amino groups and then turn them into urea molecules in the liver. Some key reactions are used to remove amino groups from amino acids. Therefore, the amino group on glutamate (which comes from amino acids) must be transferred to the liver (if it is not already there) and into the creation of the urea molecule. Nitrogen from amino groups in the liver in the form of glutamate can be released as ammonia (principally the ammonium ion NH_4^+) in the *glutamate dehydrogenase* reaction.

$$\text{glutamate} + H_2O + NAD^+ \leftrightarrow \text{α-ketoglutarate} + NADH + H^+ + NH_4^+$$

As shown, the coenzyme is NAD^+, but $NADP^+$ can also be used. The freely reversible glutamate dehydrogenase reaction takes place only in the mitochondrial matrix, whereas most aminotransferase enzymes exist in both the mitochondrial matrix and the cell cytosol. The glutamate dehydrogenase reaction, as it proceeds to the right, is a deamination reaction, since it removes the amino group. Because NADH is also formed, it is an oxidative deamination reaction. The glutamate dehydrogenase reaction removes nitrogen from glutamate, which can be used to make urea. As we will discuss later, exercising muscle releases ammonia (as the NH_4^+ ion). The glutamate dehydrogenase reaction in muscle is one source of this ammonia. Ammonia is also produced by the deamination of AMP in the AMP deaminase (adenylate deaminase) reaction, as discussed in chapter 4. This latter reaction is generally not significant unless muscle is working at a high intensity. Overall, the production of ammonium ion and its release from muscle are proportional to the intensity of the exercise.

✓ KEY POINT

Ammonia is a toxic substance, but it is produced in small quantities by some tissues (kidney and muscle). Some is excreted in the urine. We have been using the term *ammonium ion*—the principal form we would find in the blood—rather than *ammonia*, although the former is simply protonated ammonia. Ammonia (ammonium ion) released from muscle is taken up by the liver (primarily) and kidneys. In the liver, ammonia can be used to make urea.

Glutamine

Glutamine is a special amino acid even though it is not essential for most of us. The content of the free amino acid glutamine is high in a variety of cells and in the blood, in concentrations beyond that of any other amino acid. It represents about 60% of the amino acid pool. We obtain some glutamine in our diets, but most is synthesized in the body from glutamate using the enzyme *glutamine synthetase* (see figure 8.3). Glutamine is an important fuel for the gut and immune system. Notice that the additional nitrogen in glutamine is an amide of the side chain carboxylate of glutamate. Skeletal muscle is an important site for glutamine synthesis, but

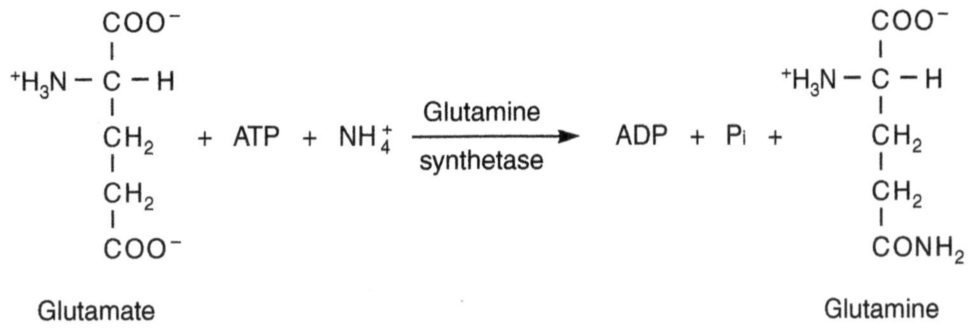

$$\text{Glutamate} \quad + \text{ATP} + NH_4^+ \xrightarrow{\text{Glutamine synthetase}} \text{ADP} + P_i + \text{Glutamine}$$

Figure 8.3 The glutamine synthetase reaction. This reaction is very active in skeletal muscle, transferring ammonium ions to glutamate to create glutamine.

glutamine is also made in the brain, adipose tissue, and the heart. As already mentioned, ammonia is toxic, so converting it to glutamine is a good strategy. Glutamine content in skeletal muscle and other tissues appears to have a regulatory role in whole-body protein synthesis. During a variety of catabolic conditions, its content is decreased. During anabolic states, intracellular glutamine content is elevated. Glutamine is also an important fuel for macrophages and lymphocytes, cells that are important to our health and appropriate immune responses.

The last reaction to consider is the removal of the amino group from the side chain of glutamine, which is the reverse of the glutamine synthetase reaction. *Glutaminase*, found mainly in liver mitochondria, catalyzes this deamination reaction:

$$\text{glutamine} + H_2O \rightarrow \text{glutamate} + NH_4^+$$

> ### ✓ KEY POINT
>
> Some studies have demonstrated that athletes who overtrain have an intracellular glutamine content that resembles catabolic states. These people are also more prone to infections, particularly of the respiratory tract. Although much of the intracellular glutamine is formed by the glutamine synthetase reaction, supplemental protocols with glutamine in critically ill patients have been attempted with mixed results. Besides being unstable in solution, much of the supplemental glutamine is oxidized by tissues of the intestinal tract or taken up by the liver and kidney. For this reason, glutamine is often provided intravenously as a dipeptide, attached to either the amino acid glycine or alanine. In the kidney, dipeptides are hydrolyzed, releasing free glutamine (Van de Poll et al. 2004).

Branched-Chain Amino Acids

The branched-chain amino acids leucine, isoleucine, and valine are the most common essential amino acids in proteins. Unlike the other 17 amino acids, the BCAAs are metabolized mainly in skeletal muscle, although they can also be metabolized in liver. These essential amino acids are treated as a group because the first three reactions in their catabolism are catalyzed by the same enzymes: *branched-chain amino acid aminotransferase*, *branched-chain keto acid dehydrogenase*, and *acyl CoA dehydrogenase* (see figure 8.4). After these first three reactions, carbon skeletons remaining from the three amino acids take their separate reaction directions. Both valine and isoleucine produce a *succinyl CoA* as a product. Since this is a citric acid cycle intermediate, valine and leucine are glucogenic amino acids. The products from leucine catabolism, acetoacetate and acetyl CoA, cannot be converted to glucose, so leucine, like lysine, is considered a ketogenic amino acid. Recall that acetoacetate is a ketone body (see chapter 7).

The first reaction in the catabolism of the BCAAs is a transamination in which the amino group on each BCAA is transferred to α-ketoglutarate, making glutamate. While transamination reactions are reversible, the branched-chain amino acid aminotransferase goes only one way because the products undergo an oxidative decarboxylation catalyzed by branched-chain keto acid dehydrogenase (BCKAD). This mitochondrial enzyme is a multienzyme complex, regulated by phosphorylation and dephosphorylation mechanisms. This is reminiscent of the regulation of pyruvate dehydrogenase (PDH) by phosphorylation and dephosphorylation. Indeed, PDH and BCKAD have an identical subunit. Phosphorylation of BCKAD by BCKAD kinase renders the enzyme inactive. A BCKAD phosphatase removes the phosphate group, leading to an active BCKAD.

In chapter 6, figure 6.24 (p. 192) showed that the amino groups on the BCAAs (transferred to α-ketoglutarate, making glutamate) can be transferred from glutamate to pyruvate by the action of alanine aminotransferase. Alanine is released from muscle to the blood, where its amino group is available for urea synthesis. In this way, the amino groups of the BCAAs are removed from muscle to be disposed of by the liver. Leucine and its α-keto acid, α-ketoisocaproate, can suppress protein degradation in skeletal muscle. Leucine can also enhance protein synthesis by stimulating the initiation of translation (Rennie and Tipton 2000). Leucine is available as a supplement at many nutrition retail suppliers.

UREA CYCLE

Ammonia is quite toxic, especially to the brain. However, two safe forms of ammonia exist: the amino group of glutamate and the side chain amide nitrogen in glutamine. Although we can temporarily store ammonia in these innocuous forms, we must eliminate the nitrogen we cannot use. Mammals convert the nitrogen to urea, whereas birds and reptiles eliminate amino groups by converting nitrogen to uric acid. Urea, whose structure is shown next, is a simple molecule formed in the liver and excreted from the kidney when urine is formed:

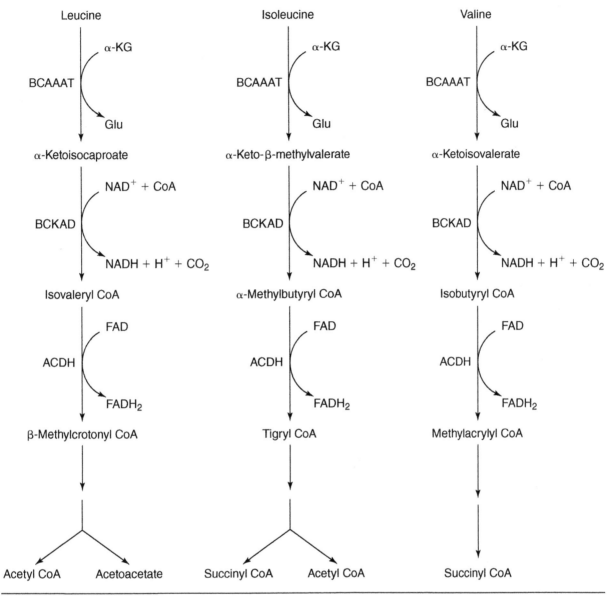

Figure 8.4 Catabolism of the three branched-chain amino acids leucine, isoleucine, and valine. The process starts with the same three enzymes: BCAAAT (branched-chain amino acid aminotransferase), BCKAD (branched-chain keto acid dehydrogenase), and ACDH (acyl CoA dehydrogenase). Isoleucine and valine produce succinyl CoA as a final product, which can be converted to glucose in the liver by the process of gluconeogenesis.

$$H_2N - \overset{\overset{O}{\|}}{C} - NH_2$$

The two amino groups in urea allow the body to rid itself of nitrogen. The carbonyl group comes from carbon dioxide. The nitrogen comes from the ammonium ion and from the amino group of aspartate. Figure 8.5 summarizes the path taken by nitrogen from amino acids in the body, using only three organs as examples. Muscle contains more protein than any other tissue. Amino acids released when muscle protein is degraded can be reused to make new protein, which is partially catabolized in muscle, as in the case of the BCAAs, or can be released to the blood. Amino groups from the BCAAs are first passed to α-ketoglutarate to make glutamate. Some glutamate transfers its amino group to pyruvate to make alanine. Some glutamate combines with ammonia to create glutamine. We have seen that ammonia production increases in muscle during exercise, especially intense exercise.

Muscle releases alanine and glutamine, which contain much of the muscle nitrogen. As we saw in figure 6.24, the source of the nitrogen on alanine is BCAAs. In the liver, the amino group on alanine is

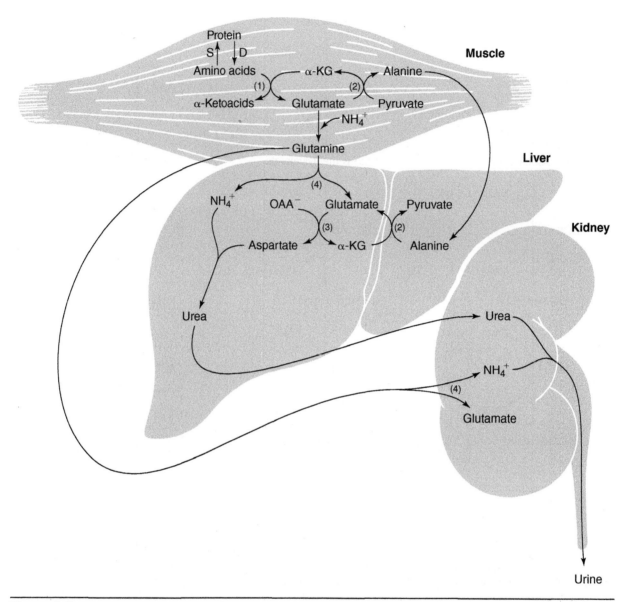

Figure 8.5 The interorgan transport of nitrogen, involving skeletal muscle, the liver, and kidneys. Branched-chain and other amino acids generated by the degradation (D) of protein and not used in protein synthesis (S) can lose their amino groups by transfer to α-ketoglutarate (α-KG) to make glutamate. Amino groups on glutamate can be transferred to pyruvate to make alanine. Glutamate can also be converted to glutamine using ammonia (NH_4^+). Muscle releases alanine and glutamine. Alanine amino groups can end up on aspartate, and glutamine can be converted into glutamate and ammonia. Urea is synthesized in liver using amino groups from ammonia and aspartate. The kidney releases most nitrogen as urea and a smaller amount as ammonia. Reactions: (1) branched-chain amino acid aminotransferase or other aminotransferase; (2) alanine aminotransferase; (3) aspartate aminotransferase; (4) glutaminase.

transferred to α-ketoglutarate to make glutamate, catalyzed by alanine aminotransferase. The glutamine released from muscle results from the amination of glutamate in the glutamine synthetase reaction. The liver is the major site for nitrogen removal from most of the amino acids (except the BCAAs). The amino groups end up on glutamate. Muscle also releases ammonium ions, especially during exercise. Urea synthesis requires the ammonium ion that comes from ammonia taken up by liver from the blood, from glutamine via the glutaminase reaction, or from glutamate via the glutamate dehydrogenase reaction. Aspartate, which provides the other urea nitrogen, comes from a transamination reaction in which oxaloacetate undergoes transamination from glutamate to make aspartate and α-ketoglutarate. The enzyme *aspartate aminotransferase* catalyzes the following reaction.

glutamate + oxaloacetate ↔ aspartate + α-ketoglutarate

The urea cycle was one of the first metabolic pathways to be fully described when it was elucidated in the early 1930s. Dr. Hans Krebs, who later also described the Krebs (or citric acid) cycle, was the scientist primarily responsible for elucidation of the urea cycle. The urea cycle consists of four enzymatic steps that take place in the liver cells, in both the mitochondrial matrix and cytosol. Before the actual cycle begins, carbamoyl phosphate must be synthesized from ammonia (as ammonium ion) and carbon dioxide (as bicarbonate), in a reaction catalyzed by *carbamoyl phosphate synthetase* (see figure 8.6). This reaction takes place in the mitochondrial matrix. It is driven by the simultaneous hydrolysis of two ATP and involves the formation of carbamoyl phosphate, catalyzed by carbamoyl phosphate synthetase. The carbamoyl phosphate synthetase reaction, the committed step in urea synthesis, controls the overall rate of the urea cycle. Carbamoyl phosphate synthetase activity is allosterically activated by *N-acetyl glutamate*, which is formed from acetyl CoA and glutamate by the enzyme *N-acetylglutamate synthase*. When amino acid degradation is accelerated, glutamate increases in the liver. By its conversion to N-acetylglutamate and subsequent allosteric activation of carbamoyl phosphate synthetase, it is signaling a need to speed up the urea cycle to get rid of the amino groups of the degraded amino acids.

The actual urea cycle is shown in figure 8.7. The carbamoyl phosphate enters the urea cycle in the mitochondrial matrix, joining with *ornithine* to form *citrulline* in a reaction catalyzed by *ornithine transcarbamoylase*. The citrulline exits the mitochondrial matrix. The second nitrogen, in the form of the *aspartate* amino group, enters the cycle when aspartate combines with citrulline to form *argininosuccinate*. This step involves ATP hydrolysis and is catalyzed by *argininosuccinate synthetase*. The argininosuccinate is then cleaved to arginine and

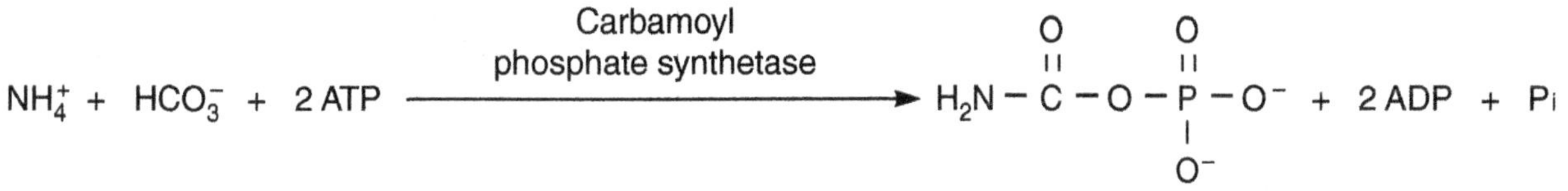

$$NH_4^+ + HCO_3^- + 2\,ATP \xrightarrow{\text{Carbamoyl phosphate synthetase}} H_2N-\underset{\underset{O^-}{|}}{\overset{O}{\overset{\|}{C}}}-O-\overset{O}{\overset{\|}{P}}-O^- + 2\,ADP + P_i$$

Figure 8.6 The reaction controlling the rate of urea synthesis, catalyzed by carbamoyl phosphate synthetase in the matrix of liver mitochondria.

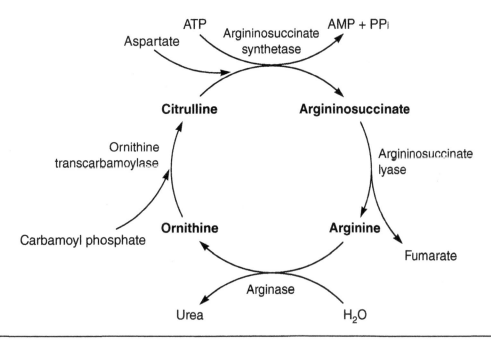

Figure 8.7 The urea cycle. This cycle is a sequence of four enzyme-catalyzed reactions that produce urea from nitrogen on carbamoyl phosphate and aspartate. There is no net consumption or formation of ornithine or other intermediates in the cycle, shown in bold letters.

fumarate in a reaction catalyzed by *argininosuccinate lyase*. In the final step, arginine is cleaved by *arginase* to yield ornithine and urea. The ornithine is now regenerated and enters the mitochondrion to combine again with carbamoyl phosphate.

Urea is extremely water soluble. Thus, it leaves the liver via the blood and enters the kidneys, where it is filtered out. The urea cycle is metabolically expensive because the ATP cost to make one urea molecule is four. Two are used to make carbamoyl phosphate. Can you think where the other two ATP come from? (Hint: How many phosphate groups are removed from ATP during the condensation of citrulline and aspartate?)

FATE OF AMINO ACID CARBON SKELETONS

After the amino groups are removed from the amino acids, carbon skeletons remain, in many cases in α-keto acids, such as pyruvate, oxaloacetate, or α-ketoglutarate (see figure 8.8). These carbon skeletons have a variety of fates, such as gluconeogenesis; 18 of the 20 amino acids can be a source of glucose. These are described as glucogenic amino acids. Note in figure 8.8 that leucine and lysine do not produce either pyruvate or citric acid (TCA) cycle intermediates. Leucine and lysine are described as ketogenic amino acids because their carbon skeletons produce only acetoacetyl CoA and acetyl CoA. Note that leucine degradation initially produces acetoacetate and acetyl CoA (figure 8.4). However, as with the ketone body acetoacetate, a CoA is transferred from succinyl CoA, making acetoacetyl CoA and succinate. The carbon skeletons of all amino acids can also be used for immediate oxidation because they form citric acid cycle intermediates, or acetyl CoA. Finally, all are potential sources of carbon to make new fatty acids because all are also potential sources of acetyl CoA. As already mentioned, amino acid

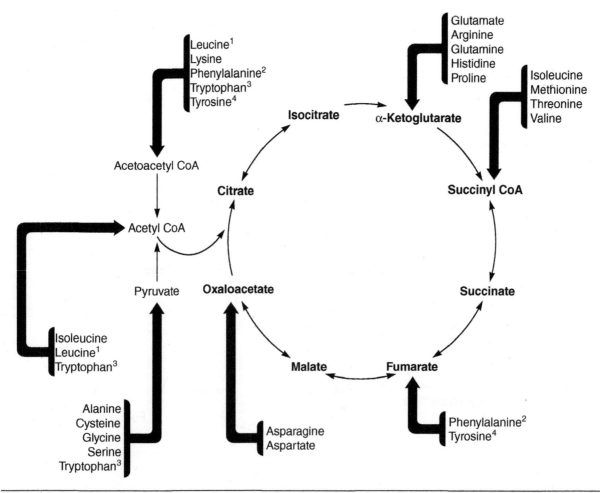

Figure 8.8 Conversion of amino acid carbon skeletons. Following removal of nitrogen groups on amino acids through deamination or transamination, the amino acid carbon skeletons are converted into citric acid cycle intermediates (shown in bold) or related molecules. The large arrows show the major paths taken by the carbon skeletons. The carbon atoms of some large amino acids, identified with superscript numbers, give rise to two or three different molecules.

carbon skeletons are not a very significant source for de novo fatty acid synthesis.

Figure 8.8 illustrates what the carbon skeletons of the amino acids have in common with the citric acid cycle or substances directly related to this cycle. The amino acids shown can generate the citric acid cycle intermediates or related molecules by simple removal of their amino groups via transamination (alanine, glutamate, and aspartate) or through a number of steps not shown. As we discovered in chapter 6, the formation of citric acid cycle intermediates through removal of the amino groups from amino acids is known as anaplerosis. It is an important mechanism to maintain citric acid cycle intermediate concentration, especially during exercise, when demand on the citric acid cycle is high.

AMINO ACID METABOLISM DURING EXERCISE

During exercise, carbohydrate and lipid supply most of the energy needs of the body, as already discussed. However, metabolism of amino acids also takes place at an accelerated rate in a manner designed both to transfer amino acid nitrogen (amino groups) out of muscle and to move their carbon skeletons from muscle to liver. To get a proper perspective on the effects of exercise, we will look at low- to moderate-intensity exercise and high-intensity exercise, because the effects of the two are quite different. First, it is important to understand how we learn about amino acid metabolism in muscle.

The study of metabolism of amino acids during exercise is complex. In humans, it is carried out through measurement of the concentrations of amino acids in the arterial and venous blood across an exercising muscle. From concentration differences and measures of muscle blood flow, we can determine which amino acids are being taken up and released from muscle and the rate at which this is taking place. We can then describe the release of an amino acid from muscle in terms of the concentration difference in arterial and venous blood times the blood flow, producing such numbers as micromoles of amino acid released per minute. In addition, muscle biopsies can be taken to determine the concentrations of amino acids and a variety of other metabolic intermediates in the muscle samples obtained before, during, and after exercise. For example, the lack of release of an amino acid or metabolite from muscle may mean that nothing is happening with it compared to what occurs in the inactive condition, or that its concentration is increasing within the muscle

and that is why it is not being released. Thus, it is necessary to measure both changes within muscle and changes in exchange across the muscle.

Even when good measures can be obtained of muscle metabolites, arterial and venous concentration differences, and blood flow, it may still be difficult to get an accurate picture of amino acid metabolism in muscle during exercise. The challenge is to estimate with reasonable accuracy the mass of muscle involved in the actual exercise. In other words, we need to be able to describe changes taking place in muscle amino acids by describing releases or uptakes in terms of micromoles per minute per unit mass of muscle (i.e., micromoles per minute per kilogram of muscle). Infusion of **stable isotopes** of specific amino acids is now also used to determine rates of amino acid uptake and use in muscle protein synthesis. The rate of appearance of these isotopes in skeletal muscle proteins, determined from muscle biopsies, can provide further information regarding rate of protein uptake and synthesis. One of the interesting developments seen from this type of research was that infusion of essential amino acids will stimulate muscle protein synthesis, but infusion of nonessential amino acids will not (Miller 2007). These findings have implications for exercise and nutrition strategies that are discussed further in the Next Stage section of this chapter.

Moderate-Intensity Exercise

Figure 8.9 summarizes some of the important changes in amino acid metabolism that take place when exercising skeletal muscle. During exercise in the postabsorptive state, skeletal muscle is in a net protein catabolic state, in which protein breakdown exceeds the rate of protein synthesis. Most of the amino acids produced by the net protein catabolism are released from the muscle. The major exception is glutamate, because there is a net uptake of glutamate when the body is at rest. This uptake increases even more during exercise. Glutamate is needed as a precursor for glutamine, which, along with alanine, is released in increasing amounts with exercise intensity and duration. The release of glutamine and alanine is far greater than their content in skeletal muscle proteins, suggesting that both amino acids are being formed in skeletal muscle at an accelerated rate during exercise. Figure 8.9 illustrates the derivation of the ammonia, alanine, and glutamine.

We have already discussed the fact that during glycolysis, approximately 1% of the pyruvate produced is converted to alanine. Transamination of pyruvate with glutamate by alanine aminotransferase produces alanine and α-ketoglutarate. As shown previously, the alanine leaves the muscle and travels in the blood

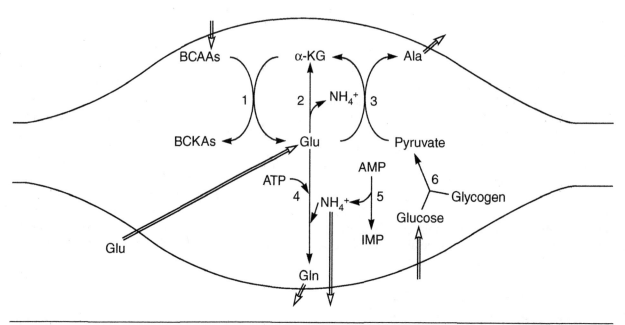

Figure 8.9 Net release of amino acids from skeletal muscle during exercise, which occurs with most amino acids. The exceptions are glutamate and the branched-chain amino acids (BCAAs), for which there is a net uptake. Alanine (Ala) and glutamine (Gln), formed by reactions 3 and 4, respectively, are released from muscle at rates far beyond those for the other amino acids. Alanine is synthesized from pyruvate and an amino group donated by the BCAAs. The source of the pyruvate is glycolysis (6), using glucose or glycogen as a precursor. Glutamine is synthesized from glutamate (Glu) using glutamine synthetase (4). Glutamate may also be deaminated by glutamate dehydrogenase (2), forming α-ketoglutarate (α-KG). Active in muscle during vigorous exercise is AMP deaminase (5), which produces ammonium ion (NH_4^+) and IMP (inosine monophosphate). Exercising muscle releases NH_4^+. Reaction 1 is catalyzed by branched-chain amino acid aminotransferase. Reaction 3 is catalyzed by alanine aminotransferase. BCKAs are branched-chain keto acids.

to the liver (see figure 6.24 on p. 192). In the liver, alanine aminotransferase converts alanine to pyruvate and α-ketoglutarate. The pyruvate is a source of carbon to make glucose by gluconeogenesis. If the glucose produced in liver is released to the blood and catabolized to pyruvate in muscle, a cycle is produced that is known as the *glucose-alanine cycle*.

The principal source of the amino group that is released from muscle in alanine is from the BCAAs. As mentioned earlier, these amino acids are preferentially transaminated in skeletal muscle by branched-chain amino acid aminotransferase. This reaction converts α-ketoglutarate to glutamate and the deaminated products of the BCAAs, known as branched-chain keto acids. These may undergo further oxidative metabolism within the muscle or may be released to the blood for metabolism in the liver. The carbon skeleton of leucine can be completely oxidized in the citric acid cycle since one product is acetyl CoA and the other, acetoacetate, gives rise to two acetyl CoAs. Degradation of isoleucine and valine generates succinyl CoA, which can help to maintain citric acid cycle intermediates.

During low- to moderate-intensity exercise, muscle releases alanine at a rate far in excess of the rate at rest. As shown in figure 8.9, the source of the pyruvate for alanine formation is from production during glycolysis from glucose and, more importantly, stored glycogen. We also know from chapter 6 that the contribution of glycogen to ATP production during prolonged exercise declines as the glycogen stores are gradually reduced. Although this is partially compensated by an increased use of blood glucose, it is typically noted that the respiratory exchange ratio during constant-intensity exercise declines over time. With glucose and glycogen as the sources of pyruvate to make alanine, we would expect that the release of alanine declines with exercise duration, and this is what is observed.

As shown in figure 8.9, skeletal muscle also releases glutamine, especially during exercise. Amination of glutamate, using the glutamine synthetase reaction (shown in figure 8.3), is the major source of glutamine, because muscle proteins are not rich in this amino acid. The ammonium groups needed to form glutamine from glutamate can be derived from glutamate via the glutamate dehydrogenase reaction, as well as from the deamination of AMP by the enzyme AMP deaminase. With its net uptake from the blood, glutamate is the source for glutamine synthesis and for some of the ammonia produced. Indeed, so much glutamate is needed to make glutamine that the concentration of glutamate actually declines during exercise, unlike that of the other amino acids. Interestingly, when subjects were given large doses of monosodium glutamate before exercise, so that plasma levels of glutamate were greatly elevated, the release of both alanine and glutamine increased, but release of ammonia was attenuated during the exercise (Mourtzakis and Graham 2003).

> **KEY POINT**
>
> If we consider the net effect of the glucose-alanine cycle, the carbon atoms of glucose are simply recycled from liver to muscle (as glucose) and muscle to liver (in alanine). However, this cycle also allows the muscle to transfer nitrogen from branched-chain amino acids to the liver for urea synthesis. When muscle glycogen levels are low, the consequent decrease in glycolysis during exercise reduces pyruvate formation. Consequently, the formation of alanine in muscle also declines.

The changes just noted for moderate-intensity exercise can be modulated depending on the nutritional state of the exercising person, as well as the concentration of muscle glycogen. In a fasted, low-glycogen state, less pyruvate is formed in glycolysis to produce alanine. In this state, protein breakdown is likely much greater, with a corresponding increase in the release of most amino acids from muscle. Early exercise studies also suggested that the rate of catabolism of amino acids by the liver increases in a fasted, low muscle glycogen state, as evidenced by increased formation of urea and larger concentrations of urea in sweat. Some studies have also noted an increase in muscle IMP and a decrease in **total adenine nucleotide** (TAN) content in untrained subjects during prolonged moderate-intensity exercise when muscle glycogen concentration is severely reduced. Apparently the steep decline in muscle glycogen leads to an increase in ADP concentration and, thereby, AMP concentration. This can result in a modest flux through AMP deaminase by a mass action effect, producing IMP even though conditions favoring complete activation of AMP deaminase are not present.

Purine Nucleotide Cycle

During exercise, there are significant increases in cytosolic ADP concentration, even though the level of ATP is reasonably well preserved by the three energy systems discussed in chapter 4. The ADP can be converted into ATP and AMP in a freely reversible reaction that is catalyzed by the enzyme *adenylate kinase*. As shown in table 4.3 (p. 98), the increase in AMP concentration during exercise is directly related to exercise intensity.

$$2\ ADP \leftrightarrow ATP + AMP$$

Chapter 4 also discusses the deamination of AMP, catalyzed by the enzyme *adenylate deaminase*. This enzyme converts AMP to IMP and ammonia, as shown in the following:

$$AMP + H_2O \rightarrow IMP + NH_3$$

Proton addition to the base NH_3 generates ammonium ion (NH_4^+). As discussed in chapter 4, the adenylate deaminase reaction results in an actual decrease in the TAN content of the muscle.

Adenylate deaminase is more active in Type II (fast-twitch) than Type I (slow-twitch) muscle fibers. Therefore, it is unlikely to be activated significantly in low-intensity exercise, since the force production will be handled to a greater extent by motor units containing Type I muscle fibers. Furthermore, adenylate deaminase activity is low in rested or slowly contracting muscle, increasing only if the pH is decreased and the concentrations of AMP and ADP are increased. Type I fibers not only have less adenylate deaminase but also are unlikely to experience the rise in AMP and decrease in pH that are needed to activate the enzyme, since these fibers have a higher capacity for oxidative metabolism and lower glycolytic activity. Therefore, we can anticipate that adenylate deaminase is unlikely to be significant during low to moderate levels of exercise.

> **KEY POINT**
>
> Many athletes ingest fluids that contain carbohydrate during training. While they may be thinking that these liquids provide fuel for their training, they will also help maintain protein balance in the muscle by modulating the rate of protein breakdown as they attenuate the depletion of carbohydrate stores. Therefore, the need for gluconeogenesis from protein sources, a process normally activated by strenuous prolonged exercise, is lessened.

Since the adenylate deaminase can reduce TAN levels by up to 50% in very severe exercise, there must be a way to regenerate the adenine nucleotides. The process for doing this is known as the purine nucleotide cycle and is shown in figure 8.10. In reaction 1, which is catalyzed by adenylate deaminase, IMP and NH_3 are produced. As already noted, this reaction is active during intense muscle activity. Adenosine monophosphate is regenerated by two additional reactions. In the first, aspartate combines with IMP at the cost of a guanosine triphosphate (GTP), forming adenylosuccinate. This is catalyzed by *adenylosuccinate synthetase*. In the final step, adenylosuccinate is split into AMP and fumarate, catalyzed by *adenylosuccinate lyase*. The net reaction shows that it costs energy to regenerate AMP in the form of a GTP. In the process, an aspartate is deaminated, producing fumarate, a TCA cycle intermediate. The regeneration of AMP by adenylosuccinate synthetase and adenylosuccinate lyase takes place predominantly after exercise because the activities of these enzymes are too low to produce appreciable AMP during exercise.

High-Intensity Exercise

When the intensity of exercise is high, there are likely to be only modest increases in the release of glutamine and alanine from muscle. Glutamate uptake by muscle also appears to be little influenced by increased exercise intensity. Glutamine synthesis consumes ATP, and its release represents a loss of glutamate from muscle. Since glutamate is a source for the citric acid cycle intermediate α-ketoglutarate, accelerated formation and release of glutamine may not be a productive reaction when ATP demand is high. The key reaction that is accelerated by an increase in exercise intensity is the adenylate deaminase reaction. In particular, the metabolic changes in an intensely contracting fast-twitch muscle fiber (increase in AMP and H^+) favor activation of adenylate deaminase, with the attendant formation of IMP and ammonia. As a phosphorylated molecule, IMP is trapped within the fiber, but ammonia is released at an accelerated rate. Studies generally show a positive relationship between lactate formation and release from skeletal muscle and release of ammonia.

> ## ✓ KEY POINT
>
> During intense exercise, a primary purpose of the adenylate kinase reaction, which transforms two ADP into an ATP and an AMP molecule, is to enhance the cellular signaling to activate the ATP regenerating metabolic pathways. This is because increased presence of AMP is an important signaling molecule for activation of enzymes that regulate glycolytic and aerobic metabolism. In addition, by helping stabilize ADP levels in intensely exercising muscle, this reaction also helps to maintain higher [ATP]/[ADP] ratios. As noted in chapter 4, these are important to maximize the free energy release of ATP hydrolysis, so they are essential for driving the ATPase reactions associated with muscle contraction.

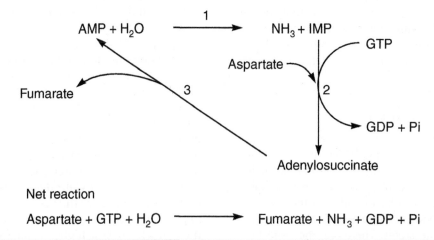

Figure 8.10 The purine nucleotide cycle. In this simple pathway, AMP is deaminated during intense muscle activity, catalyzed by adenylate deaminase (1). The next two steps regenerate AMP. First, IMP combines with aspartate, catalyzed by adenylosuccinate synthetase (2). Secondly, adenylosuccinate is split into AMP and fumarate, catalyzed by adenylosuccinate lyase. The net reaction shows that the regeneration of AMP costs a GTP, and the amino acid aspartate is deaminated and a fumarate is produced.

ADDITIONAL ROLES FOR AMINO ACIDS

We have seen in previous chapters that amino acids have purposes beyond simply serving as the building blocks of proteins. Table 8.1 summarizes roles for amino acids other than as components of body proteins.

Creatine is created in a two-step process from the amino acid *glycine* and part of the amino acid *arginine*. A methyl group must be transferred, and this is accomplished through the use of *S-adenosylmethionine* (SAM), a derivative of the essential amino acid *methionine*. *Carnitine*, the essential molecule used in the transfer of long-chain fatty acyl groups across the mitochondrial inner membrane, begins as a *lysine* residue that is methylated on its side chain amino group with SAM. We have also discussed the essential role played by glutathione in maintaining cellular redox status. Glutathione is a tripeptide formed from the amino acids *glutamate*, *cysteine*, and *glycine*. Biosynthesis of pyrimidines, such as uracil, thymine, and cytosine, depends on aspartate and glutamine. For purine biosynthesis, including the bases adenine and guanine, the amino acids glutamine, glycine, and aspartate are needed. Heme proteins, such as hemoglobin, myoglobin, and the cytochromes, have a heme group with a central iron ion. The heme group is built from a structure class known as **porphyrins**, which are synthesized in the body using carbons from succinyl CoA and the amino acid glycine.

We generally acknowledge a total of about 20 different essential and nonessential amino acids in our proteins. This total does not include a number of modifications, such as **hydroxylation**, which is the addition of an OH group, to certain amino acids. Hydroxylation of some amino acids occurs when specific proteins are completing their tertiary, or three-dimensional, structural modifications. Examples of hydroxylated amino acids are hydroxyproline and hydroxylysine, which are hydroxylated versions of proline and lysine. These modifications of proline and lysine are of particular interest in humans, since they are essential steps in collagen protein synthesis, and hydroxylation of these amino acids is part of the process that converts procollagen

Table 8.1 Roles for Amino Acids Other Than as Components of Proteins

Precursor amino acid	Special product	Role of product
Alanine and histidine	Carnosine	Dipeptide involved as an antioxidant
Arginine	Nitric oxide	Inter- and intracellular signaling
Arginine, glycine, and methionine	Creatine/creatine phosphate	Energy transfer reactions
Aspartate, glutamine, and glycine	Purines and pyrimidines	Nucleic acids, nucleoside phosphates
Aspartate	Aspartate	Excitatory neurotransmitter
Cysteine, glutamate, and glycine	Glutathione	Maintain cellular redox
Glutamate	Glutamate	Excitatory neurotransmitter
Glutamate	Tetrahydrofolic acid (THFA)	One-carbon transfer reactions
Glycine	Porphyrins	Heme groups in proteins
Glycine	Glycine	Inhibitory neurotransmitter
Histidine	Histamine	Neurotransmitter
Lysine and methionine	Carnitine	Fatty acyl transfer
Methionine	Homocysteine	Association with circulatory diseases
Methionine	S-adenosylmethionine	Methyl group transfers
Proline and lysine	Hydroxyproline and hydroxylysine	Collagen and connective tissue
Tryptophan	Serotonin	Neurotransmitter
Tyrosine	Dopamine, norepinephrine, and epinephrine	Neurotransmitters, hormone

into functional collagen. Essentially, the presence of hydroxyproline and hydroxylysine allows for proper three-dimensional configuration and folding of the collagen proteins into their stable tertiary and quaternary structures so that they can function effectively in the formation of connective tissues. The synthesis of hydroxyproline and hydroxylysine from proline and lysine relies on the presence of vitamin C to act as a coenzyme. Unlike most other mammals, who can synthesize vitamin C or ascorbic acid from glucose, humans lack the critical enzyme to run this synthesis pathway. Hence, humans require vitamin C in their diet to supply this particular requirement. When vitamin C is lacking in our diets, humans develop an ultimately fatal condition known as scurvy, which is characterized by the breakdown of connective tissue that is made with collagen. Without vitamin C, humans lack the ability to hydroxylate proline and lysine into hydroxyproline and hydroxylysine, and procollagen cannot be converted to functional collagen. Hence, in situations where dietary vitamin C is unavailable, collagen cannot be synthesized and replaced. In previous centuries, sailors on long ocean voyages, who did not have access to fresh fruits and vegetables, regularly showed symptoms of scurvy.

The amino acid *tyrosine* can also be created by hydroxylation of the essential amino acid phenylalanine. Tyrosine is the precursor for three important molecules, the neurotransmitters *dopamine* and *norepinephrine* and the adrenal medulla hormone *epinephrine*. Tyrosine is first converted to dopamine, a neurotransmitter in the brain. Hydroxylation of dopamine creates norepinephrine, and methylation of norepinephrine produces epinephrine. *Histamine*, another neurotransmitter, is synthesized from the essential amino acid histidine by decarboxylation of its carboxyl group. In a simple two-step process, the essential amino acid tryptophan is converted to the brain neurotransmitter *serotonin*.

Homocysteine is a form of the amino acid cysteine with an additional methyl group that is actually synthesized from methionine through the removal of a methyl group. Although the physiological and biochemical mechanisms behind the relationship are not yet fully defined, high blood levels of homocysteine have been epidemiologically associated with increased incidence of cardiovascular disease. Low levels of B vitamins are thought to be associated with abnormal blood levels of homocysteine. Dietary supplements of B vitamins, such as folic acid, vitamin B6, and vitamin B12, have been associated with reductions in blood homocysteine levels, since these vitamins are important coenzymes involved in homocysteine metabolism. However, a number of studies have cast doubt on a direct link between reduction of blood homocysteine levels and any reduction in risk of cardiovascular disease. The relationship between homocysteine elevations and cardiovascular diseases continues to be controversial as well (Bazzano et al. 2006).

Some special amino acids, found in specific tissues, also exist. *Carnosine* (β-alanylhistidine) is a dipeptide formed from a modified form of the amino acid *alanine* (β-alanine) and *histidine*. It is found, along with two related forms, *anserine* and *homocarnosine*, in skeletal and cardiac muscle and in the brain in particular. These novel amino acid derivatives are thought to play roles as antioxidants. Carnosine has been tested as a supplement to attenuate the development of cataracts in rats. However, insufficient positive data exist at this time to recommend dietary supplements of carnosine for this purpose in humans.

NEXT STAGE

Effects of Diet and Exercise on Protein Degradation and Synthesis

Resistance training has long been known to induce muscle hypertrophy as well as increase muscle strength. Increased muscle hypertrophy results from an overall positive net protein balance (NPB), where muscle protein synthesis (MPS) exceeds muscle protein breakdown (MPB). The molecular regulation of MPS and MPB, both under basal conditions and in response to resistance training, are discussed in chapter 3. Here, we will discuss research related to how diet and exercise affect muscle protein accretion.

A review by Burd and colleagues (2009) covered research developments related to dietary intake of specific amino acids and proteins, and to their timing relative to optimization of positive NPB. As previously noted, protein and amino acid turnover in skeletal muscle is dynamic and rapid. Following a meal, amino acids are absorbed into muscle and MPS exceeds MPB. Between meals, the reverse is true (see figure 8.11a). Generally, these two processes cancel each other out, and relatively little net gain or loss of muscle mass occurs. However, as depicted in figure 8.11b, resistance exercise stimulates MPS to the extent that a positive NPB is achieved. Feeding of protein also accentuates the MPS signal arising from resistance training, which can last for up to 48 h following a single bout of resistance exercise.

The timing, amino-acid type, and protein source of the feedings that optimize MPS in conjunction with resistance exercise have been subjects of recent research. Findings from these studies, as reviewed by Burd and colleagues (2009), indicate that the essential amino acids are necessary for the stimulation of MPS with optimal amounts obtained from approximately 20 g of complete proteins (or 8-9 g of essential amino acids). Ingestion of protein up to 2 h following resistance training is also much more potent in stimulating MPS than ingestion of protein prior to exercise (Fujita et al. 2009). Delaying of feeding beyond 2 h postexercise will also reduce the rate of training-induced MPS and muscle hypertrophy.

Rapidly digested proteins, such as whey and soy, can quickly but transiently augment MPS following resistance training. Milk proteins, which include whey, also promote greater net MPS following training than soy proteins do. This is possibly due to milk's high content of the essential amino acid leucine and its rapid digestion and absorption (Burd et al. 2009). Milk proteins such as casein may also improve muscle NPB following resistance exercise by inhibiting MPB.

Strength training in the elderly is particularly dependent on timely postexercise provision of protein to optimize muscle mass gains. Young men and women appear not to differ appreciably in rates of postexercise MPS and in the effects of diet on MPS. This indicates that local muscle control factors may be more important in regulating muscle hypertrophy than circulating hormone levels. However, elderly women seem to have greater impairment of MPS following protein feeding and resistance exercise than elderly men (Burd et al. 2009). This may be due to their loss of estrogen, which has been demonstrated to stimulate factors

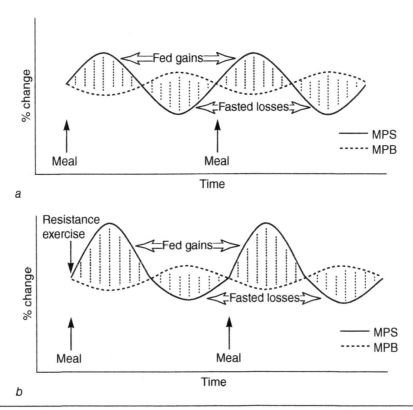

Figure 8.11 (a) Changes in muscle protein synthesis (MPS) and muscle protein breakdown (MPB) in response to feeding (i.e., amino acids). (b) Changes in MPS and MPB in response to resistance exercise and feeding. Chronic application of these anabolic stimuli, as in (b), results in muscle hypertrophy.

Reprinted from N.A. Burd et al., 2009, "Exercise training and protein metabolism: Influences of contraction, protein intake, and sex-based differences," *Journal of Applied Physiology* 106: 1692-1701. Used with permission.

associated with muscle hypertrophy, such as satellite-cell activation and proliferation (Enns and Tiidus 2008). Research is continuing on optimizing the effects of exercise and diet for regulating protein and amino-acid metabolism to maximize muscle hypertrophy, particularly for the elderly, who lose significant amounts of muscle mass and strength as they age.

SUMMARY

The content of protein in the adult body remains remarkably constant. Because most adults take in about 10% to 15% of their dietary energy in the form of protein, an equivalent amount of amino acids must be lost each day. The body cannot store excess amino acids. Those not needed to make protein lose their nitrogen groups, and the carbon skeletons are used to make glucose (gluconeogenesis) or are converted to acetyl CoA, which can feed into the citric acid cycle or can be used to make fatty acids. The first step in disposing of excess amino acids is to remove the amino groups, which are primarily transferred to α-ketoglutarate by aminotransferase enzymes to produce glutamate and α-keto acids. This process occurs in the liver for most amino acids. However, the branched-chain amino acids (leucine, isoleucine, and valine) lose their amino groups primarily in muscle through the action of branched-chain amino acid aminotransferase. Skeletal muscle contains a high concentration of glutamine, synthesized by glutamine synthetase from glutamate and ammonium ions. During exercise, muscle releases most amino acids, particularly alanine, ammonium ions, and glutamine, which are released at accelerated rates. During exercise, branched-chain amino acids and glutamate are taken up by muscle. Amino groups on amino acids are removed from the body in the form of urea, a molecule synthesized in liver using the urea cycle and excreted in the urine. The simple urea cycle eliminates amino groups that cannot be oxidized. The purine nucleotide cycle is active in skeletal muscle. During vigorous exercise, AMP deaminase is activated, changing AMP into IMP and ammonia. Regeneration of AMP, and thus the other adenine nucleotides, is accomplished by two reactions of this cycle. Many amino acids have functions beyond their role as building blocks of proteins. A number of amino acids can be converted into special molecules that function as hormones.

REVIEW QUESTIONS

1. What do the terms *energy balance* and *protein balance* mean?

2. The nitrogen content of proteins is considered to average 16% by mass of the body's proteins. The excretion of nitrogen can be measured in a simple manner. If a young adult excretes 16 g of nitrogen on average each day for a week, what is the average daily intake of protein if we assume that this person is in nitrogen balance?

3. You determine that the content of urea in the urine averages 40 g per day. What is the approximate intake of protein assuming that the person is in energy balance?

4. The measurement of the excretion of 3-methylhistidine has been taken as an index of skeletal muscle protein breakdown. If you want to estimate muscle protein degradation, you need to determine the concentration of 3-methylhistidine in the urine. What else do you need to measure?

5. A young woman in energy and nitrogen balance takes in 8,000 kJ of food energy per day. She excretes 30 g per day of urea. Assuming that each gram of amino acids when oxidized yields 18 kJ, what percentage of her daily food intake comes from proteins?

6. The data in the following table were obtained from subjects at rest and during exercise on a cycle ergometer at 70% of $\dot{V}O_2$max for 1 h. Blood flow was measured, and exchange of amino acids across the leg was determined from blood samples obtained from the femoral artery and femoral vein. Leg blood flow averaged 0.3 L/min during rest and 8.0 L/min during exercise. The average concentration difference between arterial and venous blood for glutamate, glutamine, alanine, and leucine is shown for rest and exercise conditions. The molecular weight (MW) for each amino acid is also shown.

Amino acid	MW	Arterial minus venous blood concentration difference per minute (μmol/min)	
		Rest	Exercise
Glutamate	146	12	40
Glutamine	146	−25	−50
Alanine	89	−30	−100
Leucine	133	−1	20

a. What does a negative arterial minus venous concentration difference mean?

b. What was the total volume of blood flow through the leg during the 60 min of exercise?

c. What was the fold increase in blood flow from rest to exercise?

d. How many millimoles of glutamate were exchanged by the leg during the exercise period?

e. What are the two major sources of carbon for the alanine released from the muscle?

f. How many grams of alanine were released from the muscle?

g. Assuming that all of the alanine released during the exercise period is taken up by the liver, how many moles of pyruvate would be produced?

h. If the alanine released from muscle during the 60 min of exercise was entirely converted to glucose in the liver, how many millimoles of glucose would this produce?

7. Discuss how the timing and form of dietary protein intake can help optimize muscle protein synthesis and muscle hypertrophy following resistance training.

appendix

Answers to Review Questions

Chapter 1

1. a. 11
   b. N-terminus is glycine and C-terminus is leucine.
   c. Zero: arg and lys have +1 each, and asp and glu have net −1 each.
   d. Cysteine only
   e. Nonpolar amino acids are gly, phe, val, trp, leu.
   f. Phe, val, trp, leu
   g. Yes, there are two cys residues.
   h. GRCEDKFVWCL

2. For cysteine to carry a net negative charge, the side-chain sulfhydryl group must have a net negative charge, while the ammonium group carries its charge. This happens only when the pH is greater than the pK_a for the sulfhydryl group, but less than the pK_a for the amino group, which would be about pH 10. Thus, the pH must between 8.2 and 10.0.

3. At pH of 6.0, the $H_2PO_4^-$ (acid form) predominates, whereas at pH 8.0, it is HPO_4^{2-}.

4. No, only those proteins made up of two or more polypeptide chains (subunits) have quaternary structure.

Chapter 2

1. Plot the data using a Lineweaver–Burk plot. These are actual data, so your line will need to be adjusted to reflect them fairly. Make sure you extend the line into the other quadrant to get an intercept on the X-axis. K_m should be about 15 to 16 mmol/L, and Vmax should be about 33 mmol · ml^{-1} · min^{-1}.

2. Assuming that the enzyme behaves traditionally, then a 10 °C rise in temperature should double the Vmax.

3. This was a common problem in early studies because labs were not careful about temperature control. Too often, samples sat at room temperature and then were measured in the spectrophotometer or fluorometer. Different labs could have different temperatures. A second possibility lies in calibration using solutions of NADH, especially with the use of a fluorometric assay. Finally, samples may have less or more water depending on how they are treated. Since water is about 77% of a wet muscle sample, this could introduce a large error.

4. a. 10 mmol · kg^{-1} · min^{-1}
   b. 10 IU/g

5. The K_m was the same, but the Vmax was lower (27 mmol · ml^{-1} · min^{-1}). This means that the inhibitor was noncompetitive.

6. PDH phosphatase actually activates PDH, since the phosphorylated form of PDH is the inactive form.

Chapter 3

1. Complementary strand: 3'-TCAGGTCG-CAATCTGGCTTCA-5'

2. Base sequence of mRNA: 5'-AGUCCAGC-GUUAGACCGAAGU-3'

3. Amino-acid sequence: N-Ser-Pro-Ala-Leu-Asp-Arg-Ser-C

4. The sequence of the sense strand would be the same as the mRNA sequence except that you would have a T instead of a U. Therefore, go to the genetic code and work from

amino acid to base sequence. You would get some redundancy, but this is also correct: 5'-GAAAAAATGGCTGGTCATACT-3'.

5. This is a bit of a tricky question because you are asked two different things. The coding region would begin with AUG, the start codon, and proceed to a stop codon. Thus, the minimum number of nucleotides would be 3 × 200 if the protein began with methionine, or 3 × 201 if the protein did not begin with methionine because the initial methionine coded by AUG would be removed. The open reading frame is the area between the start and stop codon, so it would contain a minimum of 600 nucleotides.

6. Calpains are activated by Ca^{2+}. Since autolysis increases the sensitivity of calpains to Ca^{2+}, it will increase the proteolytic activity of calpains at a given $[Ca^{2+}]$.

7. Mitochondrial biogenesis is regulated by at least four different signaling pathways including CaMK, CaN, AMPK, and p38γ MAPK pathways. Calcium is a key signal that activates both CaMK and CaN pathways; increased [AMP]/[ATP] activates AMPK; and ROS activates p38γ MAPK. Collectively, calcium, [AMP]/[ATP], and ROS are key primary signals associated with endurance exercise that result in activation of these signaling pathways, increasing transcription of genes that encode mitochondrial proteins. Mitochondrial biogenesis with endurance training is the result of the cumulative effects of repeated endurance exercise and activation of these signaling pathways.

Chapter 4

1. Γ rest is 5.85×10^{-3} mM; Γ mild fatigue is 3.53×10^{-2} mM; Γ severe fatigue is 0.629 mM; and Γ following marathon is 3.75×10^{-2} mM.

2. a. 4.0 μmol/g wet weight
   b. 17.4 mmol/kg dry weight
   c. 5.7 mM

3. K_{eq} is 25.6.

4. ΔG for ATP hydrolysis = 10.5 kcal/mol.

5. Approximately 10.2 mM for total magnesium concentration

6. ΔG for ATP hydrolysis = 7.0 kcal/mol.

7. ATP + H_2O → ADP + Pi
   PCr + ADP → ATP + Cr
   PCr + H_2O → Cr + Pi

8. During very intense exercise, the rate of ATP utilization by the contracting muscles exceeds the maximum rate of ATP resynthesis of oxidative phosphorylation. Hence, in order to continue to exercise (at least for a limited time) at very high intensities, increased ATP-resynthesis capacity must be utilized. This can be accomplished by rapid hydrolysis of PCr stores and by upregulation of the glycolytic pathway.

9. PCr stores are limited in muscle. There is no need to utilize PCr for ATP resynthesis during rest or low-intensity activity, since ATP utilization rates are low and can be easily matched by aerobic metabolism (oxidative phosphorylation).

10. Dietary PCr supplementation can boost muscle PCr stores, thereby potentially enhancing ability to perform repeated, short bursts of high-intensity activities. Ergogenic benefits to a single bout of high-intensity activity may also exist. Recent research has also suggested that there may be an anabolic component to elevated muscle PCr content such that it enhances muscle size and protein accretion, particularly in older adults. Few of these benefits can be seen without concurrent power or resistance training.

11. 93

Chapter 5

1. The number of ATP formed maximally from the complete oxidation of a pyruvate would be 12.5. Four NADH are generated (4 × 2.5), one $FADH_2$ (1.5), and one GTP (equivalent to an ATP). Using the earlier numbers, the ATP yield would be 15.

2. To reduce one oxygen molecule requires 4 electrons. To reduce four would require 16 electrons. Forty protons would be pumped.

3. The same, 2.0 L/min

4. Rotenone inhibits complex I. Therefore, any substrate that generates NADH will not work. A good substrate would be succinate because you would bypass complex I.

5. The $\dot{V}O_2$ of 0.4 L/min would represent the person's resting metabolism. Since the subject is sitting and not doing much in terms of muscular work, the metabolism would represent mostly metabolism of the heart, respiratory muscles, kidney, brain, and liver.

6. Use the relationship that a $\dot{V}O_2$ of 1.0 L/min is equivalent to an energy expenditure of 5 kcal/min. Then multiply by 120 min and use 3.0 L/min. You should get 1,800 kcal, or 7,525 kJ.

7. The level-off phenomenon occurs with people who are trained or well trained. It means that they continue to exercise past the point of any increase in $\dot{V}O_2$, a fact attributable to their $\dot{V}O_2$ being limited by the ability to pump oxygen-rich blood from the heart. Untrained people often stop without a plateau in $\dot{V}O_2$. This is taken to mean that their weak link lies in the ability to use the oxygen in blood pumped from the heart. The limitation would be in the citric acid cycle, in the electron transport chain, or both.

8. Using table 5.4, determine that the $\Delta E^{\circ\prime}$ for the algebraic sum of $FADH_2$ oxidation and reduction of oxygen is 1.04 V. The standard free energy change from this would be 200 kJ/mol.

9. Dehydroascorbic acid can donate electrons only to substances below it in table 5.4. Therefore, the answer is both oxygen and the oxidized form of cytochrome c.

10. The $\Delta E^{\circ\prime}$ is –0.14 V.

11. The ΔE is –0.13 V.

12. Antioxidants may be able to reduce the amount of oxidative damage to muscle caused by RONS produced during exercise. However, since RONS are important signaling factors that can induce positive adaptations to training as well as stimulate physiological adaptations that benefit health, too much suppression of RONS production during exercise by dietary antioxidant supplementation can diminish the positive responses to exercise training.

Chapter 6

1. Glycogen by itself has a fuel value of 4 kcal/g (17 kJ/g). However, since each gram of glycogen is associated with 3 g of water, the fuel value is diluted to 1 kcal/g, or 4.2 kJ/g.

2. Activities that involve vertically lifting your body weight may require anaerobic glycolysis, but if the activity is relatively brief, having a huge glycogen store does not provide an advantage. Rock climbing or rope climbing would use glycogen at a high rate but for brief periods of time, so large stores would be just added weight. Storing a lot of glycogen could also add extra weight, so wrestlers should be careful.

3. A lactate concentration of 990 mg/dL is equivalent to 11 mM. A glucose concentration of 108 mg/dL is equivalent to 6 mM.

4. Using the calorie equivalents of 4 kcal/g for carbohydrate and 9 kcal/g for fat, the runner's rate of energy expenditure is 20.5 kcal/min (86 kJ/min). The proportion of energy from carbohydrate is 16/20.5, or 0.78. The total time he runs is total energy expended divided by rate of energy expended. The answer is 39.9 min. At 4 min/km, he covers just under 10 km.

5. A glycogen concentration of 500 mmol/kg dry weight is equivalent to 115 mmol/kg wet weight if the water content is 77%. Converting this to millimolar units, you would divide 115 mmol/kg wet weight by 0.7, because each kilogram of muscle would contain 0.7 kg of water, or 0.7 L of intracellular water.

6. The FT fiber is metabolically endowed to produce lactate, whereas the ST fiber is better endowed to use lactate as a fuel. An elevated concentration of extracellular lactate leads to fatigue, whereas lactate may help maintain force within cells. Therefore, the FT fiber should have the high K_m transporter, while the ST fiber should have the low K_m transporter.

7. Lactate production will result from upregulation of glycolysis, which provides a rapid resynthesis of ATP that allows intense exercise to be continued. Lactate can also be utilized as a fuel source, both within the working muscle as well as by other muscles that can convert lactate back to pyruvate. The pyruvate can then be used for oxidative phosphorylation in the mitochondria. Lactate can also be converted by the liver to glucose via gluconeogenic pathways. Lactate itself does not contribute to metabolic acidosis or cause muscle soreness.

Chapter 7

1.

Arachidonic acid, showing all carbon atoms and cis orientation for each double bond.

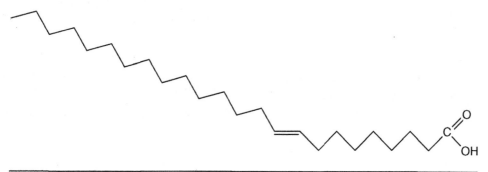

Elaidic acid, showing all carbon atoms and the orientation about the double bond.

2. A 3:1 ratio tells us that 100% of fatty acids released by lipolysis are released. Therefore a 1.5:1 ratio means that 50% of the fatty acids are being reesterified.

3. Energy expended during 60 min is 60 min × 2.4 L O_2/min × 5.043 kcal/L O_2 = 726 kcal (3,037 kJ). The number of grams of carbohydrate oxidized is 0.585 × 726 kcal divided by 4 kcal/g = 106.2 g carbohydrate and 33.4 g fat. Because this is a nonprotein RER, we are excluding the contribution made by protein oxidation. This is usually less than 5%.

4. Each mole of triacylglycerol has 1 mol of glycerol. The molecular formula for glycerol is $C_3H_8O_3$, and its molecular weight is 92. During the formation of a triacylglycerol, three water molecules are released for each fatty acid attached. If we ignore that glycerol contributes to this water loss, we can determine that glycerol contributes 92/860 × 100, or 10.7%.

5. The $\dot{V}O_2$ due entirely to the active kicking muscles was 0.75 L/min (three times that of rest, which must then be 0.25 L/min). The kilocalories of energy devoted entirely to kicking are 60 min × 0.75 L O_2/min × 5.043 kcal/L O_2 = 227 kcal. Weight of intramuscular TAG used is first calculated using the relationship that if there are 860 g/mol of TAG, then there are 860 mg/mmol. Now, intramuscular triacylglycerol used = (12 mmol/kg − 10 mmol/kg) × 3 kg of muscle used × 860 mg/mmol TAG = 5,160 mg, or 5.16 g. The energy equivalent would be 5.16 g × 9 kcal/g, or 46.4 kcal. The mass of fat oxidized is given from the RQ relationship as described in question 3. With an RER of 0.87, 41.5% of the energy comes from fat. This amounts to 227 kcal × 0.415 = 94 kcal. We can account for 46 kcal from IMTG. Thus, the remainder is from FFA and perhaps VLDL. Obviously, the balance of the remaining energy would be derived from carbohydrate, notably muscle glycogen.

6. See the Next Stage section in this chapter for a complete answer to this question. Note that a negative caloric balance needs to be achieved in order to increase 24 h fat oxidation regardless of macronutrient content of diet or exercise during that time period.

Chapter 8

1. *Energy balance* means that food energy intake is equal to total daily energy expenditure. *Protein balance* means that no net protein is stored in the body, so that protein intake is matched by the complete degradation of protein.

2. Nitrogen balance means that nitrogen intake and nitrogen excretion are equivalent. If excretion of nitrogen is 16 g/day, then intake is 100/16 × 16 g, or 100 g/day.

3. Reviewing the chemical structure of urea, you can determine that its molecular formula is CH_4N_2O. Its molecular weight is 60, and the

contribution to this molecular weight based on nitrogen is 28/60. Therefore, the nitrogen content in 40 g of excreted urea is 40 × 28/60, or 18.7 g. To convert this to a protein value, simply multiply by 100/16. This gives us 117 g of protein intake per day.

4. You would need to measure the total volume of urine produced over 24 h.

5. As you did in question 2, you would determine nitrogen excretion as 28/60 × 30 g urea per day, or 14 g N per day. Multiply this by 100/16 to get total protein lost, or 87.5 g protein. Because she is in nitrogen balance (protein balance), she must take in 87.5 g of protein. At 18 kJ/g, the energy equivalent of protein is 18 kJ/g × 87.5 g, or 1,575 kJ. This represents 1,575/8,000 × 100%, or 19.7% protein.

6.
   a. A negative arterial concentration difference means that the venous value is greater than the arterial value.
   b. 8 L/min × 60 min = 480 L of blood.
   c. Fold increase = 8 L/min divided by 0.3 L/min, or 26.7-fold greater.
   d. Glutamate exchange would be uptake of 40 mmol/min. Multiply this by 60 min to get total exchange during exercise (2,400 mmol).
   e. Muscle glycogen and glucose taken up from the blood.
   f. Release of alanine is 100 mmol/min × 60 min × 89 mg/mmol. Divide by 1,000 to get grams = 534 g.
   g. The MW of alanine is 89, so 534 g represents 6 mol of alanine. Now, you need to know that each mole of alanine is converted to 1 mol of pyruvate. Further, you need to know that it takes 2 mol of pyruvate to make 1 mol of glucose (from chapter 5). Thus, the 6 mol of alanine would be converted into 6 mol of pyruvate, which would be converted into 3 mol of glucose, or 3,000 mmol of glucose.
   h. 100 mmol of alanine released per minute would be 6,000 mmol released for 1 h. As shown in answer g, it takes 2 mmol of alanine to make 1 mmol of glucose. Therefore, there would be the potential to make 3,000 mmol of glucose from the alanine.

7. Dietary intake of at least 20 g of protein rich in essential amino acids (8-9 g) taken within 2 h of resistance training will optimize muscle protein synthesis.

Glossary

acceptor control—Regulation of the rate of oxidative phosphorylation by the availability of ADP.

acid—A compound that can donate a proton.

acid dissociation constant—The constant K_a, which describes the ability of an acid to donate a proton. The larger the value, the stronger the acid.

acidosis—A metabolic condition in which the production of protons is elevated and the capacity to buffer these is diminished. The blood pH may be reduced.

actin—The principal protein of the thin filament in muscle. It can bind with myosin and enhance myosin's ability to hydrolyze ATP.

activator—A DNA-binding protein that enhances the expression of a gene, or an allosteric effector that increases the activity of an enzyme.

active site—The region of an enzyme protein that binds substrate and converts it to product.

active transport—Transport across a membrane against a concentration gradient, requiring energy.

adenine—A purine base found in DNA and RNA.

adenosine—The molecule formed when the base adenine binds to the sugar ribose.

adenosine diphosphate—See ADP.

adenosine monophosphate—See AMP.

adenosine triphosphate—See ATP.

adenylyl (adenylate) cyclase—A membrane-bound enzyme that converts ATP into cyclic AMP.

adipocyte—A single fat cell.

adipokine—A cytokine secreted from adipose tissue.

adipose tissue—A collection of adipocytes embedded in a connective tissue network.

ADP—Adenosine 5'-diphosphate, an adenine nucleotide produced when ATP is hydrolyzed, used as a substrate in reactions producing ATP.

adrenergic receptors—Membrane-bound receptors, categorized as alpha or beta, that can bind epinephrine and norepinephrine.

aerobic—Requiring or taking place in the presence of oxygen.

alkalosis—A condition in which the blood or intracellular pH becomes too high due to a decrease in the proton concentration.

allosteric enzyme—An enzyme that plays a regulatory role in metabolism because, in addition to binding its substrate or substrates at its active site, it binds small molecules at allosteric sites, modulating the enzyme activity.

allosteric site—The site on an allosteric enzyme where effector molecules bind.

alpha-helix—A type of secondary structure of a protein in which the polypeptide backbone forms coils with maximum hydrogen bonding between amino acids.

amino acid pool—The collection of free amino acids in a compartment such as the bloodstream.

amino acid residues—The amino acids in a peptide or protein.

aminotransferases—A class of enzymes that remove an amino group from one amino acid and transfer it to another, usually α-ketoglutarate.

AMP—Adenosine 5'-monophosphate, an adenine nucleotide.

amphipathic—Referring to a molecule that has both hydrophilic and hydrophobic properties.

amphoteric—Capable of accepting or donating protons and thus able to act as a base or acid, respectively.

anabolism—The part of metabolism in which larger molecules are made from smaller molecules.

anaerobic—Occurring without air or oxygen.

anaplerosis—Reactions that can generate TCA cycle intermediates.

anhydride bond—A type of energy-rich bond formed when two acids are joined together with the elimination of water.

anion—A negatively charged ion.

anticodon—A specific sequence of three bases in tRNA that is complementary to an mRNA codon.

antiparallel—Referring to two polynucleotide chains that are opposite in polarity.

antiport—A membrane transport protein that simultaneously moves two molecules in opposite directions across the membrane.

apoptosis—Programmed cell death in which a regulated process leads through a series of steps to the ultimate death of a cell.

ATP—Adenosine 5'-triphosphate, an adenine nucleotide used as the energy currency in metabolism. The free energy released when ATP is hydrolyzed is used to drive reactions in cells.

ATP–ADP antiport (adenine nucleotide translocase)—An inner membrane protein that transports one ATP out of the matrix while simultaneously transferring an ADP into the matrix.

ATPase—An enzyme that can hydrolyze ATP, producing ADP and inorganic phosphate.

ATP synthase—An enzyme complex in the inner mitochondrial membrane that uses the free energy release in oxidative phosphorylation to combine Pi with ADP to make ATP.

autocrine—Pertaining to the effect exerted on a cell by a substance produced by that cell.

(a-v̄)O₂ difference—The average difference in oxygen content between the arterial blood and the mixed venous blood, usually expressed as milliliters of oxygen contained in 100 mL of arterial and venous blood.

base—A compound that can accept a proton.

base pair—Two nucleotides in polynucleotide chains that can be paired through hydrogen bonding between their bases (for example, base A with base T).

beta-oxidation—The process through which fatty acids attached to coenzyme A are broken down into a sequence of four steps to produce acetyl CoA units.

beta-sheet—A type of secondary structure of a polypeptide chain in which hydrogen bonding between components of the chain creates a zigzag structure.

binding site—The part of the active site of an enzyme where substrate is bound, prior to its conversion to product.

biotin—A B vitamin involved as a coenzyme in some carboxylation reactions.

branched-chain amino acids—The name given to three essential, hydrophobic amino acids: leucine, isoleucine, and valine.

Brønsted–Lowry—The surnames of scientists who described acids as proton donors and bases as proton acceptors.

brown adipose tissue (BAT)—A form of adipose tissue in which much of the reduction of oxygen to make water is not coupled with ADP phosphorylation to make ATP. Therefore, most of the energy release appears as heat.

buffer—A system that can resist changes in the pH of a solution.

calmodulin—A ubiquitous calcium-binding protein in cells, capable of binding four calcium ions.

carnitine—A small molecule that aids in transport of long-chain fatty acids across the inner mitochondrial membrane.

catabolism—That part of metabolism in which larger molecules are broken down into smaller components and energy is released.

catalase—Enzyme found in peroxisomes that degrades hydrogen peroxide.

catalytic site—The part of the active site of an enzyme molecule that actually participates in the bond making or breaking (or both) that converts a substrate into a product.

ceramide—A bioactive lipid metabolite that is synthesized from the precursors palmitate and serine.

chaperonins—A class of proteins involved in post-translational processing of polypeptide chains.

chemiosmotic hypothesis—The proposal currently accepted to describe how the free energy released during electron transport from reduced coenzymes to oxygen is coupled with ATP formation in oxidative phosphorylation.

chromatin—A complex of DNA, histone, and nonhistone proteins from which chromosomes emerge during cell division.

chromosome—A discrete structure, observed as such during cell division, consisting of a long DNA molecule with associated proteins.

chylomicrons—Lipoproteins formed in intestinal cells and released into the bloodstream from the lymphatic system. These contain lipids obtained from the diet and some proteins to maintain their particulate form.

citric acid cycle—A sequence of reactions taking place in mitochondria where acetyl units attached to CoA are degraded to carbon dioxide and the electrons produced are transferred to the coenzymes NAD^+ and FAD; also known as the *TCA* or *Krebs cycle*.

codon—Three adjacent nucleotides that specify a particular amino acid. We generally consider these to be in mRNA.

coenzyme—A molecule generally derived from a B vitamin that is essential to a particular enzyme in its catalytic role.

coenzyme A (CoA)—A particular coenzyme, formed in part from the B vitamin pantothenic acid, used to carry acyl groups such as acetyl groups or fatty acyl groups from fatty acids or carbohydrates.

coenzyme Q (also known as ubiquinone)—A small molecule involved in the electron transport chain, leading to ATP formation.

competitive inhibitor—An enzyme inhibitor that resembles the normal substrate and blocks the enzyme by reversibly binding and releasing from the active site. Its effects can be overcome by an excess of substrate.

complementary—Referring to two bases or two nucleotide chains containing purine and pyrimidine bases that recognize each other by their unique structures and that can form stable hydrogen bonds with each other.

cooperativity—Property of some proteins with multiple subunits in which binding of a substrate or ligand to one subunit alters the affinity of the other subunits for the substrate or ligand.

Cori cycle—A cycle in which glucose is broken down to lactate in a muscle fiber. The lactate is released to the blood, taken up by the liver, and converted back to glucose, and then released again to be used by muscle to form lactate.

cristae—Inwardly projecting folds of the inner mitochondrial membrane.

crossbridge—A part of the myosin molecule with ATPase- and actin-binding activity. During ATP hydrolysis, a myosin crossbridge can go through a cycle of attachment to actin, force generation, and then detachment.

crossover concept—The point during graded exercise when the contribution of fat to ATP production is matched by that of carbohydrate. At intensities beyond this, carbohydrate becomes increasingly important.

C-terminus—The end of a polypeptide chain with a free alpha carboxylate group.

cyclic AMP (cAMP)—Adenosine 3', 5'-monophosphate in which the phosphate group forms an ester bond with OH groups attached to both the 3' and 5' carbon atoms of the sugar ribose.

cytochromes—Heme-containing proteins in the electron transport chain that can be alternately in an oxidized or a reduced state.

cytokines—Polypeptide hormones secreted by a variety of tissues that bind to cell membrane receptors and signal such cell activities as proliferation or differentiation.

cytosine—A pyrimidine base found in DNA and RNA.

deamination—An enzymatic process in which an amino acid loses an amino group.

degenerate—A term to describe the fact that there is more than one codon specifying each amino acid.

degradation—Breakdown of something, as when a protein is changed to its component amino acids.

denaturation—Disruption of secondary and tertiary structure of a biological polymer such as a protein. With this comes a loss of normal activity.

de novo lipogenesis—Synthesis of fatty acids in the body starting from acetyl CoA.

dephosphorylation—Removal of a phosphate group from a molecule to which it is covalently attached.

diabetes mellitus—Type 1 or insulin-dependent diabetes, caused by destruction of the insulin-producing and -secreting beta cells of the pancreas, and type 2 or insulin-independent or insulin-resistant diabetes.

diacylglycerol (also known as a diglyceride)—A glycerol molecule with two fatty acids attached through ester bonds.

disaccharide—A molecule composed of two monosaccharides joined together by a covalent bond.

dismutation reaction—Reaction that occurs when two identical chemical species react together and one species donates an electron to the other.

dissociation—The process in which an acid dissolved in water gives up a proton.

disulfide bond—A covalent bond joining two sulfhydryl (SH) groups on cysteine side chains in the same or different polypeptide chains.

domain—A distinct section of a protein molecule with a distinct three-dimensional structure. It may or may not have a special function.

downstream—For a gene, representing the direction in which RNA polymerase II moves during transcription, considered to be in the 3' direction.

electrochemical gradient—The energy required to separate a charge and concentration difference across a membrane.

electron transport chain—A group of four protein-lipid complexes in the inner mitochondrial membrane that transfer electrons from reduced coenzymes to oxygen.

endergonic reaction—A reaction or process that requires free energy in order to occur.

enhancers—Regulatory base sequences in DNA to which specific protein transcription factors may bind to influence the expression of a gene. Generally, these lie a considerable distance from the gene they regulate.

enthalpy change—The total energy liberated in a reaction or process.

entropy change—A change in the degree of randomness or disorder in a system.

epigenetics—The study of heritable change in gene expression in the absence of changes to the sequence of the genome.

epinephrine—Hormone released from the adrenal medulla in response to physical stress.

EPOC—Excess postexercise oxygen consumption; refers to the amount of oxygen consumed at the end of exercise in excess of oxygen consumption at rest.

equilibrium—The state of a system when there is no net change in energy.

equilibrium constant (K_{eq})—The ratio of the product of the concentrations of products of a reaction to the product of the concentrations of reactants when the system is in equilibrium.

equilibrium reaction—A reversible enzyme-catalyzed reaction that, over time, results in an equilibrium mixture of products and substrates.

essential amino acids—Amino acids that must be obtained in the diet because the body cannot synthesize them or cannot synthesize them in adequate amounts. Examples are leucine, isoleucine, valine, threonine, phenylalanine, histidine, tryptophan, lysine, and methionine.

essential fatty acids—Fatty acids that the body requires but cannot synthesize in amounts sufficient to meet our physiological needs. Examples are linoleic acid and linolenic acid.

esterification—The process of triacylglycerol synthesis, which takes place in the cytosol of fat cells and other fat-containing cells.

estrogens—Female sex hormones that stimulate and maintain female secondary sex characteristics and also have metabolic effects.

euglycemia—A normal range of blood glucose concentration.

eukaryotic—Refers to species, including all plants and animals and many single-celled microorganisms, whose cells have complex systems of internal membranes or organelles such as nuclei and mitochondria. Prokaryotic organisms, such as bacteria, are different from eukaryotes because they generally have only one type of membrane, the plasma membrane, which forms the boundary of the cell as it does in eukaryotic cells.

exergonic reaction—A reaction that can proceed spontaneously with the release of free energy.

exon—Part of the primary transcript that will be retained in the functional mRNA molecule.

exothermic reaction—A reaction that liberates energy when it proceeds from reactants to products.

facilitated diffusion—Diffusion in which a substance crosses a membrane, down its concentration gradient, with the aid of a specific carrier.

FAD (flavin adenine dinucleotide)—An oxidation-reduction coenzyme that oxidizes a class of substrate molecules, transferring their electrons to coenzyme Q.

Faraday constant (F)—The number (96.5 kJ per mole per volt) that allows one to convert a change in redox potential, expressed in volts, into units of energy.

feedback inhibition—Inhibition of an allosteric enzyme by a product of its reaction or a product of a pathway in which the enzyme participates.

Fenton reaction—The nonenzymatic reaction of ferrous iron with hydrogen peroxide to produce hydroxyl radical.

free energy change—The fraction of the total energy released in a reaction or process that can be used to do useful work.

free fatty acid (FFA)—A fatty acid, bound noncovalently to albumin, circulating in the blood.

free radical—A chemical species containing an unpaired electron.

gene—The segment of DNA that provides the information for the sequence of amino acids in a polypeptide chain.

genetic code—The set of rules that specify which three-nucleotide codons correspond to which amino acids or termination signals in the mRNA sequence (Roth 2007).

genome—The total genetic information carried within a cell nucleus.

ghrelin—A polypeptide secreted by gastric cells. It can signal the need to form new fat cells.

glucagon—Polypeptide hormone secreted from the alpha cells of the pancreas in response to a decrease in blood glucose concentration.

glucogenic amino acids—Those amino acids whose carbon skeletons can be used to make glucose in the liver.

glucose transporters—A family of membrane proteins that transport glucose across cell membranes, down their concentration gradient.

glutathione—A tripeptide, containing a sulfhydryl group (-SH), that participates in a variety of intracellular reduction reactions.

glutathione peroxidase—An enzyme that breaks down hydrogen peroxide or other peroxides and uses the reducing power of glutathione.

glycemic index—Rating of the potential for foods to elevate blood glucose levels.

glyceroneogenesis—The formation of glycerol from pyruvate or carbon skeletons of some amino acids.

glycogen—A large branched polymer, formed entirely of glucose, that is the principal means of storing carbohydrate in liver and muscle.

glycogenesis—The formation of glycogen from glucose.

glycogenin—A self-glycosylating polypeptide that is the precursor for glycogen synthesis.

glycogenolysis—The process in which glycogen is broken down in liver and muscle to yield glucose 6-phosphate.

glycolysis—A catabolic pathway that breaks down glucose 6-phosphate, derived from glucose or glycogen, into lactate, generating ATP in the process.

glycosome—A molecule of glycogen along with the enzymes involved both in synthesizing and in degrading glycogen.

G protein—A GTP-binding membrane protein that participates in the process of signal transduction, linking the binding of an external ligand, its membrane receptor, into an action within the cell.

growth factors—A family of extracellular polypeptides that can bind to specific receptors on the cell membrane, influencing the growth and differentiation of that cell.

growth hormone—A polypeptide hormone synthesized in and secreted from the anterior portion of the pituitary gland.

guanine—A purine base found in DNA and RNA.

half-life—The time taken for the disappearance or decay of one-half of the population of chemical substances.

heat shock proteins—A family of proteins originally shown to be synthesized in response to cellular or organism stress. Among the members of this expanded family are the chaperonins, which participate in posttranslational processing of new polypeptides.

heme—A prosthetic group containing a central iron ion.

Henderson-Hasselbalch equation—The equation relating the solution pH, the pK_a, and the ratio of undissociated to dissociated acid concentrations in a solution.

Hill slope—Quantification method for determining steepness of the enzyme-substrate concentration curve, which characterizes the degree of cooperativity of multiple substrate binding sites on an enzyme.

histones—A family of nuclear proteins carrying net positive charges because of a preponderance of basic amino acids. These are associated with the polynucleotide backbone of DNA in structures known as nucleosomes.

holoenzyme—A catalytically active enzyme containing the polypeptide subunit and any necessary cofactors.

hormone receptor—A protein located in a cell membrane or within a cell that recognizes and binds a specific hormone.

hybridization—Association of two complementary polynucleotide chains because of base pairing through hydrogen bonding between complementary bases.

hydride ion—A hydrogen atom with an additional electron so that it carries a negative charge.

hydrogen bond—A weak, noncovalent bond between a hydrogen atom, covalently attached to an electronegative atom, and another electronegative atom in the same or a different molecule.

hydrophilic—Referring to polar molecules or charged ions that can interact with the polar water molecule.

hydrophobic—Referring to nonpolar groups or molecules that cannot interact with water.

hydroxylation—Addition of an OH group to specific amino acids.

hydroxyl radical—The most reactive free radical normally produced in our bodies.

hyperglycemia—Elevated blood glucose concentration.

hypoglycemia—Very low level of blood glucose concentration.

inorganic phosphate—Ionized phosphate group that contains a net negative charge at neutral pH.

inorganic pyrophosphate—An anhydride containing two phosphate groups joined together; it is hydrolyzed by the enzyme inorganic pyrophosphatase.

insulin—A polypeptide hormone produced in and secreted from the beta cells of the pancreas in response to an elevation in blood glucose concentration.

integral membrane proteins—Proteins that are permanently attached to biological membranes.

intermembrane space—The space between the outer and inner mitochondrial membranes.

intermyofibrillar mitochondria—Rows of mitochondria located between adjacent myofibrils in a muscle fiber.

international unit—The amount of enzyme that converts 1 μm of substrate to product in 1 min.

intron—Part of the primary transcript that will be removed prior to the formation of the functional mRNA molecule.

isoelectric point (pI)—The pH at which a molecule or ion has no net charge but contains at least one positive and one negative charge.

isozymes (isoenzymes)—Different molecular forms of the same enzyme, catalyzing the same reaction with the same mechanism but having different kinetic parameters.

ketogenic amino acids—The carbon skeletons of amino acids that cannot be converted into glucose in the liver. Two examples are lysine and leucine.

ketone bodies—A group of small molecules, including acetoacetate, 3-hydroxybutyrate, and acetone that are formed in the body during carbohydrate deprivation or uncontrolled diabetes mellitus.

ketonemia—A condition of elevated ketone-body concentration in the blood.

ketonuria—Loss of ketone bodies in the urine.

kinetic parameters—The Vmax and K_m are considered to define parameters of an enzyme.

Krebs cycle—See citric acid cycle.

leptin—A polypeptide synthesized in and secreted by adipose tissue, especially when the fat cells are enlarged from storing more triacylglycerol.

ligand—A small molecule that binds to a specific larger one, such as a receptor.

Lineweaver–Burk plot—A transformation of the Michaelis–Menten equation in which reciprocals of substrate concentration are plotted, yielding a straight line whose intercepts can produce values for the kinetic parameter.

lipoic acid—An organic coenzyme, made in the body, that works with two mitochondrial dehydrogenase enzymes.

lipolysis—The enzymatic breakdown of triacylglycerol molecules, catalyzed by a class of enzymes known as lipases.

lipoproteins—A family of protein–lipid particles found in the blood.

long-chain fatty acids—Fatty acids with 14 or more carbon atoms; may be saturated or unsaturated.

macrophage—Tissue-based phagocytic cell, derived from blood monocyte cells, that plays a role in immune responses within tissues.

mammalian target of rapamycin (mTOR)—A protein kinase that regulates cell growth, protein synthesis, and transcription and is activated by resistance exercise.

mass action ratio—The ratio of the product of the concentrations of the products of a reaction to the product of the concentrations of reactants at any time during the course of a chemical reaction.

matrix—That part of the mitochondrion that is bound by the inner membrane.

maximum velocity (Vmax)—The greatest possible velocity of an enzyme-catalyzed reaction.

medium-chain triacylglycerols—Triacylglycerol molecules containing fatty acids with a carbon chain length of 6 to 10.

messenger RNA (mRNA)—A class of RNA molecules, complementary to a portion of a strand of DNA, that provides the information for the synthesis of a polypeptide chain.

metabolic pathway—A sequence of enzyme-catalyzed reactions in which a starting substrate is converted into a product after undergoing a number of steps.

metabolic syndrome—A condition in which a person has three or more of the following five disorders: fasting hyperglycemia; obesity, particularly in the abdominal area; high blood pressure; elevated blood triglyceride levels; and decreased levels of high-density lipoprotein (HDL) cholesterol.

metabolon—A cluster of enzymes in which the product of one enzyme is immediately passed to the next enzyme because of the physical association.

Michaelis constant (K_m)—The concentration of substrate needed to produce one-half the maximum velocity for an enzyme-catalyzed reaction.

mitochondrial biogenesis—The process by which new mitochondria are formed in the cell.

mitogen-activated protein kinases (MAPK)—A family of protein kinases that respond to cell stimulation by mitogens and other signals and regulate various cellular functions such as gene expression.

monocarboxylate transporter (MCT)—A membrane protein that transports small anions against their concentration gradient. Usually refers to transport of lactate and pyruvate.

monosaccharide—A simple sugar that cannot be digested or broken down further without having its structure destroyed.

mutation—A change in the nucleotide sequence of DNA that can be passed on.

myeloperoxidase—An enzyme found in a variety of phagocytic cells (neutrophils, macrophages) that reacts hydrogen peroxide with a chloride ion to generate hypochlorite.

myogenic regulatory factors—A class of transcription factors that, when bound to their cognate cis elements, regulate the expression of genes specific for development and maintenance of muscle properties.

myosin—The principal protein of the thick filament of muscle fibers. It has ATPase- and actin-binding activity.

myosin heavy chain—The largest subunit in the oligomeric protein, myosin.

NAD$^+$ (nicotinamide adenine dinucleotide)—The oxidized form of the more common coenzyme involved in redox reactions in the cell.

NADH—Reduced form of NAD$^+$. Transfers its electrons to oxygen in the electron transport chain.

NADPH (nicotinamide adenine dinucleotide phosphate)—The reduced form for the coenzyme involved in redox reactions in which larger molecules are built out of smaller components. The oxidized form is NADP$^+$.

NADPH oxidase—Enzyme found in immune cells and endothelial cells that can use electrons from NADPH to reduce oxygen to superoxide.

Nernst equation—An equation to determine the redox potential of a reaction using known values for standard redox potential, temperature, and the concentrations of the oxidized and reduced forms of the substrates and products.

neutrophil—A class of white blood cells that attack foreign cells such as bacteria.

NFAT (nuclear factor of activated T cells)—A transcription factor first identified in T lymphocytes but now known to operate in other cells.

nitric oxide (NO)—Nitrogen monoxide, a free radical-signaling molecule, produced by nitric oxide synthase and derived from the amino acid arginine.

nitric oxide synthase—A class of enzyme that converts arginine to nitric oxide and citrulline.

nitrogen balance—The relationship between nitrogen intake into the body in the form of food protein and nitrogen loss from the body in urine, feces, and sweat.

noncompetitive inhibitor—An agent that causes irreversible inhibition of an enzyme and loss of enzyme function. This kind of inhibition decreases Vmax but has no effect on K_m.

norepinephrine (also known as noradrenaline)—The chief neurotransmitter in the sympathetic nervous system.

N-terminus—The start of a polypeptide chain with a free alpha ammonium group.

nucleoside—Product that results when a purine or pyrimidine base is attached to a ribose sugar.

nucleoside triphosphate—A sugar plus base (nucleoside) with one or more attached phosphate groups.

nucleosome—A basic structural component of chromatin, consisting of a DNA strand wound around a core of histone proteins.

nucleotide—A nucleoside with one or more attached phosphate groups.

oligomeric—Form of small polymers, usually of nucleotides (oligonucleotide), amino acids (oligopeptide), or monosaccharides (oligosaccharide).

open reading frame—A sequence of nucleotide codons that code for amino acids in the absence of stop codons.

osmotic pressure—The pressure that must be applied to a solution to prevent solvent (water) from flowing into it.

oxidation—A chemical reaction in which a substrate loses electrons.

oxidative phosphorylation—The formation of ATP from ADP and Pi in association with the transfer of electrons from substrate to coenzymes to oxygen.

oxidative stress—The condition in a cell or organism in which the formation of reactive oxygen and nitrogen species exceeds the capacity to quench them.

oxygen consumption ($\dot{V}O_2$)—The utilization of oxygen (measured in L/min) to accept electrons and produce ATP in oxidative phosphorylation.

pancreatic lipase—A lipase produced in and secreted by the pancreas. Its function is to break down dietary triacylglycerol.

pantothenic acid—A B vitamin that forms a major part of coenzyme A.

paracrine—A term to describe the mechanism of signaling between neighboring cells in which a growth factor diffuses from one cell to its close neighbors.

pentose phosphate pathway—A pathway to interconvert hexoses and pentoses; a source of NADPH for synthetic reactions. It is also called the *phosphogluconate pathway*.

peptide bond—A specialized form of amide bond that joins two amino acids together.

perilipins—A family of proteins that coat adipose tissue droplets in fat cells so that the triacylglycerol molecules are inaccessible to enzymes unless the cell is stimulated.

peroxidation—RONS-induced degradation of fats and lipids.

phosphagen—A molecule such as phosphocreatine that can transfer its phosphate group to ADP to make ATP.

phosphate transporter—Protein that transfers a phosphate group across the inner membrane while simultaneously transferring a hydroxide group the opposite way, to the intermembrane space.

phosphatidic acid—A diglyceride with a phosphate group attached in ester linkage to the third OH of glycerol.

phosphatidylinositols—A family of phospholipids in which inositol or phosphorylated derivatives are attached to phosphatidic acid.

phosphocreatine—A phosphagen capable of rapidly transferring its phosphate group to ADP to quickly regenerate ATP.

phospholipid—A lipid containing one or more phosphate groups.

phosphoprotein phosphatase—An enzyme that removes a phosphate group attached in ester linkage to a protein.

phosphorylation—Covalent attachment of a phosphate group to a molecule.

phosphorylation potential—The ratio of the concentration of ATP, divided by the concentrations of ADP and Pi (i.e., [ATP]/[ADP] × [Pi]).

phytochemical—Name given to a wide variety of compounds from plant sources that can have an antioxidant effect in our bodies.

pK_a—The negative logarithm of the acid-dissociation constant; equals the pH of a solution in which one-half of the molecules are dissociated and one-half are not dissociated (i.e., neutral).

polypeptide—A molecule composed of 20 or more amino acids joined together by peptide bonds.

polyribosome—An mRNA molecule with two or more ribosomes that carry out protein synthesis on it.

polyunsaturated fatty acids (PUFAs)—Fatty acids with two or more double bonds, which include linoleic acid, α-linolenic acid, and omega-3 fatty acids.

P/O ratio—The ratio of the number of ATP molecules produced per pair of electrons transferred from a substrate to oxygen to reduce the oxygen to water.

porphyrin—A heterocyclic molecule, made in the body, that forms the basic structure for heme groups in hemoglobin, myoglobin, and the cytochromes.

postabsorptive state—The metabolic state following absorption of dietary nutrients, when blood nutrient levels are maintained through sources other than absorption from the gut.

primary gene transcript—The initial product of transcription.

primary structure—The sequence of amino acids in a peptide or protein, starting from the N-terminus.

promoter—The region upstream of the start site of a gene to which transcription factors can bind to influence the initiation of transcription.

prosthetic group—A cofactor (metal ion or organic compound) that is covalently attached to a protein and is necessary for that protein to function.

proteasome—A cylindrical-shaped protein complex responsible for much of the degradation of cellular proteins.

proteinases—Protein-degrading enzymes.

protein balance—The relationship between protein synthesis and protein degradation, which affects the net content of protein in an organism. This is often inferred from nitrogen balance.

protein kinase—An enzyme that attaches a phosphate group from ATP to the side chain of a serine, threonine, or tyrosine residue in the protein.

protein tyrosine kinase—An enzyme that attaches the terminal phosphate of ATP to the hydroxyl group in the side chain of tyrosine residues in a protein.

proteomics—The study of large compilations of proteins, such as those expressed in a single cell.

proton motive force (pmf)—The combination of the membrane electric potential difference and the pH gradient that provides the driving force for protons to move from the intermembrane space to the matrix of mitochondria.

purine—A nitrogen-containing heterocyclic base found in nucleic acids. Principal examples are adenine and guanine, cytosine, thymine, and uracil.

pyrimidine—See purine.

quaternary structure—The three-dimensional organization of a protein that is composed of subunits, dealing with the way the subunits interact with each other.

rate limiting—Referring to the step in a reaction or pathway that limits the overall rate of the reaction or pathway.

reactive oxygen species (ROS)—A molecule or ion of oxygen that is highly reactive in accepting or donating one or more electrons, or an oxygen species with one or more unpaired electrons.

redox potential—The tendency of a substance to be reduced, measured in volts.

redox reaction—A reaction in which electrons are transferred from one substrate (oxidation) to another substrate (reduction).

reducing equivalents—A pair of electrons that can be used to reduce something. Generally transferred as a hydride ion or as a pair of hydrogen atoms.

reduction—What occurs when a substrate gains one or more electrons.

repressor—A protein that binds to a regulatory element associated with a gene and blocks its transcription.

respiratory burst—A process by which a reactive oxygen species such as a superoxide is produced in macrophages or other phagocytic cells when stimulated.

respiratory exchange ratio (RER)—The ratio of carbon dioxide produced to oxygen consumed as measured by collection of expired air at a subject's mouth.

respiratory quotient (RQ)—The ratio of carbon dioxide produced to oxygen consumed as measured in an isolated system or across a tissue or organ.

ribonuclease—A class of enzymes that can hydrolyze the sugar-phosphate backbone of RNA.

RNA polymerase II—The enzyme that transcribes a section of DNA, producing an RNA molecule complementary to the copied strand of DNA.

sarcolemma—The cell membrane of a muscle cell or fiber.

sarcopenia—Age-related loss of muscle mass due primarily to the loss of muscle cells or muscle fibers through cell death.

sarcoplasmic reticulum—A specialized form of endoplasmic reticulum, found in striated muscle, that holds, releases, and takes up calcium.

satellite cells—Muscle precursor cells located around the periphery of muscle fibers that are critical for muscle repair and hypertrophy.

secondary structure—A regular, repeating structure assumed by the polypeptide backbone of a polypeptide. Examples are the alpha-helix and the beta-sheet.

SERCA pump—ATPase that provides the energy to reabsorb calcium into the sarcoplasmic reticulum during muscular relaxation.

signal transduction—The process through which an extracellular signal is amplified and converted to a response within the cell.

small nuclear RNA (snRNA)—Small RNA molecules in the nucleus that play a role in RNA processing, such as splicing out introns.

sodium–calcium antiport—A membrane protein that transports calcium against its gradient using the energy produced when sodium flows down its gradient.

stable isotopes—Isotopes of common elements that are not radioactive.

standard free energy change ($\Delta G°'$)—The free energy change for a reaction under standard conditions at pH 7.0 and with all reactants and products at 1 M concentration.

standard redox potential ($\Delta E°'$)—The tendency for a substance to be reduced, measured in volts and based on standard conditions.

start codon—The codon AUG, which signals the start of translation; also the codon for methionine.

start site—The nucleotide on the sense strand of DNA that indicates the beginning of transcription.

state 3 respiration—Maximally stimulated oxidative phosphorylation when all substrates and components are present in adequate amounts.

state 4 respiration—The level of electron transport in mitochondria when the substrate ADP is inadequate.

steady state—A nonequilibrium condition of a system in which there is a constant flux, yet the intermediates within the system are at constant concentration.

stereoisomers—Isomers that have the same composition of components yet have different molecular arrangements.

steroid hormones—A class of hormones based on the structure of their precursor, cholesterol.

stop (termination) codons—Three codons, UAG, UAA, and UGA, that signal the end to translation.

stroke volume—The volume of blood ejected from the left ventricle at each beat of the heart.

subsarcolemmal mitochondria—Mitochondria located immediately beneath the cell membrane (sarcolemma) of a muscle fiber.

substrate—The specific compound that is chemically modified by an enzyme.

substrate-level phosphorylation—Process whereby ADP is converted into ATP by any mechanism other than oxidative phosphorylation.

superoxide—The product of the addition of a single electron to an oxygen molecule.

superoxide dismutase—An enzyme that catalyzes the reaction between two superoxide anions, in which an electron is donated by one to the other, creating hydrogen peroxide.

TATA box—A sequence of bases located about 25 to 35 base pairs upstream of the start of many eukaryotic genes that helps to locate RNA polymerase II.

TCA (tricarboxylic acid) cycle—See citric acid cycle.

template DNA strand—The strand of DNA that is copied in a complementary way by RNA polymerase II.

thermodynamics—The study of the principles of energy exchange in metabolism.

thiamine pyrophosphate (TPP)—A coenzyme involved in several mitochondrial oxidative decarboxylation reactions.

thick filament—A filament in the middle of the sarcomere of a striated muscle fiber, composed almost entirely of myosin molecules.

thin filament—A filament in striated muscle consisting of the proteins actin, tropomyosin, and troponin.

thymine—A pyrimidine base found in DNA.

total adenine nucleotide (TAN)—The sum of the concentrations of ATP, ADP, and AMP.

transamination—Transfer of an amino group from an amino acid to a keto acid to make a new amino acid and a new keto acid.

transcription—The enzymatic reaction whereby a segment of a DNA strand is copied through the making of RNA.

transcription factor—Proteins that regulate DNA transcription; any protein that binds to a cis element near a gene and increases the frequency of initiation of transcription.

translation—The process whereby the sequence of codons on an mRNA molecule is used to direct the sequence of amino acids in a polypeptide chain.

triacylglycerol (triglyceride)—A triester consisting of the alcohol glycerol and three fatty acid molecules.

troponin—A calcium-binding protein in the thin filament of striated muscle.

T-tubules—Invaginations in the sarcolemma of striated muscle.

turnover number—The number of molecules of substrate that an enzyme active site can convert to product in 1 s.

ubiquinol (reduced coenzyme Q or QH_2)—The reduced form of an electron carrier in the mitochondrial inner membrane.

ubiquinone (oxidized coenzyme Q or Q)—The oxidized form of an electron carrier in the mitochondrial inner membrane.

ubiquitin—Small protein found in all cells that becomes covalently attached to proteins destined for degradation by the proteasome complex.

ubisemiquinone—An intermediate in the reduction of ubiquinone or the oxidation of ubiquinol in which there is an unpaired electron that can be passed to oxygen.

uncoupling protein—Inner membrane protein that allows protons to flow down their electrochemical gradient without ADP phosphorylation.

uniport—A membrane protein that allows a polar molecule or ion to flow across the membrane down a concentration gradient.

upstream—Representing a direction with respect to a gene—that is, in the 5' direction—based on the sense strand of DNA.

uracil—A pyrimidine base found in RNA.

urea—Small molecule produced in the liver in the urea cycle; the major form in which the body expels nitrogen groups from amino acids.

urea cycle—A sequence of reactions in the liver in which urea is synthesized.

very-low-density lipoprotein (VLDL)—A triacylglycerol-rich lipoprotein fraction, synthesized in liver.

wobble—The relatively loose base pairing between the third (3') base in a codon and the first base (5') of an anticodon.

xanthine oxidase—An enzyme that can convert hypoxanthine to xanthine and xanthine to uric acid; in the process, it generates superoxide.

xenobiotics—Man-made foreign molecules that may or may not be harmful to humans if they enter their bodies.

zwitterion—A molecule with one positive and one negative charge.

References

Ahmadian, M., R.E. Duncan, and H.S. Sul. 2009. Skinny on fat metabolism: Lipolysis and fatty acid utilization. *Trends in Endocrinology and Metabolism* 20: 424-428.

Allen, D.A., and H. Westerblad. 2004. Lactic acid—the latest performance-enhancing drug. *Science* 305: 1112-1113.

Allen, D.G., G.D. Lamb, and H. Westerblad. 2008. Skeletal muscle fatigue: Cellular mechanisms. *Physiological Reviews* 88: 287-332.

Alsted, T.J., L. Nybo, M. Schweiger, C. Fledeilius, P. Jacobsen, R. Zimmerman, R. Zechner, and B. Kiens. 2009. Adipose triglyceride lipase in human skeletal muscle is upregulated by exercise training. *American Journal of Physiology: Endocrinology and Metabolism* 296: E445-E453.

Angus, D.J., M. Hargreaves, J. Dancey, and M.A. Febbraio. 2000. Effect of carbohydrate or carbohydrate plus medium chain triglyceride ingestion on cycling time trial performance. *Journal of Applied Physiology* 88: 113-119.

Arkinstall, M.J., C.R. Bruce, S.A. Clark, C.A. Rickards, L.M. Burke, and J.A. Hawley. 2004. Regulation of fuel metabolism by preexercise muscle glycogen content and exercise intensity. *Journal of Applied Physiology* 97: 2275-2283.

Armitage, J.A., I.Y. Khan, P.D. Taylor, P.W. Nathanielsz, and L. Poston. 2004. Developmental programming of the metabolic syndrome by maternal nutritional imbalance: How strong is the evidence from experimental models in mammals? *Journal of Physiology* 561: 355-377.

Austin, K., and B. Seebohar. 2011. *Performance nutrition: Applying the science of nutrient timing.* Champaign, IL: Human Kinetics.

Baar, K. 2010. Epigenetic control of skeletal muscle fibre type. *Acta Physiologica* 199: 477-487.

Bailey, D.M., B. Davies, I.S. Young, M.J. Jackson, G.W. Davison, R. Isaacson, and R.S. Richardson. 2003. EPR spectroscopic detection of free radical outflow from an isolated muscle bed in exercising humans. *Journal of Applied Physiology* 94: 1714-1718.

Baldwin, J., R.J. Snow, M.J. Gibala, A. Garnham, K. Howarth, and M.A. Febbraio. 2003. Glycogen availability does not affect the TCA cycle or TAN pools during prolonged, fatiguing exercise. *Journal of Applied Physiology* 94: 2181-2187.

Baldwin, K.M. 2000. Research in the exercise sciences: Where do we go from here? *Journal of Applied Physiology* 88: 332-336.

Barazzoni, R., A. Bosutti, M. Stebel, M.R. Cattlin, E. Roder, L. Visinin, L. Cattin, G. Biolf, M. Zanetti, and G. Guarnieri. 2005. Ghrelin regulates mitochondrial-lipid metabolism gene expression and tissue fat distribution in liver and skeletal muscle. *American Journal of Physiology* 288: E228-E235.

Barclay, C.J., R.C. Woledge, and N.A. Curin. 2007. Energy turnover for Ca^{2+} cycling in skeletal muscle. *Journal of Muscle Research and Cell Motility* 28: 259-274.

Bazzano, L.A., K. Reynolds, K.N. Holder, and J. He. 2006. Effect of folic acid supplementation on risk of cardiovascular diseases: A meta-analysis of randomized control studies. *Journal of the American Medical Association* 296: 2720-2726.

Belcastro, A.N., L.D. Shewchuk, and D.A. Raj. 1998. Exercise-induced muscle injury: A calpain hypothesis. *Molecular and Cellular Biochemistry* 179: 135-145.

Berg, J.M., J.L. Tymoczko, and L. Stryer. 2002. *Biochemistry*. 5th ed. New York: Freeman.

Berggren, J.R., C.J. Tanner, T.R. Koves, D.M. Muoio, and J.A. Houmard. 2005. Glucose uptake in muscle cell cultures from endurance-trained men. *Medicine and Science in Sports and Exercise* 37: 579-584.

Bilan, P.J., V. Samokhvalov, A. Koshkina, J.D. Schertzer, M.C. Samaan, and A. Klip. 2009. Direct and macrophage-mediated actions of fatty acids causing insulin resistance in muscle cells. *Archives of Physiology and Biochemistry* 115: 176-190.

Blackstock, W.P, and M.P. Weir. 1999. Proteomics: Quantitative and physical mapping of cellular proteins. *Trends in Biotechnology* 17: 121-127.

Boeger, H., D.A. Bushnell, R. Davis, J. Griesenbeck, Y. Lorch, J.S. Strattan, K.D. Westover, and R.D. Kornberg. 2005. Structural basis of eukaryotic gene transcription. *FEBS Letters* 579: 899-903.

Bonen, A., A. Chabowski, J.J. Luiken, and J.F. Glatz. 2007. Mechanisms and regulation of protein-mediated cellular fatty acid uptake: Molecular, biochemical and physiological evidence. *Physiology* 22: 15-28.

Booth, F.W., and P.D. Neufer. 2005. Exercise controls gene expression. *American Scientist* 93: 28-35.

Boveris, A., L.E. Cosa, E. Cadenas, and J.J. Poderoso. 1999. Regulation of mitochondrial respiration by adenosine diphosphate, oxygen, and nitric oxide. *Methods in Enzymology* 301: 188-198.

Bradley, N.S., G.J.F. Heigenhauser, B.D. Roy, E.M. Staples, J.G. Inglis, P.J. LeBlanc, and S.J. Peters. 2008. Acute effects of differential dietary fatty acids on human skeletal muscle pyruvate dehydrogenase activity. *Journal of Applied Physiology* 104: 1-9.

Bray, G.A. 2007. How bad is fructose? *American Journal of Clinical Nutrition* 86: 895-896.

Brooks, G.A. 1997. Importance of the "crossover" concept in exercise metabolism. *Clinical and Experimental Pharmacology and Physiology* 24: 889-895.

Brooks, G.A. 2007. Genome, proteome, and transcriptomes: The new systems approach to research. *Exercise and Sport Science Reviews* 35: 41-42.

Bülow, J. 2003. Lipid mobilization and utilization. In *Principles of exercise biochemistry*, 3rd ed., ed. J.R. Poortmans. Basel: Karger.

Burd, N.A., J.E. Tang, D.R. Moore, and S.M. Phillips. 2009. Exercise training and protein metabolism: Influences of contraction, protein intake, and sex-based differences. *Journal of Applied Physiology* 106: 1692-1701.

Burniston, J.G. 2009. Adaptation of the rat cardiac proteome in response to intensity-controlled endurance exercise. *Proteomics* 9(1): 106-115.

Burtis, C.A., and E.R. Ashwood, eds. 2001. *Tietz fundamentals of clinical chemistry*. 5th ed. New York: Saunders.

Bushati, N., and S.M. Cohen. 2007. MicroRNA functions. *Annual Review of Cell and Developmental Biology* 23: 175-205.

Callis, T.E., Z. Deng, J.F. Chen, and D.Z. Wang. 2008. Muscling through the microRNA world. *Experimental Biology and Medicine* 233: 131-138.

Canto, C., Z. Gerhart-Hines, J.N. Feige, M. Lagouge, L. Noriega, J.C. Milne, P.J. Elliott, P. Puigserver, and J. Auwerx. 2009. AMPK regulates energy expenditure by modulating NAD^+ metabolism and SIRT1 activation. *Nature* 458: 1056-1062.

Chakravarthy, S., Y.-J. Park, J. Chodaparambil, R.S. Edayathumangalam, and K. Luger. 2005. Structure and dynamic properties of nucleosome core particles. *FEBS Letters* 579: 895-898.

Chanutin, A. 1926. The fate of creatine when administered to man. *Journal of Biological Chemistry* 67: 29-43.

Chin, E.R. 2010. Intracellular Ca^{2+} signaling in skeletal muscle: Decoding a complex message. *Exercise and Sport Science Reviews* 38: 76-85.

Chinetti-Gbaguidi, G., J.-C. Fruchart, and B. Staels. 2005. Role of the PPAR family of nuclear receptors in the regulation of metabolic and cardiovascular homeostasis: New approaches to therapy. *Current Opinion in Pharmacology* 5: 177-183.

Coffey, V.G., and J.A. Hawley. 2007. The molecular bases of training adaptation. *Sports Medicine* 37: 737-763.

Coles, L., J. Litt, H. Hatta, and A. Bonen. 2004. Exercise rapidly increases expression of the monocarboxylate transporters MCT1 and MCT4. *Journal of Physiology* 561: 253-261.

Conaway, J.W., L. Florens, S. Sata, C. Tomomori-Sato, T.J. Parmely, T. Yao, S.K. Swanson, C.A.S. Banks, M.P. Washburn, and R.C. Conaway. 2005. The mammalian mediator complex. *FEBS Letters* 579: 904-908.

Costa, L.E., G. Mendez, and A. Boveris. 1997. Oxygen dependence of mitochondrial function measured by high-resolution respirometry in long-term hypoxic rats. *American Journal of Physiology* 273: C852-C858.

Crowther, G.J., M.F. Carey, W.F. Kemper, and K.E. Conley. 2002a. Control of glycolysis in contracting skeletal muscle. II. Turning it off. *American Journal of Physiology* 282: E74-E79.

Crowther, G.J., W.F. Kemper, M.F. Carey, and K.E. Conley. 2002b. Control of glycolysis in contracting skeletal muscle. I. Turning it on. *American Journal of Physiology* 282: E67-E73.

Davidsen, P.K., I.J. Gallagher, J.W. Hartman, M.A. Tarnopolsky, F. Dela, J.W. Helge, J.A. Timmons, and S.M. Phillips. 2011. High responders to resistance exercise training demonstrate differential regulation of skeletal muscle microRNA expression. *Journal of Applied Physiology* 110: 309-317.

de Glisezinski, I., C. Moro, F. Pillard, F. Marion-Latard, I. Harant, M. Meste, M. Berlan, F. Crampes, and D. Rivière. 2003. Aerobic training improves exercise-induced lipolysis in SCAT and lipid utilization in overweight men. *American Journal of Physiology* 285: E984-E990.

Didier, L., R.S. Hundal, A. Dresner, T.B. Price, S.M. Vogel, F. Falk Petersen, and G.I. Shulman. 2000. Mechanism of muscle glycogen autoregulation in humans. *American Journal of Physiology* 278: E663-E668.

Dirksen, R.T. 2009. Sarcoplasmic reticulum-mitochondrial through-space coupling in skeletal muscle. *Applied Physiology Nutrition and Metabolism* 34: 389-395.

Du, J., Z. Hu, and W.E. Mitch. 2005. Molecular mechanisms activating muscle protein degradation in chronic kidney disease and other catabolic conditions. *European Journal of Clinical Investigation* 35: 157-163.

Duchen, M.R. 2004. Mitochondria in health and disease: Perspectives on a new mitochondrial biology. *Molecular Aspects of Medicine* 25: 365-451.

Dulloo, A.G., M. Gubler, J.P. Montani, J. Seydoux, and G. Solinas. 2004. Substrate cycling between de novo lipogenesis and lipid oxidation: A thermogenic mechanism against skeletal muscle lipotoxicity and glucolipotoxicity. *International Journal of Obesity* 28: S29-S37.

Duteil, S., C. Bourrilhon, J.S. Raynaud, C. Wary, R.S. Richardson, A. Leroy-Willig, J.C. Jouanin, C.Y. Guezennec, and P.G. Carlier. 2004. Metabolic and vascular support for the role of myoglobin in humans: A multiparametric NMR study. *American Journal of Physiology* 287: R1441-R1449.

Dyck, D.J. 2005. Leptin sensitivity in skeletal muscle is modulated by diet and exercise. *Exercise and Sport Sciences Reviews* 33(4): 189-194.

Dyck, D.J., G.J. Heigenhauser, and C.R. Bruce. 2006. The role of adipokines as regulators of skeletal muscle fatty acid metabolism and insulin sensitivity. *Acta Physiologica* 186: 5-16.

Elosua, R., L. Molina, M. Fito, A. Arquer, J.L. Sanchez-Quesada, M.I. Covas, J. Ordoñez-Llanos, and J. Arruga. 2003. Response of oxidative stress biomarkers to a 16-week aerobic physical activity program and to acute physical activity, in healthy young men and women. *Atherosclerosis* 167: 327-334.

Enns, D.L., and P.M. Tiidus. 2008. Estrogen influences satellite cell activation and proliferation following downhill running in rats. *Journal of Applied Physiology* 104: 347-353.

Ferrington, D.A., A.G. Krainev, and D.J. Bigelow. 1998. Altered turnover of calcium regulatory proteins of the sarcoplasmic reticulum in aged skeletal muscle. *Journal of Biological Chemistry* 273: 5885-5891.

Flatt, J.P. 1995. Use and storage of carbohydrate and fat. *American Journal of Clinical Nutrition* 61: 952S-959S.

Flaus, A., and T. Owen-Hughes. 2004. Mechanisms for ATP-dependent chromatin remodeling: Farewell to the tuna-can octamer. *Current Opinion in Genetics and Development* 14: 165-173.

Frøsig, C., S.B. Jørgensen, D.G. Hardie, E.A. Richter, and J.F.P. Wojtaszewski. 2004. 5'-AMP-activated protein kinase activity and protein expression are regulated by endurance training in human skeletal muscle. *American Journal of Physiology* 286: E411-E417.

Fueger, P.T., S. Heikkinen, D.P. Bracy, C.M. Malabanan, R.R. Pencek, M. Laakso, and D.H. Wasserman. 2003. Hexokinase II partial knockout impairs exercise-stimulated glucose uptake in oxidative muscles of mice. *American Journal of Physiology* 285: E958-E963.

Fujita S., H.C. Dreyer, M.J. Drummond, E.L. Glynn, E. Volpi, and B.B. Rasmussen. 2009. Essential amino acid and carbohydrate ingestion before resistance exercise does not enhance postexercise muscle protein synthesis. *Journal of Applied Physiology* 106: 1730-1739.

Gailly, P., F. De Backer, M. Van Schoor, and J.M. Gillis. 2007. In situ measurements of calpain activity in isolated muscle fibres from normal and dystrophin-lacking mdx mice. *Journal of Physiology* 582: 1261-1275.

Geiger, P.C., and A.A. Gupte. 2011. Heat shock proteins are important mediators of skeletal muscle insulin sensitivity. *Exercise and Sport Sciences Reviews* 39: 32-42.

Gomez-Cabrera, M.C., C. Borras, F.V. Pallardo, J. Sastre, L.L. Ji, and J. Vina. 2005. Decreased xanthine oxidase-mediated oxidative stress prevents useful cellular adaptations to exercise in rats. *Journal of Physiology* 567: 113-120.

Gomez-Cabrera, M.C., E. Domenech, M. Romagnioli, A. Arduini, C. Borras, F.V. Pallardo, J. Sastre, and J. Vina. 2008. Oral administration of vitamin C decreases muscle mitochondrial biogenesis and hampers training-induced adaptations in endurance performance. *American Journal of Clinical Nutrition* 87: 142-149.

Graham, T.E., Z. Yuan, A.K. Hill, and R.J. Wilson. 2010. The regulation of muscle glycogen: The granule and its proteins. *Acta Physiologica* 199: 489-498.

Grassi, B. 2001. Regulation of oxygen consumption at exercise onset: Is it really controversial? *Exercise and Sport Sciences Reviews* 29: 134-138.

Gropper, S.S., J.L. Smith, and J.L. Groff. 2005. *Advanced nutrition and human metabolism.* Belmont, CA: Thomson Wadsworth.

Gross, L.S., L. Li, E.S. Ford, and S. Liu. 2004. Increased consumption of refined carbohydrates and the epidemic of type 2 diabetes in the United States: An ecological assessment. *American Journal of Clinical Nutrition* 79: 774-779.

Guilherme, A., J.B. Virbasius, V. Puri, and M.P. Czech. 2008. Adipocyte dysfunctions linking obesity to insulin resistance and type 2 diabetes. *Nature Reviews: Molecular Cell Biology* 9: 367-377.

Gundersen, K. 2011. Excitation-transcription coupling in skeletal muscle: The molecular pathways of exercise. *Biological Reviews* 86(3): 564-600.

Halliwell, B., and J.M.C. Gutteridge. 1999. *Free radicals in biology and medicine.* 3rd ed. Oxford: Clarendon Press.

Hamilton, M.T., D.G. Hamilton, and T.W. Zderic. 2004. Exercise physiology versus inactivity physiology: An essential concept for understanding lipoprotein lipase regulation. *Exercise and Sport Sciences Reviews* 32: 161-166.

Hardie, D.G. 2004. AMP-activated protein kinase: A key system mediating metabolic responses to exercise. *Medicine and Science in Sports and Exercise* 36: 28-34.

Harper, M.E., K. Green, and M.D. Brand. 2008. The efficiency of cellular energy transduction and its implications for obesity. *Annual Review of Nutrition* 28: 13-33.

Hawley, J.A., and L.M. Burke. 2010. Carbohydrate availability and training adaptation: effects on cell metabolism. *Exercise and Sport Science Reviews* 38: 152-160.

Helge, J.W. 2010. Arm and leg substrate utilization and muscle adaptation after prolonged low-intensity training. *Acta Physiologica* 199: 519-528.

Hepple, R.T. 2009. Why eating less keeps mitochondria working in aged skeletal muscle. *Exercise and Sport Sciences Reviews* 37: 23-28.

Hickner, R.C., J.S. Fisher, P.A. Hansen, S.B. Racette, C.M. Mier, M.J. Turner, and J.O. Holloszy. 1997. Muscle glycogen accumulation after endurance exercise in trained and untrained individuals. *Journal of Applied Physiology* 83: 897-903.

Hittel, D.S., Y. Hathout, and E.P. Hoffman. 2007. Proteomics and systems biology in exercise and sport sciences research. *Exercise and Sport Science Reviews* 35: 5-11.

Holloway, G.P., S.S. Jain, V. Bezaire, X.X. Han, J.F. Glatz, J.J. Luiken, M.E. Harper, and A. Bonen. 2009. FAT/CD36-null mice reveal that mitochondrial FAT/CD36 is required to upregulate mitochondrial fatty acid oxidation in contracting muscle. *American Journal of Physiology: Regulatory Integrative and Comparative Physiology* 297: R960-R967.

Holm, C. 2003. Molecular mechanisms regulating hormone-sensitive lipase and lipolysis. *Biochemical Society Transactions* 31: 1120-1124.

Hood, D.A. 2001. Contractile activity-induced mitochondrial biogenesis in skeletal muscle. *Journal of Applied Physiology* 90: 1137-1157.

Horowitz, J.F., A.E. Kaufman, A.K. Fox, and M.P. Harber. 2005. Energy deficit without reducing dietary carbohydrate alters resting carbohydrate oxidation and fatty acid availability. *Journal of Applied Physiology* 98: 1612-1628.

Hughson, R.L. 2005. Regulation of $\dot{V}O_2$ on kinetics by O_2 delivery. In *Oxygen uptake kinetics in health and disease*, ed. A.M. Jones and D.C. Poole. London: Routledge.

Hughson, R.L., M.E. Tschakovsky, and M.E. Houston. 2001. Regulation of oxygen consumption at the onset of exercise. *Exercise and Sport Sciences Reviews* 29: 129-133.

Humphery-Smith, I. 2004. A human proteome project with a beginning and an end. *Proteomics* 4: 2519-2521.

Ivy, J.L., H.W. Goforth Jr., B.M. Damon, T.R. McCauley, E.C. Parsons, and T.B. Price. 2002. Early postexercise muscle glycogen recovery is enhanced with a carbohydrate protein supplement. *Journal of Applied Physiology* 93: 1337-1344.

Jaworski, K., E. Sarkadi-Nagy, R.E. Duncan, M. Ahmadian, and H.S. Sul. 2007. Regulation of triglyceride metabolism IV. Hormonal regulation of lipolysis in adipose tissue. *American Journal of Physiology: Gastrointestinal and Liver Physiology* 293: G1-G4.

Jennissen, H.P. 1995. Ubiquitin and the enigma of intracellular protein degradation. *European Journal of Biochemistry* 231: 1-30.

Jeukendrup, A.E. 2003. Modulation of carbohydrate and fat utilization by diet, exercise and environment. *Biochemical Society Transactions* 31: 1270-1273.

Jeukendrup, A.E., W.H.M. Saris, and A.J.M. Wagenmakers. 1998. Fat metabolism during exercise: A review. Part II: Regulation of metabolism and the effects of training. *International Journal of Sports Medicine* 19: 293-302.

Jocken, J.W., E. Smit, G.H. Goossens, Y.P. Essers, M.A. van Baak, M. Mensink, W.H. Saris, and E.E. Blaak. 2008. Adipose triglyceride lipase (ATGL) expression in human skeletal muscle is type I (oxidative) fiber specific. *Histochemistry and Cell Biology* 129: 535-538.

Jones, D.P. 2008. Radical-free biology of oxidative stress *American Journal of Physiology: Cell Physiology* 295: C849-C868.

Karelis, A.D., J.W. Smith, D.H. Passe, and F. Peronnet. 2010. Carbohydrate administration and exercise performance: What are the potential mechanisms involved? *Sports Medicine* 40: 747-763.

Katz, A., and K. Sahlin. 1990. Role of oxygen in regulation of glycolysis and lactate production in human skeletal muscle. *Exercise and Sport Sciences Reviews* 18: 1-28.

Keller, P., N. Vollaard, J. Babraj, D. Ball, D.A. Sewell, and J.A. Timmons. 2007. Using systems biology to define the essential biological networks responsible for adaptation to endurance exercise training. *Biochemical Society Transactions* 35: 1306-1309.

Kendrew, J.C., R.E. Dickerson, B.E. Strandberg, R.G. Hart, D.R. Davies, D.C. Phillips, and V.C. Shore. 1960. Structure of myoglobin: A three-dimensional Fourier synthesis at 2 Å resolution. *Nature* 185: 422-427.

Kislinger, T., and A.O. Gramolini. 2010. Proteome analysis of mouse model systems: A tool to model human disease and for the investigation of tissue-specific biology. *Journal of Proteomics* 73: 2205-2218.

Kjaer, M., K. Howlett, J. Langfort, T. Zimmerman-Belsing, J. Lorentsen, J. Bülow, J. Ihlemann, U. Feldt-Rasmussen, and H. Galbo. 2000. Adrenaline and glycogenolysis in skeletal muscle during exercise: A study in adrenalectomised humans. *Journal of Physiology* 528: 371-378.

Lamb, G.D. 2009. Mechanisms of excitation-contraction uncoupling relevant to activity-induced muscle fatigue. *Applied Physiology Nutrition and Metabolism* 34: 368-372.

Lanza, I.R., and K.S. Nair. 2010. Mitochondrial function as a determinant of life span. *Pflugers Archives* 459: 277-289.

Leary, S.C., C.N. Lyons, A.G. Rosenberger, J.S. Ballantyne, J. Stillman, and C.D. Moyes. 2003. Fiber-type differences in muscle mitochondrial profiles. *American Journal of Physiology* 285: R817-R826.

LeBlanc, P.J., K.R. Howarth, M.J. Gibala, and G.J.F. Heigenhauser. 2004. Effects of 7 wk of endurance training on human skeletal muscle metabolism during submaximal exercise. *Journal of Applied Physiology* 97: 2148-2153.

Lecarpentier, Y. 2007. Physiological role of free radicals in skeletal muscles. *Journal of Applied Physiology* 103: 1917-1918.

Little, J.P., A. Safdir, G.P. Wilkin, M.A. Tarnopolsky, and M.J. Gibala. 2010. A practical model of low-volume, high interval training induces mitochondrial biogenesis in human skeletal muscle: Potential mechanisms. *Journal of Physiology* 588: 1011-1022.

Lyon, C.J., R.E. Law, and W.A. Hsieh. 2003. Adiposity, inflammation, and atherogenesis. *Endocrinology* 144: 2195-2200.

MacIntosh, B., P. Gardiner, and A.J. McComas. 2004. *Skeletal muscle: Form and function*. 2nd ed. Champaign, IL: Human Kinetics.

MacLennan, D.H., M. Abu-Abed, and C.-H. Kang. 2002. Structure–function relationships in Ca^{2+} cycling proteins. *Journal of Molecular and Cellular Cardiology* 34: 897-918.

Maher, A.C., M. Akhtar, J. Vockley, and M.A. Tarnopolsky. 2010. Women have higher protein content of β-oxidation enzymes in skeletal muscle than men. *PLoS One* 5: e12025.

Marcinek, D.J., M.J. Kushmerick, and K.E. Conley. 2010. Lactic acidosis in vivo: Testing the link between lactate generation and H^+ accumulation in ischemic mouse muscle. *Journal of Applied Physiology* 108: 1479-1486.

Martínez-Ruiz, A., and S. Lamas. 2007. Signaling by NO-induced protein S-nitrosylation and S-glutathionylation: Convergences and divergences. *Cardiovascular Research* 75: 220-228.

McCarthy, J.J., and K.A. Esser. 2007. MicroRNA-1 and microRNA-133a expression are decreased during skeletal muscle hypertrophy. *Journal of Applied Physiology* 102: 306-313.

McGee, S.L., and M. Hargreaves. 2011. Histone modifications and exercise adaptations. *Journal of Applied Physiology* 110: 258-263.

McGee, S.L., B.J. van Denderen, K.F. Howlett, J. Mollica, J.D. Schertzer, B.E. Kemp, and M. Hargreaves. 2008. AMP-activated protein kinase regulates GLUT-4 transcription by phosphorylating histone deacetylase 5. *Diabetes* 57: 860-867.

McGinley, C., A. Shafat, and A.E. Donnelly. 2009. Does antioxidant vitamin supplementation protect against muscle damage? *Sports Medicine* 39: 1011-1032.

Melanson, E.L., W.S. Gozansky, D.W. Barry, P.S. MacLean, G.K. Grunwald, and J.O. Hill. 2009. When energy balance is maintained, exercise does not induce negative fat balance in lean sedentary, obese sedentary, or lean endurance-trained individuals. *Journal of Applied Physiology* 107: 1847-1856.

Melanson, E.L., P.S. MacLean, and J.O. Hill. 2009. Exercise improves fat metabolism in muscle but does not increase 24-h fat oxidation. *Exercise and Sport Sciences Reviews* 37: 93-101.

Miller, B.F. 2007. Human muscle protein synthesis after physical activity and feeding. *Exercise and Sport Sciences Reviews* 35: 50-55.

Mittendorfer, B., D.A. Fields, and S. Klein. 2004. Excess body fat in men decreases plasma fatty acid availability and oxidation during endurance exercise. *American Journal of Physiology* 286: 354-362.

Mooren, F.C., A. Lechtermann, and K. Völker. 2004. Exercise-induced apoptosis of lymphocytes depends on training status. *Medicine and Science in Sports and Exercise* 36: 1476-1483.

Morio, B., U. Holmbäck, D. Gore, and R.R. Wolfe. 2004. Increased VLDL-TAG turnover during and after acute moderate intensity exercise. *Medicine and Science in Sports and Exercise* 36: 801-806.

Mourtzakis, M., J. Gonzalez-Alonso, T.E. Graham, and B. Saltin. 2004. Hemodynamics and O_2 uptake during maximal knee extension exercise in untrained and trained human quadriceps muscle: effects of hyperoxida. *Journal of Applied Physiology* 97: 1796-1802.

Mourtzakis, M., and T.E. Graham. 2003. Glutamate ingestion and its effects at rest and during exercise in humans. *Journal of Applied Physiology* 93: 1251-1259.

Moyes, C.D., and D.A. Hood. 2003. Origins and consequences of mitochondrial variation in vertebrate muscle. *Annual Reviews of Physiology* 65: 177-201.

Nguyen, H.X., and J.G. Tidball. 2003. Expression of a muscle-specific, nitric oxide synthase transgene prevents muscle membrane injury and reduces muscle inflammation during modified muscle use in mice. *Journal of Physiology* 550: 347-356.

Novak, M.L., W. Billich, S.M. Smith, K.B. Sukhija, T.J. MocLoghlin, T.A. Hornberger, and T.J. Koh. 2009. COX-2 inhibitor reduces skeletal muscle hypertrophy in mice. *American Journal of Physiology: Regulatory Integrative and Comparative Physiology* 296: R1132-R1139.

Olsen, S., P. Aagaard, F. Kadi, G. Tufekovic, J. Verney, J.L. Olesen, C. Suetta, and M. Kjaer. 2006. Creatine supplementation augments the increase in satellite cell and myonuclei number in human skeletal muscle induced by strength training. *Journal of Physiology* 572: 525-534.

Oosthuyse, T., and A.N. Bosch. 2010. The effect of menstrual cycle on exercise metabolism. *Sports Medicine* 207-227.

Ouzounian, M., D.S. Lee, A.O. Gramolini, A. Emili, M. Fukuoka, and P.P. Liu. 2007. Predict, prevent and personalize: Genomic and proteomic approaches to cardiovascular medicine. *Canadian Journal of Cardiology* 23: S28A-S33A.

Peterson, J.M., R.W. Bryner, J.C. Frisbee, and A.E. Always. 2008. Effects of exercise and obesity on UCP3 content in rat hindlimb muscles. *Medicine and Science in Sports and Exercise* 40: 1616-1622.

Phillips, D., A.M. Aponte, S.A. French, D.J. Chess, and R.S. Balaban. 2009. Succinyl-CoA synthetase is a phosphate target for the activation of mitochondrial metabolism. *Biochemistry* 48: 7140-7149.

Phillips, S.M., H.J. Green, M.A. Tarnopolsky, G.J.F. Heigenhauser, and S.M. Grant. 1996. Progressive effect of endurance training on metabolic adaptations in working skeletal muscle. *American Journal of Physiology* 270: E265-E272.

Philp, A., D.L. Hamilton, and K. Baar. 2011. Signals mediating skeletal muscle remodeling by resistance exercise: PI3-kinase independent activation of mTORC1. *Journal of Applied Physiology* 110: 561-568.

Pourova, J., M. Kottova, M. Voprsalova, and M. Pour. 2010. Reactive oxygen and nitrogen species in normal physiological processes. *Acta Physiologica* 198: 15-35.

Powers, S.K., J. Duarte, A.N. Kavazis, and E.E. Talbert. 2010. Reactive oxygen species are signalling molecules for skeletal muscle adaptation. *Experimental Physiology* 95: 1-9.

Price, T.B., D. Laurent, K.F. Petersen, D.L. Rothman, and G.I. Shulman. 2000. Glycogen loading alters muscle glycogen resynthesis after exercise. *Journal of Applied Physiology* 88: 698-704.

Pruchnic, R., A. Katsiaras, J. He, D.E. Kelley, C. Winters, and B.H. Goodpaster. 2004. Exercise training increases intramyocellular lipid and oxidative capacity in older subjects. *American Journal of Physiology* 287: E857-E862.

Randle, P.J., P.B. Garland, C.N. Hales, and E.A. Newsholme. 1963. The glucose fatty-acid cycle: Its role in insulin sensitivity and the metabolic disturbances of diabetes mellitus. *Lancet* 1: 785-789.

Randle, P.J., E.A. Newsholme, and P.B. Garland. 1964. Regulation of glucose uptake by muscle. *Biochemistry Journal* 93: 652-665.

Rennie, M.J., and K.D. Tipton. 2000. Protein and amino acid metabolism during and after exercise and the effects of training. *Annual Reviews of Nutrition* 20: 457-483.

Reshef, L., Y. Olswang, H. Cassuto, B. Blum, C.M. Croniger, S.C. Halhan, S.M. Tilghman, and R.W. Hanson. 2003. Glyceroneogenesis and the triglyceride/fatty acid cycle. *Journal of Biological Chemistry* 278: 30413-30416.

Richardson, R.S., J.S. Leigh, P.D. Wagner, and E.A. Noyszewski. 1999. Cellular PO_2 as a determinant of maximal mitochondrial O_2 consumption in trained human skeletal muscle. *Journal of Applied Physiology* 87: 325-331.

Richardson, R.S., E.A. Noyszewski, L.J. Haseler, S. Blumi, and L.R. Frank. 2002. Evolving techniques for the investigation of muscle bioenergetics and oxygenation. *Biochemical Society Transactions* 30: 232-237.

Rider, M.H., L. Bertrand, D. Vertommen, P.A. Michels, G.G. Rousseau, and L. Hue. 2004. 6-Phosphofructo-2-kinase/fructose-2,6-bisphosphatase: Head-to-head with a bifunctional enzyme that controls glycolysis. *Biochemical Journal* 381: 561-579.

Ristow, M., K. Zarse, A. Oberback, N. Klotikng, M. Birringer, M. Kiehntopt, M. Sumvoll, C.R. Kahn, and M. Bluher. 2009. Antioxidants prevent health-promoting effects of physical exercise in humans. *Proceedings of the National Academy of Sciences* 106: 8665-8670.

Robergs, R.A., F. Ghiasvand, and D. Parker. 2004. Biochemistry of exercise-induced metabolic acidosis. *American Journal of Physiology* 287: R502-R516.

Robergs, R.A., and S.O. Roberts. 1997. *Exercise physiology: Exercise performance and clinical applications*. St. Louis: Mosby.

Rocchini, A.P. 2002. Childhood obesity and the diabetes epidemic. *New England Journal of Medicine* 346: 854-855.

Roden, M. 2004. How free fatty acids inhibit glucose utilization in human skeletal muscle. *News in Physiological Sciences* 19: 92-96.

Rodgers, J.T., C. Lerin, W. Haas, S.P. Gygl, B.M. Spiegelman, and P. Pulgserver. 2005. Nutrient control of glucose homeostasis through a complex of PGC-1α and SIRT1. *Nature* 434: 113-118.

Roeder, R.G. 2005. Transcriptional regulation and the role of diverse coactivators in animal cells. *FEBS Letters* 579: 909-915.

Roepstorff, C., M. Donsmark, M. Thiele, B. Vitisen, G. Stewart, K. Vissing, P. Schjerling, D.G. Hardie, H. Galbo, and B. Kiens. 2006. Sex differences in hormone-sensitive lipase expression, activity, and phosphorylation in skeletal muscle at rest and during exercise. *American Journal of Physiology: Endocrinology and Metabolism* 291: E1106-E1114.

Roepstorff, C., N. Halberg, T. Hillis, A.K. Saha, N.B. Ruderman, J.F.P. Wojtaszewski, E.A. Richter, and B. Kiens. 2005. Malonyl-CoA and carnitine in regulation of fat oxidation in human skeletal muscle during exercise. *American Journal of Physiology* 288: E133-E142.

Rossi, A.E., S. Boncompagni, and R.T. Dirksen. 2009. Sarcoplasmic reticulum-mitochondrial symbiosis: Bidirectional signaling in skeletal muscle. *Exercise and Sport Sciences Reviews* 37: 29-35.

Roth, S.M. 2007. *Genetics primer for exercise science and health*. Champaign, IL: Human Kinetics.

Roth, S.M. 2011. MicroRNAs: Playing a big role in explaining skeletal muscle adaptation? *Journal of Applied Physiology* 110: 301-302.

Rush, J.W.E., S.G. Denniss, and D.A. Graham. 2005. Vascular nitric oxide and oxidative stress: Determinants of endothelial adaptations to cardiovascular disease and to physical activity. *Canadian Journal of Applied Physiology* 30: 442-474.

Rush, J.W.E., and L.L. Spriet. 2001. Skeletal muscle glycogen phosphorylase kinetics: Effects of adenine nucleotides and caffeine. *Journal of Applied Physiology* 91: 2071-2078.

Safdar, A., A. Abadi, M. Akhtar, B.P. Hettinga, and M.A. Tarnopolsky. 2009. MiRNA in the regulation of skeletal muscle adaptation to acute endurance exercise in C57Bl/6J male mice. *PLoS One* 4: e5610.

Safdar, A., M.J. Hamadeh, J.J. Kaczor, S. Raha, J. deBeer, and M.A. Tarnopolsky. 2010. Aberrant mitochondrial homeostasis in skeletal muscle of sedentary older adults. *PLoS One* 5: e10778.

Safdar, A., N. Yardley, R.J. Snow, S. Melov, and M.A. Tarnopolsky. 2004. Genomic and protein expression in skeletal muscle of men following acute creatine monohydrate supplementation. *Canadian Journal of Applied Physiology* 29: S77.

Sahlin, K., M. Fernström, M. Svensson, and M. Tonkonogi. 2002. No evidence of an intracellular lactate shuttle in rat skeletal muscle. *Journal of Physiology* 541: 569-574.

Saltin, B., A.P. Gagge, and J.A.J. Stolwijk. 1968. Muscle temperature during submaximal exercise in man. *Journal of Applied Physiology* 25: 679-688.

Sambandam, N., M. Steinmetz, A. Chu, J.Y. Altarejos, J.R.B. Dyck, and G.D. Lopaschuk. 2004. Malonyl CoA decarboxylase (MCD) is differentially regulated in subcellular compartments by 5' AMP-activated protein kinase (AMPK). *European Journal of Biochemistry* 271: 2831-2840.

Sandström, M.E., S.J. Zhang, J. Bruton, J.P. Silva, M.B. Reid, H. Westerblad, and A. Katz. 2006. Role of reactive oxygen species in contraction-mediated glucose transport in mouse skeletal muscle. *Journal of Physiology* 575: 251-262.

Shearer, J., and T. Graham. 2002. New perspectives on muscle glycogen storage and utilization during exercise. *Canadian Journal of Applied Physiology* 27(2): 179-203.

Smith, S.R. 2009. Beyond the bout: New perspectives on exercise and fat oxidation. *Exercise and Sports Sciences Reviews* 37: 58-59.

Snow, R.J., and R.M. Murphy. 2003. Factors influencing creatine loading into human skeletal muscle. *Exercise and Sport Sciences Reviews* 31: 154-158.

Spencer, M.K., Z. Yan, and A. Katz. 1992. Effect of low glycogen on carbohydrate and energy metabolism in human muscle during exercise. *American Journal of Physiology* 262: C975-C979.

Stich, V., I. de Glisezinski, F. Crampes, J. Hejnova, J.M. Cottet-Emard, J. Galitzky, M. Lafontan, D. Rivière, and M. Berlan. 2000. Activation of α_2 adrenergic receptors impairs exercise-induced lipolysis in SCAT of obese subjects. *American Journal of Physiology* 279: R499-R504.

Stojanovski, D., A.J. Johnston, I. Streimann, N.J. Hoogenraad, and M.T. Ryan. 2003. Import of nuclear encoded proteins into mitochondria. *Experimental Physiology* 88: 57-64.

Strandberg, B. 2009. Chapter 1: Building the ground for the first two protein structures: Myoglobin and haemoglobin. *Journal of Molecular Biology* 392: 2-10.

Strawford, A., F. Antero, M. Christiansen, and M.K. Hellerstein. 2004. Adipose tissue triglyceride turnover, de novo lipogenesis, and cell proliferation in humans measured with ^2H_2O. *American Journal of Physiology* 286: E577-E588.

Taillandier, D., L. Combaret, M.-N. Pouch, S.E. Samuels, D. Béchet, and D. Attaix. 2004. The role of ubiquitin-proteasome-dependent proteolysis in the remodeling of skeletal muscle. *Proceedings of the Nutrition Society* 63: 357-361.

Talanian, J.L., G.P. Holloway, L.A. Snook, G.J. Heigenhouser, A. Bonen, and L.L. Spreit. 2010. Exercise training increases sarcolemmal and mitochondrial fatty acid transport proteins in human skeletal muscle. *American Journal of Physiology: Endocrinology and Metabolism* 299: E180-E188.

Tauler, P., M.D. Ferrer, A. Sureda, P. Pujol, F. Drobnic, J.A. Tur, and A. Pons. 2008. Supplementation with an antioxidant cocktail containing coenzyme A prevents plasma oxidative damage induced by soccer. *European Journal of Applied Physiology* 104: 777-785.

Tiidus, P.M., and J.K. Shoemaker. 1995. Effleurage massage, muscle blood flow and long-term post-exercise strength recovery. *International Journal of Sports Medicine* 16: 478-483.

Tong, X., A. Evangelista, and R.A. Cohen. 2010. Targeting the redox regulation of SERCA in vascular physiology and disease. *Current Opinion in Pharmacology* 10: 133-138.

Toyoshima, C., M. Nakasako, H. Nomura, and H. Ogawa. 2000. Crystal structure of the calcium pump of sarcoplasmic reticulum at 2.6 A resolution. *Nature* 405: 647-655.

Trimmer, J.K., J.-M. Schwarz, G.A. Casazza, M.A. Horning, N. Rodriguez, and G.A. Brooks. 2002. Measurement of gluconeogenesis in exercising men by mass isotopomer distribution analysis. *Journal of Applied Physiology* 93: 233-241.

Trumpower, B.L. 1990. The protonmotive Q cycle. Energy transduction by coupling of proton translocation to electron transfer by the cytochrome bc1 complex. *Journal of Biological Chemistry* 265: 11409-11412.

Tseng, Y.H., A.M. Cypress, and R. Khan. 2010. Cellular bioenergetics as a target for obesity therapy. *Nature Reviews: Drug Discovery* 9: 465-481.

Tucker, R.M., C. May, R. Bennett, J. Hymer, and B. McHaney. 2004. A gym-based wellness challenge for people with type 2 diabetes: Effect on weight loss, body composition and glycemic control. *Diabetes Spectrum* 17: 176-180.

Tupling, A.R. 2004. The sarcoplasmic reticulum in muscle fatigue and disease: Role of the sarco(endo)plasmic reticulum Ca^{2+}-ATPase. *Canadian Journal of Applied Physiology* 29: 308-329.

Tupling, A.R., C. Vigna, R.J. Ford, S.C. Tsuchiya, D.A. Graham, S.G. Denniss, and J.W.E. Rush. 2007. Effects of buthionine sulfoximine treatment on diaphragm contractility and SR Ca^{2+} pump function in rats. *Journal of Applied Physiology* 103: 1921-1928.

Uguccioni, G., D. D'souza, and D.A. Hood. 2010. Regulation of PPARγ coactivator-1α function and expression in muscle: Effect of exercise. *PPAR Research*: 937123, Epub.

Vandenboom, R. 2004. The myofibrillar complex and fatigue: A review. *Canadian Journal of Applied Physiology* 29: 330-356.

Van de Poll, M.C.G., P.B. Soeters, N.E.P. Deutz, K.C.H. Fearon, and C.H.D. Dejong. 2004. Renal metabolism of amino acids: Its role in interorgan amino acid exchange. *American Journal of Clinical Nutrition* 79: 185-197.

van Hall, G. 2010. Lactate kinetics in human tissues at rest and during exercise. *Acta Physiologica* 199: 499-508.

van Rooij, E., N. Liu, and E.N. Olson. 2008. MicroRNAs flex their muscles. *Trends in Genetics* 24: 159-166.

Venables, M.C., J. Achten, and A.E. Jeukendrup. 2005. Determinants of fat oxidation during exercise in healthy men and women: A cross-sectional study. *Journal of Applied Physiology* 98: 160-167.

Vincent, H.K., J.W. Morgan, and K.R. Vincent. 2004. Obesity exacerbates oxidative stress levels after acute exercise. *Medicine and Science in Sports and Exercise* 36: 772-779.

Vogt, M., A. Puntschart, H. Howard, B. Mueller, C. Mannhart, L. Feller-Tuescher, P. Mullis, and H. Hoppeler. 2003. Effects of dietary fat on muscle substrates, metabolism, and performance in athletes. *Medicine and Science in Sports and Exercise* 35: 952-960.

Wang, C., W.S. Harris, M. Chung, A.H. Lichtenstein, E.M. Balk, B. Kupelnick, H.S. Jordan, and J. Lau. 2006. N-3 fatty acids from fish or fish-oil supplements, but not α-linolenic acid, benefit cardiovascular disease outcomes in primary and secondary prevention studies: A systematic review. *American Journal of Clinical Nutrition* 84: 5-17.

Wasserman, K., B.J. Whipp, S.N. Koyal, and W.L. Beaver. 1973. Anaerobic threshold and respiratory gas exchange during exercise. *Journal of Applied Physiology* 35: 236-245.

Watt, M.J., G.J. Heigenhauser, P.J. LeBlanc, J.G. Inglis, L.L. Spriet, and S.J. Peters. 2004. Rapid upregulation of pyruvate dehydrogenase kinase activity in human skeletal muscle during prolonged exercise. *Journal of Applied Physiology* 97: 1261-1267.

Watt, M.J., G.J.F. Heigenhauser, and L.L. Spriet. 2002. Intramuscular triacylglycerol utilization in human skeletal muscle during exercise: Is there a controversy? *Journal of Applied Physiology* 93: 1185-1195.

Watt, M.J., A.G. Holmes, S.K. Pinnamaneni, A.P. Garnham, G.R. Steinberg, B.E. Kemp, and M.A. Febbraio. 2006. Regulation of HSL serine phosphorylation in skeletal muscle and adipose tissue. *American Journal of Physiology: Endocrinology and Metabolism* 290: E500-E508.

Watt, M.J., P. Krustrup, N.H. Secher, B. Saltin, B.K. Pedersen, and M.A. Febbraio. 2004. Glucose ingestion blunts hormone-sensitive lipase activity in contracting human skeletal muscle. *American Journal of Physiology* 286: E144-E150.

Weltan, S.M., A.N. Bosch, S.C. Dennis, and T.D. Noakes. 1998. Preexercise muscle glycogen content affects metabolism during exercise despite maintenance of hyperglycemia. *American Journal of Physiology* 274: E83-E88.

Westerblad, H., D.C. Allen, and J. Lännergren. 2002. Muscle fatigue: Lactic acid or inorganic phosphate the major cause? *News in Physiological Sciences* 17: 17-21.

Wilfred, B.R., W.X. Wang, and P.T. Nelson. 2007. Energizing miRNA research: A review of the role of miRNAs in lipid metabolism, with a prediction that miR-103/107 regulates human metabolic pathways. *Molecular Genetics and Metabolism* 91: 209-217.

Wilmore, J.H., and D.L. Costill. 2004. *Physiology of sport and exercise.* 3rd ed. Champaign, IL: Human Kinetics.

Wiltshire, E.V., V. Poitras, M. Pak, T. Hong, J. Rayner, and M.E. Tschakovsky. 2010. Massage impairs postexercise muscle blood flow and "lactic acid" removal. *Medicine and Science in Sports and Exercise* 42: 1062-1071.

Xu, X., C.J. Chen, E.A. Arriaga, and L.V. Thompson. 2010. Asymmetric superoxide release inside and outside the mitochondria in skeletal muscle under conditions of aging and disuse. *Journal of Applied Physiology* 109: 1133-1139.

Yan, Z., M. Okutsu, Y.N. Akhtar, and V.A. Lira. 2011. Regulation of exercise-induced fiber type transformation, mitochondrial biogenesis, and angiogenesis in skeletal muscle. *Journal of Applied Physiology* 110: 264-274.

Yeaman, S.J. 2004. Hormone-sensitive lipase—new roles for an old enzyme. *Biochemical Journal* 379: 11-22.

Yfanti, C., T. Akerstrom, S. Nielson, A.R. Nielson, R. Mounier, O.H. Mortensen, J. Lykkesfeldt, A.J. Rose, C.P. Fischer, and B.K. Pedersen. 2010. Antioxidant supplementation does not alter endurance training adaptations. *Medicine and Science in Sports and Exercise* 42: 1388-1395.

Zderic, T.W., C.J. Davidson, S. Schenk, L.O. Byerley, and E.F. Coyle. 2004. High fat diet elevates resting intramuscular triglyceride concentration and whole body lipolysis during exercise. *American Journal of Physiology* 286: E217-E225.

Index

Note: The italicized *f* and *t* following page numbers refer to figures and tables, respectively.

A

acceptor control model 136
acetic acid 5-6
acetoacetate 223, 223f-225f
acetone 223, 223f
acetylation 31, 52
acetylcholine 82
acetyl CoA 115-116, 131-132, 139, 159, 182, 196, 226, 256
acetyl CoA carboxylase 33, 225, 227, 229, 237f
acid(s) 4-5
acid dissociation constant 5, 9f
acidosis 4
aconitase 116
actin 15
actin-activated myosin ATPase 80
activating transcriptional factor 2 67
activation domains 48
activator protein-1 49-50
activators 47-48
active site 19
active transport 27, 27f
acylcarnitine 219-220
acyl carrier protein 227
acyl CoA carboxylase 227, 229, 236
acyl CoA synthetase 210, 219
adenine 39, 40f, 85
adenine nucleotides 86
adenosine 85, 215
adenosine diphosphate. *See* ADP
adenosine 5'-monophosphate 40, 41f
adenosine triphosphate. *See* ATP
adenylate deaminase. *See* AMP deaminase
adenylate kinase 86, 96-97, 199, 259-260
adenylosuccinate lyase 260
adenylyl cyclase 51, 171, 213
adipocytes 107, 209, 210f, 217f, 241f
adipokines 240
adiponectin 241
adipose tissue
  description of 126-127, 205
  as endocrine tissue 240-242
  fatty acids from 216-217, 230
  lipolysis regulation in 213-215
ADP
  ATP regeneration from 96, 112, 119
  description of 80
  formation of 129
  free 86
  mitochondrial transport of 127-128
  phosphorylation of 129
ADP:ATP ratio 87
β-adrenergic receptor 171
aerobic glycolysis 91, 159
aerobic metabolism 98, 114
Akt 70, 214
alanine 8f-10f, 253, 257
alanine aminotransferase 250
aldolase 162
alkalosis 4
allosteric effectors 30, 164, 165f, 171t, 173-174
allosteric enzymes 30-31
allosteric sites 30
alpha-helix 13
amino acid(s)
  acid–base properties of 7, 8f-9f
  acids 4-5
  bases 4-5
  basic 7
  branched-chain 191, 191f, 250, 252-253, 258
  buffers 5-6
  carbon skeletons 256f, 256-257
  D- 9, 10f
  degradation of 249-252
  diversity of 7

essential 248
functions of 4
glucogenesis role of 190-191
glucogenic 190
ketogenic 190
L- 9, 10f
monoamino-dicarboxylic 7
nonessential 248, 261
pH 10
release of 3
roles of 261t, 261-262
sequence of 11
side chains of 12
skeletal muscle 249, 258f
sources of 247
stereoisomerism of 9, 10f
structure of 6-7, 6f-7f
amino acid metabolism
  during exercise 257-260
  in high-intensity exercise 260
  in moderate-intensity exercise 257-259
  overview of 247-249, 248f
amino acid pool 247
amino acid residues 10
amino acid transporters 249
aminoacyl-tRNA 59-60, 62
aminoacyl-tRNA synthetase 60
5-amino-imidazole-4-carboxamide-riboside 199
aminotransferase enzymes 250
aminotransferases 186
ammonia 5, 96, 251-252
AMP-activated protein kinase 67, 198-199, 216, 229, 234, 237
AMP deaminase 96-97, 259
amphipathic 13
amphoteric 7
AMP kinase kinase 199, 237
amyloglucosidase 168
anabolism 75
anaerobic alactic system 92
anaerobic glycolysis 91, 97, 159, 176, 180
anaplerosis 191
anhydride bonds 85
anions 205
anserine 262
anticodons 58, 61
antioxidants 142-143, 146-148
antiparallel strands 41
antiport 114, 127
apoenzyme 25
apoptosis 110, 114
arachidonic acid 207t
arginine 8f
ascorbic acid 141
asparagine 8f
aspartate aminotransferase 250, 254
aspartate–glutamate transporter 186
aspartic acid 7, 8f-9f
ATP
  ADP and 87, 119, 130f
  cellular concentration of 87
  description of 27, 32
  energy released from hydrolysis of 85-86
  formation of 75
  functions of 77
  hydrolysis of 83-85, 88, 91, 101-103, 128-129, 130f, 142
  mitochondrial transport of 127-128
  molecular structure of 85, 85f
  muscle use of 77, 83-84
  skeletal muscle use of 77, 84f
  turnover of 77
ATP–ADP antiport 127
ATPases
  SERCA 84, 88

  types of 88, 89f
ATP-citrate lyase 226
ATP/O ratio 112, 126
ATP regeneration
  ADP as source of 96
  creatine kinase for 92
  description of 87-88
  during exercise 97
  glycolysis for 88, 90-91, 160, 164. *See also* glycolysis
  oxidative phosphorylation for 88, 128-129. *See also* oxidative phosphorylation
  phosphocreatine. *See* phosphocreatine
ATP synthase 110, 125f, 125-126
AUG codon 43t, 44
autocrine effect 51
(a-v̄)O₂ difference 134

B

base pairs 40-41
bases 4-5, 14
bases (mRNA) 43
basic amino acids 7
basic helix-loop-helix motif 47
basic zipper motif 47
beta-oxidation 109, 220-222
beta-sheet 13
binding site 19
bioenergetics 99-103
biological redox reactions 140-142
biotin 188
2,3-bisphosphate glycerate 163
blood
  creatine concentrations in 93-94
  glucose concentrations in 156, 157f, 166, 186, 194, 229
  lactate concentrations in 176f, 178, 186
  pH of 4
body composition 234
bonds
  anhydride 85
  disulfide 11-13, 12f
  hydrogen 12, 12f
  peptide 10
  protein structure 11-12
branched-chain amino acids 191, 191f, 250, 252-253, 258
branching enzyme 168
Brønsted-Lowry 4-5
brown adipose tissue 126-127, 205
buffers 5-6
B vitamins 25, 25t, 188, 262

C

Ca²⁺-ATPase 3, 13
calcineurin 67
calcitonin 57
calcitonin gene-related peptide 57
calcium
  in glycogenolysis regulation 172
  inner mitochondrial membrane transport of 128f
  mitochondrial uptake of 113-114
calcium channels 113
calcium uniport 113
calmodulin 172
calorie restriction 149
calpains 66
cAMP phosphodiesterase 214
carbamoyl phosphate synthetase 255
carbohydrates
  classification of 153-155
  description of 153
  dietary amounts of 167
  disaccharides 154

carbohydrates (continued)
  glycogen. See glycogen
  monosaccharides 153, 154f
  muscle fatigue affected by 201-202
  oxidation of 235
  storage sites for 107
α-carbon 9, 12
carbonic anhydrase 23
cardiac muscle 88
cardiac muscle cells 135, 238f
cardiac output 133, 134t
carnitine 219, 239f, 261
carnitine acetyl transferase 239
carnitine-acylcarnitine translocase 219
carnitine palmitoyl transferase I/II 219-220, 236
carnosine 262
catabolism 75, 84
catalase 144
catalytic constant 23
catalytic site 19
cathepsins 66
cell
  ATP concentration in 87
  calcium benefits for 114
  energy pathways in 108f
  free energy in 102-103
  glucose uptake by 155-157
  phosphates in 84-86
  reactive oxygen and nitrogen species effects on 144-146
cell-membrane receptors 52
cell signaling 51-52, 147-148
cellular respiration. See oxidative phosphorylation
ceramides 240, 242f
cervonic acid 207t
chaperones 64
chaperonins 64
chemical proteomics 16
chemiosmotic hypothesis 125
cholesterol 242-243, 243f
chromatin 41-43, 52
chromosomes 39
chylomicrons 210
citrate 164, 236
citrate synthase 116, 131, 226
citric acid cycle
  chemical structures 117f
  description of 28, 107, 109, 112
  discovery of 114, 255
  electron transport chain and 129, 131
  function of 114
  NADH from 129
  overview of 114-116, 115f
  oxaloacetate for 119-120
  reactions of 116-120
  regulation of 129, 131, 131t
  schematic diagrams of 117f-118f
coactivators 48
codons 43-44, 58
coenzyme(s) 25, 119
coenzyme A 114-115
coenzyme Q 118, 121-123, 184
coenzyme Q-cytochrome c oxidoreductase 123
cofactors, enzyme 25-26
competitive inhibitors 24, 24f
complementary strands 40
cooperativity 30
corepressors 48
Cori cycle 186
cortisol 194, 202, 215, 218
coupled phosphorylation 125f, 125-126
covalent modification of enzymes 31-34
creatine. See also phosphocreatine
  blood concentrations of 93-94
  description of 27, 27f
  food sources of 93, 94f
  formation of 261
  properties of 93-95
  skeletal muscle metabolism of 94f
  sources of 93, 94f
  supplementation of 93, 95-96
  total concentration in muscle 94-95, 95t
creatine kinase 88, 92, 103, 129
creatinine 94
cristae 109, 111

crossbridge cycle 82f
crossbridges 14, 79-82
crossover concept 231
C-terminus 10
cyclic AMP 52, 52f
cyclic AMP response element binding protein 52, 194
cyclooxygenase-1 25
cyclooxygenase-2 25
cysteine 8f-9f
cystic fibrosis transmembrane conductance regulator protein 11
cystolic redox state 181
cytidine triphosphate 85
cytochromes
  b 123
  c 121, 123, 143
  c1 123
  c oxidase 123-124, 135
  description of 123
cytokines 51, 240
cytoplasmic NADH 184-186
cytosine 39, 40f
cytosolic glycerol phosphate dehydrogenase 185

D

D-amino acids 9, 10f
deamination 250-251
degenerate 43
dehydrogenation 28f
denaturation 15-16
de novo lipogenesis 225, 227, 229, 236
deoxyadenosine 40, 41f
deoxyribonucleic acid. See DNA
2-deoxyribose 40f, 153
dephosphorylation 32-33, 51
depolarization 82
desnutrin/ATGL 215-216
α-D-fructose 154f
α-D-galactose 154f
α-D-glucose 154f
diabetes mellitus 155-157, 194
diacylglycerols 207, 209
dichloroacetate 133
diet
  fuel utilization during exercise affected by 232-233
  high-fat 232-233, 236
  high-protein 191
  protein degradation affected by 262-263
dihydrogen 6
dipeptide 10, 10f
disaccharides 154
dismutation reaction 143
dissociation 5
disulfide bonds 11-13, 12f
DNA
  base pairs 40-41
  chromatin 41-43, 52
  description of 3
  double-helix structure of 39, 42f
  genetic code 43t, 43-44
  mitochondrial 44
  nucleosomes 41-43, 42f
  strands of 40-41, 44f
  structure of 39-41, 42f
  transcription. See transcription
DNA-binding domain 47
domains 13
downstream 44

E

E box 51
eicosanoids 207
18S ribosomal RNA transcript 58
80S ribosomal RNA transcript 61f
elderly 263
electrochemical gradient 112
electron paramagnetic resonance spectroscopy 147
electron transport chain
  citric acid cycle and 129, 131
  coenzyme Q-cytochrome c oxidoreductase 123
  complexes involved in 120f, 120-124
  cytochrome c oxidase 123-124

description of 109, 112, 120-122
in inner mitochondrial membrane 121f, 121-122
NADH-coenzyme Q oxidoreductase 122
oxygen availability to 139
schematic diagram of 119f
succinate-coenzyme Q oxidoreductase 122-123
summary of 124
endergonic reaction 101
endothelial nitric oxide synthase 144
endothermic reactions 100
endurance training
  carbohydrate availability during 202
  frequency in 66
  fuel utilization during exercise affected by 233-234
  gene expression regulation in 66-68
  glycolytic capacity affected by 201
  lactate metabolism affected by 183-184
  metabolic stress caused by 199
  PDK4 enzyme increases secondary to 133
energy systems
  glycolysis. See glycolysis
  overview of 88-89
  oxidative phosphorylation. See oxidative phosphorylation
  phosphocreatine 92-95
enhancers 47, 49
enolase 163
enoyl CoA hydratase 222
enthalpy change 100
entropy change 100
enzymatic reactions
  enzyme concentration effects on 22
  pH effect on 23, 23f
  rates of 20-25
  substrate concentration effects on 20-22, 21f
  temperature effects on 23, 23f
enzyme(s)
  allosteric 30-31
  as catalysts 19-20
  classification of 26, 26t
  cofactors 25-26
  concentration of 22
  covalent modification of 31-34
  definition of 19
  glycolytic 160-164
  inhibition of 24f, 24-25
  mineral effects on 25-26
  mitochondrial 35
  rate-limiting 31
enzyme activity
  measurement of 35
  modification of 32f
  regulation of 30-34
enzyme-catalyzed reactions
  description of 19-20
  direction of 21
  illustration of 20f
  kinetic parameters for 21
  rate of 30
enzyme kinetics 20
enzyme velocity 22f
epigenetics 104
epinephrine 32, 171-172, 213
E proteins 51
equilibrium 101-102
equilibrium constant 20
equilibrium reactions 20
essential amino acids 248
essential fatty acids 206-207
essential light chains 80
esterification 212
estradiol 51
estrogen 51
euchromatin 43
euglycemia 155
eukaryotic 47
eukaryotic elongation factors 62
eukaryotic initiation factors 60
excess postexercise oxygen consumption 92, 93f
excitation-transcription coupling 66, 67f
exercise. See also endurance training; resistance training
  amino acid metabolism during 257-260
  ATP regeneration during 97

cell signaling in 147-148
creatine supplementation effects on 95
feeding during 233
free fatty acids and 229-231, 232f
glucose uptake regulated by 200-201
glycogen synthesis after 169
lactic acid formation during 91
lipids as fuel for 229-239
metabolism during 97-99, 231-239
oxidative phosphorylation regulation at onset of 136-138
oxidative stress and 146-147
oxygen consumption during 92
protein degradation affected by 262-263
reactive oxygen and nitrogen species formation during 147-148
exercise intensity
  fat oxidation affected by 243
  lactate metabolism affected by 183-184
  metabolism affected by 231-232
exercise performance 201-202
exercise training
  adaptive response to 70-71
  carbohydrate ingestion during 259
  epigenetic changes in muscle-gene expression affected by 104
  mitochondrial volume increased by 111
  muscle mitochondria affected by 148-149
exergonic reaction 101
exons 54, 57
exothermic reactions 100
expression proteomics 16
extracellular fluids 9

F

facilitated diffusion 26
$FADH_2$ 29, 112, 114, 119, 121
Faraday constant 142
fast-twitch muscle fibers 81, 91, 181, 181t
fat oxidation 243-244
fatty acid-binding protein 219
fatty acids. See also lipids
  beta-oxidation of 139, 220-222, 221f
  description of 108, 205-207
  dietary sources of 210, 225
  essential 206-207
  intracellular transport of 219
  odd-carbon 222
  oxidation of 218-222, 236-239
  peroxidation of 146
  polyunsaturated 145, 206, 233
  saturated, beta-oxidation of 220-222
  shorthand notation for 205-206
  structure of 206f
  synthesis of 210, 225-229, 228f
  unsaturated, oxidation of 222
fatty-acid synthase 227
fatty acyl CoA 220, 223
feedback inhibition 31
feed-forward mechanism 172
Fenton reaction 144, 145f
Fick equation 133
5S ribosomal RNA transcript 58
flavin adenine dinucleotide 28, 89, 114, 116
flavin mononucleotide 122
flavoproteins 122-123
flux 160
fold 14
40S ribosomal RNA transcript 58, 60
45S ribosomal RNA transcript 58
Fos 49
free ADP 86
free energy
  for ATP hydrolysis 142
  in cell 102-103
  description of 99-100
  electron transfer in cytochrome oxidase as source of 124
  quantitative values for 101-102
free energy change 100-101
free fatty acids 205, 218, 229-231, 232f, 235
free radical 143, 145
fructokinase 158
fructose 154f, 158-159, 163
fructose 1,6-bisphosphatase 189, 194, 196
fructose 2,6-bisphosphate 165, 180
fumarase 118

fumarate 118
functional proteomics 16

G

GC-rich regions 48
gene
  control region of 50f
  definition of 44, 47
  description of 3
  promoter region of 49f
gene expression
  in endurance training 66-68
  epigenetic changes in 104
  gluconeogenesis control by alterations in 194-196
  in resistance training 68-70
general transcription factors 48
gene regulatory proteins 47-48
gene repression 48
genetic code 43t, 43-44
genome 39
ghrelin 241-242
glucagon 11, 174, 192, 195f
glucocorticoids 212
glucogenic amino acids 190
glucokinase 22, 22f, 158
glucokinase regulatory protein 158
gluconeogenesis
  amino acids' role in 190-191
  definition of 186
  description of 163, 186-187
  enzyme activity control used to regulate 196
  gene expression alterations for control of 194-196
  glycerol as precursor for 190
  insulin effects on 193
  in liver 193f
  reactions of 187-190
  regulation of 191-196
glucose
  blood concentrations of 156, 157f, 166, 186, 194, 229
  cellular uptake of 155-157
  description of 108
  functions of 186
  metabolism of 159f
  oxidation of 111
  phosphorylation of 157-159
  skeletal muscle uptake of 156, 200-201
glucose-alanine cycle 191, 192f, 258-259
glucose-fatty acid cycle 231, 234-239
glucose-6-phosphatase 166, 189-190
glucose-6-phosphate 157-160, 161f, 168, 176, 180
glucose-6-phosphate dehydrogenase 197
glucose-phosphate isomerase reaction 162
glucose transporters 155t, 155-156, 158-159, 199-200
glutamate 257
glutamate dehydrogenase 251
glutamic acid 7, 8f
glutaminase 252
glutamine 8f, 251-252, 258
glutamine synthetase 251, 251f
glutaredoxins 34
glutathione 10, 33, 141, 144, 261
glycemic index 159
glyceraldehyde 3-phosphate dehydrogenase 162, 197
glycerol kinase 190, 217-218
glycerol 3-phosphate 217-218
glycerol-phosphate shuttle 184f, 184-185, 211
glyceroneogenesis 218
glycine 8f, 10f
glycogen
  breakdown of 166, 173
  composition of 165
  description of 90, 107
  dietary supercompensation of 169
  energy from 174
  liver storage of 166-167, 168t, 174, 209
  metabolism of 165-166, 170-176
  skeletal muscle storage of 166-168, 174
  storage of 166-169
  structure of 166f
  synthesis of 166-169, 168f
glycogenesis 168, 175-176

glycogenin 165
glycogenolysis
  allosteric effector regulation of 173-174
  definition of 166
  liver regulation of 174-175
  regulation of 171t, 171-175
glycogen phosphorylase 166, 170, 172-174
glycogen phosphorylase kinase 173f
glycogen synthase 169, 170f, 175, 175f
glycolysis
  aerobic 91, 159
  anaerobic 91, 97, 159, 176, 180
  ATP production from 91, 160, 164
  definition of 159
  description of 88, 90-91
  enzyme activity control used to regulate 196
  enzymes of 160-164
  fuel sources for 107
  functions of 160
  in liver 193f
  pyruvate generation from 160
  reactions of 160-164, 162t
  regulation of 164-165
glycosidic bond 165
glycosome 165
glycylalanine 10, 10f
G protein 52, 215
G protein-coupled receptors 52
growth factors 11
growth factor-signaling pathway 69f
growth hormone 11
guanine 39, 40f
guanosine triphosphate 85-86, 115-116, 118

H

half-life 3
heat shock proteins 64, 157
heavy chains 80
heme 13
hemoglobin 15, 134-136
Henderson–Hasselbalch equation 6
hepatocyte nuclear factor-4 195
heterochromatin 42
heterogeneous nuclear RNA 54
heterogeneous ribonucleoprotein particles 54
hexokinase 22, 22f, 31, 158, 164
hexose bisphosphate 162
high-fat diet 232-233, 236
Hill coefficient 30
Hill slope 30
histamine 262
histidine 8f-9f
histone 31, 42, 52, 53f
histone acetylases 53
histone acetyltransferases 53
histone deacetylases 53, 67
histone tails 43, 52-53
holoenzyme 25
homeodomain motif 47
homocarnosine 262
homocysteine 262
hormone(s)
  gluconeogenesis regulation by 192-194
  as peptides 11
  steroid 51
hormone receptors 51
housekeeping genes 48
hybridization 54
hydride ion 28
hydrogen bonds 12, 12f
hydrogen peroxide 144
hydrolases 26t
hydrolysis 80
hydrophilic side chains 13
hydrophobic 12, 12f
3-hydroxyacyl CoA dehydrogenase 222
3-hydroxybutyrate dehydrogenase 224, 225f
hydroxylation 261
hyperglycemia 155, 236
hypochlorite 144, 145f
hypoglycemia 155, 201
hypoxanthine 97

I

initiator 48
inorganic phosphate
  description of 6, 88, 173

inorganic phosphate *(continued)*
 formation of 129
 mitochondrial transport of 127-128
 muscle fatigue caused by 178
 phosphocreatine breakdown effects on 103, 129
inorganic pyrophosphatase 85, 210
inorganic pyrophosphate 46, 60, 169
inosine monophosphate 96-97, 99
inositol 209f
insulin 11, 156, 157f, 193, 200, 214-215
insulin-like growth factor-1 69
insulin receptor 200
insulin resistance 194, 218
integral membrane proteins 26
intermembrane space 109
intermyofibrillar mitochondria 109
international unit 35
interval training 150
intracellular fluids 9
intramuscular triacylglycerol 109, 231, 234
introns 54, 56f, 56-57
ionic interactions 12, 12f
ischemia 177
isocitrate dehydrogenase 116, 131, 131t
isoelectric point 7, 9f
isoleucine 8f, 253f
isomerases 26t
isozymes 22

J
Jun 49

K
K_a. *See* acid dissociation constant
k_{cat}. *See* turnover number
K_{eq}. *See* equilibrium constant
ketoacidosis 224
ketogenic amino acids 190
α-ketoglutarate 116, 185, 250, 250f, 252, 254
α-ketoglutarate dehydrogenase 116, 131, 131t, 138
ketone bodies 222-225, 223f, 235
ketonemia 224
ketonuria 224-225
ketosis 224-225
kinetic energy 99
K_m. *See* Michaelis constant
Krebs cycle. *See* citric acid cycle

L
lactate
 accumulation of 177-180
 acidification of 177-180
 anaerobic glycolysis production of 180
 blood concentrations of 176f, 178, 186
 description of 28-29, 103
 dietary effects on 183-184
 disappearance of 177-180
 exercise intensity effects on 183-184
 formation of 177-180
 functions of 163
 lipolysis inhibition by 230
 metabolism of 176-184
 pyruvate oxidation of 187
 red blood cell production of 180
 sources of, during exercise 96
 transport of 182f, 182-183
lactate: NAD⁺ oxidoreductase 26
lactate dehydrogenase 26, 28, 29f, 90, 177f, 181
lactate shuttle 182f, 182-183, 187f
lactic acid
 description of 28, 176
 exercise-induced formation of 91
 muscle fatigue caused by 4, 178
 pK_a of 91
laforin 169
L-amino acids 9, 10f
lecithin 208
leptin 216-217, 240
leucine 8f, 252, 253f, 256
leucine zipper motif 47
ligands 30
ligases 26t
light chains 80
Lineweaver-Burk plot 21, 24f

linoleic acid 206, 207t
lipids
 exercise use of 229-239
 fatty acids. *See* fatty acids
 metabolism of 209-218
 oxidation of 243-244
 phospholipids 208-209, 208f-209f
 storage of 209-218
 triacylglycerols. *See* triacylglycerols
 types of 205-209
lipoic acid 116, 132
lipolysis
 definition of 212
 fatty acid production by 216
 lactate effects on 230
 regulation of 213-216
 skeletal muscle regulation of 215-216
lipoprotein lipase 210, 212, 218, 242
liver
 amino acid metabolism 247, 249
 glucagon's roles in 195f
 gluconeogenesis in 186, 193f
 glucose release from 186
 glucose uptake by 155-156
 glycogen in 107-108, 166, 168t, 174, 209
 glycogenolysis regulation in 174-175, 193f
 glycolysis in 193f
long-chain fatty acids 205, 218, 237
lyases 26t
lysine 7, 8f-9f, 52, 53f, 256
lysosomes 66

M
macroglycogen 169
macrophages 144
magnesium 26
magnetic resonance imaging 97
malate 118, 141
malate-aspartate shuttle 185f, 185-186
malate dehydrogenase 119
malic enzyme 226
malin 169
malonyl CoA 225, 226f, 227, 229, 236-239, 237f
malonyl CoA decarboxylase 237
maltose 154
mammalian target of rapamycin 70
mass action 30
mass action ratio 102
matrix, mitochondria 110, 110f
maximum velocity 20
mediator 48
medium-chain triacylglycerols 220
messenger RNA
 bases in 43
 codons in 58
 definition of 53-54
 description of 39
 5' end of, capping of 54, 55f
 formation of 4f, 54-57
 introns 54, 56f, 56-57
 lifetime of 63
 model of 60f
 poly A tail of 54, 56, 63
 pre-mRNA 54, 55f-56f
 translation role of 60
metabolic pathway 90
metabolic syndrome 153, 157
metabolism
 adenylate kinase's role in 96-97
 amino acid 247-249, 248f
 AMP deaminase's role in 96-97
 body composition effects 234
 definition of 75
 in exercise 97-99, 231-239
 glucose 159f
 glycogen. *See* glycogen, metabolism of
 glycolysis. *See* glycolysis
 lactate 176-184
 overview of 88-89, 107-109, 108f
 oxidative phosphorylation. *See* oxidative phosphorylation
 oxygen consumption test as indicator of 90
 phosphocreatine. *See* phosphocreatine
 sex differences 234
metabolites 97-98, 98t
metabolon 119

methionine 8f
methionyl-tRNA 58, 60
Michaelis constant 21
Michaelis-Menten kinetics 21-22, 22f, 30, 155
microRNAs 70-71
minerals 25, 146
mitochondria
 ADP transport by 127-128
 aging effects on 148-149
 ATP transport by 127-128
 calcium uptake in 113-114
 cristae 109, 111
 cross-sectional view of 110f
 description of 64
 exercise training effects on 148-149
 inner membrane of 109-110, 110f, 121f, 128f
 inorganic phosphate transport by 127-128
 intermembrane space of 109
 matrix of 110, 110f
 outer membrane of 109-110, 110f
 oxidative phosphorylation in 109-114
 oxygen delivery to 134-135, 135f
 in skeletal muscle 109
 superoxide production in 143
 volume of, in muscle 110, 111
mitochondrial biogenesis 67
mitochondrial DNA 44, 110
mitochondrial glycerol phosphate dehydrogenase 185
mitochondrial nitric oxide synthase 144
mitochondrial partial pressure 136
mitochondrial proteins 149
mitochondrial redox state 138
mitogen-activated protein kinase 67, 215
mmHg 134
monoacylglycerol lipase 212
monoacylglycerols 207
monoamino-dicarboxylic amino acids 7
monocarboxylate transporter 180, 182
monocyte chemoattractant protein-1 240
monosaccharides 153, 154f
muscle. *See also* skeletal muscle
 ATP use by 77, 83-84
 contraction of 82-83, 83f
 energy turnover in 87
 fatty-acid oxidation in 236-239
 glycogen storage in 166-168, 174
 metabolites in 98, 98t
 total creatine concentration in 94-95, 95t
muscle fatigue
 carbohydrates and 201-202
 delays in 202
 description of 103
 inorganic phosphate's role in 178
 lactic acid's role in 4, 178
 purpose of 202
muscle fibers
 activation of 97
 description of 69
 epigenetic control of 104
 fast-twitch 81, 91, 181, 181t
 schematic diagram of 79f
 skeletal 77-78, 212
 slow-twitch 81, 91, 181, 181t
 thick filaments of 78-79, 80f
 Type I 97, 103
 Type II 96-97, 103
muscle hypertrophy 69, 262
muscle protein synthesis 262-263, 263f
muscle soreness 179
muscle-specific atrophy F box 65
muscle-specific really-interesting-novel-gene finger protein 1 65
mutases 163
myeloperoxidase 144
myocyte-enhancer factor 2 67, 104
myofibrils 77, 78f, 80f
myogenic precursor cells 69
myogenic regulatory factors 51
myoglobin 13, 13f, 135
myokinase 96
myosin
 crossbridges in 14-15, 81
 description of 64
 properties of 14
 structure of 15f, 80-81
myosin heavy chains 15, 80-81, 81t

myosin isozymes 81
myosin light chains 80

N

N-acetyl glutamate 255
NAD+ 28, 89, 185
NADH
  cytoplasmic 184-186
  description of 28-29, 35, 89-90, 112, 121, 124, 128-129, 138, 180
NADH–coenzyme Q oxidoreductase 122
NADPH 196
NADPH oxidase 144, 145f
Na$^+$-K$^+$ ATPase pump 83, 202
Nernst equation 142
neuromuscular junction 78
neuronal nitric oxide synthase 144
neutrophils 144, 147
NFAT 67
NFκB 49
nicotinamide adenine dinucleotide 28, 114
nitric oxide 35-36, 144
nitric oxide synthase 36, 144
nitrogen 254f
nitrogen balance 190
nonapeptide 10
noncompetitive inhibitors 24, 24f
nonessential amino acids 248, 261
nonsteroidal anti-inflammatory drugs 25
norepinephrine 166
N-terminus 10-11
nuclear pores 56-57
nuclear receptor superfamily 51
nuclear respiratory factors 67
nucleolus 57
nucleoside 40
nucleoside diphosphate 85-86
nucleoside diphosphate kinase 86
nucleoside monophosphate 85
nucleoside triphosphates 44, 85-86
nucleosomes 41-43, 42f
nucleotides 40, 84-85

O

odd-carbon fatty acids 222
oleic acid 206f
oligomeric proteins 15
oligopeptide 10
omega-3-polyunsaturated fatty acids 206-207, 240
open reading frame 60
ornithine decarboxylase 3
ornithine transcarbamoylase 255
osmotic pressure 166
oxaloacetate 119-120, 188-191
oxidants 142-143
oxidation
  description of 27
  fatty acids 218-222, 236-239
  glucose 111
  ketone bodies 222-225
  pyruvate 132-133
oxidation-reduction reactions 27-29
oxidative phosphorylation 29
  ATP production through 88-90, 128-129
  citric acid cycle. See citric acid cycle
  coupled. see coupled phosphorylation
  definition of 109, 128
  description of 107
  at exercise onset 136-138
  fuel molecules used in 89
  fuel sources for 107
  in high-altitude conditions 139
  limiting factors 139
  mechanism of 111-112, 113f
  in mitochondria 109-114
  mitochondrial creatine kinase subunit involvement in 92
  overview of 89-90, 113f
  pyruvate oxidation 132-133
  rate of 89-90, 133, 137f
  regulation of 128-139
  in rested muscle 136
  in steady-state exercise 138
  summary of 89f, 130f
  uncoupled 126-127
oxidative stress 146-147

oxidoreductases 26t
oxygen consumption
  excess postexercise 92, 93f
  measurement of 89-90
oxygen delivery
  description of 129
  in high-altitude conditions 139
  to mitochondria 134-135, 135f
  regulation of 133-136

P

palmitic acid 111-112, 206f, 225-227
pancreatic lipase 212
pantothenic acid 115
paracrine effect 51
partial pressure of oxygen 134
P-domain 13, 14f
pentose phosphate pathway 153, 196-198, 198f
peptide(s)
  characteristics of 10-11
  formation of 10, 10f
  hormones as 11
  primary structure of 10
peptide bond 10
perilipins 213
peroxisome proliferator–activated receptor γ 50, 240
peroxisome proliferator–activated receptor γ coactivator-1α 149-150
peroxisome proliferator α coactivator 1 70
peroxisome proliferator γ coactivator-1 48, 50, 67, 194-195
pH
  amino acids 10
  blood 4
  enzymatic reactions affected by 23, 23f
  extracellular and intracellular fluids 9
phenylalanine 8f
phosphagens 87-88
phosphatase 114
phosphates
  in cells 84-86
  inorganic. See inorganic phosphate
phosphate transporter 127
phosphatidic acid 208, 208f
phosphatidylcholine 208f
phosphatidylinositol 4,5-bisphosphate 209
phosphatidylinositol 3-kinase 69
phosphatidylinositols 209
phosphatidylinositol 3,4,5-trisphosphate 200
phosphocreatine. See also creatine
  ATP production from 91-92
  creatine supplementation effects on 95
  definition of 87
  inorganic phosphate increases caused by breakdown of 103, 129
  regeneration of, during exercise recovery 92, 93f
  resynthesis of 92
phosphocreatine shuttle 129
phosphoenolpyruvate 163
phosphoenolpyruvate carboxykinase 188-189, 194, 218
phosphofructokinase-1 31, 160, 162, 164, 165t, 235
phosphofructokinase-2 165
6-phosphofructo-2-kinase/fructose 2,6-bisphosphatase 196
phosphoglucomutase 166, 168
6-phosphogluconate dehydrogenase 197
phosphoglycerate kinase 163
phosphoglycerate mutase 163
phospholipids 208-209, 208f-209f
phosphoprotein phosphatases 32, 171-172
phosphorylation
  description of 32-33, 51
  oxidative. See oxidative phosphorylation
  substrate-level 91
phosphorylation potential 138
pH scale 4
phytochemical 146
pI. See isoelectric point
pK$_a$ 5-6, 91
plasma proteins 64
^{31}P nuclear magnetic resonance spectroscopy 97, 137
poly A polymerase 56

poly A tail 54, 56, 63
polynucleotide 41f
polypeptides
  description of 10, 13-14
  posttranslational processing of 63-64
polyribosome 62, 62f
polysaccharides 154
polyubiquitination 65
polyunsaturated fatty acids 145, 206, 233
P/O ratio 112, 126
porins 109
porphyrins 261
posttranslation modification proteomics 16
potential energy 99
pre-mRNA 54, 55f-56f
primary gene transcript 44
primary-miRNAs 70
primary structure of proteins/peptides 10-11
proglycogen 169
proline 6, 8f
promoter 45, 49f
propionate 190
protein(s)
  amino acids 247
  bonds responsible for structure of 11-12
  chaperones 64
  conformations of 13
  definition of 3
  denaturation of 15-16
  diets with high levels of 191
  domains 13
  fold of 14
  gene regulatory 47-48
  half-life of 3
  histone 53
  integral membrane 26
  lifetime of 4f
  myoglobin 13, 13f
  oligomeric 15
  phosphorylation of 32-33, 51
  as polypeptide 10
  primary structure of 10-11
  quaternary structure of 15
  redox control of 33, 34f
  secondary structure of 12-13
  structure of 11-16, 47
  tertiary structure of 13-15
  transcription regulation by 47
  transmembrane 26
proteinases 64
protein balance 190
protein degradation
  calpain system of 66
  description of 3
  diet effects on 262-263
  equation for 64
  exercise effects on 262-263
  lysosomal system of 66
  pathways of 64-65
  steps involved in 65f
  ubiquitin-proteasome pathway of 65-66
protein disulfides 34
protein kinases
  A 171, 193, 214-215
  C 200, 240
  description of 32
  G 36
protein transporters 26-27
protein turnover 3, 145
protein tyrosine kinase 200
proteomics 11, 16
proton 4
proton motive force 112
proton pumping 112, 120, 125
proximal promoter 47
purine bases 39, 40f, 85
purine nucleotide cycle 97, 259-260, 260f
pyrimidine bases 39, 40f, 85, 261
pyruvate
  description of 90-91
  formation of 181
  glucose-6-phosphate conversion to 160, 161f
  lactate oxidation to 187
  oxidation of 132-133
pyruvate carboxylase 188
pyruvate dehydrogenase 33, 114, 132-133, 235
pyruvate dehydrogenase kinase 133

pyruvate dehydrogenase phosphatase 132
pyruvate kinase 196

Q
Q cycle 123
quantitative proteomics 16
quaternary structure of proteins 15

R
rate-limiting enzymes 31
reactive oxygen and nitrogen species
  cellular damage from 144-146
  exercise-related formation of 147-148
  fatty acid peroxidation by 146
  rested muscle production of 147
  types of 143-144
reactive oxygen species
  description of 67, 103, 143
  types of 143-144
reading frame 44
redox potential 138, 140-142
redox reactions 27, 139-142
reducing equivalents 122
reduction 27
regulatory light chains 80
repressors 47-49
resistance training
  creatine supplementation during 95
  definition of 68
  gene expression regulation in 68-70
resistin 241
respiratory burst 144
respiratory exchange ratio 111, 230-231, 233
respiratory quotient 111
response elements 47-48, 50f, 52
rested muscle 136, 147
resting metabolic rate 247
retinoic acid 50
retinoid X receptor 50
reversible oxidation 33
ribonucleases 63
ribonucleic acid. See RNA
ribose 40f, 153
ribose 5-phosphate 196
ribosomal RNA 39, 54, 57-58
rigor mortis 88
RNA
  composition of 53
  messenger. See messenger RNA
  microRNA 70-71
  posttranscriptional modifications of 53-58
  ribosomal 39, 54, 57-58
  small nuclear 54
  transfer. See transfer RNA
  types of 39, 53-54
RNA polymerase II 44-45, 48

S
S-adenosylmethionine 261
sarcolemma 77, 182
sarcopenia 95
sarcoplasmic reticulum 3, 78, 82
satellite cells 25, 69, 180
scurvy 262
secondary structure of proteins 12-13
sense strand 44
SERCA 13, 14f, 35-36, 83, 103, 148
SERCA ATPase 84, 88
serine 8f
serotonin 262
S-glutathionylation 33
side chains 12
signaling pathways 198-201
signal transduction 51, 198
silent information regulator-1 149, 196, 199
SI units 154
60S ribosomal RNA transcript 58, 60
skeletal muscle
  amino acids in 249, 258f
  AMP-activated protein kinase activation in 67
  ATP use by 77, 84f
  Ca^{2+}-dependent transcriptional pathways in 67
  contraction of 81, 83f
  creatine kinase activity in 88
  creatine metabolism in 94f
  energy metabolism in 179f

energy requirements of 77-84
exercise-induced adaptation in 68f
fatty acid consumption by 236
fibers of 77-78, 81, 103-104, 212. See also muscle fibers
glucose uptake by 156, 200-201, 238f
glutamine synthesis in 251
glycogen storage in 166-168
lipolysis regulation in 215-216
metabolism 165
metabolite concentrations in 98, 98t
mitochondria in 67, 109
myofibrils 77, 78f, 80f
myosin heavy chains 81, 81t
structure of 77-79, 78f-79f
transcription factors in 50-51
skeletal muscle cells 135
slow-twitch muscle fibers 81, 91, 181, 181t
small nuclear RNA 54
sodium-calcium antiport 114
spliceosome 56
splicing 56-57
standard free energy change 101
standard redox potentials 140, 141t
start codon 44
start site 45
state 3 respiration 136
state 4 respiration 136
steady state 90
steady-state exercise 138
stearic acid 207t
stereoisomerism 9, 10f
stereoisomers 9
steroid hormones 51
stop codons 44, 60
stroke volume 133-134, 134t
strong acids 4-5
structural proteomics 16
subsarcolemmal mitochondria 109
substrate
  concentration of 20-22, 21f
  definition of 19
  glycogenolysis regulation by 173-174
substrate-level phosphorylation 91
succinate-coenzyme Q oxidoreductase 122-123
succinate dehydrogenase 29f, 118, 122-123, 185
succinyl CoA synthetase 116, 118
sucrose 158
sulfenic acid 33
sulfinic acid 33
sulfonic acid 33
superoxide 122, 143
superoxide dismutase 144
Svedberg units 58

T
TATA-binding protein 48
TATA box 47, 49
temperature, enzymatic reactions affected by 23, 23f
template DNA strand 44-45
tertiary structure of proteins 13-15
testosterone 215
thermodynamics 100-101
thiamine pyrophosphate 116, 132
thick filaments 78-79, 80f
thin filaments 79, 80f
thiolase 222-224
thiols 33-34
threonine 8f
thymine 39, 40f
timnodonic acid 207t
torr 134
total adenine nucleotide 259-260
transamination 249-251, 257
transcription
  basal apparatus 48
  cell signaling 51-52
  definition of 3, 44
  DNA organization and 52-53
  elongation phase of 45f, 45-46
  higher levels of control 49-51
  initiation phase of 45, 45f, 52
  phases of 45f
  regulation of 46-53, 63
  response elements 47-48

RNA chain in 46, 46f
schematic diagram of 47f
steps in 45f, 45-46
termination phase of 45f, 46
transcription factors 47-50, 67, 150
transferases 26t
transfer RNA
  aminoacyl-tRNA 59-60
  definition of 39, 54
  formation of 57-58
translation
  aminoacyl-tRNA formation 59-60, 62
  definition of 3
  elongation phase of 58-59, 61-62, 62f
  initiation phase of 58, 60-61
  mRNA's role in 60
  regulation of 63
  steps in 58-62
  termination phase of 59, 62
transmembrane electric potential 112
transmembrane proteins 26
transport proteins 27
triacylglycerols
  breakdown of 212f
  description of 205, 207-208
  formation of 210-212, 217
  intramuscular 109, 215-216, 231, 234
  medium-chain 220
  recycling of, in adipocyte 217f
  storage of 211
  structure of 208f
  summary of 216
  synthesis of 213
  turnover of 212-216
tricarboxylic acid cycle. See citric acid cycle
triose phosphate isomerase 162
tropomyosin 79
troponins 79
tryptophan 8f
T-tubules 82, 114, 156
tuberosclerosis complex 70
turnover number 23-24
type 1 diabetes mellitus 156
type 2 diabetes mellitus 156-157, 194
tyrosine 8f-9f, 262

U
ubiquinol 122, 143
ubiquinone 122, 123f, 184
ubiquitin 65
ubiquitin-proteasome pathway 65-66
ubisemiquinone 122, 142
uncoupled oxidative phosphorylation 126-127
uncoupling protein 1 126-127
uniport 113
unsaturated fatty acids 222
upstream 44
uracil 39, 40f
urea 16, 186, 255f, 256
urea cycle 186, 252-256, 255f
uridine diphosphate-glucose pyrophosphorylase 168-169
uridine monophosphate kinase 86
uridine triphosphate 85-86

V
valine 8f, 253f
very-low-density lipoproteins 210, 218
visceral fat 209-210
vitamin C 147, 262
vitamin E 147
Vmax 20, 21f, 35

W
weak acids 4-5
weight loss 243-244
wobble 58
women 234

X
xanthine oxidase 144, 145f
xenobiotics 146
x-ray crystallography 13

Z
zinc 25
zinc finger motif 47
zwitterion 7

About the Authors

Peter M. Tiidus, PhD, is a professor and former chair of the department of kinesiology and physical education at Wilfrid Laurier University in Waterloo, Ontario, Canada. For more than 30 years, he has focused his research on the physiological mechanisms of and practical interventions in muscle damage and repair, employing both animal models and human subjects.

Tiidus has authored more than 80 publications and presented his research in multiple lectures and conference presentations on estrogen and muscle damage, inflammation, and repair and the influence of treatment interventions on muscle recovery from damage and physiological responses. He currently serves as an editorial board member for *Medicine & Science in Sports & Exercise*. He is also a former member of the board of directors of the Canadian Society for Exercise Physiology.

A. Russell Tupling, PhD, is an associate professor in the department of kinesiology at the University of Waterloo in Ontario, Canada. His research program, which is funded by the Natural Sciences and Engineering Research Council of Canada and the Canadian Institutes of Health Research, is dedicated to the understanding of the regulation of sarcoplasmic reticulum (SR) function in muscle and understanding how defects in the function of SR proteins that occur with oxidative stress contribute to fatigue, weakness, and disease. In 2009, he received an Early Research Award from the Government of Ontario to conduct research examining a potential link between Ca^{2+} pump energetics in muscle and metabolic disorders.

Tupling has 49 peer-reviewed publications in scholarly journals and over 70 conference abstracts based on his research. In 2010, he won the Award of Excellence in Graduate Supervision, which was established by the University of Waterloo in recognition of exemplary faculty members who have demonstrated excellence in graduate student supervision. Tupling is a member of the American Physiological Society and the Canadian Society for Exercise Physiology (CSEP). He was invited to give the inaugural Mike Houston Tutorial Lecture in Skeletal Muscle at the CSEP conference in 2009.

Michael E. Houston, PhD, received his undergraduate training in biochemistry from the University of Toronto and his PhD in biochemistry from the University of Waterloo. A superb athlete and lifelong exercise fanatic, he was able to integrate his training in biochemistry with his love of exercise sciences and to forge a career as a teacher and scientist in the field of kinesiology. For almost 40 years during his career, he authored more than 100 refereed publications and taught courses on the biochemistry of exercise to many undergraduate and graduate students. In 2003, he was presented with the Honour Award from the Canadian Society for Exercise Physiology in acknowledgment of his lifetime contribution to research and education in exercise science.

Houston was the author of the first three editions of *Biochemistry Primer for Exercise Science*. This fourth edition, which is built on his body of work, still incorporates a major portion of his third edition. Dr. Houston passed away in 2008.